Student Solutions Manual

Beginning Algebra

EIGHTH EDITION

Richard N. Aufmann
Palomar College

Joanne S. Lockwood
Nashua Community College

Prepared by

Rhoda Oden
Gadsden State Community College

Australia • Brazil • Japan • Korea • Mexico • Singapore • Spain • United Kingdom • United States

For product information and technology assistance, contact us at
Cengage Learning Customer & Sales Support, 1-800-354-9706

For permission to use material from this text or product, submit all requests online at **www.cengage.com/permissions**
Further permissions questions can be emailed to
permissionrequest@cengage.com

ISBN-13: 978-1-133-11224-2
ISBN-10: 1-133-11224-2

Brooks/Cole
20 Davis Drive
Belmont, CA 94002-3098
USA

Cengage Learning is a leading provider of customized learning solutions with office locations around the globe, including Singapore, the United Kingdom, Australia, Mexico, Brazil, and Japan. Locate your local office at: **www.cengage.com/global**

Cengage Learning products are represented in Canada by Nelson Education, Ltd.

To learn more about Brooks/Cole, visit **www.cengage.com/brookscole**

Purchase any of our products at your local college store or at our preferred online store **www.cengagebrain.com**

Printed in the United States of America
2 3 4 5 6 7 15 14 13

Contents

CHAPTER 1 Prealgebra Review

Chapter 1 Prep Test

1. $127.1649 \approx 127.16$

2. $\begin{array}{r} \scriptstyle{1\,1} \\ 49{,}147 \\ 596 \\ \hline 49{,}743 \end{array}$

3. $\begin{array}{r} \scriptstyle{4\ 10\ 10\ 14} \\ 5\,0\,0\,4 \\ -4\,8\,7 \\ \hline 4\,5\,1\,7 \end{array}$

4. $\begin{array}{r} 407 \\ \times 28 \\ \hline 3256 \\ 8140 \\ \hline 11{,}396 \end{array}$

5. $19\overline{)456} = 24$; 38, 76, 76, 0

6. $8 = 2^3$

$12 = 2^2 \cdot 3$

$LCM(8,12) = 2^3 \cdot 3 = 24$

7. $16 = 2^4$

$20 = 2^2 \cdot 5$

$GCF(16,20) = 2^2 = 4$

8. $21 = 3 \cdot 7$

9. $\frac{4}{10} = \frac{2}{5}$

Section 1.1 Introduction to Integers

1. The statement is **sometimes true**. The absolute value of a number is positive unless the number is zero. The absolute value of zero is zero, which is neither positive nor negative.

3. The statement is **always true** because the absolute value of a number is either a positive number or zero, both of which are greater than -2.

5. **a.** -12 is a negative integer.

b. 18 is a positive integer.

c. -7 is a negative integer.

d. 0 is neither positive nor negative.

e. $\frac{3}{4}$ is neither a positive integer nor a negative integer.

f. 365 is a positive integer.

7. The whole numbers include the number zero (0), but the natural numbers do not.

9. The inequality $-5 < -1$ is read "negative 5 is less than negative one."

11. $-2 > -5$ because -2 lies to the right of -5 on the number line.

13. $-16 < 1$ because -16 lies to the left of 1 on the number line.

15. $3 > -7$ because 3 lies to the right of -7 on the number line.

17. $0 > -3$ because 0 lies to the right of -3 on the number line.

19. $-42 < 27$ because -42 lies to the left of 27 on the number line.

21. $21 > -34$ because 21 lies to the right of -34 on the number line.

23. $-27 > -39$ because -27 lies to the right of -39 on the number line.

25. $-131 < 101$ because -131 lies to the left of 101 on the number line.

27. If n is to the right of 5 on the number line, then n must be a positive number because all numbers to the right of 5 are positive numbers greater than 5. Only statement ***i*** is true.

29. Yes, the inequalities do represent the same order relation. The statement $6 \geq 1$ says that 6 lies to the right of 1on the number line. The statement $1 \leq 6$ says that 1 lies to the left of 6 on the number line.

31. The natural numbers less than 9:
{1, 2, 3, 4, 5, 6, 7, 8}

33. The positive integers less than or equal to 8:
{1, 2, 3, 4, 5, 6, 7, 8}

35. The negative integers greater than -7:
{-6, -5, -4, -3, -2, -1}

37. The only element of A greater than 2 is the element 5.

39. The elements of D that are less than -8 are -23 and -18.

41. The elements of E that are greater than -10 are 21 and 37.

43. The elements of B that are less than or equal to 0 are -52, -46 and 0.

45. The elements of C that are greater than or equal to -17 are -17, 0, 4 and 29.

47. The elements of A that are greater than or equal to 5 are 5, 6, 7, 8, and 9.

49. The elements of D that are less than -4 are -10, -9, -8, -7, -6 and -5.

51. The equation $|-5| = 5$ is read "the absolute value of negative five is five."

53. The opposite of 22 is -22.

55. The opposite of -31 is 31.

57. The opposite of -168 is 168.

59. The opposite of 630 is -630.

61. $-(-18) = 18$

63. $-(49) = -49$

65. $|16| = 16$

67. $|-12| = 12$

69. $-|29| = -29$

71. $-|-14| = -14$

73. $-|0| = 0$

75. $-|34| = -34$

77. A = {-8, -5, -2, 1, 3}

a. Opposite of each element of A: 8, 5, 2, -1, -3

b. Absolute value of each element: 8, 5, 2, 1, 3

79. True. The absolute value of a negative number *n* is greater than *n* because the absolute value of a negative number is a positive number and any positive number is greater than any negative number.

81. $|-83| > |58|$ because 83 > 58.

83. $|43| < |-52|$ because 43 < 52.

85. $|-68| > |-42|$ because 68 > 42.

87. $|-45| < |-61|$ because 45 < 61.

89. From least to greatest: $-19, -|-8|, |-5|, 6$

91. From least to greatest: $-22, -(-3), |-14|, |-25|$

93. a. From the table, a temperature of 5 °F with a 20 mph wind feels like -15 °F. A temperature of 10F with a 15 mph wind feels like -7 °F. So 5 °F with a 20 mph wind feels colder.

b. From the table, a temperature of -25 °F with a 10 mph wind feels like -47 °F. A temperature of -15 °F with a 20 mph wind feels like -42 °F. So -25 °F with a 10 mph wind feels colder.

95. On the number line, the two points that are four units from 0 are 4 and -4.

97. On the number line, the two points that are seven units from 4 are 11 and -3.

99. If *a* is a positive number, then $-a$ is a negative number.

101. -5 < 3 because -5 is to the left of 3 on the number line. 3 > -5 because 3 is to the right of –5 on the number line.

–5 –4 –3 –2 –1 0 1 2 3

103. The opposite of the additive inverse of 7 is **7**.

105. The opposite of the absolute value of 8 is **-8**.

Section 1.2 Operations with Integers

1. It is **sometimes true** that the sum of two integers is larger than either of the integers being added. If two nonnegative integers are added the sum is larger than either addend.

3. It is **always true** that the quotient of two integers with different signs is negative.

5. It is **always true** that if x is an integer and $4x = 0$ then $x = 0$. The only way to get a result of zero when multiplying is if there is a factor of zero.

7. In -6 – 2 the first "-" is a **negative** and the second "-" is a **minus**.

9. To add two numbers with the same sign, add the absolute values of the numbers. The sum will have the sign of the addends.

11. In the addition equation $8+(-3)=5$, the addends are 8 and -3 and the sum is 5.

13. $-3+(-8)=-11$

15. $-4+(-5)=-9$

17. $6+(-9)=-3$

19. $-6+7=1$

21. $2+(-3)+(-4)=-1+(-4)=-5$

23. $-3+(-12)+(-15)=-15+(-15)=-30$

25. $-17+(-3)+29=-20+29=9$

27. $-3+(-8)+12=-11+12=1$

29. $13+(-22)+4+(-5)=-9+4+(-5)$
$=-5+(-5)=-10$

31. The sum $812+(-537)$ is positive because the positive addend has the larger absolute value.

33. The word "minus" refers to the operation of subtraction. The word "negative" refers to the sign of a number.

35. $-10-4=-10+(-4)=-14$

37. $16-8=16+(-8)=8$

39. $7-14=7+(-14)=-7$

41. $3-(-4)=3+4=7$

43. $-4-(-2)=-4+2=-2$

45. $-12-16=-12+(-16)=-28$

47. $4-5-12=4+(-5)+(-12)=-1+(-12)=-13$

49. $-12-(-3)-(-15)=-12+3+15=-9+15=6$

51. $13-7-15=13+(-7)+(-15)=6+(-15)=-9$

53. $-30-(-65)-29-4=-30+65+(-29)+(-4)$
$=35+(-29)+(-4)$
$=6+(-4)=2$

55. The difference $-25-52$ will be negative. Rewriting as an addition problem yields $-25+(-52)$, the sum of two negatives, which is negative.

57. **a.** The operation in 8(-7) is multiplication because there is no operation symbol between the 8 and the left parentheses.

b. The operation in 8 – 7 is subtraction because there is a space before and after the minus sign.

c. The operation in 8 – (- 7) is subtraction because there is a space before and after the minus sign.

d. The operation in $-xy$ is multiplication because there is no operation symbol between the x and the y.

e. The operation in $x(-y)$ is multiplication because there is no operation symbol between the x and the parentheses.

f. The operation in $-x - y$ is subtraction because there is a space before and after the minus sign.

59. In the equation (-10)(7)= -70, the factors are <u>-10 and 7</u> and the product is <u>70</u>.

61. For the product (-4)(-12), the signs of the factors are the same. The sign of the product is <u>positive</u>. The product is <u>48</u>.

63. $14 \cdot 3 = 42$

65. $5(-4) = -20$

67. $-8(2) = -16$

69. $(-5)(-5) = 25$

71. $(-7)(0) = 0$

73. $-24 \cdot 3 = -72$

75. $6(-17) = -102$

77. $-4(-35) = 140$

79. $5 \cdot 7(-2) = 35(-2) = -70$

81. $(-9)(-9)(2) = 81(2) = 162$

83. $-5(8)(-3) = (-40)(-3) = 120$

85. $-1(4)(-9) = -4(-9) = 36$

87. The product of three negative integers is negative because an odd number of negative factors yields a negative.

89. Using a division symbol $\frac{-15}{3} = -15 \div 3$.

91. Division problem: $\frac{-36}{-12} = 3$.

Related multiplication problem: $3(-12) = -36$.

93. Division problem: $\frac{-55}{11} = -5$.

Related multiplication problem: $-5(11) = -55$.

95. $12 \div (-6) = -2$

97. $(-72) \div (-9) = 8$

99. $0 \div (-6) = 0$

101. $45 \div (-5) = -9$

103. $-36 \div 4 = -9$

105. $-81 \div (-9) = 9$

107. $72 \div (-3) = -24$

109. $-60 \div 5 = -12$

111. $78 \div (-6) = -13$

113. $-72 \div 4 = -18$

115. $-114 \div (-6) = 19$

117. $-130 \div (-5) = 26$

119. The quotient $-\frac{520}{-13}$ is positive.

121. The word *drop* indicates a decrease in temperature, so at 10:00 P.M. the temperature is (85 – 20) degrees Fahrenheit, choice **ii**.

123. $-6°C + 9°C = 3°C$

125. $10°C - (-4)°C = 10°C + 4°C = 14°C$
(high temperature – low temperature)

127. $360°C - (-39)°C = 360°C + 39°C = 399°C$
(boiling temperature – freezing temperature)

129. $5642 - (-28) = 5642 + 28 = 5670\,meters$
(Mt. Elbrus – Valdez Peninsula)

131. $5895 - (-156) = 5895 + 156 = 6051\,meters$
(Mt. Kilimanjaro – Lake Assal)

133. $8850 - (-411) = 8850 + 411 = 9261\,meters$
(Mt. Everest – Dead Sea)

135.
$$\begin{aligned} average &= \frac{sum}{7} \\ &= \frac{4+(-5)+8+0+(-9)+(-11)+(-8)}{7} \\ &= \frac{-21}{7} = -3°C \end{aligned}$$

137. $45°F - (-4)°F = 45°F + 4°F = 49°F$

139. $16°F - (-70)°F = 16°F + 70°F = 86°F$

141. Lee Westwood:
$$\begin{aligned} -5+(-3)+(-4)+(-1) &= -8+(-4)+(-1) \\ &= -12+(-1) = -13 \end{aligned}$$
Anthony Kim:
$$\begin{aligned} -4+(-2)+1+(-7) &= -6+1+(-7) \\ &= -5+(-7) = -12 \end{aligned}$$
K.J. Choi:
$$\begin{aligned} -5+(-1)+(-2)+(-3) &= -6+(-2)+(-3) \\ &= -8+(-3) = -11 \end{aligned}$$

143. $|13-(-4)| = |13+4| = |17| = 17$

145. $|18-21| = |-3| = 3$

147. -4, -9, -14 (subtract 5)

149. -16, 4, -1 (divide by -4)

151. For a number to be divisible by 4, the last two digits must form a number divisible by 4. We can eliminate numbers that do not contain the digits 4, 5, 6 and 3. So our only choices are 4536, 5436, 3456, 4356, 5346, 5364. The largest of those is **5436**.

153. Statement **b** is false because $|3-4|=|-1|=1$ and $|3|-|4|=3-4=-1$.

155. Statement **a** is true for all real numbers.

157. If the product -4x is a positive integer, then x must be a negative integer because a product is positive only when the two factors have like signs.

159. 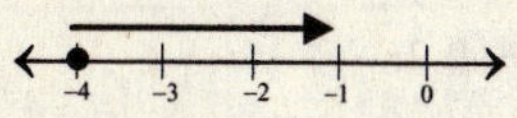

$-4+3=-1$

161. 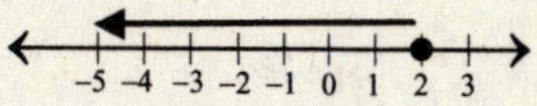

$2+(-7)=-5$

163. 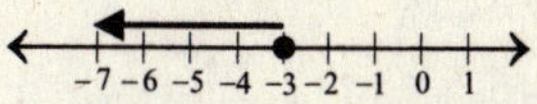

$-3+(-4)=-7$

165. To model -7 + 4, place 7 red chips and 4 blue chips in a circle. Pair as many red and blue chips as possible. There are 3 red chips remaining, or -3. For -2 + 6, use 2 red chips and 6 blue chips. After pairing, there are 4 blue chips remaining, or +4. For -5 + (-3), use 5 red chips and then 3 more red chips. There are no red/blue pairs, so there are 8 red chips. The solution is -8.

167. Answers will vary. For example, -16 – (-8) = -8 or -25 – (-17) = -8. The difference between the absolute values of the numbers being subtracted must be 8.

Section 1.3 Rational Numbers

1. This statement is **never true**. To multiply fractions, simply multiply the numerators together and multiply the denominators together.

3. It is **always true** that an irrational number is a real number.

5. It is **never true** that to write a decimal as a percent, the decimal is multiplied by $\frac{1}{100}$.

7. To write $\frac{2}{3}$ as a decimal, divide $\underline{2}$ by $\underline{3}$. The quotient is 0.6666…, which is a $\underline{\text{repeating}}$ decimal.

9. $$\begin{array}{r} 0.33 \\ 3\overline{)1.00} \\ \underline{-9} \\ 10 \\ \underline{-9} \\ 1 \end{array} \qquad \frac{1}{3}=0.\overline{3}$$

11. $$\begin{array}{r} 0.25 \\ 4\overline{)1.00} \\ \underline{-8} \\ 20 \\ \underline{-20} \\ 0 \end{array} \qquad \frac{1}{4}=0.25$$

13. $$\begin{array}{r} 0.4 \\ 5\overline{)2.0} \\ \underline{20} \\ 0 \end{array} \qquad \frac{2}{5}=0.4$$

15.
$$\begin{array}{r} 0.166 \\ 6\overline{)1.000} \\ \underline{-6} \\ 40 \\ \underline{-36} \\ 40 \\ \underline{-36} \\ 4 \end{array} \qquad \frac{1}{6}=0.1\overline{6}$$

17.
$$\begin{array}{r} 0.125 \\ 8\overline{)1.000} \\ \underline{-8} \\ 20 \\ \underline{-16} \\ 40 \\ \underline{40} \\ 0 \end{array} \qquad \frac{1}{8}=0.125$$

19.
$$\begin{array}{r} 0.22 \\ 9\overline{)2.00} \\ \underline{18} \\ 20 \\ \underline{18} \\ 2 \end{array} \qquad \frac{2}{9}=0.\overline{2}$$

21.
$$\frac{5}{11}=0.\overline{45} \qquad \begin{array}{r} 0.4545 \\ 11\overline{)5.0000} \\ \underline{44} \\ 60 \\ \underline{55} \\ 50 \\ \underline{44} \\ 6 \end{array}$$

23.
$$\begin{array}{r} 0.5833 \\ 12\overline{)7.0000} \\ \underline{60} \\ 100 \\ \underline{-96} \\ 40 \\ \underline{36} \\ 40 \\ \underline{36} \\ 4 \end{array} \qquad \frac{7}{12}=0.58\overline{3}$$

25.
$$\begin{array}{r} 0.266 \\ 15\overline{)4.000} \\ \underline{30} \\ 100 \\ \underline{-90} \\ 100 \\ \underline{90} \\ 10 \end{array} \qquad \frac{4}{15}=0.2\overline{6}$$

27.
$$\begin{array}{r} 0.4375 \\ 16\overline{)7.0000} \\ \underline{64} \\ 60 \\ \underline{48} \\ 120 \\ \underline{112} \\ 80 \\ \underline{80} \\ 0 \end{array} \qquad \frac{7}{16}=0.4375$$

29.
$$\begin{array}{r} 0.24 \\ 25\overline{)6.00} \\ \underline{50} \\ 100 \\ \underline{100} \\ 0 \end{array} \qquad \frac{6}{25}=0.24$$

31.
$$40\overline{)9.000}$$
0.225
80
100
80
200
200
0

$\frac{9}{40}=0.225$

33. $\frac{15}{22}=0.6\overline{81}$

$$22\overline{)15.00000}$$
0.68181
132
180
176
40
22
180
176
40
22
18

35. The product of 1.762 and -8.4 will have four decimal places because the factors have a total of four decimal places.

37. $\frac{1}{2}\left(-\frac{3}{4}\right)=-\frac{1\cdot 3}{2\cdot 4}=-\frac{3}{8}$

39. $\left(-\frac{3}{8}\right)\left(-\frac{4}{15}\right)=\frac{\overset{1}{\cancel{3}}\cdot\overset{1}{\cancel{2}}\cdot\overset{1}{\cancel{2}}}{\underset{1}{\cancel{2}}\cdot\underset{1}{\cancel{2}}\cdot 2\cdot\underset{1}{\cancel{3}}\cdot 5}=\frac{1}{10}$

41. $\left(\frac{1}{2}\right)\left(-\frac{3}{4}\right)\left(-\frac{5}{8}\right)=\frac{15}{64}$

43. $\frac{3}{8}\div\frac{1}{4}=\frac{3}{8}\cdot\frac{4}{1}=\frac{3\cdot\overset{1}{\cancel{2}}\cdot\overset{1}{\cancel{2}}}{\underset{1}{\cancel{2}}\cdot\underset{1}{\cancel{2}}\cdot 2}=\frac{3}{2}$

45. $-\frac{5}{12}\div\frac{15}{32}=-\frac{5}{12}\cdot\frac{32}{15}=-\frac{\overset{1}{\cancel{5}}\cdot\overset{1}{\cancel{2}}\cdot\overset{1}{\cancel{2}}\cdot 2\cdot 2\cdot 2}{\underset{1}{\cancel{2}}\cdot\underset{1}{\cancel{2}}\cdot 3\cdot 3\cdot\underset{1}{\cancel{5}}}=-\frac{8}{9}$

47. $-\frac{4}{9}\div\left(-\frac{2}{3}\right)=-\frac{4}{9}\cdot\left(-\frac{3}{2}\right)=\frac{\overset{1}{\cancel{2}}\cdot 2\cdot\overset{1}{\cancel{3}}}{\underset{1}{\cancel{3}}\cdot 3\cdot\underset{1}{\cancel{2}}}=\frac{2}{3}$

49.
3.47
×1.2
694
3470
4.164

(1.2)(3.47) = 4.164

51.
1.89
×2.3
567
3780
4.347

(-1.89)(-2.3) = 4.347

53.
1.06
×3.8
848
3180
4.028

(1.06)(-3.8) = -4.028

55. **a.** The product is negative because there are an odd number of negative factors.

b. The quotient is positive because the quotient of two numbers with like signs is positive.

57. $\dfrac{-1.27}{-1.7}=\dfrac{12.7}{17}\approx 0.75$

$$\begin{array}{r} 0.747 \\ 17\overline{)12.70} \\ \underline{119} \\ 80 \\ \underline{68} \\ 120 \\ \underline{119} \\ 1 \end{array}$$

59. $\dfrac{-354.2086}{0.1719}=\dfrac{-3{,}542{,}086}{1719}\approx -2060.55$

$$\begin{array}{r} 2060.550 \\ 1719\overline{)3542086.000} \\ \underline{3438} \\ 10408 \\ \underline{10314} \\ 9460 \\ \underline{8595} \\ 8650 \\ \underline{8595} \\ 550 \end{array}$$

61. $\dfrac{3}{14}=\dfrac{3\cdot 2}{14\cdot 2}=\dfrac{6}{28}$

63. $-\dfrac{1}{4}+\dfrac{3}{4}=\dfrac{-1+3}{4}=\dfrac{2}{4}=\dfrac{1}{2}$

65. $-\dfrac{5}{6}-\dfrac{1}{6}=\dfrac{-5-1}{6}=\dfrac{-6}{6}=-1$

67. $-\dfrac{5}{6}-\dfrac{5}{9}=-\dfrac{15}{18}+\left(-\dfrac{10}{18}\right)=-\dfrac{25}{18}$

69. $-\dfrac{7}{12}+\dfrac{5}{8}=-\dfrac{14}{24}+\dfrac{15}{24}$

$=\dfrac{-14+15}{24}=\dfrac{1}{24}$

71. $\dfrac{1}{3}+\dfrac{5}{6}-\dfrac{2}{9}=\dfrac{6}{18}+\dfrac{15}{18}-\dfrac{4}{18}=\dfrac{6+15-4}{18}=\dfrac{17}{18}$

73. $-\dfrac{3}{8}-\dfrac{5}{12}-\dfrac{3}{16}=-\dfrac{18}{48}-\dfrac{20}{48}-\dfrac{9}{48}$

$=\dfrac{-18-20-9}{48}=-\dfrac{47}{48}$

75. $\dfrac{1}{2}-\dfrac{3}{8}-\left(-\dfrac{1}{4}\right)=\dfrac{4}{8}-\dfrac{3}{8}+\dfrac{2}{8}=\dfrac{4-3+2}{8}=\dfrac{3}{8}$

77. $\dfrac{1}{3}-\dfrac{1}{4}-\dfrac{1}{5}=\dfrac{20}{60}-\dfrac{15}{60}-\dfrac{12}{60}$

$=\dfrac{20-15-12}{60}=-\dfrac{7}{60}$

79. $\dfrac{5}{16}+\dfrac{1}{8}-\dfrac{1}{2}=\dfrac{5}{16}+\dfrac{2}{16}-\dfrac{8}{16}=\dfrac{5+2-8}{16}=-\dfrac{1}{16}$

81. $\dfrac{1}{8}-\dfrac{11}{12}+\dfrac{1}{2}=\dfrac{3}{24}-\dfrac{22}{24}+\dfrac{12}{24}$

$=\dfrac{3-22+12}{24}=-\dfrac{7}{24}$

83.
$$\begin{array}{r} 1.09 \\ \underline{+6.20} \\ 7.29 \end{array}$$

85.
$$\begin{array}{r} 8.179 \\ \underline{-5.13} \\ 3.049 \end{array}$$

$5.13-8.179=-3.049$

87.
$$\begin{array}{r} 3.60 \\ \underline{-2.54} \\ 1.06 \end{array}$$

$2.54-3.6=-1.06$

89. $\begin{array}{r} 16.92 \\ +6.925 \\ \hline 23.845 \end{array}$

$-16.92-6.925=-23.845$

91. $\begin{array}{r} 17.6920 \\ -6.9027 \\ \hline 10.7893 \end{array}$

$6.9027-17.692=-10.7893$

93. $\begin{array}{r} 16.4 \\ +3.09 \\ \hline 19.49 \end{array}$ $16.4+3.09=19.49$

$\begin{array}{r} 19.49 \\ -7.93 \\ \hline 11.56 \end{array}$ $19.49-7.93=11.56$

95. $\begin{array}{r} 19 \\ +3.72 \\ \hline 22.72 \end{array}$ $19-(-3.72)=22.72$

$\begin{array}{r} 82.75 \\ -22.72 \\ \hline 60.03 \end{array}$ $22.72-82.75=-60.03$

97. $\begin{array}{r} 3.09 \\ +4.6 \\ \hline 7.69 \end{array}$ $-3.09-4.6=-7.69$

$\begin{array}{r} 27.3 \\ +7.69 \\ \hline 34.99 \end{array}$ $-7.69-27.3=-34.99$

99. **a.** $\frac{7}{8}+\frac{4}{5}\approx 1+1=2$

b. $\frac{1}{3}+\left(-\frac{1}{2}\right)\approx 0+0=0$

c. $-0.125+1.25\approx 0+1=1$

d. $-1.3+0.2\approx -1+0=-1$

101. **a.** To convert a decimal to a percent, multiply the decimal by 100%.

b. To convert a percent to a decimal, remove the percent sign and divide by 100.

103. To write 80% as a fraction, remove the percent sign and multiply by $\frac{1}{100}$: $80\%=80\cdot\frac{1}{100}=\frac{4}{5}$

105. To write $\frac{3}{10}$ as a percent, multiply by 100%:

$\frac{3}{10}=\frac{3}{10}\cdot(100\%)=\frac{300}{10}\%=30\%$

107. $75\%=75\left(\frac{1}{100}\right)=\frac{75}{100}=\frac{3}{4}$

$75\%=75(0.01)=0.75$

109. $50\%=50\left(\frac{1}{100}\right)=\frac{50}{100}=\frac{1}{2}$

$50\%=50(0.01)=0.5$

111. $64\%=64\left(\frac{1}{100}\right)=\frac{64}{100}=\frac{16}{25}$

$16\%=16(0.01)=0.16$

113. $175\%=175\left(\frac{1}{100}\right)=\frac{175}{100}=1\frac{3}{4}$

$175\%=175(0.01)=1.75$

115. $19\% = 19\left(\frac{1}{100}\right) = \frac{19}{100}$

$19\% = 19(0.01) = 0.19$

117. $5\% = 5\left(\frac{1}{100}\right) = \frac{5}{100} = \frac{1}{20}$

$5\% = 5(0.01) = 0.05$

119. $450\% = 450\left(\frac{1}{100}\right) = \frac{450}{100} = 4\frac{1}{2}$

$450\% = 450(0.01) = 4.5$

121. $8\% = 8\left(\frac{1}{100}\right) = \frac{8}{100} = \frac{2}{25}$

$8\% = 8(0.01) = 0.08$

123. $11\frac{1}{9}\% = \frac{100}{9}\left(\frac{1}{100}\right) = \frac{1}{9}$

125. $31\frac{1}{4}\% = \frac{125}{4}\left(\frac{1}{100}\right) = \frac{5 \cdot \overset{1}{\cancel{25}}}{4 \cdot 4 \cdot \underset{1}{\cancel{25}}} = \frac{5}{16}$

127. $\frac{1}{2}\% = \frac{1}{2}\left(\frac{1}{100}\right) = \frac{1}{200}$

129. $6\frac{1}{4}\% = \frac{25}{4}\left(\frac{1}{100}\right) = \frac{\overset{1}{\cancel{25}}}{4 \cdot 4 \cdot \underset{1}{\cancel{25}}} = \frac{1}{16}$

131. $7.3\% = 7.3(0.01) = 0.073$

133. $15.8\% = 15.8(0.01) = 0.158$

135. $9.15\% = 9.15(0.01) = 0.0915$

137. $18.23\% = 18.23(0.01) = 0.1823$

139. $0.15 = 0.15(100\%) = 15\%$

141. $0.05 = 0.05(100\%) = 5\%$

143. $0.175 = 0.175(100\%) = 17.5\%$

145. $1.15 = 1.15(100\%) = 115\%$

147. $0.008 = 0.008(100\%) = 0.8\%$

149. $0.065 = 0.065(100\%) = 6.5\%$

151. $\frac{27}{50} = \frac{27}{50}(100\%) = \frac{2700}{50}\% = 54\%$

153. $\frac{1}{3} = \frac{1}{3}(100\%) = \frac{100}{3}\% \approx 33.3\%$

155. $\frac{4}{9} = \frac{4}{9}(100\%) = \frac{400}{9}\% \approx 44.4\%$

157. $2\frac{1}{2} = \frac{5}{2} = \frac{5}{2}(100\%) = \frac{500}{2}\% = 250\%$

159. $\frac{3}{8} = \frac{3}{8}(100\%) = \frac{300}{8}\% = 37\frac{1}{2}\%$

161. $\frac{5}{14} = \frac{5}{14}(100\%) = \frac{500}{14}\% = 35\frac{5}{7}\%$

163. $1\frac{1}{4} = \frac{5}{4} = \frac{5}{4}(100\%) = \frac{500}{4}\% = 125\%$

165. $1\frac{5}{9} = \frac{14}{9} = \frac{14}{9}(100\%) = \frac{1400}{9}\% = 155\frac{5}{9}\%$

167. The fraction $\frac{4}{3}$ represents a number <u>greater than 100%</u> because the numerator is greater than the denominator.

169. Internet: $40\% = \frac{40}{100} = \frac{2}{5}$

171. Newspaper ad: 22% represents less than one-quarter because 25% is one quarter and 22% < 25%.

173. The number 28 is a natural number, an integer, a positive integer, a rational number, and a real number.

175. The number -7.707 is a rational number and a real number.

177. The number 0.171771777... is an irrational number and a real number.

179. **a.** $112.1°F-(-87.9°F)=112.1°F+87.9°F$
$=200.0°F$

b. $44.5°C-(-66.6°C)=44.5°C+66.6°C$
$=111.1°C$

181. Difference in the deficits in 1980 and 1985:
$-73.830-(-212.308)=-73.830+212.308$
$=\$138.478\,billion$

183. $\frac{-212.308}{-53.242}=3.987...\approx 4$ times greater

185. $-3.2°C-0.4°C=-3.2°C+(-0.4°C)$
$=-3.6°C$

187. $x-0.30x=\mathbf{0.70x}$

189. A common denominator allows the fractions to be written as like terms that can then be added or subtracted.
It is not necessary to have a common denominator when multiplying two fractions. Multiplication does not require that we have like objects.

191.

$\frac{2}{3}$	$-\frac{1}{6}$	0
$-\frac{1}{2}$	$\frac{1}{6}$	$\frac{5}{6}$
$\frac{1}{3}$	$\frac{1}{2}$	$-\frac{1}{3}$

193. Answers will vary. For example:
$-\frac{1}{2}+\left(-\frac{1}{4}\right)=-\frac{3}{4}; \frac{1}{2}+\frac{1}{4}=\frac{3}{4};$
$\frac{3}{4}+\left(-\frac{1}{4}\right)=\frac{1}{2}.$

Section 1.4 Exponents and the Order of Operations Agreement

1. nine to the fifth power: 9^5

3. seven to the *n*th power: 7^n

5. False. $(-5)^2 = 25$, $-5^2 = -25$, and $-(5)^2 = -25$.

7. True. To evaluate an expression means to determine what one number it is equal to.

9. In the expression $(-5)^2$, -5 is called the <u>base</u> and 2 is called the <u>exponent</u>. To evaluate $(-5)^2$, find the product <u>(-5)(-5)</u> = <u>25</u>.

11. $6^2 = (6)(6) = 36$

13. $-7^2 = -(7)(7) = -49$

15. $(-3)^2 = (-3)(-3) = 9$

17. $(-3)^4 = (-3)(-3)(-3)(-3) = 81$

19. $\left(\frac{1}{2}\right)^2 = \frac{1}{2}\cdot\frac{1}{2} = \frac{1}{4}$

21. $(0.3)^2 = (0.3)(0.3) = 0.09$

23. $(-3)2^2 = (-3)(2)(2) = -12$

25. $2^3 \cdot 3^3 \cdot (-4) = 8 \cdot 27 \cdot (-4) = -864$

27. $\left(\frac{2}{3}\right)^2 \cdot 3^3 = \frac{4}{\cancel{9}_1} \cdot \cancel{27}^3 = 12$

29. $(0.3)^3 \cdot 2^3 = 0.027(8) = 0.216$

31. $\left(\frac{2}{3}\right)^2 \cdot \frac{1}{4} \cdot 3^3 = \frac{\cancel{4}^1}{\cancel{3}_1 \cdot \cancel{3}_1} \cdot \frac{1}{\cancel{4}_1} \cdot \frac{\cancel{3}^1 \cdot \cancel{3}^1 \cdot 3}{1} = \frac{3}{1} = 3$

33. The fifth power of negative eighteen is **negative** because the product of an odd number of negative factors is negative.

35. We need an Order of Operations Agreement to ensure that there is only one way in which an expression can be correctly simplified.

37. $2(3^3) = 2\ \underline{(27)} = \underline{54}$

39. $4 - 8 \div 2 = 4 - 4 = 0$

41. $2(3-4) - (-3)^2 = 2(-1) - (-3)^2$
$= 2(-1) - 9$
$= -2 - 9 = -11$

43. $24 - 18 \div 3 + 2 = 24 - 6 + 2 = 20$

45. $16 + 15 \div (-5) - 2 = 16 + (-3) - 2 = 11$

47. $3 - 2[8 - (3-2)] = 3 - 2[8-1]$
$= 3 - 2[7] = 3 - 14$
$= -11$

49. $6 + \frac{16-4}{2^2+2} - 2 = 6 + \frac{12}{4+2} - 2$
$= 6 + \frac{12}{6} - 2$
$= 6 + 2 - 2 = 6$

51. $96 \div 2[12 + (6-2)] - 3^3 = 96 \div 2[12+4] - 3^3$
$= 96 \div 2[16] - 3^3$
$= 96 \div 2[16] - 27$
$= 48[16] - 27$
$= 768 - 27 = 741$

53. $16 \div 2 - 4^2 - (-3)^2 = 16 \div 2 - 16 - 9$
$= 8 - 16 - 9 = -17$

55. $16 - 3(8-3)^2 \div 5 = 16 - 3(5)^2 \div 5$
$= 16 - 3(25) \div 5$
$= 16 - 75 \div 5$
$= 16 - 15 = 1$

57. $\frac{(-10)+(-2)}{6^2 - 30} \div |2-4| = \frac{-12}{36-30} \div |-2|$
$= \frac{-12}{6} \div 2 = -2 \div 2 = -1$

59. $0.3(1.7 - 4.8) + (1.2)^2 = 0.3(-3.1) + 1.44$
$= -0.93 + 1.44 = 0.51$

61. $\frac{3}{8}\div\left|\frac{5}{6}+\frac{2}{3}\right|=\frac{3}{8}\div\left|\frac{5}{6}+\frac{4}{6}\right|=\frac{3}{8}\div\left|\frac{9}{6}\right|$

$=\frac{3}{8}\div\frac{9}{6}=\frac{3}{8}\cdot\frac{6}{9}$

$=\frac{\cancel{3}^{1}}{\cancel{2}_{1}\cdot 2\cdot 2}\cdot\frac{\cancel{2}^{1}}{\cancel{3}_{1}}=\frac{1}{4}$

63. Statement **iii** because

$9-2^2(1-5)=9-2^2(-4)=9-4(-4)$.

65. To simplify Exercise 55, first perform the subtraction inside the parentheses. Then simplify the exponential expression. Next multiply and divide from left to right. Finally, subtract.

67. $(-3)^3 > (-2)^5$ because -27 > -32.

69. $\frac{10^{12}\ operations}{76.4\times 10^{9}\ operations\ per\ \text{sec}}\approx 13\ \text{sec}$

71. Answers may vary.

a. possible answer: r = ½

b. possible answer: $r = 0$ or $r = 1$

c. possible answer: $r = 2$ or $r = 4$

73. Since $4^2 = 16$, the last digit in 34^{202} will be 6.

75. Since $7^2 = 49$, the last digit in 27^{622} will be 9.

Section 1.5 Concepts from Geometry

1. True. Perpendicular lines form four 90° angles.

3. True. Every square is a parallelogram.

5. An angle whose measure is 58° is **less than** the measure of a right angle.

7. If $\angle A$ and $\angle B$ are complementary angles, then $m\angle A + m\angle B = \underline{90}°$.

9. If $\angle E$ is a right angle, then $m\angle E = \underline{90}°$.

11. At 3 o'clock the measure of the smaller angle between the hands of the analog clock is 90°, which is **less than** 120°.

13. Complement of 62° = 90° - 62° = 28°

15. Supplement of 48° = 180° - 48° = 132°

17. Complement of 7° = 90° - 7° = 83°

19. Supplement of 89° = 180° - 89° = 91°

21. $m\angle AOC = m\angle AOB - m\angle COB$
$=180°-48°=132°$

23. $m\angle x = 90° - 39° = 51°$

25. $m\angle AOB = m\angle AOC + m\angle COB$
$= 32° + 45° = 77°$

27. $m\angle AOC = m\angle AOB - m\angle COB$
$= 138° - 59° = 79°$

29. $m\angle A = 360° - 68° = 292°$

31. A plane figure that has three sides is called a <u>triangle</u>.

33. $perimeter = side\ 1 + side\ 2 = side\ 3$
$= 2.51 + 4.08 + 3.12 = 9.71 cm$

35. $perimeter = 2 \cdot length + 2 \cdot width$
$= 2(4\,ft\ 8in) + 2(2\,ft\ 5in)$
$= 8\,ft\ 16in + 4\,ft\ 10in$
$= 12\,ft\ 26in = 14\,ft\,2in$

37. $perimeter = 4 \cdot side = 4(13) = 52in$

39. $circumference = 2 \cdot \pi \cdot radius$
$= 2(3.14)21 = 131.88cm$

41. $radius = \frac{1}{2} \cdot diameter = \frac{1}{2} \cdot 1.2 = 0.6m$
$circumference = 2 \cdot \pi \cdot radius$
$= 2(3.14)0.6 = 3.768m$

43. $cost\ for\ framing = (perimeter\ in\ ft)(cost\ per\ ft)$
$perimeter = 2 \cdot length + 2 \cdot width$
$= 2 \cdot 5 + 2 \cdot 3 = 16\,feet$
$price = (16\,feet)(\$4.81 / ft) = \$76.96$

45. $cost\ of\ binding = (circumference\ in\ ft)(cost / ft)$
$circumference = 2 \cdot \pi \cdot radius$
$= 2 \cdot (3.14)3 = 18.84\,feet$
$price = (18.84\,feet)(\$1.05 / ft) = \$19.78$

47. The first square has a perimeter of 8 units, the second has a perimeter of 10 units, the third has a perimeter of 12 units, the next four squares have perimeters of 14, 16, 18, 20 units. The eighth square has a perimeter of 22 units.

49. The formula for the area of a rectangle is Area = length · width.

51. $area = length \cdot width = 4 \cdot 8 = 32\,ft^2$

53. $area = base \cdot height = 14 \cdot 27 = 378cm^2$

55. $area = \pi \cdot radius^2 = 3.14(4^2)$
$= 3.14(16) = 50.24in^2$

57. $area = side^2 = 4.1^2 = 16.81m^2$

59. $area = \frac{1}{2} \cdot base \cdot height = \frac{1}{2} \cdot 7 \cdot 15 = 52.5cm^2$

61. $area = \pi \cdot radius^2 = 3.14(8.5^2)$
$= 3.14(72.25) = 226.865in^2$

63. $cost\ of\ carpeting = (area\ of\ floors\ in\ yd^2)(cost / yd^2)$
$area = 400{,}000\,ft^2 \cdot \left(\frac{1yd^2}{9\,ft^2}\right)$
$= 44{,}444.4yd^2$
$price = (44{,}444.4yd^2)(\$36 / yd^2)$
$= \$1{,}600{,}000$

65. The room is a square with sides measuring 18 feet or 6 yards.
$area = side^2 = 6^2 = 36yd^2$

67. $cost\ of\ window = (area\ in\ ft^2)(cost\ per\ ft^2)$
$area = \pi \cdot radius^2$
$area = (3.14)2^2 = 12.56 square\ feet$
$price = (12.56 square\ feet)(\$48 / square\ foot)$
$price \approx \$603$

69. $area = length \cdot width = 150 \cdot 70 = 10{,}500mi^2$

71. By trial-and-error, we find that the rectangle with perimeter of 20 units with the greatest possible area measures **5 units by 5 units**.

73. $perimeter = sum\ of\ lengths\ of\ sides$
$$= 60+12+42+16+18+28 = 176\,ft$$

$$area = sum\ of\ areas\ of\ smaller\ rectangles$$
$$= 60(12)+18(16) = 1008\,ft^2$$

75. The area of the shaded portion is the area of the square minus the sum of the areas of the four triangles in the corners.

$$area\ of\ square = side^2$$
$$= 4^2 = 16\,ft^2$$
$$area\ of\ corners = 4\cdot\frac{1}{2}\cdot 2\cdot 2$$
$$= 8\,ft^2$$
$$shaded\ area = 16-8 = 8\,ft^2$$

77. Cutting out the triangle and placing it on the other side of the parallelogram does not change the area of the figure. The new figure is a rectangle whose length is *b* and whose width is *h*. The area of the rectangle and of the parallelogram is *bh*.

Chapter 1 Review Exercises

1. Natural numbers less than 7: {1, 2, 3, 4, 5, 6}

2. $\frac{5}{8} = 0.625 = 62.5\%$

3. $-|-4| = -4$

4. $16-(-30)-42 = 16+30+(-42) = 4$

5. $area = \frac{1}{2}\cdot base\cdot height = \frac{1}{2}\cdot 4\cdot 9 = 18\,cm^2$

6. $\frac{7}{9} = 0.777... = 0.\overline{7}$

7. $(6.02)(-0.89) = -5.3578$

8. $\frac{-10+2}{2+(-4)} \div 2+6 = \frac{-8}{-2} \div 2+6 = 4\div 2+6 = 2+6 = 8$

9. The opposite of -4 is **4**.

10. $16-30 = 16+(-30) = -14$

11. $0.672 = 67.2\%$

12. $79\frac{1}{2}\% = \frac{159}{2}\cdot\frac{1}{100} = \frac{159}{200}$

13. $-72\div 8 = -9$

14. $\frac{17}{20} = 17\div 20 = 0.85$

15. $\frac{5}{12}\div\left(-\frac{5}{6}\right) = \frac{\overset{1}{\cancel{5}}}{\underset{2}{\cancel{12}}}\cdot\left(-\frac{\overset{1}{\cancel{6}}}{\underset{1}{\cancel{5}}}\right) = -\frac{1}{2}$

16. $3^2-4+20\div 5 = 9-4+20\div 5 = 9-4+4 = 9$

17. $m\angle AOB = m\angle AOC - m\angle BOC$
$$= 82° - 45° = 37°$$

18. -22 + 14 + (-8) = -8 + (-8) = -16

19. $(-5)(-6)(3) = (30)(3) = 90$

20. $6.039-12.92 = 6.039+(-12.92) = -6.881$

21. Elements of A less than or equal to -3: -5 and 3

22. $7\% = 0.07$

23. $\frac{3}{4}(4)^2 = \frac{3}{4}(16) = 12$

24. -2 > -40 because -2 lies to the right of -40 on the number line.

25. 13 + (-16) = -3

26. Complement of 56° = 90° - 56° = 34°

27. $-\frac{2}{5}+\frac{7}{15}=-\frac{6}{15}+\frac{7}{15}=\frac{1}{15}$

28. $(-3^3)\cdot 2^2 = -27(4) = -108$

29. $2\frac{7}{9}=\frac{25}{9}\cdot 100\% = \frac{2500}{9}\%$
$= 277.777...\% \approx 277.8\%$

30. 240% = 2.4

31. Supplement of 28° = 180° - 28° = 152°

32. $96 \div (-12) = -8$

33. $area = \pi \cdot radius^2 = 3.14(3^2) = 28.26m^2$

34. $2^3 \div 4 - 2(2-7) = 2^3 \div 4 - 2(-5)$
$= 8 \div 4 - 2(-5) = 2 + 10 = 12$

35. $|-3| = 3$

36. $1\frac{2}{3}=\frac{5}{3}\cdot 100\% = \frac{500}{3}\% = 166\frac{2}{3}\%$

37. **a.** Opposites: 12, 8, 1, -7
b. Absolute values: 12, 8, 1, 7

38. $\frac{7}{11} = 7 \div 11 = 0.6363... = 0.\overline{63}$

39. $$\begin{array}{r} 11.53 \\ 23\overline{)265.40} \\ \underline{23} \\ 35 \\ \underline{23} \\ 124 \\ \underline{115} \\ 90 \\ \underline{69} \\ 21 \end{array}$$ so $\frac{0.2654}{-0.023}=\frac{265.4}{-23}\approx -11.5$

40. $(7-2)^2 - 5 - 3\cdot 4 = 5^2 - 5 - 3\cdot 4 = 25 - 5 - 12 = 8$

41. -12 + 8 + (-4) = -8

42. $-\frac{5}{8}+\frac{1}{6}=-\frac{15}{24}+\frac{4}{24}=-\frac{11}{24}$

43. Elements of *D* greater than -19: -17, -9, 0, 4

44. 0.002 = 0.2%

45. $-4^2\cdot\left(\frac{1}{2}\right)^2 = -16\cdot\left(\frac{1}{4}\right) = -4$

46. -1.329 + 4.89 = 3.561

47. $-|17| = -17$

48. $-5 - 22 - (-13) - 19 - (-6)$
$= -5 + (-22) + 13 + (-19) + 6$
$= -27$

49. $\left(\frac{1}{\cancel{3}_1}\right)\left(-\frac{\cancel{2}^1\cdot\cancel{2}^1}{5}\right)\left(\frac{\cancel{3}^1}{\cancel{2}_1\cdot\cancel{2}_1\cdot 2}\right) = -\frac{1}{10}$

50. -43 < -34 because -43 is to the left of -34 on the number line.

51. $perimeter = 2 \cdot length + 2 \cdot width$
$= 2(12) + 2(10) = 42in$

52. $(-2)^3 \cdot 4^2 = -8(16) = -128$

53. $0.075 = 7.5\%$

54. $\frac{19}{35} \cdot 100\% = \frac{1900}{35} = 54\frac{10}{35}\% = 54\frac{2}{7}\%$

55. 14 + (-18) + 6 + (-20) = -18

56. -4(-8)(12)(0) = 0

57. $2^3 - 7 + 16 \div (-3+5) = 2^3 - 7 + 16 \div 2$
$= 8 - 7 + 8 = 9$

58. $\frac{3}{4} + \frac{1}{2} - \frac{3}{8} = \frac{6}{8} + \frac{4}{8} - \frac{3}{8} = \frac{7}{8}$

59. $-128 \div (-8) = 16$

60. -57 < 28 because -57 lies to the left of 28 on the number line.

61. $\left(-\frac{1}{3}\right)^3 \cdot 9^2 = -\frac{1}{27} \cdot 81 = -3$

62. $-7 + (-3) + (-12) + 16 = -6$

63. 5(-2)(10)(-3)=300

64. Negative integers greater than -4: {-3, -2, -1}

65. *cost of flower bed = (area in ft²)(cost per ft²)*
$area = length \cdot width = (8)12 = 96 square\ feet$
$price = (96 square\ feet)(\$2.51 / square\ foot)$
$= \$240.96$

66. $-6°C + 14°C = 8°C$

67. $average = \frac{sum}{3} = \frac{-8+7+(-5)}{3} = \frac{-6}{3} = -2°C$

68. $record\ high - record\ low = 63°F - (-45°F)$
$= 63°F + 45°F = 108°F$

69. $-13°C + 7°C = -6°C$

70. $Venus - Pluto = 480°C - (-234°C)$
$= 480°C + 234°C = 714°C$

Chapter Test

1. $55\% = \frac{55}{100} = \frac{11}{20}$

2. Elements of *B* less than -5: -8 and -6

3. $m\angle x = 94° - 47° = 47°$

4. $\frac{3}{20} = 3 \div 20 = 0.15$

5. $\frac{\cancel{3}^{1}}{\cancel{2}_{1} \cdot 2}\left(-\frac{\cancel{2}^{1}}{\cancel{3}_{1} \cdot 7}\right) = -\frac{1}{14}$ **6.** $-75 \div 5 = -15$

7. $\left(-\frac{2}{3}\right)^3 3^2 = -\frac{2 \cdot 2 \cdot 2}{\cancel{3} \cdot \cancel{3} \cdot 3} \cdot \cancel{3} \cdot \cancel{3} = -\frac{8}{3}$

8. -7 + (-3) + 12 = 2

9. Positive integers less than or equal to 6:

$\{1, 2, 3, 4, 5, 6\}$

10. $(1.59)100\% = 159\%$

11. $|-29| = 29$

12. -47 > -68 because -47 lies to the right of -68 on the number line.

13. $-\frac{4}{9}-\frac{5}{6} = -\frac{8}{18}+\left(-\frac{15}{18}\right) = -\frac{23}{18}$

14. $area = base \cdot height = 10(9) = 90cm^2$

15. $8+\frac{12-4}{3^2-1}-6 = 8+\frac{8}{8}-6 = 8+1-6 = 3$

16. $-\frac{5}{8} \div \left(-\frac{3}{4}\right) = -\frac{5}{\overset{1}{\cancel{2}}\cdot\overset{1}{\cancel{2}}\cdot 2}\cdot\left(-\frac{\overset{1}{\cancel{2}}\cdot\overset{1}{\cancel{2}}}{3}\right) = \frac{5}{6}$

17. $\frac{3}{13}\cdot 100\% = \frac{300}{13}\% \approx 23.1\%$

18. $6.2\% = 0.062$

19. $13 - (-5) - 4 = 13 + 5 + (-4) = 14$

20. $\frac{13}{30} = 0.4333... = 0.4\overline{3}$

21. $(-0.9)(2.7) = -2.43$

22. Complement of 28° = 90° - 28° = 62°

23. $2^2 \cdot (-4)^2 \cdot 10 = 4 \cdot 16 \cdot 10 = 640$

24. $-|-34| = -34$

25. $circumference = \pi \cdot diameter$

$= (3.14)27 = 84.78in$

26. **a.** Opposites: 17, 6, -5, -9

b. Absolute values: 17, 6, 5, 9

27. $\frac{16}{23}\cdot 100\% = \frac{1600}{23}\% = 69\frac{13}{23}\%$

28. $-18.354 + 6.97 = -11.384$

29. $-4(8)(-5) = 160$

30.
$$\begin{aligned} 9(-4) \div [2(8-5)^2] &= 9(-4) \div [2\cdot 3^2] \\ &= 9(-4) \div [2\cdot 9] \\ &= 9(-4) \div 18 \\ &= -36 \div 18 = -2 \end{aligned}$$

31. $-8°C + 12°C = 4°C$

32.
$$\begin{aligned} average = \frac{sum}{4} &= \frac{-61+(-58)+(-49)+(-24)}{4} \\ &= \frac{-192}{4} = -48°F \end{aligned}$$

33. *cost of fencing = (perimeter in ft)(cost per ft²)*

$$\begin{aligned} perimeter &= 2\cdot length + 2\cdot width \\ &= 2(150) + 2(200) = 700\ feet \end{aligned}$$

$price = (700\ feet)(\$6.52 / foot) = \$4564$

CHAPTER 2 Variable Expressions

Chapter 2 Prep Test

1. $-12-(-15)=-12+15=3$

2. $-36\div(-9)=4$

3. $-\frac{3}{4}+\frac{5}{6}=-\frac{3}{4}\cdot\frac{3}{3}+\frac{5}{6}\cdot\frac{2}{2}$
$=-\frac{9}{12}+\frac{10}{12}$
$=\frac{-9+10}{12}=\frac{1}{12}$

4. Reciprocal of $-\frac{9}{4}$: $-\frac{4}{9}$

5. $\left(-\frac{3}{4}\right)\div\left(-\frac{5}{2}\right)=\left(-\frac{3}{\cancel{4}_2}\right)\cdot\left(-\frac{\cancel{2}^1}{5}\right)=\frac{3}{10}$

6. $-2^4=-2\cdot2\cdot2\cdot2=-16$

7. $\left(\frac{2}{3}\right)^3=\frac{2}{3}\cdot\frac{2}{3}\cdot\frac{2}{3}=\frac{8}{27}$

8. $3\cdot4^2=3\cdot4\cdot4=3\cdot16=48$

9. $7-2\cdot3=7-6=1$

10. $5-7(3-2^2)=5-7(3-4)$
$=5-7(-1)$
$=5+7$
$=12$

Section 2.1 Evaluating Variable Expressions

1. It is **always true** that $3x^2$ is a variable expression.

3. It is **sometimes true** that for the expression x^5, the value of x is 1.

5. It is **always true** that the result of evaluating a variable expression is a single number.

7. $mn^2-m=(-2)(5)^2-(-2)$ Replace m with -2 and n with 5.
$=(-2)(25)-(-2)$
$=-50-(-2)$
$=-50+2$
$=-48$

9. Terms: $-3n^2$, $-4n$, $\underline{7}$

11. Variable terms: $9\underline{b^2}$, $-4\underline{ab}$, $\underline{a^2}$

13. Variable terms: $-8\underline{n}$, $-3\underline{n^2}$

15. Coefficients: 12, -8, -1

17. The phrase "evaluate a variable expression" means to substitute the given values of the variables into the variable expression and then use the Order of Operations Agreement to simplify the resulting expression.

19. $3a+2b=3(2)+2(3)$
$=6+6=12$

21. $-a^2=-(2)^2=-4$

23. $-3a+4b=-3(2)+4(3)$
$=-6+12=6$

25. $b^2 - 3 = (3)^2 - 3 = 9 - 3 = 6$

27. $16 \div (2c) = 16 \div [2(-4)]$
$= 16 \div (-8) = -2$

29. $bc \div (2a) = (3)(-4) \div [2(2)]$
$= -12 \div 4 = -3$

31. $a^2 - b^2 = (2)^2 - (3)^2$
$= 4 - 9 = -5$

33. $(a+b)^2 = (2+3)^2 = 5^2 = 25$

35. $2a - (c+a)^2 = 2(2) - [(-4)+(2)]^2$
$= 2(2) - [-2]^2$
$= 2(2) - 4$
$= 4 - 4 = 0$

37. $b^2 - \dfrac{ac}{8} = (3)^2 - \dfrac{(2)(-4)}{8}$
$= (3)^2 - \dfrac{-8}{8}$
$= 9 + 1 = 10$

39. $(b-2a)^2 + bc = [(3) - 2(2)]^2 + (3)(-4)$
$= [3-4]^2 + (3)(-4)$
$= [-1]^2 + (3)(-4)$
$= 1 - 12 = -11$

41. $\dfrac{d-b}{c} = \dfrac{(3)-(4)}{-1} = \dfrac{-1}{-1} = 1$

43. $\dfrac{b+2d}{b} = \dfrac{(4)+2(3)}{4}$
$= \dfrac{4+6}{4} = \dfrac{10}{4}$
$= \dfrac{5}{2}$

45. $\dfrac{2c-d}{-ad} = \dfrac{2(-1)-(3)}{-(-2)(3)}$
$= \dfrac{-2-3}{6} = -\dfrac{5}{6}$

47. $(d-a)^2 - 3c = [(3)-(-2)]^2 - 3(-1)$
$= 5^2 - 3(-1)$
$= 25 + 3 = 28$

49. $(b-c)^2 \div 5 = [(4)-(-1)]^2 \div 5$
$= 5^2 \div 5 = 25 \div 5 = 5$

51. $a^2 - 5a - 6 = (-2)^2 - 5(-2) - 6$
$= 4 + 10 - 6 = 8$

53. $\dfrac{2ac}{b} \div (-c) = \dfrac{2(-2)(-1)}{(4)} \div [-(-1)]$
$= \dfrac{4}{4} \div 1 = 1 \div 1 = 1$

55. $3(b-a) - bc = 3[(4)-(-2)] - (4)(-1)$
$= 3(6) - 4(-1)$
$= 18 + 4 = 22$

57. $\dfrac{b^2 - a}{ad + 3c} = \dfrac{(4)^2 - (-2)}{(-2)(3) + 3(-1)}$
$= \dfrac{16+2}{-6-3} = \dfrac{18}{-9}$
$= -2$

59. $\dfrac{5}{8}a^4 - c^2 = \dfrac{5}{8}(-2)^4 - (-1)^2$
$= \dfrac{5}{8}(16) - 1 = 10 - 1 = 9$

61. $\dfrac{abc}{b-d} = \dfrac{(-2)(4)(-1)}{(4)-(3)} = \dfrac{8}{1} = 8$

63. $d^3 - 3d - 9 = (3)^3 - 3(3) - 9$
$= 27 - 3(3) - 9$
$= 27 - 9 - 9 = 9$

65. $c^2 - ab = (-0.8)^2 - (2.7)(-1.6)$
$= 0.64 - (2.7)(-1.6)$
$= 0.64 + 4.32$
$= 4.96$

67. $\frac{b^3}{c} - 4a = \frac{(-1.6)^3}{(-0.8)} - 4(2.7)$
$= \frac{-4.096}{-0.8} - 4(2.7)$
$= 5.12 - 10.8$
$= -5.68$

69. a. No, V cannot be a whole number because it is the product of a whole number and an irrational number, which is irrational.

b. No, the exact volume cannot be V cm^2 because volume is measured in cubic units.

71. $V = \pi r^2 h$
$= \pi(1.25)^2(5.25)$
$= \pi(1.5625)(5.25)$
$= 25.8 in^3$

73. $A = \frac{1}{2}h(b + B)$
$= \frac{1}{2}(6.75)(17.5 + 10.25)$
$= \frac{1}{2}(6.75)(27.75)$
$= 93.7 cm^2$

75. $V = \pi r^2 h$
$= \pi(3.5)^2(12.6)$
$\approx 484.9 m^3$

77. $a = 3x^2 - 4x - 5$
$= 3(-2)^2 - 4(-2) - 5$
$= 3(4) - 4(-2) - 5$
$= 12 + 8 - 5 = 15$

$3a - 4 = 3(15) - 4$
$= 45 - 4 = 41$

79. $d = 3w^2 - 2v = 3(-1)^2 - 2(3)$
$= 3(1) - 2(3) = 3 - 6 = -3$

$d^2 - 4d = (-3)^2 - 4(-3) = 9 + 12 = 21$

81. $|-4ab| = |-4(-2)(-3)| = |-24| = 24$

83. $3^x - x^3 = 3^2 - 2^3 = 9 - 8 = 1$

85. $z^y = (-2)^3 = -8$

87. $x^x - y^y = 2^2 - 3^3 = 4 - 27 = -23$

89. a. **4** is the first natural number for which 3^x is greater than x^3.

b. **5** is the first natural number for which 4^x is greater than x^4.

c. **6** is the first natural number for which 5^x is greater than x^5.

d. **7** is the first natural number for which 6^x is greater than x^6.

Conjecture: $n^x > x^n$ if $x \geq n + 1$

91. Recorded grade = $G + 8$

93. Niece's age = $A - 16$

Section 2.2 Simplifying Variable Expressions

1. The only pair of terms that are like terms are in part **iv**. The variable parts are exactly the same.

3. The statement is **true** because of the Multiplication Property of One.

5. The statement is **false** because of the Inverse Property of Multiplication.

7. The statement is **false** because like terms must have exactly the same variable parts, not necessarily the same coefficients.

9. The fact that two terms can be added in either order is called the Commutative Property of Addition.

11. The Inverse Property of Multiplication tells us that the product of a number and its reciprocal (or multiplicative inverse) is one.

13. Commutative Property of Multiplication: $2 \cdot 5 = 5 \cdot 2$

15. Associative Property of Multiplication: $(4 \cdot 5) \cdot 6 = 4 \cdot (5 \cdot 6)$

17. The Distributive Property: $2(4+3) = 8+6$

19. The Inverse Property of Addition: $8+(-8) = 0$

21. The Multiplication Property of One: $-4 \cdot 1 = -4$

23. The Inverse Property of Addition

25. The Commutative Property of Addition

27. The Associative Property of Addition

29. The Commutative Property of Multiplication

31. The Associative Property of Multiplication

33. *Like terms* are terms that have exactly the same variable parts. The terms $5xy$ and $-3xy$ are like terms. The terms $5xy$ and $3x^2$ are not like terms.

35. The terms $3x$ and $5x$ are like terms because they have the same variable parts.

37. $$\begin{aligned} 5a - 8a &= 5a + (-8a) \\ &= [5 + (-8)]a \quad \textit{(Distributive)} \\ &= -3a \end{aligned}$$

39. $6x + 8x = (6+8)x = 14x$

41. $9a - 4a = 9a + (-4)a = [9 + (-4)]a = 5a$

43. $4y - 10y = 4y + (-10y) = [4 + (-10)]y = -6y$

45. $-3b - 7$

47. $-12a + 17a = (-12+17)a = 5a$

49. $5ab - 7ab = 5ab + (-7ab) = [5 + (-7)]ab = -2ab$

51. $-12xy + 17xy = (-12+17)xy = 5xy$

53. $-3ab + 3ab = (-3+3)ab = 0ab = 0$

55. $-\frac{1}{2}x-\frac{1}{3}x=-\frac{1}{2}x+\left(-\frac{1}{3}x\right)=\left[-\frac{1}{2}+\left(-\frac{1}{3}\right)\right]x$

$=\left[-\frac{3}{6}+\left(-\frac{2}{6}\right)\right]x=-\frac{5}{6}x$

57. $\frac{3}{8}x^2-\frac{5}{12}x^2=\left[\frac{3}{8}+\left(-\frac{5}{12}\right)\right]x^2$

$=\left[\frac{9}{24}+\left(-\frac{10}{24}\right)\right]x^2=-\frac{1}{24}x^2$

59. $3x+5x+3x=(3+5+3)x=11x$

61. $5a-3a+5a=(5-3+5)a=7a$

63. $-5x^2-12x^2+3x^2=[-5+(-12)+3]x^2=-14x^2$

65. $7x-8x+3y=[7+(-8)]x+3y=-x+3y$

67. $7x-3y+10x=7x+10x-3y$

$=(7+10)x-3y=17x-3y$

69. $3a-7b-5a+b=3a-5a-7b+b$

$=[3+(-5)]a+(-7+1)b$

$=-2a-6b$

71. $3x-8y-10x+4x=3x-10x+4x-8y$

$=[3+(-10)+4]x-8y$

$=-3x-8y$

73. $x^2-7x-5x^2+5x=x^2-5x^2-7x+5x$

$=[1+(-5)]x^2+(-7+5)x$

$=-4x^2-2x$

75. **a.** The coefficient of a^2 will be **negative**.

b. The coefficient of a will be **positive.**

c. The constant term will be **0**.

77. $4(-12x)=[4(-12)]x=-48x$

Use the **Associative** Property of Multiplication to group the factors.

79. $4(3x)=(4\cdot 3)x=12x$

81. $-3(7a)=(-3\cdot 7)a=-21a$

83. $-2(-3)y=[(-2)(-3)]y=6y$

85. $(4x)2=2(4x)=(2\cdot 4)x=8x$

87. $(3a)(-2)=(-2)(3a)=(-2\cdot 3)a=-6a$

89. $(-3b)(-4)=(-4)(-3b)=[(-4)(-3)]b=12b$

91. $-5(3x^2)=(-5\cdot 3)x^2=-15x^2$

93. $\frac{1}{3}(3x^2)=\left(\frac{1}{3}\cdot 3\right)x^2=x^2$

95. $\frac{1}{8}(8x)=\left(\frac{1}{8}\cdot 8\right)x=x$

97. $-\frac{1}{7}(-7n)=\left[\left(-\frac{1}{7}\right)(-7)\right]n=n$

99. $\frac{12x}{5}\cdot\frac{5}{12}=\left(\frac{5}{12}\cdot\frac{12}{5}\right)x=x$

101. $(-10n)\left(-\frac{1}{10}\right)=\left(-\frac{1}{10}\right)(-10n)$

$=\left[\left(-\frac{1}{10}\right)\cdot(-10)\right]n$

$=n$

103. $\frac{1}{7}(14x)=\left(\frac{1}{7}\cdot 14\right)x=2x$

105. $-\frac{1}{7}(16x)=\left(-\frac{1}{8}\cdot 16\right)x=-2x$

107. $-\frac{5}{8}(24a^2)=\left(-\frac{5}{8}\cdot 24\right)a^2=-15a^2$

109. $-\frac{3}{4}(-8y)=\left[\left(-\frac{3}{4}\right)(-8)\right]y=6y$

111. $(33y)\left(\frac{1}{11}\right)=\left(\frac{1}{11}\right)(33y)=\left(\frac{1}{11}\cdot 33\right)y=3y$

113. $(-10x)\left(\frac{1}{5}\right)=\left(\frac{1}{5}\right)(-10x)=\left[\frac{1}{5}(-10)\right]x=-2x$

115. After multiplying by a proper fraction, the coefficient will be **less than one** because the product of two fractions that are each less than one will be a number that is also less than one.

117. $2(4x-3)=(2)(4x)-(2)(3)=8x-6$

119. $-2(a+7)=(-2)(a)+(-2)(7)=-2a-14$

121. $-3(2y-8)=(-3)(2y)-(-3)(8)=-6y+24$

123. $-(x+2)=(-1)(x)+(-1)(2)=-x-2$

125. $(5-3b)7=(5)(7)-(3b)(7)=35-21b$

127. $-3(3-5x)=(-3)(3)-(-3)(5x)=-9+15x$

129. $3(5x^2+2x)=(3)(5x^2)+(3)(2x)=15x^2+6x$

131. $-2(-y+9)=(-2)(-y)+(-2)(9)=2y-18$

133. $(-3x-6)5=(-3x)(5)-(6)(5)=-15x-30$

135. $2(-3x^2-14)=(2)(-3x^2)-(2)(14)=-6x^2-28$

137. $-3(2y^2-7)=(-3)(2y^2)-(-3)(7)=-6y^2+21$

139. $3(x^2+2x-6)=(3)(x^2)+(3)(2x)-(3)(6)$
$=3x^2+6x-18$

141. $-2(y^2-2y+4)=(-2)(y^2)-(-2)(2y)+(-2)(4)$
$=-2y^2+4y-8$

143. $2(-a^2-2a+3)=(2)(-a^2)-(2)(2a)+(2)(3)$
$=-2a^2-4a+6$

145. $-5(-2x^2-3x+7)$
$=(-5)(-2x^2)-(-5)\cdot 3x+(-5)\cdot 7$
$=10x^2+15x-35$

147. $-(3a^2+5a-4)$
$=(-1)(3a^2)+(-1)(5a)-(-1)(4)$
$=-3a^2-5a+4$

149. After being multiplied by a negative the constant term will be **positive**. The constant term in the original expression is negative, so multiplying by a negative yields a positive product.

151. When simplifying $4(3a - 7) + 2(a - 3)$, the first step is to use the Distributive Property to remove parentheses:

$$4(3a - 7) + 2(a - 3) = \underline{12}a - \underline{28} + \underline{2}a - \underline{6}$$

153. $4x - 2(3x + 8) = 4x - 6x - 16 = -2x - 16$

155. $9 - 3(4y + 6) = 9 - 12y - 18 = -12y - 9$

157. $5n - (7 - 2n) = 5n - 7 + 2n = 7n - 7$

159. $3(x + 2) - 5(x - 7) = 3x + 6 - 5x + 35 = -2x + 41$

161. $3(a - b) - 4(a + b) = 3a - 3b - 4a - 4b = -a - 7b$

163. $4[x - 2(x - 3)] = 4[x - 2x + 6]$
$= 4[-x + 6] = -4x + 24$

165. $-3[2x - (x + 7)] = -3[2x - x - 7]$
$= -3[x - 7]$
$= -3x + 21$

167. $2x - 3[x - 2(4 - x)] = 2x - 3[x - 8 + 2x]$
$= 2x - 3[3x - 8]$
$= 2x - 9x + 24$
$= -7x + 24$

169. $2x + 3(x - 2y) + 5(3x - 7y) = 2x + 3x - 6y + 15x - 35y$
$= 20x - 41y$

171. $12 - 7(y - 9) = 12 - 7y + 63$

Expression **iii** is equivalent.

173. A number that has no reciprocal is $\underline{0}$.

175. The additive inverse of $a - b$ is
$\underline{-(a - b) = -a + b}$.

177. **a. False.** $10 \div 5 \neq 5 \div 10$

b. False. $(10 \div 5) \div 2 \neq 10 \div (5 \div 2)$

c. False. $(9 - 5) - 2 \neq 9 - (5 - 2)$

d. False. $10 - 5 \neq 5 - 10$

e. True.

179. Reading from the addition table, Δ + ‡ = ◊.

181. Inspection of the addition table reveals that ◊ is the additive identity element because each sum with ◊ as an addend yields the other addend.

183. The sum of opposites is the additive identity element, ◊. Inspection of the table reveals that the sum ‡ + Δ = ◊. So, -Δ = ‡.

185. From problem #183, we know that -Δ = ‡. The sum then becomes ‡+ ‡ - ◊. Moving from left to right: Δ - ◊. From the table, -◊ = ◊ and Δ + ◊ = Δ.

Section 2.3 Translating Verbal Expressions into Variable Expressions

1. The statement is **false**. The correct translation of "five less than *n*" is "$n - 5$."

3. The statement is **false**. If the sum of two numbers is 12 and one of the numbers is x, then the other number is correctly expressed as $12 - x$.

5. The statement is **true.**

7. *Less than* indicated subtraction and *quotient* indicated division.

9. *Subtracted from* indicates subtraction, *product* indicates multiplication, and *cube* indicates an exponent.

11. $6+c$

13. $w+55$

15. $16+y$

17. b^2-30

19. $\frac{4}{5}m+18$

21. $9+\frac{t}{5}$

23. $7(r+8)$

25. $a(a+13)$

27. $\frac{1}{2}z^2+14$

29. $9m^3+m^2$

31. $s-\frac{s}{2}$

33. $c^2-(c+14)$

35. $8(b+5)$

37. $13-n$

39. $\frac{3}{7}n$

41. $\frac{2n}{5}$

43. $n(n+10)$

45. $\frac{3}{4+n}$

47. n^2+3n

49. $7n^2-4$

51. n^3-12n

53. $32-\frac{a}{7}$ translates into all three of the phrases.

55. The phrase "the difference between twelve times a number and fifteen times the number" can be translated as $12n-15n$. This expression simplifies to $-3n$.

57. $5n+n$ simplifies to $6n$.

59. $(n+11)+8$ simplifies to $n+19$.

61. $(n+9)+4$ simplifies to $n+13$.

63. $7(5n)$ simplifies to $35n$.

65. $17n + 2n$ simplifies to $19n$.

67. $n + 12n$ simplifies to $13n$.

69. $3(n^2 + 4)$ simplifies to $3n^2 + 12$.

71. $\frac{3}{4}(16n + 4)$ simplifies to $12n + 3$.

73. $16 - (n + 9)$ simplifies to $-n + 7$.

75. $6(n + 8)$ simplifies to $6n + 48$.

77. $7 - (n + 2)$ simplifies to $-n + 5$.

79. $\frac{1}{3}(n + 6n)$ simplifies to $\frac{7n}{3}$.

81. $(n - 6) + (n + 12)$ simplifies to $2n + 6$.

83. $(n - 20) + (n + 9)$ simplifies to $2n - 11$.

85. The sum of two numbers is 25. To express both numbers in terms of the same variable, let x be one number. Then the other number is $\underline{25 - x}$.

87. The width of a rectangle is one-half the length. To express the length and width in terms of the same variable, let L be the length. Then the width is $\frac{1}{2}L$.

89. Let x be one number; the other number is $18 - x$.

91. Let d be the diameter of Dione; the other diameter is $d + 253$.

93. Let G be the number of genes in the roundworm genome; the genes in the human genome is $G + 11{,}000$.

95. Let N be the total number of Americans; then Americans who want to explore Mars is $\frac{3}{4}N$.

97. Let s be the number of points awarded for a safety; then the points awarded for a touchdown is $3s$.

99. Let B be the attendance at major league Basketball games; then $B + 50{,}000{,}000$ is the attendance at major league baseball games.

101. Let L be the measure of the largest angle; then the smallest angle is $\frac{1}{2}L - 10$.

103. Let h be the number of hours of labor; then the amount of the bill is $238 + 89h$.

105. Let n be either the number of nickels or the number of dimes; the other is then $35 - n$.

107. There are twice as many hydrogen atoms. This is expressed as $2x$.

109. The distance that the rope moves is $\frac{3}{5}x$.

111. Variables are used to represent unknown quantities and the value can change according to the context of the problem.

113. Answers will vary. Examples include: the sum of p and 8; p increased by 8; the total of p and 8; p more than 8; 8 more than p.

115. Answers will vary. Examples include: the product of 4 and c; 4 times c; 4 multiplied by c.

117. a. Complete the table.

Figure Number	Number of Tiles
1	5
2	7
3	9
4	11
5	13
6	15
7	17

b. The number of tiles in the nth figure will be $n+n+3$ or $2n+3$.

Chapter 2 Review Exercises

1. $-7y^2+6y^2-(-2y^2)=(-7+6+2)y^2=y^2$

2. $(12x)\left(\frac{1}{4}\right)=\left(\frac{1}{4}\right)(12x)=\left(\frac{1}{4}\cdot 12\right)x=3x$

3. $\frac{2}{3}(-15a)=\left[\left(\frac{2}{3}\right)(-15)\right]a=-10a$

4. $-2(2x-4)=-4x+8$

5. $5(2x+4)-3(x-6)=10x+20-3x+18=7x+38$

6. $a^2-3b=(2)^2-3(-4)=4+12=16$

7. $-9+\underline{9}=0$

8. $-4(-9y)=[(-4)(-9)]y=36y$

9. $-2(-3y+9)=6y-18$

10.
$$\begin{aligned}3[2x-3(x-2y)]+3y&=3[2x-3x+6y]+3y\\&=3[-x+6y]+3y\\&=-3x+18y+3y\\&=-3x+21y\end{aligned}$$

11. $-4(2x^2-3y^2)=-8x^2+12y^2$

12. $3x-5x+7x=(3-5+7)x=5x$

13. $b^2-3ab=(-2)^2-3(3)(-2)=4+18=22$

14. $\frac{1}{5}(10x)=\left(\frac{1}{5}\cdot 10\right)x=2x$

15. $5(3-7b)=15-35b$

16.
$$\begin{aligned}2x+3[4-(3x-7)]&=2x+3[4-3x+7]\\&=2x+3[11-3x]\\&=2x+33-9x\\&=-7x+33\end{aligned}$$

17. The Commutative Property of Multiplication

18. $3(8-2x)=24-6x$

19. $-2x^2-(-3x^2)+4x^2=(-2+3+4)x^2=5x^2$

20. $-3x-2(2x-7)=-3x-4x+14=-7x+14$

21. $-3(3y^2-3y-7)=-9y^2+9y+21$

22. $-2[x-2(x-y)]+5y=-2[x-2x+2y]+5y$
$=-2[-x+2y]+5y$
$=2x-4y+5y$
$=2x+y$

23. $\dfrac{-2ab}{2b-a}=\dfrac{-2(-4)(6)}{2(6)-(-4)}=\dfrac{48}{16}=3$

24. $(-3)(-12y)=36y$

25. $4(3x-2)-7(x+5)=12x-8-7x-35=5x-43$

26. $(16x)\left(\dfrac{1}{8}\right)=2x$

27. $-3(2x^2-7y^2)=-3(2x^2)-(-3)(7y^2)$
$=-6x^2+21y^2$

28. $3(a-c)-2ab=3[2-(-4)]-2(2)(3)$
$=3[2+4]-2(2)(3)$
$=3(6)-2(2)(3)$
$=18-12=6$

29. $2x-3(x-2)=2x-3x+6=-x+6$

30. $2a-(-3b)-7a-5b=2a+3b-7a-5b$
$=-5a-2b$

31. $-5(2x^2-3x+6)=-10x^2+15x-30$

32. $3x-7y-12x=-9x-7y$

33. $\dfrac{1}{2}(12a)=\left(\dfrac{1}{2}\cdot 12\right)a=6a$

34. $2x+3[x-2(4-2x)]=2x+3[x-8+4x]$
$=2x+3[5x-8]$
$=2x+15x-24$
$=17x-24$

35. $3x+(-12y)-5x-(-7y)=3x+(-12y)+(-5x)+7y$
$=-2x-5y$

36. $\left(-\dfrac{5}{6}\right)(-36b)=\left(-\dfrac{5}{6}(-36)\right)b=30b$

37. $(6+3)7=42+21$

38. $4x^2+9x-6x^2-5x=-2x^2+4x$

39. $-\dfrac{3}{8}(16x^2)=\left(-\dfrac{3}{8}\cdot 16\right)x^2=-6x^2$

40. $-3[2x-(7x-9)]=-3[2x-7x+9]$
$=-3[-5x+9]$
$=15x-27$

41. $-(8a^2-3b^2)=-8a^2+3b^2$

42. The Multiplication Property of Zero

43. "*b* decreased by the product of seven and *b*" translates to $\boldsymbol{b-7b}$.

44. "The sum of a number and twice the square of the number" translates to $n+2n^2$.

45. "Three less than the quotient of six and a number" translates to $\dfrac{6}{n}-3$.

46. "Ten divided by the difference between y and two" translates to $\frac{10}{y-2}$.

47. "Eight times the quotient of twice a number and sixteen" translates to $8\left(\frac{2n}{16}\right)=n$.

48. "The product of four and the sum of two and five times a number" translates to $4(2+5n)=8+20n$.

49. Let h be the height of the triangle; then the base is $h+15$.

50. Let b be the amount of either espresso beans or mocha java beans; then the other is $20-b$.

Chapter 2 Test

1. $(9y)4=4(9y)=(4\cdot 9)y=36y$

2. $7x+5y-3x-8y=(7x-3x)+(5y-8y)=4x-3y$

3. $8n-(6-2n)=8n-6+2n=10n-6$

4. $$\begin{aligned}3ab-(2a)^2&=3(-2)(-3)-[2(-2)]^2\\&=3(-2)(-3)-[-4]^2\\&=3(2)(-3)-16\\&=18-16=2\end{aligned}$$

5. The Multiplication Property of One

6. $-4(-x+10)=4x-40$

7. $$\begin{aligned}\frac{2}{3}x^2-\frac{7}{12}x^2&=\left(\frac{2}{3}-\frac{7}{12}\right)x^2\\&=\left(\frac{8}{12}-\frac{7}{12}\right)x^2=\frac{1}{12}x^2\end{aligned}$$

8. $$\begin{aligned}(-10x)\left(-\frac{2}{5}\right)&=\left(-\frac{2}{5}\right)(-10x)\\&=\left[-\frac{2}{5}(-10)\right]x\\&=4x\end{aligned}$$

9. $(-4y^2+8)6=-24y^2+48$

10. $-19+19=0$

11. $\frac{-3ab}{2a+b}=\frac{-3(-1)(4)}{2(-1)+4}=\frac{12}{-2+4}=\frac{12}{2}=6$

12. $$\begin{aligned}5(x+y)-8(x-y)&=5x+5y-8x+8y\\&=-3x+13y\end{aligned}$$

13. $6b-9b+4b=(6-9+4)b=b$

14. $13(6a)=(13\cdot 6)a=78a$

15. $3(x^2-5x+4)=3x^2-15x+12$

16. $$\begin{aligned}4(b-a)+bc&=4(-3-2)+(-3)(4)\\&=4(-5)+(-3)(4)\\&=-20-12=-32\end{aligned}$$

17. $$\begin{aligned}&6x-3(y-7x)+2(5x-y)\\&=6x-3y+21x+10x-2y\\&=37x-5y\end{aligned}$$

18. "The quotient of eight more than n and seventeen" translates to $\frac{n+8}{17}$.

19. "The difference between the sum of a and b and the square of b" translates to $(a+b)-b^2$

20. "The sum of the square of a number and the product of the number and eleven" translates to $n^2 + 11n$.

21. "Twenty times the sum of a number and nine" is $20(n+9) = 20n + 180$.

22. "Two more than a number added to the difference between the number and three" is $(n-3)+(n+2) = 2n-1$.

23. "A number minus the product of one-fourth and twice the number" is $n - \frac{1}{4}(2n) = n - \frac{1}{2}n = \frac{1}{2}n$.

24. Let d be the distance from Earth to the sun; then the distance from Neptune to the sun is $30d$.

25. Let L be the length of one piece; then the length of the other piece is $9 - L$.

Cumulative Review Exercises

1. $-4+7+(-10) = 3+(-10) = -7$

2. $-16-(-25)-4 = -16+25+(-4) = 9+(-4) = 5$

3. $(-2)(3)(-4) = (-6)(-4) = 24$

4. $-60 \div 12 = -5$

5. $1\frac{1}{4} = \frac{5}{4} = 1.25$

$$\begin{array}{r} 1.25 \\ 4\overline{)5.00} \\ \underline{4\,0} \\ 100 \\ \underline{100} \\ 0 \end{array}$$

6. $60\% = 60 \cdot \frac{1}{100} = \frac{\overset{1}{\cancel{20}} \cdot 3}{1} \cdot \frac{1}{\underset{1}{\cancel{20}} \cdot 5} = \frac{3}{5}$

$60\% = 60(0.01) = 0.6$

7. Negative integers greater than or equal to -4:

{-4, -3, -2, -1}

8. $\frac{2}{25} = \frac{2}{25} \cdot 100\% = \frac{2}{\underset{1}{\cancel{25}}} \cdot \frac{\overset{4}{\cancel{100}}}{1}\% = 8\%$

9. $\frac{7}{12} - \frac{11}{16} - \left(-\frac{1}{3}\right) = \frac{28}{48} - \frac{33}{48} + \frac{16}{48} = \frac{11}{48}$

10. $\frac{5}{12} \div \left(\frac{3}{2}\right) = \frac{5}{\underset{6}{\cancel{12}}} \cdot \frac{\overset{1}{\cancel{2}}}{3} = \frac{5}{18}$

11. $\left(-\frac{9}{16}\right)\left(\frac{8}{27}\right)\left(-\frac{3}{2}\right) = \frac{\overset{1}{\cancel{3}} \cdot \overset{1}{\cancel{3}} \cdot \overset{1}{\cancel{2}} \cdot \overset{1}{\cancel{2}} \cdot \overset{1}{\cancel{2}} \cdot \overset{1}{\cancel{3}}}{\underset{1}{\cancel{2}} \cdot \underset{1}{\cancel{2}} \cdot \underset{1}{\cancel{2}} \cdot 2 \cdot \underset{1}{\cancel{3}} \cdot \underset{1}{\cancel{3}} \cdot \underset{1}{\cancel{3}} \cdot 2} = \frac{1}{4}$

12. $-3^2 \cdot \left(-\frac{2}{3}\right)^3 = -\overset{1}{\cancel{9}}\left(-\frac{8}{\underset{3}{\cancel{27}}}\right) = \frac{8}{3}$

13. $-2^5 \div (3-5)^2 - (-3) = -2^5 \div (-2)^2 - (-3)$
$= -32 \div 4 - (-3)$
$= -8 + 3 = -5$

14. $\left(-\frac{3}{4}\right)^2-\left(\frac{3}{8}-\frac{11}{12}\right)=\left(-\frac{3}{4}\right)^2-\left(\frac{9}{24}-\frac{22}{24}\right)$
$=\left(-\frac{3}{4}\right)^2-\left(-\frac{13}{24}\right)$
$=\frac{9}{16}+\frac{13}{24}$
$=\frac{27}{48}+\frac{26}{24}=\frac{53}{24}$

15. $a-3b^2=4-3(-2)^2$
$=4-3(4)$
$=4-12$
$=-8$

16. $-2x^2-(-3x^2)+4x^2=-2x^2+3x^2+4x^2=5x^2$

17. $8a-12b-9a=8a-9a-12b=-a-12b$

18. $\frac{1}{3}(9a)=\left(\frac{1}{3}\cdot 9\right)a=3a$

19. $\left(-\frac{5}{8}\right)(-32b)=\left(-\frac{5}{8}(-32)\right)b=20b$

20. $5(4-2x)=20-10x$

21. $-3(-2y+7)=6y-21$

22. $-2(3x^2-4y^2)=-6x^2+8y^2$

23. $-4(2y^2-5y-8)=-8y^2+20y+32$

24. $-4x-3(2x-5)=-4x-6x+15=-10x+15$

25. $3(4x-1)-7(x+2)=12x-3-7x-14=5x-17$

26. $3x+2[x-4(2-x)]=3x+2[x-8+4x]$
$=3x+2[5x-8]$
$=3x+10x-16$
$=13x-16$

27. $3[4x-2(x-4y)]+5y=3[4x-2x+8y]+5y$
$=3[2x+8y]+5y$
$=6x+24y+5y$
$=6x+29y$

28. "The difference between six and the product of a number and twelve" translates to $6-12n$.

29. "The total of five and the difference between a number and seven" translates to $5+(n-7)$.

30. Let w be the speed of the wildebeest; then the speed of the peregrine falcon is $4w$.

CHAPTER 3 Solving Equations and Inequalities

Chapter 3 Prep Test

1. $8-12=8+(-12)=-4$

2. $-\frac{\overset{1}{\cancel{3}}}{\underset{1}{\cancel{4}}}\left(-\frac{\overset{1}{\cancel{4}}}{\underset{1}{\cancel{3}}}\right)=1$

3. $-\frac{5}{8}(16)=-\frac{5}{\underset{1}{\cancel{8}}}\cdot\frac{\overset{2}{\cancel{16}}}{1}=-10$

4. $90\%=0.90$

5. $0.75=75\%$

6. $3x^2-4x-1;\ x=-4$
$$3(-4)^2-4(-4)-1$$
$$3(16)+16-1$$
$$48+16-1=63$$

7. $3x-5+7x=3x+7x-5=10x-5$

8. $8x-9-8x=8x-8x-9=-9$

9. $6x-3(6-x)=6x-18+3x=9x-18$

Section 3.1 Introduction to Equations

1. **a.** This is an **equation.**

   b. This is an **expression** – there is no equals sign.

   c. This is an **expression** – there is no equals sign.

   d. This is an **equation.**

   e. This is an **expression** – there is no equals sign.

3. Both **a** and **d** are in the form $x+a=b$. To solve equations of this form, subtract a from both sides of the equation.

5. It is **sometimes true** that both sides of an equation can be multiplied by the same number without changing the solution of the equation.

7. It is **never true** that the Multiplication Property of Equations is used to remove a term from one side of an equation. This property is used to remove a factor from one side of an equation.

9. To determine whether -3 is a solution of the equation $9-5x=-3-9x$ replace the variable by the given number, <u>-3</u>. Evaluate the numerical expressions. Compare the results.
$$9-5x=-3-9x$$
$$9-5(-3)=-3-9(-3)$$
$$9+15=-3+27$$
$$24=24$$
The results are equal; therefore, -3 is a solution of the equation $9-5x=-3-9x$.

11.
$$3a-4=10$$
$$3(-2)-4=10$$
$$-12-4=10$$
$$-16\neq 10$$
No, -2 is not a solution.

13.
$$7-3n=2$$
$$7-3(2)=2$$
$$7-6=2$$
$$1\neq 2$$
No, 2 is not a solution.

15. $3y-4=2y$
$3(4)-4=2(4)$
$12-4=8$
$8=8$

Yes, 4 is a solution.

17. $4-3b=4-5b$
$4-3(0)=4-5(0)$
$4-0=4-0$
$4=4$

Yes, 1 is a solution.

19. $2x^2-1=4x-1$
$2(2)^2-1=4(2)-1$
$8-1=8-1$
$7=7$

Yes, 2 is a solution.

21. $2a(a-1)=3a+3$
$2(3)(3-1)=3(3)+3$
$6(2)=9+3$
$12=12$

Yes, 3 is a solution.

23. $4y+1=3$
$4\left(\frac{1}{2}\right)+1=3$
$2+1=3$
$3=3$

Yes, $\frac{1}{2}$ is a solution.

25. $8x-1=12x+3$
$8\cdot\frac{3}{4}-1=12\cdot\frac{3}{4}+3$
$6-1=9+3$
$5\neq 12$

No, $\frac{3}{4}$ is not a solution.

27. $c^2-3c=4c-8.25$
$1.5^2-3(1.5)=4(1.5)-8.25$
$2.25-4.5=6-8.25$
$-2.25=-2.25$

Yes, 1.5 is a solution.

29. To solve the equation $18+n=25$, subtract 18 from each side of the equation. The solution is 7.

31. The value of x is greater than $\frac{19}{24}$. To solve the equation, $\frac{11}{16}$ is added to each side of the equation, and $\frac{19}{24}+\frac{11}{16}>\frac{19}{24}$.

33. $y+3=9$
$y+3-3=9-3$
$y=6$

Check: $6+3\stackrel{?}{=}9$
$9=9$ *True*

35. $z-6=10$
$z-6+6=10+6$
$z=16$

Check: $16-6\stackrel{?}{=}10$
$10=10$ *True*

37. $5+x=12$
$5+x-5=12-5$
$x=7$

Check: $5+7\stackrel{?}{=}12$
$12=12$ *True*

39. $t+12=10$
$t+12-12=10-12$
$t=-2$

Check: $-2+12\stackrel{?}{=}10$
$10=10$ *True*

41. $x-6=-5$
$x-6+6=-5+6$
$x=1$

Check: $1-6\stackrel{?}{=}-5$
$-5=-5$ *True*

43. $y-5=-5$
$y-5+5=-5+5$
$y=0$

Check: $0-5\overset{?}{=}-5$
$-5=-5$ *True*

45. $x-6=-3$
$x-6+6=-3+6$
$x=3$

Check: $3-6\overset{?}{=}-3$
$-3=-3$ *True*

47. $n+11=1$
$n+11-11=1-11$
$n=-10$

Check: $-10+11\overset{?}{=}1$
$1=1$ *True*

49. $8+x=5$
$8+x-8=5-8$
$x=-3$

Check: $8+(-3)\overset{?}{=}5$
$5=5$ *True*

51. $10+y=-4$
$10+y-10=-4-10$
$y=-14$

Check: $10+(-14)\overset{?}{=}-4$
$-4=-4$ *True*

53. $t-6=-4$
$y-6+6=-4+6$
$t=2$

Check: $2-6\overset{?}{=}-4$
$-4=-4$ *True*

55. $-8=n+1$
$-8-1=n+1-1$
$-9=n$

Check: $-8\overset{?}{=}-9+1$
$-8=-8$ *True*

57. $-6=y-5$
$-6+5=y-5+5$
$-1=y$

Check: $-6\overset{?}{=}-1-5$
$-6=-6$ *True*

59. $-9=5+x$
$-9-5=5+x-5$
$-14=x$

Check: $-9\overset{?}{=}5+(-14)$
$-9=-9$ *True*

61. $-7=-2+x$
$-7+2=-2+x+2$
$-5=x$

Check: $-7\overset{?}{=}-2+(-5)$
$-7=-7$ *True*

63. $c+\frac{3}{4}=-\frac{1}{4}$
$c+\frac{3}{4}-\frac{3}{4}=-\frac{1}{4}-\frac{3}{4}$
$c=-\frac{4}{4}$
$c=-1$

Check: $-1+\frac{3}{4}\overset{?}{=}-\frac{1}{4}$
$-\frac{4}{4}+\frac{3}{4}\overset{?}{=}-\frac{1}{4}$
$-\frac{1}{4}=-\frac{1}{4}$ *True*

65. $x-\frac{2}{5}=\frac{3}{5}$
$x-\frac{2}{5}+\frac{2}{5}=\frac{3}{5}+\frac{2}{5}$
$x=\frac{5}{5}$
$x=1$

Check: $1-\frac{2}{5}\overset{?}{=}\frac{3}{5}$
$\frac{5}{5}-\frac{2}{5}\overset{?}{=}\frac{3}{5}$
$\frac{3}{5}=\frac{3}{5}$ *True*

67. $\frac{4}{9}+a=-\frac{2}{9}$
$\frac{4}{9}+a-\frac{4}{9}=-\frac{2}{9}-\frac{4}{9}$
$a=-\frac{6}{9}$
$a=-\frac{2}{3}$

Check: $\frac{4}{9}+\left(-\frac{2}{3}\right)\overset{?}{=}-\frac{2}{9}$
$\frac{4}{9}-\frac{6}{9}\overset{?}{=}-\frac{2}{9}$
$-\frac{2}{9}=-\frac{2}{9}$ *True*

69. $-\frac{1}{4}=c-\frac{2}{3}$
$-\frac{1}{4}+\frac{2}{3}=c-\frac{2}{3}+\frac{2}{3}$
$-\frac{3}{12}+\frac{8}{12}=c$
$\frac{5}{12}=c$

Check: $-\frac{1}{4}\overset{?}{=}\frac{5}{12}-\frac{2}{3}$
$-\frac{1}{4}\overset{?}{=}\frac{5}{12}-\frac{8}{12}$
$-\frac{1}{4}\overset{?}{=}-\frac{3}{12}$
$-\frac{1}{4}=-\frac{1}{4}$ *True*

71. $\frac{5}{9}=b-\frac{1}{3}$

$\frac{5}{9}+\frac{1}{3}=b-\frac{1}{3}+\frac{1}{3}$

$\frac{5}{9}+\frac{3}{9}=b$

$\frac{8}{9}=b$

Check: $\frac{5}{9}\overset{?}{=}\frac{8}{9}-\frac{1}{3}$

$\frac{5}{9}\overset{?}{=}\frac{8}{9}-\frac{3}{9}$

$\frac{5}{9}=\frac{5}{9}$ *True*

73. $d+1.3619=2.0148$

$d+1.3619-1.3619=2.0148-1.3619$

$d=0.6529$

Check: $0.6529+1.3619\overset{?}{=}2.0148$

$2.0148=2.0148$ *True*

75. $-0.813+x=-1.096$

$-0.813+x+0.813=-1.096+0.813$

$x=-0.283$

Check: $-0.813+(-0.283)\overset{?}{=}-1.096$

$-1.096=-1.096$ *True*

77. The Multiplication Property of Equations is used to remove a coefficient from a variable term by multiplying each side of the equation by the reciprocal of the coefficient.

79. To solve the equation $\frac{2}{3}w=-18$, use the Multiplication Property of Equations to multiply each side of the equation by $\frac{3}{2}$. The solution is <u>-27</u>.

81. The value of x is greater than zero because to find x we must divide a negative number by a negative number which results in a positive number.

83. $5x=15$

$\frac{5x}{5}=\frac{15}{5}$

$x=3$

Check: $5(3)\overset{?}{=}15$

$15=15$ *True*

85. $3b=-12$

$\frac{3b}{3}=\frac{-12}{3}$

$b=-4$

Check: $3(-4)\overset{?}{=}-12$

$-12=-12$ *True*

87. $-3x=6$

$\frac{-3x}{-3}=\frac{6}{-3}$

$x=-2$

Check: $-3(-2)\overset{?}{=}6$

$6=6$ *True*

89. $-3x=-27$

$\frac{-3x}{-3}=\frac{-27}{-3}$

$x=9$

Check: $-3(9)\overset{?}{=}-27$

$-27=-27$ *True*

91. $20=4c$

$\frac{20}{4}=\frac{4c}{4}$

$5=c$

Check: $20\overset{?}{=}4(5)$

$20=20$ *True*

93. $-32=8w$

$\frac{-32}{8}=\frac{8w}{8}$

$-4=w$

Check: $-32\overset{?}{=}8(-4)$

$-32=-32$ *True*

95. $8d=0$

$\frac{8d}{8}=\frac{0}{8}$

$d=0$

Check: $8(0)\overset{?}{=}0$

$0=0$ *True*

97. $-64=8a$

$\frac{-64}{8}=\frac{8a}{8}$

$-8=a$

Check: $-64\overset{?}{=}8(-8)$

$-64=-64$ *True*

99. $\frac{x}{3}=2$

$3\cdot\frac{x}{3}=2\cdot 3$

$x=6$

Check: $\frac{6}{3}\stackrel{?}{=}2$

$2=2$ *True*

101. $-\frac{y}{2}=5$

$-2\left(-\frac{y}{2}\right)=-2(5)$

$y=-10$

Check: $-\frac{-10}{2}\stackrel{?}{=}5$

$-(-5)\stackrel{?}{=}5$

$5=5$ *True*

103. $\frac{n}{7}=-4$

$7\cdot\frac{n}{7}=7(-4)$

$n=-28$

Check: $\frac{-28}{7}\stackrel{?}{=}-4$

$-4=-4$ *True*

105. $\frac{2}{5}x=12$

$\frac{5}{2}\cdot\frac{2}{5}x=\frac{5}{2}\cdot 12$

$x=30$

Check: $\frac{2}{\not{5}_1}\cdot \not{30}^{6}\stackrel{?}{=}12$

$12=12$ *True*

107. $\frac{5}{6}y=-20$

$\frac{6}{5}\cdot\frac{5}{6}y=\frac{6}{5}(-20)$

$y=-24$

Check: $\frac{5}{\not{6}_1}(\not{-24}^{-4})\stackrel{?}{=}-20$

$-20=-20$ *True*

109. $\frac{2n}{3}=2$

$\frac{3}{2}\cdot\frac{2n}{3}=\frac{3}{2}\cdot 2$

$n=3$

Check: $\frac{2(\not{3}^{1})}{\not{3}_1}\stackrel{?}{=}2$

$2=2$ *True*

111. $\frac{-3z}{8}=9$

$-\frac{8}{3}\cdot\left(-\frac{3z}{8}\right)=-\frac{8}{3}(9)$

$z=-24$

Check: $\frac{-3(\not{-24}^{-3})}{\not{8}_1}\stackrel{?}{=}9$

$9=9$ *True*

113. $-6=-\frac{2}{3}y$

$\left(-\frac{3}{2}\right)(-6)=\left(-\frac{3}{2}\right)\left(-\frac{2}{3}y\right)$

$9=y$

Check: $-6\stackrel{?}{=}-\frac{2}{\not{3}_1}\cdot\not{9}^{3}$

$-6=-6$ *True*

115. $3n+2n=20$

$5n=20$

$\frac{5n}{5}=\frac{20}{5}$

$n=4$

Check: $3\cdot 4+2\cdot 4\stackrel{?}{=}20$

$12+8\stackrel{?}{=}20$

$20=20$ *True*

117. $10y-3y=21$

$7y=21$

$\frac{7y}{7}=\frac{21}{7}$

$y=3$

Check: $10\cdot 3-3\cdot 3\stackrel{?}{=}21$

$30-9\stackrel{?}{=}21$

$21=21$ *True*

119. $\frac{x}{1.4}=3.2$

$(1.4)\frac{x}{1.4}=(1.4)(3.2)$

$x=4.48$

Check: $\frac{4.48}{1.4}\stackrel{?}{=}3.2$

$3.2=3.2$ *True*

121. $3.4a=7.004$

$\frac{3.4a}{3.4}=\frac{7.004}{3.4}$

$a=2.06$

Check: $3.4(2.06)\stackrel{?}{=}7.004$

$7.004=7.004$ *True*

123. No. In the equation $15x=y$, if y is a positive integer, then x must also be positive because a positive product is the result of multiplying factors having like signs.

125. Yes. In the equation $-\frac{1}{4}x = y$, if y is a positive integer, then x must be negative because for a product to be positive, the factors must have the same signs.

127.
$$\frac{3y-8y}{7}=15$$
$$\frac{-5y}{7}=15$$
$$\left(-\frac{7}{5}\right)\left(\frac{-5y}{7}\right)=\left(-\frac{7}{5}\right)(15)$$
$$y=-21$$

129.
$$\frac{1}{\frac{1}{x}}+8=-19$$
$$\frac{1}{\frac{1}{x}}=-27$$
$$\left(\frac{\frac{1}{x}}{1}\right)\left(\frac{1}{\frac{1}{x}}\right)=-27\left(\frac{\frac{1}{x}}{1}\right)$$
$$x\cdot 1=-27\left(\frac{1}{x}\right)x$$
$$x=-27$$

131.
$$\frac{5}{\frac{7}{a}}-\frac{3}{\frac{7}{a}}=6$$
$$\frac{2}{\frac{7}{a}}=6$$
$$\frac{7}{a}\cdot\frac{2}{\frac{7}{a}}=6\cdot\frac{7}{a}$$
$$2=\frac{42}{a}$$
$$2a=\frac{42}{a}\cdot a$$
$$2a=42$$
$$a=21$$

133. Answers will vary.
One possible answer: $x+5=7$

135.
$$\frac{1}{a}+\frac{1}{b}=2$$
$$\frac{3}{7}+\frac{1}{b}=2$$
$$\frac{1}{b}=2-\frac{3}{7}$$
$$\frac{1}{b}=\frac{11}{7}$$
$$b=\frac{7}{11}$$

137. **a.** $x+3=8$ pairs with **iv**, what number plus 3 is equal to 8?

b. $x-5=20$ pairs with **ii,** what number minus 5 equals 20?

c. $4x=16$ pairs with **i**, 4 times what number is equal to 16?

d. $\frac{x}{7}=1$ pairs with **v**, what number divided by 7 is equal to 1?

e. $99=-9x$ pairs with **iii**, 99 is equal to -9 times what number?

Section 3.2 Applications of Equations of the Form *ax* = *b*

1. 30 is 75% of 40.
amount: 30, base: 40

3. Since Keith and Jennifer started at the same time and Keith had traveled farther when they met, **Keith** had the greater average speed.

5. Employee B will still have the highest salary after the 5% raise. Since each employee receives a raise of 5%, the employee with the greatest salary before the raise will have the greatest salary after the raise.**6.** Employee B will have the greatest salary after the raises. Since all three employees had the same salary before the raise, the employee that receive the largest percent raise with have the greatest salary.

7. Given: $B = 50$

$A = 12$

Unknown: Percent

$$PB = A$$
$$P(50) = 12$$
$$\frac{50P}{50} = \frac{12}{50}$$
$$P = \frac{6}{25}$$
$$P = 24\%$$

12 is 24% of 50.

9. Given: $P = 18\% = 0.18$

$B = 40$

Unknown: Amount

$$PB = A$$
$$0.18(40) = A$$
$$7.2 = A$$

18% of 40 is 7.2.

11. Given: $A = 48$

$P = 12\% = 0.12$

Unknown: Base

$$PB = A$$
$$0.12B = 48$$
$$\frac{0.12B}{0.12} = \frac{48}{0.12}$$
$$B = 400$$

12% of 400 is 48.

13. Given: $B = 27$

$P = 33\frac{1}{3}\% = \frac{1}{3}$

Unknown: Amount

$$PB = A$$
$$\frac{1}{3}(27) = A$$
$$9 = A$$

$33\frac{1}{3}\%$ of 27 is 9.

15. Given: $B = 12$

$A = 3$

Unknown: Percent

$$PB = A$$
$$P(12) = 3$$
$$\frac{12P}{12} = \frac{3}{12}$$
$$P = 0.25$$
$$P = 25\%$$

25% of 12 is 3.

17. Given: $P = 60\% = 0.60$

$A = 3$

Unknown: Base

$$PB = A$$
$$0.60B = 3$$
$$\frac{0.60B}{0.60} = \frac{3}{0.60}$$
$$B = 5$$

60 % of 5 is 3.

19. Given: $A = 12$

$B = 6$

Unknown: Percent

$$PB = A$$
$$P(6) = 12$$
$$\frac{6P}{6} = \frac{12}{6}$$
$$P = 2$$
$$P = 200\%$$

200 % of 6 is 12.

21. Given: $P = 5\frac{1}{4}\% = 0.0525$

$A = 21$

Unknown: Base

$$PB = A$$
$$0.0525B = 21$$
$$\frac{0.0525B}{0.0525} = \frac{21}{0.0525}$$
$$B = 400$$

$5\frac{1}{4}$% of 400 is 21.

23. Given: $P = 15.4\% = 0.154$

$B = 50$

Unknown: Amount

$$PB = A$$
$$0.154(50) = A$$
$$7.7 = A$$

15.4 % of 50 is 7.7.

25. Given: $A = 1$

$P = 0.5\% = 0.005$

Unknown: Base

$$PB = A$$
$$0.005B = 1$$
$$\frac{0.005B}{0.005} = \frac{1}{0.005}$$
$$B = 200$$

0.5 % of 200 is 1.

27. Given: $P = \frac{3}{4}\% = 0.0075$

$A = 3$

Unknown: Base

$$PB = A$$
$$0.0075B = 3$$
$$\frac{0.0075B}{0.0075} = \frac{3}{0.0075}$$
$$B = 400$$

$\frac{3}{4}$% of 400 is 3.

29. Given: $P = 125\% = 1.25$

$B = 16$

Unknown: Amount

$$PB = A$$
$$1.25(16) = A$$
$$20 = A$$

125 % of 16 is 20.

31. Given: $A = 16.4$

$B = 20.4$

Unknown: Percent

$$PB = A$$
$$P(20.4) = 16.4$$
$$\frac{20.4P}{20.4} = \frac{16.4}{20.4}$$
$$P = 0.80392...$$
$$P \approx 80\%$$

80% of 20.4 is 16.4.

33. Because of the properties of multiplication, the two expressions are **equal**.

35. Using the formula $PB = A$, $P =$ unknown, $B =$ 30, and $A =$ 24.

37. $A = 22{,}500$

$P = 1.11\% = 0.0111$

$$PB = A$$
$$0.0111(22{,}500) = A \approx 250$$

Approximately 250 seats are reserved for wheelchair accessibility.

39. $A = 13.2$, $P = 17.8\% = 0.178$

$$A = PB$$
$$13.2 = 0.178B$$
$$\frac{13.2}{0.178} = \frac{0.178B}{0.178}$$
$$74 \approx B$$

Each person uses approximately 74 gallons of water each day.

41. There is not enough information given to answer the question.

43. $B = 2227$, $A = 277$

$$A = PB$$
$$277 = P(2227)$$
$$\frac{277}{2227} = \frac{2227P}{2227}$$
$$0.12 \approx P$$
$$12\% \approx P$$

Approximately 12% of the accidental deaths were not attributed to motor vehicle accidents.

45. $A = 31.7$ billion, $P = 33\% = 0.33$

$$A = PB$$
$$31.7 = 0.33B$$
$$\frac{31.7}{0.33} = \frac{0.33B}{0.33}$$
$$96.1 \approx B$$

The total electricity used for a home in this country is approximately 96.1 billion KWH.

47. $B = 310$ million, $A = 106.5$ million

$$A = PB$$
$$106.5 = P(310)$$
$$\frac{106.5}{310} = \frac{310P}{310}$$
$$0.344 \approx P$$
$$34.4\% \approx P$$

Approximately 34.4% of the U. S. population watched Super Bowl XLIV.

49. $A = 1579.99$, $P = 111\% = 1.11$

$$A = PB$$
$$1579.99 = 1.11B$$
$$\frac{1579.99}{1.11} = \frac{1.11B}{1.11}$$
$$\$1423.41 \approx B$$

The less expensive model costs $1423.41.

51. $P = \$1200$, $T = 8$ months $= \frac{2}{3}$ year, $I = \$72$

$$I = PRT$$
$$72 = 1200(R)\left(\frac{2}{3}\right)$$
$$72 = 800R$$
$$\frac{72}{800} = \frac{800R}{800}$$
$$0.09 = R$$
$$9\% = R$$

The simple interest rate is 9%.

53. Account 1: $P = \$1000$, $R = 7.5\% = 0.075$, $T = 1$ year

$$I = PRT$$
$$I = 1000(0.075)(1) = \$75$$

Account 2: $P = \$2000$, $R = 8.25\% = 0.0825$, $T = \text{I year}$

$$I = PRT$$
$$I = 2000(0.0825)(1) = \$165$$

Total Interest = \$75 + 165 = \$240

Sal earned \$240 in interest.

55. Marlys: $P = \$850$, $T = 1$ year, $I = \$51$

$$I = PRT$$
$$51 = 850(R)1$$
$$\frac{51}{850} = \frac{850R}{850}$$
$$0.06 = R$$
$$6\% = R$$

Makana: $P = \$900$, $T = 1$ year, $R = 7\%$

$$I = PRT$$
$$I = 900(0.07)(1)$$
$$I = \$63$$

Makana earned \$63.

57. The interest rate on the combined investment is **between 6% and 9%.**

59. $Q = 5$ mL, $A = 250$ mL

$$Q = Ar$$
$$5 = 250r$$
$$\frac{5}{250} = \frac{250r}{250}$$
$$0.02 = r$$
$$2\% = r$$

The concentration is 2%.

61. Apple Dan: $Q = 8$ oz, $A = 32$ oz

$$Q = Ar$$
$$8 = (32)r$$
$$\frac{8}{32} = \frac{32r}{32}$$
$$0.25 = r$$
$$25\% = r$$

Generic: $Q = 9$ oz, $A = 40$ oz

$$Q = Ar$$
$$9 = 40r$$
$$\frac{9}{40} = \frac{40r}{40}$$
$$0.225 = r$$
$$22.5\% = r$$

Apple Dan's has a greater concentration of apple juice.

63. $r = 25\% = 0.25$, $A = 50$ g

$$Q = Ar$$
$$Q = 50(0.25)$$
$$Q = 12.5g$$

12.5 grams of the cream are not glycerin.

65. $Q = 50$ g, $A = 400$ g

$$Q = Ar$$
$$50 = 400r$$
$$\frac{50}{400} = \frac{400r}{400}$$
$$0.125 = r$$
$$12.5\% = r$$

The remaining solution is 12.5% salt.

67. A car that travels 10 mi in 30 min is traveling at a rate of 20 miles per hour.

$$\frac{10mi}{30\,\text{min}} = \frac{x}{60\,\text{min}}$$
$$10(60) = 30x$$
$$20miles = x$$

69. Flying with the wind, the plane travels $325 + 30 = 355$ mph.

Flying against the wind, the plane travels $325 - 30 = 295$ mph.

71. **a.** They start from the same place and end at the same place so they travel an equal distance.

b. Emma spends less time biking because she travels faster.

73. $d = rt$
$$d = 45(3) = 135mi$$

$$d = rt$$
$$d = 55(2) = 110mi$$

$$total = 135 + 110 = 245mi$$

The train travels 245 miles.

75. $d = 20$ mi, $t = 40$ min $= \frac{2}{3}$ hr

$$d = rt$$
$$20 = r\left(\frac{2}{3}\right)$$
$$\frac{3}{2}\cdot 20 = r\left(\frac{2}{3}\right)\cdot\frac{3}{2}$$
$$30mph = r$$

She must drive 30 mph.

77. $d = 36$ mi, $r = 12$ mph

$$d = rt$$
$$36 = 12t$$
$$\frac{36}{12} = \frac{12t}{12}$$
$$3hr = t$$

The bicycling time is 3 hours and lunch is 1 hour so the trip will take 4 hours.

79. $d = 250$ ft, $r = 8$ fps

$$d = rt$$
$$250 = 8t$$
$$\frac{250}{8} = \frac{8t}{8}$$
$$31.25\,\text{sec} = t$$

It takes 31.25 seconds.

81. $r = 12$ mph, $d = 8$ mi

$$d = rt$$
$$8 = 12t$$
$$\frac{8}{12}hr = t$$
$$\frac{2}{3}hr = 40\,\text{min} = t$$

They will meet after running for 40 minutes.

83. $r = 17$ mph, $t = 30$ min $= 0.5$ hr

$$d = rt$$
$$d = 17(.5)$$
$$d = 8.5mi$$

They will be 8.5 miles apart.

85. Train 1: $d = rt$
$$d = 45(2)$$
$$d = 90mi$$

Train 2: $d = rt$
$$d = 60(1)$$
$$d = 60mi$$

At 10:00 A.M. the trains are 30 miles apart.

87.
$$\begin{aligned} bill &= meal + tip + tax \\ tax &= 0.06x \\ tip &= 0.15(x+0.06x) \\ &= 0.15(1.06x) = 0.159x \\ 97.52 &= x + 0.159x + 0.06x \\ 97.52 &= 1.219x \\ \$80 &= x \end{aligned}$$

The meal cost $80.

89. If a quantity increases by 100%, its new value is 2 times the original value**90.** To answer the question, we must know the speed of the airplane and the distance between New York and Los Angeles.

91. Northeast: $P = \dfrac{55.8}{309} \approx 18.1\%$

Midwest: $P = \dfrac{67.4}{309} \approx 21.8\%$

South: $P = \dfrac{113.6}{309} \approx 36.8\%$

West: $P = \dfrac{72.2}{309} \approx 23.4\%$

Section 3.3 General Equations

1. **a.** The first step is **i**, add 7 to each side.
   b. The first step is **iii**, subtract 7 from each side.
   c. The first step is **ii**, add 5 to each side.
   d. The first step is **iv**, subtract 5 from each side.

3. The statement is **true**.

5. The statement is **true**.

7. The statement is **true**.

9. To clear denominators from the equation $\dfrac{x}{9} + 2 = \dfrac{1}{6}$, multiply each side of the equation by $\underline{9}$, the least common multiple of 9 and 6.

11. The solution of the equation will be positive. Subtracting 25 from both sides gives a negative equa to a negative coefficient times x. A negative divided by a negative is positive.

13.
$$\begin{aligned} 4y+3 &= 11 \\ 4y+3-3 &= 11-3 \\ 4y &= 8 \\ \frac{4y}{4} &= \frac{8}{4} \\ y &= 2 \end{aligned}$$

Check:
$$\begin{aligned} 4\cdot 2+3 &\stackrel{?}{=} 11 \\ 8+3 &\stackrel{?}{=} 11 \\ 11 &= 11 \textit{ True} \end{aligned}$$

15.
$$\begin{aligned} 5m-6 &= 9 \\ 5m-6+6 &= 9+6 \\ 5m &= 15 \\ \frac{5m}{5} &= \frac{15}{5} \\ m &= 3 \end{aligned}$$

Check:
$$\begin{aligned} 5\cdot 3-6 &\stackrel{?}{=} 9 \\ 15-6 &\stackrel{?}{=} 9 \\ 9 &= 9 \textit{ True} \end{aligned}$$

17.
$$\begin{aligned} 2 &= 5b+12 \\ 2-12 &= 5b+12-12 \\ -10 &= 5b \\ \frac{-10}{5} &= \frac{5b}{5} \\ -2 &= b \end{aligned}$$

Check:
$$\begin{aligned} 2 &\stackrel{?}{=} 5(-2)+12 \\ 2 &\stackrel{?}{=} -10+12 \\ 2 &= 2 \textit{ True} \end{aligned}$$

19.
$$\begin{aligned} 7-c &= 9 \\ 7-c-7 &= 9-7 \\ -c &= 2 \\ \frac{-c}{-1} &= \frac{2}{-1} \\ c &= -2 \end{aligned}$$

Check:
$$\begin{aligned} 7-(-2) &\stackrel{?}{=} 9 \\ 7+2 &\stackrel{?}{=} 9 \\ 9 &= 9 \textit{ True} \end{aligned}$$

21. $$\begin{aligned} 4-3w &= -2 \\ 4-3w-4 &= -2-4 \\ -3w &= -6 \\ \frac{-3w}{-3} &= \frac{-6}{-3} \\ w &= 2 \end{aligned}$$

Check: $$\begin{aligned} 4-3\cdot 2 &\stackrel{?}{=} -2 \\ 4-6 &\stackrel{?}{=} -2 \\ -2 &= -2 \text{ } \textit{True} \end{aligned}$$

23. $$\begin{aligned} 8-3t &= 2 \\ 8-3t-8 &= 2-8 \\ -3t &= -6 \\ \frac{-3t}{-3} &= \frac{-6}{-3} \\ t &= 2 \end{aligned}$$

Check: $$\begin{aligned} 8-3\cdot 2 &\stackrel{?}{=} 2 \\ 8-6 &\stackrel{?}{=} 2 \\ 2 &= 2 \text{ } \textit{True} \end{aligned}$$

25. $$\begin{aligned} -8x-3 &= -19 \\ -8x-3+3 &= -19+3 \\ -8x &= -16 \\ \frac{-8x}{-8} &= \frac{-16}{-8} \\ x &= 2 \end{aligned}$$

Check: $$\begin{aligned} -8\cdot 2-3 &\stackrel{?}{=} -19 \\ -16-3 &\stackrel{?}{=} -19 \\ -19 &= -19 \text{ } \textit{True} \end{aligned}$$

27. $$\begin{aligned} -12x+30 &= -6 \\ -12x+30-30 &= -6-30 \\ -12x &= -36 \\ \frac{-12x}{-12} &= \frac{-36}{-12} \\ x &= 3 \end{aligned}$$

Check: $$\begin{aligned} -12\cdot 3+30 &\stackrel{?}{=} -6 \\ -36+30 &\stackrel{?}{=} -6 \\ -6 &= -6 \text{ } \textit{True} \end{aligned}$$

29. $$\begin{aligned} 2 &= 7-5a \\ 2-7 &= 7-5a-7 \\ -5 &= -5a \\ \frac{-5}{-5} &= \frac{-5a}{-5} \\ 1 &= a \end{aligned}$$

Check: $$\begin{aligned} 2 &\stackrel{?}{=} 7-5\cdot 1 \\ 2 &\stackrel{?}{=} 7-5 \\ 2 &= 2 \text{ } \textit{True} \end{aligned}$$

31. $$\begin{aligned} -35 &= -6b+1 \\ -35-1 &= -6b+1-1 \\ -36 &= -6b \\ \frac{-36}{-6} &= \frac{-6b}{-6} \\ 6 &= b \end{aligned}$$

Check: $$\begin{aligned} -35 &\stackrel{?}{=} -6\cdot 6+1 \\ -35 &\stackrel{?}{=} -36+1 \\ -35 &= -35 \text{ } \textit{True} \end{aligned}$$

33. $$\begin{aligned} -3m-21 &= 0 \\ -3m-21+21 &= 0+21 \\ -3m &= 21 \\ \frac{-3m}{-3} &= \frac{21}{-3} \\ m &= -7 \end{aligned}$$

Check: $$\begin{aligned} -3(-7)-21 &\stackrel{?}{=} 0 \\ 21-21 &\stackrel{?}{=} 0 \\ 0 &= 0 \text{ } \textit{True} \end{aligned}$$

35. $$\begin{aligned} 8y+3 &= 7 \\ 8y+3-3 &= 7-3 \\ 8y &= 4 \\ \frac{8y}{8} &= \frac{4}{8} \\ y &= \frac{1}{2} \end{aligned}$$

Check: $$\begin{aligned} 8\cdot\frac{1}{2}+3 &\stackrel{?}{=} 7 \\ 4+3 &\stackrel{?}{=} 7 \\ 7 &= 7 \text{ } \textit{True} \end{aligned}$$

37. $$\begin{aligned} 3m+4 &= 11 \\ 3m+4-4 &= 11-4 \\ 3m &= 7 \\ \frac{3m}{3} &= \frac{7}{3} \\ m &= \frac{7}{3} \end{aligned}$$

Check: $$\begin{aligned} 3\cdot\frac{7}{3}+4 &\stackrel{?}{=} 11 \\ 7+4 &\stackrel{?}{=} 11 \\ 11 &= 11 \text{ } \textit{True} \end{aligned}$$

39. $$\begin{aligned} 7-8z &= 0 \\ 7-8z-7 &= 0-7 \\ -8z &= -7 \\ \frac{-8z}{-8} &= \frac{-7}{-8} \\ z &= \frac{7}{8} \end{aligned}$$

Check: $$\begin{aligned} 7-8\cdot\frac{7}{8} &\stackrel{?}{=} 0 \\ 7-7 &\stackrel{?}{=} 0 \\ 0 &= 0 \text{ } \textit{True} \end{aligned}$$

41. $7-9a=4$ **Check:** $7-9\cdot\frac{1}{3}\stackrel{?}{=}4$

$7-9a-7=4-7$ $\quad 7-3\stackrel{?}{=}4$

$-9a=-3$ $\quad 4=4$ *True*

$\frac{-9a}{-9}=\frac{-3}{-9}$

$a=\frac{1}{3}$

43. $3x-\frac{5}{6}=\frac{13}{6}$ **Check:** $3\cdot1-\frac{5}{6}\stackrel{?}{=}\frac{13}{6}$

$3x-\frac{5}{6}+\frac{5}{6}=\frac{13}{6}+\frac{5}{6}$ $\quad \frac{18}{6}-\frac{5}{6}\stackrel{?}{=}\frac{13}{6}$

$3x=\frac{18}{6}=3$ $\quad \frac{13}{6}=\frac{13}{6}$ *True*

$\frac{3x}{3}=\frac{3}{3}$

$x=1$

45. $9x+\frac{4}{5}=\frac{4}{5}$ **Check:** $9\cdot0+\frac{4}{5}\stackrel{?}{=}\frac{4}{5}$

$9x+\frac{4}{5}-\frac{4}{5}=\frac{4}{5}-\frac{4}{5}$ $\quad 0+\frac{4}{5}\stackrel{?}{=}\frac{4}{5}$

$9x=0$ $\quad \frac{4}{5}=\frac{4}{5}$ *True*

$\frac{9x}{9}=\frac{0}{9}$

$x=0$

47. $7=9-5a$ **Check:** $7\stackrel{?}{=}9-5\cdot\frac{2}{5}$

$7-9=9-5a-9$ $\quad 7\stackrel{?}{=}9-2$

$-2=-5a$ $\quad 7=7$ *True*

$\frac{-2}{-5}=\frac{-5a}{-5}$

$\frac{2}{5}=a$

49. $\frac{1}{3}m-1=5$ **Check:** $\frac{1}{3}\cdot18-1\stackrel{?}{=}5$

$\frac{1}{3}m-1+1=5+1$ $\quad 6-1\stackrel{?}{=}5$

$\frac{1}{3}m=6$ $\quad 5=5$ *True*

$3\cdot\frac{1}{3}m=3\cdot6$

$m=18$

51. $\frac{3}{4}n+7=13$ **Check:** $\frac{3}{4}\cdot8+7\stackrel{?}{=}13$

$\frac{3}{4}n+7-7=13-7$ $\quad 6+7\stackrel{?}{=}13$

$\frac{3}{4}n=6$ $\quad 13=13$ *True*

$\frac{4}{3}\cdot\frac{3}{4}n=\frac{4}{3}\cdot6$

$n=8$

53. $-\frac{3}{8}b+4=10$ **Check:** $-\frac{3}{8}(-16)+4\stackrel{?}{=}10$

$-\frac{3}{8}b+4-4=10-4$ $\quad 6+4\stackrel{?}{=}10$

$-\frac{3}{8}b=6$ $\quad 10=10$ *True*

$\left(-\frac{8}{3}\right)\left(-\frac{3}{8}\right)b=\left(-\frac{8}{3}\right)6$

$b=-16$

55. $\frac{y}{5}-2=3$ **Check:** $\frac{25}{5}-2\stackrel{?}{=}3$

$\frac{y}{5}-2+2=3+2$ $\quad 5-2\stackrel{?}{=}3$

$\frac{y}{5}=5$ $\quad 3=3$ *True*

$5\cdot\frac{y}{5}=5\cdot5$

$y=25$

57. $\dfrac{3c}{7}-1=8$

$\dfrac{3c}{7}-1+1=8+1$

$\dfrac{3c}{7}=9$

$\dfrac{7}{3}\cdot\dfrac{3c}{7}=\dfrac{7}{3}\cdot 9$

$c=21$

Check: $\dfrac{3\cdot 21}{7}-1\overset{?}{=}8$

$9-1\overset{?}{=}8$

$8=8$ *True*

59. $3-\dfrac{4}{5}w=-9$

$3-\dfrac{4}{5}w-3=-9-3$

$-\dfrac{4}{5}w=-12$

$\left(-\dfrac{5}{4}\right)\left(-\dfrac{4}{5}\right)w=\left(-\dfrac{5}{4}\right)(-12)$

$w=15$

Check: $3-\dfrac{4}{5}\cdot 15\overset{?}{=}-9$

$3-12\overset{?}{=}-9$

$-9=-9$ *True*

61. $\dfrac{2}{3}=y-\dfrac{1}{2}$

$\dfrac{2}{3}+\dfrac{1}{2}=y-\dfrac{1}{2}+\dfrac{1}{2}$

$\dfrac{4}{6}+\dfrac{3}{6}=y$

$\dfrac{7}{6}=y$

Check: $\dfrac{2}{3}\overset{?}{=}\dfrac{7}{6}-\dfrac{1}{2}$

$\dfrac{2}{3}\overset{?}{=}\dfrac{7}{6}-\dfrac{3}{6}$

$\dfrac{2}{3}\overset{?}{=}\dfrac{4}{6}$

$\dfrac{2}{3}=\dfrac{2}{3}$ *True*

63. $\dfrac{2}{3}=\dfrac{3}{4}-\dfrac{1}{2}y$

$12\left(\dfrac{2}{3}\right)=12\left(\dfrac{3}{4}-\dfrac{1}{2}y\right)$

$8=9-6y$

$8-9=9-6y-9$

$-1=-6y$

$\dfrac{1}{6}=y$

Check: $\dfrac{2}{3}\overset{?}{=}\dfrac{3}{4}-\dfrac{1}{2}\cdot\dfrac{1}{6}$

$\dfrac{2}{3}\overset{?}{=}\dfrac{3}{4}-\dfrac{1}{12}$

$\dfrac{2}{3}\overset{?}{=}\dfrac{9}{12}-\dfrac{1}{2}=\dfrac{8}{12}$

$\dfrac{2}{3}=\dfrac{2}{3}$ *True*

65. $\dfrac{2x}{15}+3=-\dfrac{1}{3}$

$15\left(\dfrac{2x}{15}+3\right)=15\left(-\dfrac{1}{3}\right)$

$2x+45=-5$

$2x+45-45=-5-45$

$2x=-50$

$\dfrac{2x}{2}=\dfrac{-50}{2}$

$x=-25$

Check: $\dfrac{2(-25)}{15}+3\overset{?}{=}-\dfrac{1}{3}$

$-\dfrac{50}{15}+3\overset{?}{=}-\dfrac{1}{3}$

$-\dfrac{10}{3}+\dfrac{9}{3}\overset{?}{=}-\dfrac{1}{3}$

$-\dfrac{1}{3}=-\dfrac{1}{3}$ *True*

67. $\dfrac{5x}{12}-\dfrac{1}{4}=\dfrac{1}{6}$

$12\left(\dfrac{5x}{12}-\dfrac{1}{4}\right)=12\cdot\dfrac{1}{6}$

$5x-3=2$

$5x-3+3=2+3$

$5x=5$

$\dfrac{5x}{5}=\dfrac{5}{5}$

$x=1$

Check: $\dfrac{5\cdot 1}{12}-\dfrac{1}{4}\overset{?}{=}\dfrac{1}{6}$

$\dfrac{5}{12}-\dfrac{3}{12}\overset{?}{=}\dfrac{1}{6}$

$\dfrac{2}{12}\overset{?}{=}\dfrac{1}{6}$

$\dfrac{1}{6}=\dfrac{1}{6}$ *True*

69. $5-\dfrac{4c}{7}=8$

$7\left(5-\dfrac{4c}{7}\right)=7\cdot 8$

$35-4c=56$

$35-4c-35=56-35$

$-4c=21$

$\dfrac{-4c}{-4}=\dfrac{21}{-4}$

$c=-\dfrac{21}{4}$

Check: $5-\dfrac{4}{7}\left(-\dfrac{21}{4}\right)\overset{?}{=}8$

$5+3\overset{?}{=}8$

$8=8$ *True*

71.
$$\begin{aligned} 5y+9+2y&=23 \\ 7y+9&=23 \\ 7y+9-9&=23-9 \\ 7y&=14 \\ \frac{7y}{7}&=\frac{14}{7} \\ y&=2 \end{aligned}$$

Check: $5\cdot 2+9+2\cdot 2\stackrel{?}{=}23$
$$10+9+4\stackrel{?}{=}23$$
$$23=23 \textit{ True}$$

73.
$$\begin{aligned} b-8b+1&=-6 \\ -7b+1&=-6 \\ -7b+1-1&=-6-1 \\ -7b&=-7 \\ \frac{-7b}{-7}&=\frac{-7}{-7} \\ b&=1 \end{aligned}$$

Check: $1-8\cdot 1+1\stackrel{?}{=}-6$
$$1-8+1\stackrel{?}{=}-6$$
$$-6=-6 \textit{ True}$$

75.
$$\begin{aligned} 8&=4n-6+3n \\ 8&=7n-6 \\ 8+6&=7n-6+6 \\ 14&=7n \\ \frac{14}{7}&=\frac{7n}{7} \\ 2&=n \end{aligned}$$

Check: $8\stackrel{?}{=}4\cdot 2-6+3\cdot 2$
$$8\stackrel{?}{=}8-6+6$$
$$8=8 \textit{ True}$$

77.
$$\begin{aligned} 1.2x-3.44&=1.3 \\ 1.2x-3.44+3.44&=1.3+3.44 \\ 1.2x&=4.74 \\ \frac{1.2x}{1.2}&=\frac{4.74}{1.2} \\ x&=3.95 \end{aligned}$$

Check: $1.2(3.95)-3.44\stackrel{?}{=}1.3$
$$4.74-3.44\stackrel{?}{=}1.3$$
$$1.3=1.3 \textit{ True}$$

79.
$$\begin{aligned} -6.5&=4.3y-3.06 \\ -6.5+3.06&=4.3y-3.06+3.06 \\ -3.44&=4.3y \\ \frac{-3.44}{4.3}&=\frac{4.3y}{4.3} \\ -0.8&=y \end{aligned}$$

Check: $-6.5\stackrel{?}{=}4.3(-0.8)-3.06$
$$-6.5\stackrel{?}{=}-3.44-3.06$$
$$-6.5=-6.5 \textit{ True}$$

81.
$$\begin{aligned} 3x+5&=-4 \\ 3x+5-5&=-4-5 \\ 3x&=-9 \\ x&=-3 \end{aligned}$$
So $2x-5=2(-3)-5=-11$

83.
$$\begin{aligned} 2-3x&=11 \\ 2-3x-2&=11-2 \\ -3x&=9 \\ x&=-3 \end{aligned}$$
So $x^2+2x-3=(-3)^2+2(-3)-3$
$$=9-6-3=0$$

85.
$$\begin{aligned} \frac{1}{6}x-7&=\frac{3}{4}x \\ (12)\left(\frac{1}{6}x-7\right)&=(12)\left(\frac{3}{4}x\right) \\ 12\left(\frac{1}{6}x\right)-(12)(7)&=9x \quad \textit{(Distributive)} \\ 2x-84&=9x \\ 2x-2x-84&=9x-2x \\ -84&=7x \\ -\frac{84}{7}&=\frac{7x}{x} \quad \textit{(Divide by 7.)} \\ -12&=x \end{aligned}$$

87. If you rewrite the equation so that it has one variable term on the left side, the coefficient of the variable will be -2.

89.

$$
\begin{aligned}
6y+2 &= y+17 \\
6y+2-2 &= y+17-2 \\
6y &= y+15 \\
6y-y &= y+15-y \\
5y &= 15 \\
\frac{5y}{5} &= \frac{15}{5} \\
y &= 3
\end{aligned}
$$

Check:
$$
\begin{aligned}
6\cdot 3+2 &\overset{?}{=} 3+17 \\
18+2 &\overset{?}{=} 20 \\
20 &= 20 \textit{ True}
\end{aligned}
$$

91.

$$
\begin{aligned}
11n+3 &= 10n+11 \\
11n+3-3 &= 10n+11-3 \\
11n &= 10n+8 \\
11n-10n &= 10n+8-10n \\
n &= 8
\end{aligned}
$$

Check:
$$
\begin{aligned}
11\cdot 8+3 &\overset{?}{=} 10\cdot 8+11 \\
88+3 &\overset{?}{=} 80+11 \\
91 &= 91 \textit{ True}
\end{aligned}
$$

93.

$$
\begin{aligned}
9a-10 &= 3a+2 \\
9a-10+10 &= 3a+2+10 \\
9a &= 3a+12 \\
9a-3a &= 3a+12-3a \\
6a &= 12 \\
\frac{6a}{6} &= \frac{12}{6} \\
a &= 2
\end{aligned}
$$

Check:
$$
\begin{aligned}
9\cdot 2-10 &\overset{?}{=} 3\cdot 2+2 \\
18-10 &\overset{?}{=} 6+2 \\
8 &= 8 \textit{ True}
\end{aligned}
$$

95.

$$
\begin{aligned}
13b-1 &= 4b-19 \\
13b-1+1 &= 4b-19+1 \\
13b &= 4b-18 \\
13b-4b &= 4b-18-4b \\
9b &= -18 \\
\frac{9b}{9} &= \frac{-18}{9} \\
b &= -2
\end{aligned}
$$

Check:
$$
\begin{aligned}
13(-2)-1 &\overset{?}{=} 4(-2)-19 \\
-26-1 &\overset{?}{=} -8-19 \\
-27 &= -27 \textit{ True}
\end{aligned}
$$

97.

$$
\begin{aligned}
7a-5 &= 2a-20 \\
7a-5+5 &= 2a-20+5 \\
7a &= 2a-15 \\
7a-2a &= 2a-15-2a \\
5a &= -15 \\
\frac{5a}{5} &= \frac{-15}{5} \\
a &= -3
\end{aligned}
$$

Check:
$$
\begin{aligned}
7(-3)-5 &\overset{?}{=} 2(-3)-20 \\
-21-5 &\overset{?}{=} -6-21 \\
-26 &= -26 \textit{ True}
\end{aligned}
$$

99.

$$
\begin{aligned}
n-2 &= 6-3n \\
n-2+2 &= 6-3n+2 \\
n &= 8-3n \\
n+3n &= 8-3n+3n \\
4n &= 8 \\
\frac{4n}{4} &= \frac{8}{4} \\
n &= 2
\end{aligned}
$$

Check:
$$
\begin{aligned}
2-2 &\overset{?}{=} 6-3\cdot 2 \\
0 &\overset{?}{=} 6-6 \\
0 &= 0 \textit{ True}
\end{aligned}
$$

101.

$$
\begin{aligned}
4y-2 &= -16-3y \\
4y-2+2 &= -16-3y+2 \\
4y &= -14-3y \\
4y+3y &= -14-3y+3y \\
7y &= -14 \\
\frac{7y}{7} &= \frac{-14}{7} \\
y &= -2
\end{aligned}
$$

Check:
$$
\begin{aligned}
4(-2)-2 &\overset{?}{=} -16-3(-2) \\
-8-2 &\overset{?}{=} -16+6 \\
-10 &= -10 \textit{ True}
\end{aligned}
$$

103.
$$\begin{aligned} m+4 &= 3m+8 \\ m+4-4 &= 3m+8-4 \\ m &= 3m+4 \\ m-3m &= 3m+4-3m \\ -2m &= 4 \\ \frac{-2m}{-2} &= \frac{4}{-2} \\ m &= -2 \end{aligned}$$

Check: $-2+4\stackrel{?}{=}3(-2)+8$

$2\stackrel{?}{=}-6+8$

$2=2$ *True*

105.
$$\begin{aligned} 6d-2 &= 7d+5 \\ 6d-2+2 &= 7d+5+2 \\ 6d &= 7d+7 \\ 6d-7d &= 7d+7-7d \\ -d &= 7 \\ \frac{-d}{-1} &= \frac{7}{-1} \\ d &= -7 \end{aligned}$$

Check: $6(-7)-2\stackrel{?}{=}7(-7)+5$

$-42-2\stackrel{?}{=}-49+5$

$-44=-44$ *True*

107.
$$\begin{aligned} 5a+7 &= 2a+7 \\ 5a+7-7 &= 2a+7-7 \\ 5a &= 2a \\ 5a-2a &= 2a-2a \\ 3a &= 0 \\ \frac{3a}{3} &= \frac{0}{3} \\ a &= 0 \end{aligned}$$

Check: $5\cdot 0+7\stackrel{?}{=}2\cdot 0+7$

$0+7\stackrel{?}{=}0+7$

$7=7$ *True*

109.
$$\begin{aligned} 10-4n &= 16-n \\ 10-4n-10 &= 16-n-10 \\ -4n &= 6-n \\ -4n+n &= 6-n+n \\ -3n &= 6 \\ \frac{-3n}{-3} &= \frac{6}{-3} \\ n &= -2 \end{aligned}$$

Check: $10-4(-2)\stackrel{?}{=}16-(-2)$

$10+8\stackrel{?}{=}16+2$

$18=18$ *True*

111.
$$\begin{aligned} 2b-10 &= 7b \\ 2b-10-2b &= 7b-2b \\ -10 &= 5b \\ \frac{-10}{5} &= \frac{5b}{5} \\ -2 &= b \end{aligned}$$

Check: $2(-2)-10\stackrel{?}{=}7(-2)$

$-4-10\stackrel{?}{=}-14$

$-14=-14$ *True*

113.
$$\begin{aligned} 9y &= 5y+16 \\ 9y-5y &= 5y+16-5y \\ 4y &= 16 \\ \frac{4y}{4} &= \frac{16}{4} \\ y &= 4 \end{aligned}$$

Check: $9\cdot 4\stackrel{?}{=}5\cdot 4+16$

$36\stackrel{?}{=}20+16$

$36=36$ *True*

115.
$$\begin{aligned} 8-4x &= 18-5x \\ 8-4x-8 &= 18-5x-8 \\ -4x &= 10-5x \\ -4x+5x &= 10-5x+5x \\ x &= 10 \end{aligned}$$

Check: $8-4\cdot 10\stackrel{?}{=}18-5\cdot 10$

$8-40\stackrel{?}{=}18-50$

$-32=-32$ *True*

117.
$$5-7m=2-6m$$
$$5-7m-5=2-6m-5$$
$$-7m=-6m-3$$
$$-7m+6m=-6m-3+6m$$
$$-m=-3$$
$$\frac{-m}{-1}=\frac{-3}{-1}$$
$$m=3$$

Check: $5-7\cdot3\stackrel{?}{=}2-6\cdot3$
$$5-21\stackrel{?}{=}2-18$$
$-16=-16$ *True*

119.
$$6y-1=2y+2$$
$$6y-1+1=2y+2+1$$
$$6y=2y+3$$
$$6y-2y=2y+3-2y$$
$$4y=3$$
$$\frac{4y}{4}=\frac{3}{4}$$
$$y=\frac{3}{4}$$

Check: $6\cdot\frac{3}{4}-1\stackrel{?}{=}2\cdot\frac{3}{4}+2$
$$\frac{18}{4}-\frac{4}{4}\stackrel{?}{=}\frac{6}{4}+\frac{8}{4}$$
$\frac{14}{4}=\frac{14}{4}$ *True*

121.
$$10x-3=3x-1$$
$$10x-3+3=3x-1+3$$
$$10x=3x+2$$
$$10x-3x=3x+2-3x$$
$$7x=2$$
$$\frac{7x}{7}=\frac{2}{7}$$
$$x=\frac{2}{7}$$

Check: $10\cdot\frac{2}{7}-3\stackrel{?}{=}3\cdot\frac{2}{7}-1$
$$\frac{20}{7}-\frac{21}{7}\stackrel{?}{=}\frac{6}{7}-\frac{7}{7}$$
$-\frac{1}{7}=-\frac{1}{7}$ *True*

123.
$$8a-2=4a-5$$
$$8a-2+2=4a-5+2$$
$$8a=4a-3$$
$$8a-4a=4a-3-4a$$
$$4a=-3$$
$$\frac{4a}{4}=\frac{-3}{4}$$
$$a=\frac{-3}{4}$$

Check: $8\left(-\frac{3}{4}\right)-2\stackrel{?}{=}4\left(-\frac{3}{4}\right)-5$
$$-6-2\stackrel{?}{=}-3-5$$
$-8=-8$ *True*

125.
$$\frac{3}{4}x=\frac{1}{12}x+2$$
$$12\cdot\frac{3}{4}x=12\left(\frac{1}{12}x+2\right)$$
$$9x=x+24$$
$$9x-x=x+24-x$$
$$8x=24$$
$$\frac{8x}{8}=\frac{24}{8}$$
$$x=3$$

Check: $\frac{3}{4}\cdot3\stackrel{?}{=}\frac{1}{12}\cdot3+2$
$$\frac{9}{4}\stackrel{?}{=}\frac{1}{4}+\frac{8}{4}$$
$\frac{9}{4}=\frac{9}{4}$ *True*

127.
$$\frac{4}{5}c-7=\frac{1}{10}c$$
$$10\left(\frac{4}{5}c-7\right)=10\cdot\frac{1}{10}c$$
$$8c-70=c$$
$$8c-70-8c=c-8c$$
$$-70=-7c$$
$$\frac{-70}{-7}=\frac{-7c}{-7}$$
$$10=c$$

Check: $\frac{4}{5}\cdot10-7\stackrel{?}{=}\frac{1}{10}\cdot10$
$$8-7\stackrel{?}{=}1$$
$1=1$ *True*

129.
$$\frac{2}{3}b = 2 - \frac{5}{6}b$$
$$6 \cdot \frac{2}{3}b = 6\left(2 - \frac{5}{6}b\right)$$
$$4b = 12 - 5b$$
$$4b + 5b = 12 - 5b + 5b$$
$$9b = 12$$
$$\frac{9b}{9} = \frac{12}{9}$$
$$b = \frac{4}{3}$$

Check: $\frac{2}{3} \cdot \frac{4}{3} \stackrel{?}{=} 2 - \frac{5}{6} \cdot \frac{4}{3}$
$$\frac{8}{9} \stackrel{?}{=} \frac{18}{9} - \frac{10}{9}$$
$$\frac{8}{9} = \frac{8}{9} \textit{ True}$$

131.
$$4.5x - 5.4 = 2.7x$$
$$4.5x - 5.4 - 4.5x = 2.7x - 4.5x$$
$$-5.4 = -1.8x$$
$$\frac{-5.4}{-1.8} = \frac{-1.8x}{-1.8}$$
$$3 = x$$

Check: $4.5(3) - 5.4 \stackrel{?}{=} 2.7(3)$
$$13.5 - 5.4 \stackrel{?}{=} 8.1$$
$$8.1 = 8.1 \textit{ True}$$

133.
$$5x = 3x - 8$$
$$5x - 3x = 3x - 8 - 3x$$
$$2x = -8$$
$$\frac{2x}{2} = \frac{-8}{2}$$
$$x = -4$$
So $4x + 2 = 4(-4) + 2$
$$= -16 + 2 = -14$$

135.
$$2 - 6a = 5 - 3a$$
$$2 - 6a - 2 = 5 - 3a - 2$$
$$-6a = 3 - 3a$$
$$-6a + 3a = 3 - 3a + 3a$$
$$-3a = 3$$
$$\frac{-3a}{-3} = \frac{3}{-3}$$
$$a = -1$$
So $4a^2 - 2a + 1 = 4(-1)^2 - 2(-1) + 1$
$$= 4 + 2 + 1 = 7$$

137. Use the Distributive Property to remove the parentheses from the equation
$4x + 3(x + 6) = 74$: $4x + \underline{3x} + \underline{18} = 74$.

139. **i.** $5 - 2(x - 1) = 5 - 2x + 2 = 7 - 2x = y$

ii. $3(x - 1) = 3x - 3 = y$

iii. $5 - 2x + 2 = 7 - 2x = y$

iv. $5 - 2x + 1 = 6 - 2x = y$

i and **iii** are the same.

141.
$$5x + 2(x + 1) = 23$$
$$5x + 2x + 2 = 23$$
$$7x + 2 = 23$$
$$7x = 21$$
$$x = 3$$

Check: $5 \cdot 3 + 2(3 + 1) \stackrel{?}{=} 23$
$$5 \cdot 3 + 2 \cdot 4 \stackrel{?}{=} 23$$
$$15 + 8 \stackrel{?}{=} 23$$
$$23 = 23 \textit{ True}$$

143.
$$9n - 3(2n - 1) = 15$$
$$9n - 6n + 3 = 15$$
$$3n + 3 = 15$$
$$3n = 12$$
$$n = 4$$

Check: $9 \cdot 4 - 3(2 \cdot 4 - 1) \stackrel{?}{=} 15$
$$9 \cdot 4 - 3(8 - 1) \stackrel{?}{=} 15$$
$$36 - 21 \stackrel{?}{=} 15$$
$$15 = 15 \textit{ True}$$

145. $7-3(3a-4)=10$
$7-9a+12=10$
$-9a+19=10$
$-9a+19-19=10-19$
$-9a=-9$
$a=1$

Check: $7-3(3\cdot 1-4)\overset{?}{=}10$
$7-3(3-4)\overset{?}{=}10$
$7-3(-1)\overset{?}{=}10$
$7+3\overset{?}{=}10$
$10=10$ *True*

147. $5(3-2y)+4y=3$
$15-10y+4y=3$
$15-6y=3$
$-6y=-12$
$y=2$

Check: $5(3-2\cdot 2)+4\cdot 2\overset{?}{=}3$
$5(3-4)+8\overset{?}{=}3$
$5(-1)+8\overset{?}{=}3$
$-5+8\overset{?}{=}3$
$3=3$ *True*

149. $10x+1=2(3x+5)-1$
$10x+1=6x+10-1$
$10x+1=6x+9$
$10x=6x+8$
$4x=8$
$x=2$

Check: $10\cdot 2+1\overset{?}{=}2(3\cdot 2+5)-1$
$20+1\overset{?}{=}2(6+5)-1$
$21\overset{?}{=}2\cdot 11-1$
$21\overset{?}{=}22-1$
$21=21$ *True*

151. $4-3a=7-2(2a+5)$
$4-3a=7-4a-10$
$4-3a=-4a-3$
$-3a=-4a-7$
$a=-7$

Check: $4-3(-7)\overset{?}{=}7-2[2(-7)+5]$
$4+21\overset{?}{=}7-2(-14+5)$
$25\overset{?}{=}7-2(-9)$
$25\overset{?}{=}7+18$
$25=25$ *True*

153. $3y-7=5(2y-3)+4$
$3y-7=10y-15+4$
$3y-7=10y-11$
$3y=10y-4$
$-7y=-4$
$y=\frac{4}{7}$

Check: $3\cdot\frac{4}{7}-7\overset{?}{=}5\left(2\cdot\frac{4}{7}-3\right)+4$
$\frac{12}{7}-\frac{49}{7}\overset{?}{=}5\left(\frac{8}{7}-\frac{21}{7}\right)+4$
$-\frac{37}{7}\overset{?}{=}5\left(-\frac{13}{7}\right)+\frac{28}{7}$
$-\frac{37}{7}\overset{?}{=}-\frac{65}{7}+\frac{28}{7}$
$-\frac{37}{7}=\frac{-37}{7}$ *True*

155. $5-(9-6x)=2x-2$
$5-9+6x=2x-2$
$-4+6x=2x-2$
$6x=2x+2$
$4x=2$
$x=\frac{2}{4}=\frac{1}{2}$

Check: $5-\left(9-6\cdot\frac{1}{2}\right)\overset{?}{=}2\cdot\frac{1}{2}-2$
$5-(9-3)\overset{?}{=}1-2$
$5-6\overset{?}{=}-1$
$-1=-1$ *True*

157. $$\begin{aligned} 3[2-4(y-1)] &= 3(2y+8) \\ 3[2-4y+4] &= 6y+24 \\ 3[6-4y] &= 6y+24 \\ 18-12y &= 6y+24 \\ -12y &= 6y+6 \\ -18y &= 6 \\ y &= -\frac{6}{18} = -\frac{1}{3} \end{aligned}$$

Check: $$\begin{aligned} 3\left[2-4\left(-\frac{1}{3}-1\right)\right] &\stackrel{?}{=} 3\left[2\left(-\frac{1}{3}\right)+8\right] \\ 3\left[2-4\left(-\frac{4}{3}\right)\right] &\stackrel{?}{=} 3\left[-\frac{2}{3}+8\right] \\ 3\left(\frac{6}{3}+\frac{16}{3}\right) &\stackrel{?}{=} 3\left(-\frac{2}{3}+\frac{24}{3}\right) \\ 3\left(\frac{22}{3}\right) &\stackrel{?}{=} 3\left(\frac{22}{3}\right) \\ 22 &= 22 \textit{ True} \end{aligned}$$

159. $$\begin{aligned} 3a+2[2+3(a-1)] &= 2(3a+4) \\ 3a+2[2+3a-3] &= 6a+8 \\ 3a+2[3a-1] &= 6a+8 \\ 3a+6a-2 &= 6a+8 \\ 9a-2 &= 6a+8 \\ 3a &= 10 \\ a &= \frac{10}{3} \end{aligned}$$

Check: $$\begin{aligned} 3\cdot\frac{10}{3}+2\left[2+3\left(\frac{10}{3}-1\right)\right] &\stackrel{?}{=} 2\left[3\left(\frac{10}{3}\right)+4\right] \\ 10+2\left[2+3\cdot\frac{7}{3}\right] &\stackrel{?}{=} 2(10+4) \\ 10+2(2+7) &\stackrel{?}{=} 2\cdot 14 \\ 10+2\cdot 9 &\stackrel{?}{=} 28 \\ 10+18 &\stackrel{?}{=} 28 \\ 28 &= 28 \textit{ True} \end{aligned}$$

161. $$\begin{aligned} -2[4-(3b+2)] &= 5-2(3b+6) \\ -2[4-3b-2] &= 5-6b-12 \\ -2[2-3b] &= -6b-7 \\ -4+6b &= -6b-7 \\ 12b &= -3 \\ b &= \frac{-3}{12} = -\frac{1}{4} \end{aligned}$$

Check:

$$\begin{aligned} -2\left[4-\left(3\left(-\frac{1}{4}\right)+2\right)\right] &\stackrel{?}{=} 5-2\left[3\left(-\frac{1}{4}\right)+6\right] \\ -2\left[4-\left(-\frac{3}{4}+\frac{8}{4}\right)\right] &\stackrel{?}{=} 5-2\left(-\frac{3}{4}+\frac{24}{4}\right) \\ -2\left[4-\frac{5}{4}\right] &\stackrel{?}{=} 5-2\left(\frac{21}{4}\right) \\ -8+\frac{5}{2} &\stackrel{?}{=} 5-\frac{21}{2} \\ -\frac{16}{2}+\frac{5}{2} &= \frac{10}{2}-\frac{21}{2} \\ -\frac{11}{2} &= -\frac{11}{2} \textit{ True} \end{aligned}$$

163. $$\begin{aligned} 0.3x-2(1.6x)-8 &= 3(1.9x-4.1) \\ 0.3x-3.2x-8 &= 5.7x-12.3 \\ -2.9x-8 &= 5.7x-12.3 \\ -8.6x &= -4.3 \\ x &= \frac{-4.3}{-8.6} = 0.5 \end{aligned}$$

Check: $$\begin{aligned} 0.3\cdot(0.5)-2(1.6)(0.5)-8 &\stackrel{?}{=} 3[1.9(0.5)-4.1] \\ 0.15-1.6-8 &\stackrel{?}{=} 3(0.95-4.1) \\ -9.45 &\stackrel{?}{=} 3(-3.15) \\ -9.45 &= -9.45 \textit{ True} \end{aligned}$$

165. $$\begin{aligned} 4-3a &= 7-2(2a+5) \\ 4-3a &= 7-4a-10 \\ 4-3a &= -3-4a \\ a &= -7 \end{aligned}$$

So $a^2+7a = (-7)^2+7(-7)$
$= 49-49 = 0$

167.

$$2z-5=3(4z+5)$$
$$2z-5=12z+15$$
$$-10z=20$$
$$z=-2$$
$$\text{So } \frac{z^2}{z-2}=\frac{(-2)^2}{-2-2}=\frac{4}{-4}=-1$$

169.

$$C=\frac{1}{4}D-45$$
$$-3=\frac{1}{4}D-45$$
$$42=\frac{1}{4}D$$
$$4\cdot 42=D$$
$$168=D$$

The car will slide 168 feet.

171.

$$P=\frac{1}{2}D+15$$
$$35=\frac{1}{2}D+15$$
$$20=\frac{1}{2}D$$
$$2\cdot 20=D$$
$$40=D$$

The depth is 40 feet.

173.

$$H=2.9L+78.1$$
$$H=2.9(36)+78.1$$
$$H=104.4+78.1$$
$$H=182.5$$

The height is 182.5 cm.

175.

$$s=16t^2+vt$$
$$80=16(2)^2+v(2)$$
$$80=64+2v$$
$$16=2v$$
$$8=v$$

The initial velocity is 8 ft/s.

177.

$$F=2.50+2.30(m-1)$$
$$14.00=2.50+2.30(m-1)$$
$$14.00=2.50+2.30m-2.30$$
$$14.00=2.30m+0.20$$
$$13.80=2.30m$$
$$\frac{13.80}{2.30}=m$$
$$6=m$$

The passenger was driven 6 miles.

179.

$$P=c+h+\frac{1}{4}s$$
$$557\frac{1}{4}=425+118+\frac{1}{4}\cdot s$$
$$557\frac{1}{4}=543+\frac{1}{4}s$$
$$14\frac{1}{4}=\frac{1}{4}s$$
$$4\cdot\frac{57}{4}=4\cdot\frac{1}{4}s$$
$$57=s$$

The average crown spread is 57 feet.

181.

$$S=\frac{W-5e}{10}$$
$$35=\frac{390-5e}{10}$$
$$350=390-5e$$
$$-40=-5e$$
$$8=e$$

There were 8 typing errors.

183.

$$Px=Cx+F$$
$$550x=325x+78750$$
$$225x=78750$$
$$x=\frac{78750}{225}$$
$$x=350$$

Break-even occurs with 350 television sets.

185. In the lever system, x represents the distance from the force F_1 to the fulcrum. d represents the length of the lever and $d-x$ represents the distance from force F_2 to the fulcrum.

187. Since the seesaw is 8 feet long, if one person sits 3 feet from the fulcrum, the other person must be 5 feet from the fulcrum.

189. The seesaw will not balance if the two people switch places.

191. $F_1 = 60\,lb,\ F_2 = 50\,lb,\ d = 3.5\ ft, x = 4.5\ ft$

$$F_1x = F_2(d-x)$$
$$60(3.5) = 50(8-3.5)$$
$$210 = 50(4.5)$$
$$210 \neq 225$$

The seesaw is not balanced.

193.
$$F_1x = F_2(d-x)$$
$$100(2) = F_2(10-2)$$
$$200 = 8F_2$$
$$\frac{200}{8} = F_2$$
$$25 = F_2$$

A force of 25 lb is necessary to balance the rock.

195.
$$F_1x = F_2(d-x)$$
$$128x = 160(18-x)$$
$$128x = 2880-160x$$
$$288x = 2880$$
$$x = 10$$

The fulcrum should be places 10 feet from the 128 pound acrobat.

197. $3(2x-1)-(6x-4) = -9$
$$6x-3-6x+4 = -9$$
$$1 = -9$$

No solution

199. $3[4(w+2)-(w+1)] = 5(2+w)$
$$3[4w+8-w-1] = 10+5w$$
$$3[3w+7] = 10+5w$$
$$9w+21 = 10+5w$$
$$4w = -11$$
$$w = -\frac{11}{4}$$

201.
$$\frac{1}{2}n = \frac{2}{3}n$$
$$6\left(\frac{1}{2}n\right) = 6\left(\frac{2}{3}n\right)$$
$$3n = 4n$$
$$0 = n$$

The number is 0.

203. There is a difference between no solution and a solution of $x = 0$. No solution means that there is not a number that satisfies the solution. A solution of 0 means that the number 0 will satisfy the equation.

205. $s = 5x-3$

$t = x+4$
$$s = 3t-1$$
$$5x-3 = 3(x+4)-1$$
$$5x-3 = 3x+12-1$$
$$5x-3 = 3x+11$$
$$5x-3+3-3x = 3x+11-3x+3$$
$$2x = 14$$
$$x = 7$$

207. Yes. Medical care cost more than three times as much because the CPI was over 300.

209. $(CPI)(base\ year\ cost) = cost\ today$

$$1.921(x) = 30{,}000$$

$$x = \frac{30000}{1.921}$$

$$x \approx \$15{,}617$$

The car cost about $15,617 in the base years.

Section 3.4 Inequalities

1. The solution set $x \leq 4$ is the set of values less than 4 and also includes 4. The solution set $x < 4$ is the set of values less than 4 but 4 itself is not included.

3. **a.** No, the direction of the inequality symbol would not be reversed because at no time during the solution process would it be necessary to multiply or divide by a negative number.

b. No, the direction of the inequality symbol would not be reversed because at no time during the solution process would it be necessary to multiply or divide by a negative number.

c. Yes, the direction of the inequality would be reversed when dividing by -3.

d. No, the direction of the inequality symbol would not be reversed because at no time during the solution process would it be necessary to multiply or divide by a negative number.

e. No, the direction of the inequality symbol would not be reversed because at no time during the solution process would it be necessary to multiply or divide by a negative number.

f. Yes, the direction of the inequality would have be reversed when multiplying by -3.

5. It is **always true** that the same variable can be subtracted from both sides of an inequality without changing the solution set of the inequality.

7. The graph of the solution set shown <u>includes</u> the number -2.

In the graph of the solution set of the inequality, $x \leq -2$.

9. 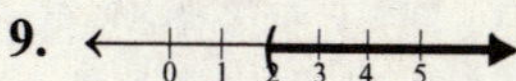

11.

13. $x + 7 \leq -3$

$$x \leq -10$$

i and **iii** are both solutions of the inequality.

15. $x + 1 < 3$

$$x < 2$$

17. $x - 5 > -2$

$$x > 3$$

19. $n + 4 \geq 7$

$$n \geq 3$$

21. $x - 6 \leq -10$

$$x \leq -4$$

23. $5 + x \geq 4$

$$x \geq -1$$

25. This solution set contains **only negative numbers** because $a - n$ will be negative since $n > a$.

27. This solution set contains **only positive numbers** because $-a + n$ will be positive since $n > a$ and a and n are both positive.

29.
$$\begin{aligned} y-3 &\geq -12 \\ y-3+3 &\geq -12+3 \\ y &\geq -9 \end{aligned}$$

31.
$$\begin{aligned} 3x-5 &< 2x+7 \\ 3x-5+5 &< 2x+7+5 \\ 3x &< 2x+12 \\ 3x-2x &< 2x+12-2x \\ x &< 12 \end{aligned}$$

33.
$$\begin{aligned} 8x-7 &\geq 7x-2 \\ 8x-7+7 &\geq 7x-2+7 \\ 8x &\geq 7x+5 \\ 8x-7x &\geq 7x+5-7x \\ x &\geq 5 \end{aligned}$$

35.
$$\begin{aligned} 2x+4 &< x-7 \\ 2x+4-4 &< x-7-4 \\ 2x &< x-11 \\ 2x-x &< x-11-x \\ x &< -11 \end{aligned}$$

37.
$$\begin{aligned} 4x-8 &\leq 2+3x \\ 4x-8+8 &\leq 2+3x+8 \\ 4x &\leq 3x+10 \\ 4x-3x &\leq 3x+10-3x \\ x &\leq 10 \end{aligned}$$

39.
$$\begin{aligned} 6x+4 &\geq 5x-2 \\ 6x+4-4 &\geq 5x-2-4 \\ 6x &\geq 5x-6 \\ 6x-5x &\geq 5x-6-5x \\ x &\geq -6 \end{aligned}$$

41.
$$\begin{aligned} 2x-12 &> x-10 \\ 2x-12+12 &> x-10+12 \\ 2x &> x+2 \\ 2x-x &> x+2-x \\ x &> 2 \end{aligned}$$

43.
$$\begin{aligned} d+\frac{1}{2} &< \frac{1}{3} \\ d+\frac{1}{2}-\frac{1}{2} &< \frac{1}{3}-\frac{1}{2} \\ d &< \frac{2}{6}-\frac{3}{6} \\ d &< -\frac{1}{6} \end{aligned}$$

45.
$$\begin{aligned} x+\frac{5}{8} &\geq -\frac{2}{3} \\ x+\frac{5}{8}-\frac{5}{8} &\geq -\frac{2}{3}-\frac{5}{8} \\ x &\geq -\frac{16}{24}-\frac{15}{24} \\ x &\geq -\frac{31}{24} \end{aligned}$$

47. $2x-\frac{1}{2}<x+\frac{3}{4}$

$4\left(2x-\frac{1}{2}\right)<4\left(x+\frac{3}{4}\right)$

$8x-2<4x+3$

$8x-2+2<4x+3+2$

$8x-2+2<4x+3+2$

$8x<4x+5$

$8x-4x<4x+5-4x$

$4x<5$

$x<\frac{5}{4}$

49. $3x+\frac{5}{8}>2x+\frac{5}{6}$

$24\left(3x+\frac{5}{8}\right)>24\left(2x+\frac{5}{6}\right)$

$72x+15>48x+20$

$24x>5$

$x>\frac{5}{24}$

51. $x+5.8\le 4.6$

$x+5.8-5.8\le 4.6-5.8$

$x\le -1.2$

53. $x-0.23\le 0.47$

$x-0.23+0.23\le 0.47+0.23$

$x\le 0.70$

55. $1.2x<0.2x-7.3$

$1.2x-0.2x<0.2x-7.3-0.2x$

$x<-7.3$

57. $-4x\le 12$

$x\ge -3$

i, ii and **iii** are solutions of the inequality.

59. The inequality -4x > 8 can be solved by dividing each side of the inequality by -4. The inequality symbol <u>is reversed</u>.

61. $8x\le -24$

$\frac{8x}{8}\le\frac{-24}{8}$

$x\le -3$

−4 −3 −2 −1 0 1

63. $24x>-48$

$\frac{24x}{24}>\frac{-48}{24}$

$x>-2$

−4 −3 −2 −1 0 1

65. $3x>0$

$\frac{3x}{3}>\frac{0}{3}$

$x>0$

−2 −1 0 1 2 3

67. $-2n\le -8$

$\frac{-2n}{-2}\ge\frac{-8}{-2}$

$n\ge 4$

0 1 2 3 4 5

69. $-4x<8$

$\frac{-4x}{-4}>\frac{8}{-4}$

$x>-2$

−4 −3 −2 −1 0 1

71. $3x < 5$

$$\frac{3x}{3} < \frac{5}{3}$$

$$x < \frac{5}{3}$$

73. $-8x \le -40$

$$\frac{-8x}{-8} \ge \frac{-40}{-8}$$

$$x \ge 5$$

75. $10x > -25$

$$\frac{10x}{10} > \frac{-25}{10}$$

$$x > -\frac{5}{2}$$

77. $-5x \ge \frac{10}{3}$

$$\left(-\frac{1}{5}\right)(-5x) \le \left(-\frac{1}{5}\right)\left(\frac{10}{3}\right)$$

$$x \le -\frac{2}{3}$$

79. $\frac{2}{3}x < -12$

$$\frac{3}{2}\cdot\frac{2}{3}x < \frac{3}{2}(-12)$$

$$x < -18$$

81. $-\frac{3}{8}x < 6$

$$\left(-\frac{8}{3}\right)\left(-\frac{3}{8}\right)x > \left(-\frac{8}{3}\right)(6)$$

$$x > -16$$

83. $-\frac{2}{11}b \ge -6$

$$\left(-\frac{11}{2}\right)\left(-\frac{2}{11}\right)b \le \left(-\frac{11}{2}\right)(-6)$$

$$b \le 33$$

85. $\frac{2}{3}n < \frac{1}{2}$

$$\frac{3}{2}\cdot\frac{2}{3}n < \frac{3}{2}\cdot\frac{1}{2}$$

$$n < \frac{3}{4}$$

87. $-\frac{2}{3}x \ge \frac{4}{7}$

$$\left(-\frac{3}{2}\right)\left(-\frac{2}{3}\right)x \le \left(-\frac{3}{2}\right)\left(\frac{4}{7}\right)$$

$$x \le -\frac{6}{7}$$

89. $-\frac{3}{4}y \ge -\frac{5}{8}$

$$\left(-\frac{4}{3}\right)\left(-\frac{3}{4}\right)y \le \left(-\frac{4}{3}\right)\left(-\frac{5}{8}\right)$$

$$y \le \frac{5}{6}$$

91. $-0.27x < 0.135$

$$\frac{-0.27x}{-0.27} < \frac{0.135}{-0.27}$$

$$x < -0.5$$

93. $8.4y \ge -6.72$

$$\frac{8.4y}{8.4} \ge \frac{-6.72}{8.4}$$

$$y \ge -0.8$$

95. $0.07x < -0.378$

$$\frac{0.07x}{0.07} < \frac{-0.378}{0.07}$$

$$x < -5.4$$

97. Statement **iv** is true. If n is positive, the product of -0.8157 and n is negative, which is less than 7.304. If n is 0, the product on the left is 0, which is less than 7.304. If n is negative, the product of n and –0.8157 is positive and can still be less than 7.304.

99. The Addition Property of Inequalities states that any real number can be added to each side of an inequality without changing the solution of the inequality. The Multiplication Property of Inequalities states that both sides of an inequality can be multiplied by a positive number without changing the solution. It also states that if both sides of an inequality are multiplied by a negative number the direction of the inequality changes.

101.

$$9-5x \geq -2x$$
$$9-5x+2x \geq -2x+2x$$
$$9-3x \geq 0$$
$$9-9-3x \geq 0-9$$

(*Subtract* 9 *from both sides.*)

$$-3x \geq -9$$
$$\frac{-3x}{-3} \leq \frac{-9}{-3}$$
$$x \leq 3$$

103.

$$3x+2 \geq 5x-8$$
$$3x+2-2 \geq 5x-8-2$$
$$3x \geq 5x-10$$
$$3x-5x \geq 5x-10-5x$$
$$-2x \geq -10$$
$$\frac{-2x}{-2} \leq \frac{-10}{-2}$$
$$x \leq 5$$

105.

$$5x-2 < 3x-2$$
$$5x-2+2 < 3x-2+2$$
$$5x < 3x$$
$$5x-3x < 3x-3x$$
$$2x < 0$$
$$\frac{2x}{2} < \frac{0}{2}$$
$$x < 0$$

107.

$$4x-8 < 2x$$
$$4x-8-4x < 2x-4x$$
$$-8 < -2x$$
$$\frac{-8}{-2} > \frac{-2x}{-2}$$
$$4 > x$$
$$x < 4$$

109.

$$2x-8 > 4x$$
$$2x-8-2x > 4x-2x$$
$$-8 > 2x$$
$$\frac{-8}{2} > \frac{2x}{2}$$
$$-4 > x$$
$$x < -4$$

111.

$$8-3x \leq 5x$$
$$8-3x+3x \leq 5x+3x$$
$$8 \leq 8x$$
$$\frac{8}{8} \leq \frac{8x}{8}$$
$$1 \leq x$$
$$x \geq 1$$

113.

$$0.1(180+x) > x$$
$$18+0.1x > x$$
$$18+0.1x-0.1x > x-0.1x$$
$$18 > 0.9x$$
$$\frac{18}{0.9} > \frac{0.9x}{0.9}$$
$$20 > x$$
$$x < 20$$

115.
$$\begin{aligned}0.15x+55&>0.10x+80\\0.15x+55-55&>0.10x+80-55\\0.15x&>0.10x+25\\0.15x-0.10x&>0.10x+25-0.10x\\0.05x&>25\\\frac{0.05x}{0.05}&>\frac{25}{0.05}\\x&>500\end{aligned}$$

117.
$$\begin{aligned}2(3x-1)&>3x+4\\6x-2&>3x+4\\6x-2+2&>3x+4+2\\6x&>3x+6\\6x-3x&>3x+6-3x\\3x&>6\\\frac{3x}{3}&>\frac{6}{3}\\x&>2\end{aligned}$$

119.
$$\begin{aligned}3(2x-5)&\ge 8x-5\\6x-15&\ge 8x-5\\6x-15+15&\ge 8x-5+15\\6x&\ge 8x+10\\6x-8x&\ge 8x+10-8x\\-2x&\ge 10\\\frac{-2x}{-2}&\le\frac{10}{-2}\\x&\le -5\end{aligned}$$

121.
$$\begin{aligned}2(2y-5)&\le 3(5-2y)\\4y-10&\le 15-6y\\4y-10+10&\le 15-6y+10\\4y&\le 25-6y\\4y+6y&\le 25-6y+6y\\10y&\le 25\\\frac{10y}{10}&\le\frac{25}{10}\\y&\le\frac{5}{2}\end{aligned}$$

123.
$$\begin{aligned}5(2-x)&>3(2x-5)\\10-5x&>6x-15\\10-5x-10&>6x-15-10\\-5x&>6x-25\\-5x-6x&>6x-25-6x\\-11x&>-25\\\frac{-11x}{-11}&<\frac{-25}{-11}\\x&<\frac{25}{11}\end{aligned}$$

125.
$$\begin{aligned}5(x-2)&>9x-3(2x-4)\\5x-10&>9x-6x+12\\5x-10&>3x+12\\5x-10+10&>3x+12+10\\5x&>3x+22\\5x-3x&>3x+22-3x\\2x&>22\\\frac{2x}{2}&>\frac{22}{2}\\x&>11\end{aligned}$$

127.
$$\begin{aligned}4-3(3-n)&\le 3(2-5n)\\4-9+3n&\le 6-15n\\-5+3n&\le 6-15n\\-5+3n+5&\le 6-15n+5\\3n&\le 11-15n\\3n+15n&\le 11-15n+15n\\18n&\le 11\\\frac{18n}{18}&\le\frac{11}{18}\\n&\le\frac{11}{18}\end{aligned}$$

129.
$$\begin{aligned}2x-3(x-4)&\ge 4-2(x-7)\\2x-3x+12&\ge 4-2x+14\\-x+12&\ge 18-2x\\-x+12-12&\ge 18-2x-12\\-x&\ge 6-2x\\-x+2x&\ge 6-2x+2x\\x&\ge 6\end{aligned}$$

131. $\frac{1}{2}(9x-10) \le -\frac{1}{3}(12-6x)$

$$6\left[\frac{1}{2}(9x-10)\right] \le 6\left[-\frac{1}{3}(12-6x)\right]$$
$$3(9x-10) \le -2(12-6x)$$
$$27x-30 \le -24+12x$$
$$27x-30+30 \le -24+12x+30$$
$$27x \le 12x+6$$
$$27x-12x \le 12x+6-12x$$
$$15x \le 6$$
$$\frac{15x}{15} \le \frac{6}{15}$$
$$x \le \frac{2}{5}$$

133. $\frac{2}{3}(9t-15)+4 < 6+\frac{3}{4}(4-12t)$

$$6t-10+4 < 6+3-9t$$
$$6t-6 < 9-9t$$
$$6t-6+6 < 9-9t+6$$
$$6t < 15-9t$$
$$6t+9t < 15-9t+9t$$
$$15t < 15$$
$$\frac{15t}{15} < \frac{15}{15}$$
$$t < 1$$

135. The solution of $3x-4 \ge 5$ is $x \ge 3$. The solution of $3x-4 > 5$ is $x > 3$. The first includes the number 3 but the second does not.

137. $-7x-2 > -4x+1$

$$-7x-2+2 < -4x+1+2$$
$$-7x < -4x+3$$
$$-3x < 3$$
$$x > -1$$

Both **iii** and **iv** are equivalent to the given inequality.

139. $5x-12 \le x+8$ $\quad$ $3x-4 \ge 2+x$

$4x \le 20$ $\quad$ $2x \ge 6$

$x \le 5$ $\quad$ $x \ge 3$

$\{3, 4, 5\}$ are in both solution sets.

141. $4(x-2) \le 3x+5$ $\quad$ $7(x-3) \ge 5x-1$

$4x-8 \le 3x+5$ $\quad$ $7x-21 \ge 5x-1$

$x \le 13$ $\quad$ $2x \ge 20$

$x \ge 10$

$\{10, 11, 12, 13\}$ are in both solution sets.

143. $x \le -2$

145. There are three possible inequalities that could apply.

$$x+18 > 10$$
$$x > -8$$

The side of a triangle cannot have a negative length.

$$x+10 > 18$$
$$x > 8$$

The unknown side must be greater than 8 inches.

$$10+18 > x$$
$$28 > x$$
$$x < 28$$

The unknown side must be less than 28 inches.

The possible values of x are between 8 inches and 28 inches.

147. 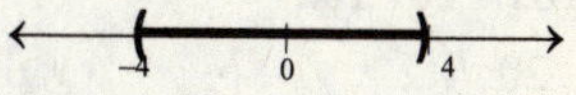

149. 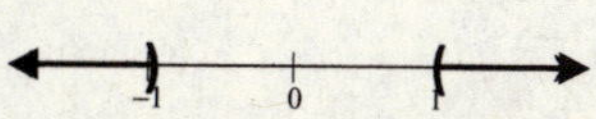

151. If $a < 0$, then it is **always true** that $a^2 > 0$.

153. If $a > b$, it is **never true** that $-a > -b$.

155. If $a \neq 0$ and $b \neq 0$ and $a > b$ it is **sometimes true** that $\frac{1}{a} > \frac{1}{b}$.

Chapter 3 Review Exercises

1.
$$5x - 2 = 4x + 5$$
$$5(3) - 2 = 4(3) + 5$$
$$15 - 2 = 12 + 5$$
$$13 \neq 17$$

No, 3 is not a solution.

2.
$$x - 4 = 16$$
$$x - 4 + 4 = 16 + 4$$
$$x = 20$$

3.
$$8x = -56$$
$$\frac{8x}{8} = \frac{-56}{8}$$
$$x = -7$$

4.
$$5x - 6 = 29$$
$$5x - 6 + 6 = 29 + 6$$
$$5x = 35$$
$$\frac{5x}{5} = \frac{35}{5}$$
$$x = 7$$

5.
$$5x + 3 = 10x - 17$$
$$5x + 3 - 3 = 10x - 17 - 3$$
$$5x = 10x - 20$$
$$5x - 10x = 10x - 20 - 10x$$
$$-5x = -20$$
$$\frac{-5x}{-5} = \frac{-20}{-5}$$
$$x = 4$$

6.
$$3(5x + 2) + 2 = 10x + 5[x - (3x - 1)]$$
$$15x + 6 + 2 = 10x + 5[x - 3x + 1]$$
$$15x + 8 = 10x + 5[-2x + 1]$$
$$15x + 8 = 10x - 10x + 5$$
$$15x + 8 = 5$$
$$15x = -3$$
$$x = -\frac{3}{15} = -\frac{1}{5}$$

7.
$$A = PB$$
$$A = 0.81(500)$$
$$A = 405$$

8.
$$A = PB$$
$$18 = 0.72B$$
$$\frac{18}{0.72} = \frac{0.72B}{0.72}$$
$$25 = B$$

9.
$$A = PB$$
$$27 = P(40)$$
$$\frac{27}{40} = \frac{40P}{40}$$
$$0.675 = P$$
$$67.5\% = P$$

10.

11.
$$x - 3 > -1$$
$$x - 3 + 3 > -1 + 3$$
$$x > 2$$

12.
$$-3x < 12$$
$$\frac{-3x}{-3} > \frac{12}{-3}$$
$$x > -4$$

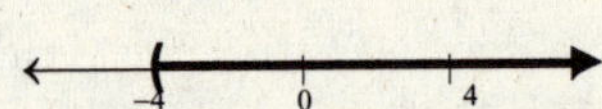

13. $3x+4\ge -8$

$3x+4-4\ge -8-4$

$3x\ge -12$

$\frac{3x}{3}\ge \frac{-12}{3}$

$x\ge -4$

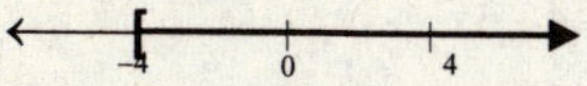

14. $7x-2(x+3)\ge x+10$

$7x-2x-6\ge x+10$

$5x-6\ge x+10$

$5x-6+6\ge x+10+6$

$5x\ge x+16$

$5x-x\ge x+16-x$

$4x\ge 16$

$x\ge 4$

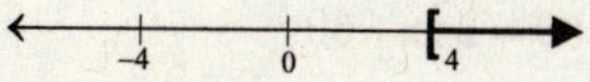

15. $x^2+4x+1=3x+7$

$2^2+4(2)+1=3(2)+7$

$4+8+1=4+7$

$13=13$

Yes, 2 is a solution.

16. $4.6=2.1+x$

$4.6-2.1=2.1+x-2.1$

$2.5=x$

17. $\frac{x}{7}=-7$

$7\cdot\frac{x}{7}=-7\cdot 7$

$x=-49$

18. $14+6x=17$

$14+6x-14=17-14$

$6x=3$

$\frac{6x}{6}=\frac{3}{6}$

$x=\frac{1}{2}$

19. $12y-1=3y+2$

$12y-1+1=3y+2+1$

$12y=3y+3$

$12y-3y=3y+3-3y$

$9y=3$

$\frac{9y}{9}=\frac{3}{9}$

$y=\frac{1}{3}$

20. $x+5(3x-20)=10(x-4)$

$x+15x-100=10x-40$

$16x-100=10x-40$

$16x-100+100=10x-40+100$

$16x=10x+60$

$16x-10x=10x+60-10x$

$6x=60$

$x=10$

21. $A=PB$

$A=\frac{2}{3}(24)$

$A=16$

22. $A=PB$

$60=0.48B$

$\frac{60}{0.48}=\frac{0.48B}{0.48}$

$125=B$

23.
$$A = PB$$
$$0.5 = P(3)$$
$$\frac{0.5}{3} = \frac{3P}{3}$$
$$\frac{1}{6} = P$$
$$16\frac{2}{3}\% = P$$

24.
$$2 + x < -2$$
$$2 + x - 2 < -2 - 2$$
$$x < -4$$

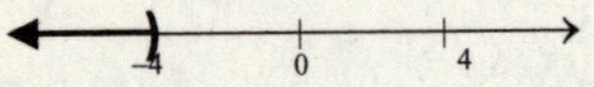

25.
$$5x \le -10$$
$$\frac{5x}{5} \le \frac{-10}{5}$$
$$x \le -2$$

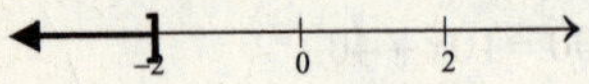

26.
$$6x + 3(2x - 1) = -27$$
$$6x + 6x - 3 = -27$$
$$12x - 3 = -27$$
$$12x - 3 + 3 = -27 + 3$$
$$12x = -24$$
$$\frac{12x}{12} = \frac{-24}{12}$$
$$x = -2$$

27.
$$a - \frac{1}{6} = \frac{2}{3}$$
$$6\left(a - \frac{1}{6}\right) = 6 \cdot \frac{2}{3}$$
$$6a - 1 = 4$$
$$6a - 1 + 1 = 4 + 1$$
$$6a = 5$$
$$a = \frac{5}{6}$$

28.
$$\frac{3}{5}a = 12$$
$$\frac{5}{3} \cdot \frac{3}{5}a = \frac{5}{3} \cdot 12$$
$$a = 20$$

29.
$$32 = 9x - 4 - 3x$$
$$32 = 6x - 4$$
$$32 + 4 = 6x - 4 + 4$$
$$36 = 6x$$
$$\frac{36}{6} = \frac{6x}{6}$$
$$6 = x$$

30.
$$-4[x + 3(x - 5)] = 3(8x + 20)$$
$$-4[x + 3x - 15] = 24x + 60$$
$$-4[4x - 15] = 24x + 60$$
$$-16x + 60 = 24x + 60$$
$$-16x + 60 - 60 = 24x + 60 - 60$$
$$-16x = 24x$$
$$-40x = 0$$
$$x = 0$$

31.
$$4x - 12 < x + 24$$
$$4x - 12 + 12 < x + 24 + 12$$
$$4x < x + 36$$
$$4x - x < x + 36 - x$$
$$3x < 36$$
$$\frac{3x}{3} < \frac{36}{3}$$
$$x < 12$$

32.
$$A = PB$$
$$A = 0.005(3000)$$
$$A = 15$$

33. $3x+7+4x=42$
$$7x+7=42$$
$$7x+7-7=42-7$$
$$7x=35$$
$$\frac{7x}{7}=\frac{35}{7}$$
$$x=5$$

34. $5x-6>19$
$$5x-6+6>19+6$$
$$5x>25$$
$$\frac{5x}{5}>\frac{25}{5}$$
$$x>5$$

35. $A=PB$
$$8=P(200)$$
$$\frac{8}{200}=\frac{200P}{200}$$
$$0.04=P$$
$$4\%=P$$

36. $6x-9<4x+3(x+3)$
$$6x-9<4x+3x+9$$
$$6x-9<7x+9$$
$$6x-9+9<7x+9+9$$
$$6x<7x+18$$
$$6x-7x<7x+18-7x$$
$$-x<18$$
$$x>-18$$

37. $5-4(x+9)>11(12x-9)$
$$5-4x-36>132x-99$$
$$-4x-31>132x-99$$
$$-4x-31+31>132x-99+31$$
$$-4x>132x-68$$
$$-4x-132x>132x-68-132x$$
$$-136x>-68$$
$$x<\frac{68}{136}$$
$$x<\frac{1}{2}$$

38. $A+B+C=180°$
$$20°+50°+C=180°$$
$$70°+C=180°$$
$$C=110°$$

The third angle measures 110°.

39. $F_1x=F_2(d-x)$
$$120(2)=F_2(12-2)$$
$$240=F_2(10)$$
$$\frac{240}{10}=\frac{10F_2}{10}$$
$$24\ lb=F_2$$

The necessary force is 24 pounds.

40. $P=2L+2W$
$$49=2(18.5)+2W$$
$$49=37+2W$$
$$12=2W$$
$$6\ ft=2W$$

The width is 6 feet.

41. $S=R-D$
$$\$198=\$239.99-D$$
$$198-239.99=239.99-D-239.99$$
$$-41.99=-D$$
$$\$41.99=D$$

The discount is \$41.99

42. $A=PB$
$$1184=0.107B$$
$$\frac{1184}{0.107}=\frac{0.107B}{0.107}$$
$$11{,}065\approx B$$

About 11,065 plants and animals are at risk.

43.

$$P = 15 + \frac{1}{2}D$$
$$55 = 15 + \frac{1}{2}D$$
$$55 - 15 = 15 + \frac{1}{2}D - 15$$
$$40 = \frac{1}{2}D$$
$$2(40) = 2 \cdot \frac{1}{2}D$$
$$80 = D$$

The depth is 80 feet.

44.

$$F_1 x = F_2(d - x)$$
$$25x = 15(8 - x)$$
$$25x = 120 - 15x$$
$$25x + 15x = 120 - 15x + 15x$$
$$40x = 120$$
$$\frac{40x}{40} = \frac{120}{40}$$
$$x = 3$$

The fulcrum should be 3 feet from the 25 pound force.

45.

$$P = 2L + 2W$$
$$84 = 2L + 2(18)$$
$$84 = 2L + 36$$
$$84 - 36 = 2L + 36 - 36$$
$$48 = 2L$$
$$\frac{48}{2} = \frac{2L}{2}$$
$$24 = L$$

The length is 24 feet long.

46.

$$d = rt$$
$$30 = 20t$$
$$\frac{30}{20} = \frac{20t}{20}$$
$$1.5 = t$$

It takes 1.5 hours.

47.

$$I = PRT$$
$$50 = P(0.08)(1)$$
$$\frac{50}{0.08} = \frac{0.08P}{0.08}$$
$$625 = P$$

She must invest \$625.

48.

$$Q = Ar$$
$$9 = 150r$$
$$\frac{9}{150} = \frac{150r}{150}$$
$$0.06 = r$$
$$6\% = r$$

The concentration of hydrochloric acid is 6%.

Chapter 3 Test

1.

$$\frac{3}{4}x = 9$$
$$\frac{4}{3} \cdot \frac{3}{4}x = \frac{4}{3} \cdot 9$$
$$x = 12$$

2.

$$6 - 5x = 5x + 11$$
$$6 - 5x - 6 = 5x + 11 - 6$$
$$-5x = 5x + 5$$
$$-5x - 5x = 5x + 5 - 5x$$
$$-10x = 5$$
$$\frac{-10x}{-10} = \frac{5}{-10}$$
$$x = -\frac{1}{2}$$

3.

$$3x - 5 = -14$$
$$3x - 5 + 5 = -14 + 5$$
$$3x = -9$$
$$\frac{3x}{3} = \frac{-9}{3}$$
$$x = -3$$

4.
$$x^2-3x=2x-6$$
$$(-2)^2-3(-2)=2(-2)-6$$
$$4+6=-4-6$$
$$10\neq-10$$

No, -2 is not a solution.

5.
$$x+\frac{1}{2}=\frac{5}{8}$$
$$8\left(x+\frac{1}{2}\right)=8\cdot\frac{5}{8}$$
$$8x+4=5$$
$$8x+4-5=5-4$$
$$8x=1$$
$$x=\frac{1}{8}$$

6.
$$5x-2(4x-3)=6x+9$$
$$5x-8x+6=6x+9$$
$$-3x+6=6x+9$$
$$-3x+6-6=6x+9-6$$
$$-3x=6x+3$$
$$-3x-6x=6x+3-6x$$
$$-9x=3$$
$$\frac{-9x}{-9}=\frac{3}{-9}$$
$$x=-\frac{1}{3}$$

7.
$$7-4x=-13$$
$$7-4x-7=-13-7$$
$$-4x=-20$$
$$\frac{-4x}{-4}=\frac{-20}{-4}$$
$$x=5$$

8.
$$11-4x=2x+8$$
$$11-4x-11=2x+8-11$$
$$-4x=2x-3$$
$$-4x-2x=2x-3-2x$$
$$-6x=-3$$
$$\frac{-6x}{-3}=\frac{-3}{-6}$$
$$x=\frac{1}{2}$$

9.
$$x-3=-8$$
$$x-3+3=-8+3$$
$$x=-5$$

10.
$$3x-2=5x+8$$
$$3x-2+2=5x+8+2$$
$$3x=5x+10$$
$$3x-5x=5x+10-5x$$
$$-2x=10$$
$$\frac{-2x}{-2}=\frac{10}{-2}$$
$$x=-5$$

11.
$$-\frac{3}{8}x=5$$
$$\left(-\frac{8}{3}\right)\left(-\frac{3}{8}\right)x=\left(-\frac{8}{3}\right)5$$
$$x=-\frac{40}{3}$$

12.
$$6x-3(2-3x)=4(2x-7)$$
$$6x-6+9x=8x-28$$
$$15x-6=8x-28$$
$$15x-6+6=8x-28+6$$
$$15x=8x-22$$
$$15x-8x=8x-22-8x$$
$$7x=-22$$
$$x=-\frac{22}{7}$$

13.
$$\begin{aligned} 6-2(5x-8) &= 3x-4 \\ 6-10x+16 &= 3x-4 \\ -10x+22 &= 3x-4 \\ -10x+22-22 &= 3x-4-22 \\ -10x &= 3x-26 \\ -10x-3x &= 3x-26-3x \\ -13x &= -26 \\ \frac{-13x}{-13} &= \frac{-26}{-13} \\ x &= 2 \end{aligned}$$

14.
$$\begin{aligned} 9-3(2x-5) &= 12+5x \\ 9-6x+15 &= 12+5x \\ -6x+24 &= 12+5x \\ -6x+24-24 &= 12+5x-24 \\ -6x &= 5x-12 \\ -6x-5x &= 5x-12-5x \\ -11x &= -12 \\ \frac{-11x}{-11} &= \frac{-12}{-11} \\ x &= \frac{12}{11} \end{aligned}$$

15.
$$\begin{aligned} 3(2x-5) &= 8x-9 \\ 6x-15 &= 8x-9 \\ 6x-15+15 &= 8x-9+15 \\ 6x &= 8x+6 \\ 6x-8x &= 8x+6-8x \\ -2x &= 6 \\ \frac{-2x}{-2} &= \frac{6}{-2} \\ x &= -3 \end{aligned}$$

16.
$$\begin{aligned} A &= PB \\ 20 &= P(16) \\ \frac{20}{16} &= \frac{16P}{16} \\ 1.25 &= P \\ 125\% &= P \end{aligned}$$

17.
$$\begin{aligned} A &= PB \\ 12 &= 0.30B \\ \frac{12}{0.30} &= \frac{0.30B}{0.30} \\ 40 &= B \end{aligned}$$

18. 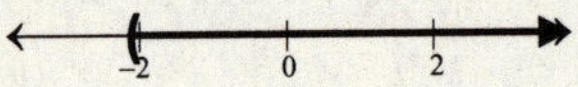

19.
$$\begin{aligned} -2+x &\le -3 \\ x &\le -1 \end{aligned}$$

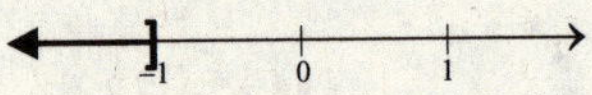

20.
$$\begin{aligned} \frac{3}{8}x &> -\frac{3}{4} \\ \frac{8}{3}\cdot\frac{3}{8}x &> -\frac{3}{4}\cdot\frac{8}{3} \\ x &> -2 \end{aligned}$$

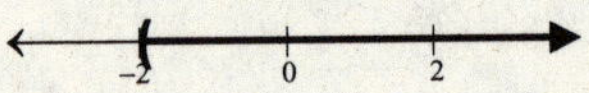

21.
$$\begin{aligned} x+\frac{1}{3} &> \frac{5}{6} \\ x+\frac{1}{3}-\frac{1}{3} &> \frac{5}{6}-\frac{1}{3} \\ x &> \frac{5}{6}-\frac{2}{6} \\ x &> \frac{3}{6} \\ x &> \frac{1}{2} \end{aligned}$$

22.
$$\begin{aligned} 3(x-7) &\ge 5x-12 \\ 3x-21 &\ge 5x-12 \\ 3x-21+21 &\ge 5x-12+21 \\ 3x &\ge 5x+9 \\ 3x-5x &\ge 5x+9-5x \\ -2x &\ge 9 \\ \frac{-2x}{-2} &\le \frac{9}{-2} \\ x &\le -\frac{9}{2} \end{aligned}$$

23. $-\frac{3}{8}x \le 6$

$\left(-\frac{8}{3}\right)\left(-\frac{3}{8}\right)x \ge \left(-\frac{8}{3}\right)6$

$x \ge -16$

24. $4x - 2(3 - 5x) \le 6x + 10$

$4x - 6 + 10x \le 6x + 10$

$14x - 6 \le 6x + 10$

$14x - 6 + 6 \le 6x + 10 + 6$

$14x \le 6x + 16$

$14x - 6x \le 6x + 16 - 6x$

$8x \le 16$

$\frac{8x}{8} \le \frac{16}{8}$

$x \le 2$

25. $3(2x - 5) \ge 8x - 9$

$6x - 15 \ge 8x - 9$

$6x - 15 + 15 \ge 8x - 9 + 15$

$6x \ge 8x + 6$

$6x - 8x \ge 8x + 6 - 8x$

$-2x \ge 6$

$\frac{-2x}{-2} \le \frac{6}{-2}$

$x \le -3$

26. $15 - 3(5x - 7) < 2(7 - 2x)$

$15 - 15x + 21 < 14 - 4x$

$36 - 15x < 14 - 4x$

$36 - 15x - 36 < 14 - 4x - 36$

$-15x < -4x - 22$

$-15x + 4x < -4x - 22 + 4x$

$-11x < -22$

$\frac{-11x}{-11} > \frac{-22}{-11}$

$x > 2$

27. $-6x + 16 = -2x$

$-6x + 16 + 6x = -2x + 6x$

$16 = 4x$

$\frac{16}{4} = \frac{4x}{4}$

$4 = x$

28. $A = PB$

$20 = \frac{5}{6}B$

$\frac{6}{5}(20) = \frac{6}{5} \cdot \frac{5}{6}B$

$24 = B$

29. $\frac{2}{3}x \ge 2$

$\frac{3}{2} \cdot \frac{2}{3}x \ge \frac{3}{2} \cdot 2$

$x \ge 3$

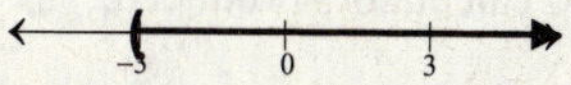

30. $x^2 + 2x + 1 = (x + 1)^2$

$5^2 + 2(5) + 1 = (5 + 1)^2$

$25 + 10 + 1 = 6^2$

$36 = 36$

Yes, 5 is a solution.

31. $A = PB$

$A = \frac{1}{6}(180)$

$A = 30$

The weight on the moon is 30 pounds.

32.
$$\begin{aligned} m_1(T_1 - T) &= m_2(T - T_2) \\ 100(80 - T) &= 50(T - 20) \\ 8000 - 100t &= 50T - 1000 \\ 8000 - 100T - 8000 &= 50T - 1000 - 8000 \\ -100T &= 50T - 9000 \\ -100T - 50T &= 50T - 9000 - 50T \\ -150T &= -9000 \\ \frac{-150T}{-150} &= \frac{-9000}{-150} \\ T &= 60 \end{aligned}$$

The final temperature is 60 °C.

33.
$$\begin{aligned} T &= UN + F \\ 5000 &= 15N + 2000 \\ 3000 &= 15N \\ \frac{3000}{15} &= \frac{15N}{15} \\ 200 &= N \end{aligned}$$

There were 200 calculators produced.

34.
$$\begin{aligned} d &= rt \\ 15 &= (4 + 3.5)t \\ 15 &= 7.5t \\ \frac{15}{7.5} &= \frac{7.5t}{7.5} \\ 2 &= t \end{aligned}$$

They meet after 2 hours.

35.
$$\begin{aligned} I &= PRT \\ I &= 750(0.062)(1) \\ I &= 46.5 \end{aligned}$$

$$\begin{aligned} I &= PRT \\ 46.5 &= P(0.05)(1) \\ \frac{46.5}{0.05} &= \frac{0.05P}{0.05} \\ 930 &= P \end{aligned}$$

He must invest \$930.

36.
$$\begin{aligned} Q &= Ar \\ 2 &= 8r \\ \frac{2}{8} &= \frac{8r}{8} \\ 0.25 &= r \\ 25\% &= r \end{aligned}$$

The milk is 25% chocolate syrup.

Cumulative Review Exercises

1. $-6 - (-20) - 8 = -6 + 20 + (-8)$
$= 14 + (-8) = 6$

2. $(-2)(-6)(-4) = 12(-4)$
$= -48$

3. $-\frac{5}{6} - \left(-\frac{7}{16}\right) = -\frac{5}{6} + \frac{7}{16}$
$= -\frac{40}{48} + \frac{21}{48}$
$= -\frac{19}{48}$

4. $-2\frac{1}{3} \div 1\frac{1}{6} = -\frac{7}{3} \div \frac{7}{6}$
$= -\frac{7}{3} \cdot \frac{6}{7} = -2$

5. $-4^2 \cdot \left(-\frac{3}{2}\right)^3 = -16\left(-\frac{27}{8}\right)$
$= 54$

6. $25 - 3 \cdot \frac{(5-2)^2}{2^3 + 1} - (-2) = 25 - 3 \cdot \frac{3^2}{8+1} + 2$
$= 25 - 3 \cdot \frac{9}{9} + 2$
$= 25 - 3 + 2$
$= 24$

7. $$\begin{aligned}3(a-c)-2ab &= 3[2-(-4)]-2(2)(3)\\ &= 3[2+4]-2(2)(3)\\ &= 3(6)-2(2)(3)\\ &= 18-12\\ &= 6\end{aligned}$$

8. $$\begin{aligned}3x-8x+(-12x) &= -5x+(-12x)\\ &= -17x\end{aligned}$$

9. $$\begin{aligned}2a-(-3b)-7a-5b &= 2a+3b+(-7a)+(-5b)\\ &= -5a-2b\end{aligned}$$

10. $$\begin{aligned}(16x)\left(\frac{1}{8}\right) &= \left(\frac{1}{8}\right)(16x)\\ &= \left(\frac{1}{8}\cdot 16\right)x\\ &= 2x\end{aligned}$$

11. $$\begin{aligned}-4(-9y) &= [(-4)(-9)]y\\ &= 36y\end{aligned}$$

12. $-2(-x^2-3x+2)=2x^2+6x-4$

13. $$\begin{aligned}-2(x-3)+2(4-x) &= -2x+6+8-2x\\ &= -4x+14\end{aligned}$$

14. $$\begin{aligned}-3[2x-4(x-3)]+2 &= -3[2x-4x+12]+2\\ &= -3[-2x+12]+2\\ &= 6x-36+2\\ &= 6x-34\end{aligned}$$

15. Negative integers greater than -8:

{-7, -6, -5, -4, -3, -2, -1}

16. $\frac{7}{8}\cdot 100\% = \frac{700}{8}\% = 87\frac{1}{2}\%$

17. $342\% = 3.42$

18. $62\frac{1}{2}\% = \frac{\overset{5}{\cancel{125}}}{2}\cdot\frac{1}{\underset{4}{\cancel{100}}} = \frac{5}{8}$

19. $$\begin{aligned}x^2+6x+9 &= x+3\\ (-3)^2+6(-3)+9 &= -3+3\\ 9-18+9 &= 0\\ 0 &= 0\end{aligned}$$

Yes, -3 is a solution.

20. $$\begin{aligned}x-4 &= -9\\ x-4+4 &= -9+4\\ x &= -5\end{aligned}$$

21. $$\begin{aligned}\frac{3}{5}x &= -15\\ \frac{5}{3}\cdot\frac{3}{5}x &= -15\cdot\frac{5}{3}\\ x &= -25\end{aligned}$$

22. $$\begin{aligned}13-9x &= -14\\ 13-9x-13 &= -14-13\\ -9x &= -27\\ \frac{-9x}{-9} &= \frac{-27}{-9}\\ x &= 3\end{aligned}$$

23. $$\begin{aligned}5x-8 &= 12x+13\\ 5x-8+8 &= 12x+13+8\\ 5x &= 12x+21\\ 5x-12x &= 12x+21-12x\\ -7x &= 21\\ \frac{-7x}{-7} &= \frac{21}{-7}\\ x &= -3\end{aligned}$$

24. $8x-3(4x-5)=-2x-11$
$$8x-12x+15=-2x-11$$
$$-4x+15=-2x-11$$
$$-4x+15-15=-2x-11-15$$
$$-4x=-2x-26$$
$$-4x+2x=-2x-26+2x$$
$$-2x=-26$$
$$\frac{-2x}{-2}=\frac{-26}{-2}$$
$$x=13$$

25.
$$-\frac{3}{4}x>\frac{2}{3}$$
$$\left(-\frac{4}{3}\right)\left(-\frac{3}{4}\right)x<\left(-\frac{4}{3}\right)\left(\frac{2}{3}\right)$$
$$x<-\frac{8}{9}$$

26.
$$5x-4\geq 4x+8$$
$$5x-4+4\geq 4x+8+4$$
$$5x\geq 4x+12$$
$$5x-4x\geq 4x+12-4x$$
$$x\geq 12$$

27.
$$3x+17<5x-1$$
$$3x+17-17<5x-1-17$$
$$3x<5x-18$$
$$3x-5x<5x-18-5x$$
$$-2x<-18$$
$$\frac{-2x}{-2}>\frac{-18}{-2}$$
$$x>9$$

28. $8-\frac{n}{12}$

29. $n+(n+2)$

30. b and 35 - b

31. L and $3-L$

32.
$$PB=A$$
$$P(1350)=229.50$$
$$\frac{P(1350)}{1350}=\frac{229.50}{1350}$$
$$P=0.17$$
$$P=17\%$$

Income tax is 17% of the salary.

33.
$$t=17.08-0.0067y$$
$$4=17.08-0.0067y$$
$$-13.08=-0.0067y$$
$$\frac{-13.08}{-0.0067}=\frac{-0.0067y}{-0.0067}$$
$$1952\approx y$$

The prediction was the year 1952.

34.
$$m_1\cdot(T_1-T)=m_2\cdot(T-T_2)$$
$$300(75-T)=100(T-15)$$
$$22{,}500-300T=100T-1{,}500$$
$$-400T=-24{,}000$$
$$T=60$$

The final temperature is 60 °C.

35.
$$F_1x=F_2(d-x)$$
$$26(12)=F_2(25-12)$$
$$312=13F_2$$
$$24=F_2$$

The required force is 24 lb.

CHAPTER 4 Solving Equations and Inequalities: Applications

Chapter 4 Prep Test

1. $R-0.35R=1R-0.35R=(1-0.35)R=0.65R$

2. $0.08x+0.05(400-x)=0.08x+20-0.05x$
$$=0.03x+20$$

3. $n+(n+2)+(n+4)=n+n+n+2+4=3n+6$

4. Let x = the number. Then the difference between 5 and twice the number is $5-2x$.

5. $0.4=40\%$

6.
$$\begin{aligned}25x+10(9-x)&=120\\25x+90-10x&=120\\15x+90&=120\\15x&=30\\x&=2\end{aligned}$$

7.
$$\begin{aligned}36&=48-48r\\-12&=-48r\\\frac{-12}{-48}&=\frac{-48r}{-48}\\\frac{1}{4}&=r\end{aligned}$$

8.
$$\begin{aligned}4(2x-5)&<12\\8x-20&<12\\8x&<32\\x&<4\end{aligned}$$

9. n = ounces of nuts; $20-n$ = ounces of pretzels

Section 4.1 Translating Sentences into Equations

1. It is **true** that when two expressions represent the same value, we say that the expressions are equal to each other.

3. It is **true** than in addition to a number, the answer to an application problem must have a unit, such as meters, dollars, minutes, or miles per hour.

5. It is **true** that given the consecutive odd integers -5 and -3, the next consecutive odd integer is -1.

7. **a.** If x is the larger number, then the smaller number is **12 –** $\boldsymbol{x}$.

b. If x is the smaller number, then the larger number is **12 –** $\boldsymbol{x}$.

9.
$$\begin{aligned}n-15&=7\\n-15+15&=7+15\\n&=22\end{aligned}$$

The number is 22.

11.
$$\begin{aligned}7n&=-21\\\frac{7n}{7}&=\frac{-21}{7}\\n&=-3\end{aligned}$$

The number is -3.

13.
$$\begin{aligned}3n-4&=5\\3n-4+4&=5+4\\3n&=9\\\frac{3n}{3}&=\frac{9}{3}\\n&=3\end{aligned}$$

The number is 3.

15. $$\begin{aligned} 4(2n+3) &= 12 \\ 8n+12 &= 12 \\ 8n+12-12 &= 12-12 \\ 8n &= 0 \\ n &= 0 \end{aligned}$$

The number is 0.

17. $$\begin{aligned} 12 &= 6(n-3) \\ 12 &= 6n-18 \\ 12+18 &= 6n-18+18 \\ 30 &= 6n \\ \frac{30}{6} &= \frac{6n}{6} \\ 5 &= n \end{aligned}$$

The number is 5.

19. $$\begin{aligned} 22 &= 6n-2 \\ 22+2 &= 6n-2+2 \\ 24 &= 6n \\ \frac{24}{6} &= \frac{6n}{6} \\ 4 &= n \end{aligned}$$

The number is 4.

21. $$\begin{aligned} 4n+7 &= 2n+3 \\ 4n+7-7 &= 2n+3-7 \\ 4n &= 2n-4 \\ 4n-2n &= 2n-4-2n \\ 2n &= -4 \\ \frac{2n}{2} &= \frac{-4}{2} \\ n &= -2 \end{aligned}$$

The number is -2.

23. $$\begin{aligned} 5n-8 &= 8n+4 \\ 5n-8+8 &= 8n+4+8 \\ 5n &= 8n+12 \\ 5n-8n &= 8n+12-8n \\ -3n &= 12 \\ \frac{-3n}{-3} &= \frac{12}{-3} \\ n &= -4 \end{aligned}$$

The number is -4.

25. $$\begin{aligned} 2(n-25) &= 3n \\ 2n-50 &= 3n \\ 2n-50-2n &= 3n-2n \\ -50 &= n \end{aligned}$$

The number is -50.

27. Let n be the smaller number.

Then $20-n$ is the larger number.

$$\begin{aligned} 3n &= 2(20-n) \\ 3n &= 40-2n \\ 3n+2n &= 40-2n+2n \\ 5n &= 40 \\ \frac{5n}{5} &= \frac{40}{5} \\ n &= 8 \\ 20-n &= 12 \end{aligned}$$

The numbers are 8 and 12.

29. Let n be the smaller number.

Then $18-n$ is the larger number.

$$\begin{aligned} 3n+2(18-n) &= 44 \\ 3n+36-2n &= 44 \\ n+36 &= 44 \\ n+36-36 &= 44-36 \\ n &= 8 \\ 18-n &= 10 \end{aligned}$$

The numbers are 8 and 10.

31. Answers will vary. Possible answer: Let n be the larger number. Then $n - 10$ is the smaller number and $n + n - 10 = 14$. If n is the smaller number, then $n + 10$ is the larger number and $n + n + 10 = 14$.

33. Integers that follow one another in order are called <u>consecutive</u> integers.

35. Let n represent the first integer. Then the second integer is $n + 1$ and the third integer is $n + 2$.

37. Let n = the first integer. Then the next two consecutive integers are $n + 1$ and $n + 2$.

$$n+n+1+n+2=54$$
$$3n+3=54$$
$$3n=51$$
$$n=17$$
$$n+1=18$$
$$n+2=19$$

The integers are 17, 18, and 19.

39. Let n = the first even integer. Then the next two consecutive even integers are $n + 2$ and $n + 4$.

$$n+n+2+n+4=84$$
$$3n+6=84$$
$$3n=78$$
$$n=26$$
$$n+2=28;\ n+4=30$$

The even integers are 26, 28, and 30.

41. Let n = the first odd integer. Then the next two consecutive odd integers are $n + 2$ and $n + 4$.

$$n+n+2+n+4=57$$
$$3n+6=57$$
$$3n=51$$
$$n=17$$
$$n+2=19$$
$$n+4=21$$

The odd integers are 17, 19, and 21.

43. Let n = the first integer. Then the next consecutive even integer is $n + 2$.

$$5n=4(n+2)$$
$$5n=4n+8$$
$$n=8$$
$$n+2=10$$

The integers are 8 and 10.

45. Let n = the first integer. Then the next consecutive odd integer is $n + 2$.

$$9n=7(n+2)$$
$$9n=7n+14$$
$$2n=14$$
$$n=7$$
$$n+2=9$$

The integers are 7 and 9.

47. Let n = the first integer. Then the next two consecutive integers are $n + 1$ and $n + 2$.

$$n+n+1+n+2=-24$$
$$3n+3=-24$$
$$3n=-27$$
$$n=-9$$
$$n+1=-8$$
$$n+2=-7$$

The integers are -9, -8, and -7.

49. Let n = the first even integer. Then the next two consecutive even integers are $n + 2$ and $n + 4$.

$$\begin{aligned} 3n &= 2(n+4)+2 \\ 3n &= 2n+8+2 \\ 3n &= 2n+10 \\ n &= 10 \\ n+2 &= 12 \\ n+4 &= 14 \end{aligned}$$

The even integers are 10, 12, and 14.

51. Let n = the first odd integer. Then the next two consecutive odd integers are $n + 2$ and $n + 4$.

$$\begin{aligned} 3(n+2) &= n+n+4+6 \\ 3n+6 &= 2n+10 \\ n &= 4 \end{aligned}$$

Since n is not an odd integer, there is no solution.

53. Statement **ii** does not represent three consecutive even integers.

55. The number of calories in a cup of low-fat milk is two-thirds the number of calories in a cup of whole milk. In this situation, let n represent the number of calories in a cup of <u>whole</u> milk, and let $\frac{2}{3}n$ represent the number of calories in a cup of <u>low-fat</u> milk.

57. $\textit{current value} = \frac{3}{5}(\textit{original value})$

$$\begin{aligned} 19{,}200 &= \frac{3}{5}v \\ \frac{5}{3}\cdot 19{,}200 &= \frac{5}{3}\cdot\frac{3}{5}v \\ \$32{,}000 &= v \end{aligned}$$

The original value of the car was $32,000.

59. $\textit{cal in an orange} = \frac{1}{5}(\textit{cal in cheese pizza})$

$$\begin{aligned} c &= \frac{1}{5}(290) \\ c &= 58\ \textit{cal} \end{aligned}$$

A medium-size orange has 58 calories.

61. Let i = the amount of iron.

Then $15i$ = the amount of mulch and

$2i$ = the amount of potassium.

$$\begin{aligned} i+15i+2i &= 18 \\ 18i &= 18 \\ i &= 1 \\ 15i &= 15lb \end{aligned}$$

There are 15 pounds of mulch.

63. Television: 70 decibels

Blender: 70 + 20 = 90 decibels

Jet Engine: 2(90) – 40 = 180 – 40 = 140 decibels

65. $$\begin{aligned} W+1.6W &= 130 \\ 2.6W &= 130 \\ \frac{2.6W}{2.6} &= \frac{130}{2.6} \\ W &= 50\,ft \\ 1.6W &= 80\,ft \end{aligned}$$

The length is 80 feet and the width is 50 feet.

67. Let h = the number of hours of labor.

$$\begin{aligned} \textit{total} &= (\textit{water pump}) + \textit{labor} \\ 820 &= 375+89h \\ 445 &= 89h \\ 5 &= h \end{aligned}$$

5 hours of labor were needed.

69. Let t = the number of tickets purchased.

$$total = 45.50 + 87.50(ticket)$$
$$833 = 45.5 + 87.5t$$
$$787.5 = 87.5t$$
$$\frac{787.5}{87.5} = \frac{87.5t}{87.5}$$
$$9 = t$$

There were 9 tickets purchased.

71. Let x = the length of the shorter piece. Then the longer piece has length $20 - x$.

$$2x = 20 - x + 4$$
$$2x = 24 - x$$
$$3x = 24$$
$$x = 8$$

The shorter piece is 8 feet long.

73. Let m = number of minutes over 900.

$$total = 80 + 0.40m$$
$$100.40 = 80 + 0.40m$$
$$20.40 = 0.40m$$
$$51 = m$$

The executive used (900 + 51) = 951 minutes of phone time.

75. Each text message over 300 text messages costs $0.15.

77. Let x = length of shorter piece = perimeter of smaller square.

Let $12 - x$ = length of longer piece = perimeter of larger square.

$$12 - x = 2x$$
$$12 = 3x$$
$$4 = x$$
$$12 - x = 8$$

The perimeter of the larger square is 8 feet.

79. Let t = the total trip time.

$$\frac{3}{5}t = 1\frac{1}{2}$$
$$\frac{3}{5}t = \frac{3}{2}$$
$$\frac{5}{3}\cdot\frac{3}{5}t = \frac{5}{3}\cdot\frac{3}{2}$$
$$t = \frac{5}{2} = 2\frac{1}{2}hr$$

The total trip takes $2\frac{1}{2}$ so there is 1 hour remaining on the trip.

81. Let n = the first even integer. Then the next three consecutive even integers are $n + 2$, $n +$ 4,and $n + 6$.

$$n + n + 2 + n + 4 + n + 6 = -36$$
$$4n + 12 = -36$$
$$4n = -48$$
$$n = -12$$
$$n + 2 = -10$$
$$n + 4 = -8$$
$$n + 6 = -6$$

The even integers are -12, -10, -8, and -6.

83. Let n = the first odd integer. Then the next two consecutive odd integers are $n + 2$ and $n + 4$.

$$n + n + 4 = 2(n + 2)$$
$$2n + 4 = 2n + 4$$
$$4 = 4$$

The equation is an identity, so any three consecutive odd integers satisfy it.

85. Answers will vary.

87. If k is an odd integer, then k - 2 is an <u>odd</u> integer.

89. If m and n are even integers, then $m - n$ is an <u>even</u> integer.

91. If m and n are odd integers, then $m + n$ is an <u>even</u> integer.

93. If m and n are odd integers, then mn is an <u>odd</u> integer.

95. If m is an even integer and n is an odd integer, then $m + n$ is an <u>odd</u> integer.

Section 4.2 Geometry Problems

1. In order from smallest to largest: acute, right, obtuse, straight

3. If $\angle D$ is obtuse, then **i**, **ii**, and **iii** cannot be D because they all measure 90° or less and **vi** cannot be D because it measures more than 180°.

5. **No,** adjacent angles cannot be vertical angles because adjacent angles must share a common side.

7. It is **true** that the formula for the perimeter of a rectangle is $P = 2L + 2W$, where L represents the length and W represents the width of a rectangle.

9. It is **true** that an isosceles triangle has two sides of equal measure and two angles of equal measure.

11. An equilateral triangle has three equal sides and three 60° angles. An isosceles triangles has two equal sides and base angles are equal.

13. **a.** If L is the length of the rectangle, then the width of the rectangle is <u>$0.25L$.</u>

b. Perimeter = $2(\underline{L}) + 2(\underline{0.25L})$

c. Perimeter = $2L + 0.5L = 2.5L$

15. Let x = the length of each of the two equal sides. Then $0.5x$ = the length of the third side.

$$x + x + 0.5x = 125$$
$$2.5x = 125$$
$$x = 50$$
$$0.5x = 25$$

The lengths of the sides are 50 feet, 50 feet, and 25 feet.

17. Let W = the width. Then length = $2W$ - 3.

$$P = 2L + 2W$$
$$42 = 2(2W - 3) + 2W$$
$$42 = 4W - 6 + 2W$$
$$42 = 6W - 6$$
$$48 = 6W$$
$$8 = W$$
$$13 = 2W - 3$$

The length is 13 meters and the width is 8 meters.

19. Let W = the width. Then length = $2W$.

$$P = 2L + 2W$$
$$120 = 2(2W) + 2W$$
$$120 = 4W + 2W$$
$$120 = 6W$$
$$20 = W$$
$$40 = 2W$$

The length is 40 feet and the width is 20 feet.

21. Let s = the length of the second side. Then the length of the first side = $2s$ and the length of the third side = $s + 30$.

$$P = sum\ of\ the\ lengths\ of\ the\ sides$$
$$110 = s + 2s + s + 30$$
$$110 = 4s + 30$$
$$80 = 4s$$
$$20 = s$$
$$40 = 2s$$
$$50 = s + 30$$

The sides measure 20 cm, 40 cm and 50 cm.

23. Let L = the length. Then width = $0.3L$.

$$P = 2L + 2W$$
$$338 = 2L + 2(0.3L)$$
$$338 = 2L + 0.6L$$
$$338 = 2.6L$$
$$130 = L$$
$$39 = 0.3L$$

The length is 130 feet and the width is 39 feet.

25. $P = 2L + 2W$

$$64 = 2(20) + 2W$$
$$64 = 40 + 2W$$
$$24 = 2W$$
$$12 = W$$

The width is 12 feet.

27. $P = 4s$

$$48 = 4s$$
$$12 = s$$

The length is 12 inches.

29. **a.** There are 90° in a right angle.

b. There are 180° in a straight angle.

c. There are 360° in one complete revolution.

d. There are between 0° and 90° in an acute angle.

e. There are between 90° and 180° in an obtuse angle.

31. $x + x + 20 = 90$

$$2x + 20 = 90$$
$$2x = 70$$
$$x = 35°$$

33. $x + 3x + 10 = 90$

$$4x + 10 = 90$$
$$4x = 80$$
$$x = 20°$$

35. $a + 127 = 180$

$$a = 53°$$

37. $a + 67 + 172 = 360$

$$a + 239 = 360$$
$$a = 121°$$

39. $4x + 6x + 2x = 180$

$$12x = 180$$
$$x = 15°$$

41. $3x + x + 36 + 4x = 180$

$$8x + 36 = 180$$
$$8x = 144$$
$$x = 18°$$

43. $x + 2x + 2x + 3x = 360$

$$8x = 360$$
$$x = 45°$$

45. $x + 131 = 180$

$$x = 49°$$

47. $7x = 4x + 36$

$$3x = 36$$
$$x = 12°$$

49. **False.** The lines must be parallel for corresponding angles to be equal.

51. **True.** If the lines are perpendicular, all of the angles formed by their intersection are equal.

53. $\angle a = 122°$ because $\angle a$ and the 122° angle are alternate interior angles.

$$\angle a + \angle b = 180$$
$$122 + \angle b = 180$$
$$\angle b = 58°$$

55. $\angle b = 136°$ because $\angle b$ and the $136°$ angle are alternate interior angles.

$$\angle a + \angle b = 180$$
$$\angle a + 136 = 180$$
$$\angle a = 44°$$

57. The statement is **true** even when the lines are not parallel.

59. $3x + 6x = 180$

$$9x = 180$$
$$x = 20°$$

61. $3x + x + 20 = 180$

$$4x + 20 = 180$$
$$4x = 160$$
$$x = 40°$$

63. $b + (90 - a) = 180$

$$b + 90 - 38 = 180$$
$$b + 52 = 180$$
$$x = 128°$$

65. $\angle y = \angle b = 70°$. The angle adjacent to $\angle a$ inside the triangle is a supplement to $\angle a$ so it measures $85°$. This means that the third angle inside the triangle measures $25°$. This angle is supplementary to $\angle x$, which must be equal to $155°$.

67. $\angle a = \angle y = 45°$. The angle inside the triangle supplementary to $\angle b$ is also equal to $45°$, so $\angle b$ must be equal to $135°$.

69. $\angle BOC + \angle AOB + x = 180$

$$\angle BOC + 90 + x = 180$$
$$\angle BOC = 90° - x$$

71. The two acute angles of a right triangle are complementary.

$$A + B = 90$$
$$A + 30 = 90$$
$$A = 60°$$

73. The sum of the interior angles of any triangle is $180°$.

$$A + B + C = 180$$
$$A + 42 + 103 = 180$$
$$A + 145 = 180$$
$$A = 35°$$

75. The sum of the interior angles of any triangle is $180°$.

$$A + B + C = 180$$
$$A + 13 + 65 = 180$$
$$A + 78 = 180$$
$$A = 102°$$

77. The sum of the interior angles of any triangle is $180°$.

Let x = the measure of each of the equal angles. Then the third angle is $3x$.

$$x + x + 3x = 180$$
$$5x = 180$$
$$x = 36°$$
$$3x = 108°$$

The angles measure $36°$, $36°$, and $108°$.

79. The sum of the interior angles of any triangle is 180 °.

Let x = the measure of each of the equal angles. Then the third angle is $2x + 16$.

$$x+x+2x+16=180$$
$$4x+16=180$$
$$4x=164$$
$$x=41°$$
$$2x+16=98°$$

The angles measure 41 °, 41 °, and 98 °.

81. False. A triangle can have only one right angle.

83. False. A triangle has six exterior angles.

85. The perimeter of the rectangle is equal to the perimeter of the triangle, which is 24 cm.

$$L=3W$$
$$P=2L+2W$$
$$24=2(3W)+2W$$
$$24=6W+2W$$
$$24=8W$$
$$3=W$$
$$9=L$$

The length is 9 cm and the width is 3 cm.

87.
$$P=2L+2W$$
$$48x=2L+2(8x)$$
$$48x=2L+16x$$
$$32x=2L$$
$$16x=L$$

The length is $16x$.

89. Let the measures of the angles be x, $x + 1$, and $x + 2$.

$$x+x+1+x+2=180$$
$$3x+3=180$$
$$3x=177$$
$$x=59°$$
$$x+1=60°$$
$$x+2=61°$$

The angles measure 59 °, 60 °, and 61 °.

91. a. Answers will vary.

b. This demonstrates that the sum of the interior angles of a triangle are equal to a straight angle, 180 °.

Section 4.3 Markup and Discount Problems

1. To determine the markup subtract the cost from the selling price.

3. To determine the markup multiply the cost times the markup rate.

5. To determine the sale price subtract the discount from the regular price.

7. Cost is the price that a business pays for a product. Selling price is the price for which the business sells the product to s customer.

9. In the formula $S = C + rC$, $S =$ $40, $C =$ unknown, and $r =$ 0.25.

11.
$$S=C+rC$$
$$S=40+(0.40)(40)$$
$$S=40+16=56$$

The selling price is $56.

13. $S = C + rC$

$S = 358 + (0.58)(358)$

$S = 358 + 207.64$

$S = \$565.64$

The selling price is $565.64.

15. $S = C + rC$

$630 = 360 + (r)(360)$

$270 = 360r$

$0.75 = r$

$75\% = r$

The markup rate is 75%.

17. $S = C + rC$

$2.45 = 1.96 + (r)(1.96)$

$0.49 = 1.96r$

$0.25 = r$

$25\% = r$

The markup rate is 25%.

19. $S = C + rC$

$168 = C + (0.40)C$

$168 = 1.40C$

$120 = C$

The cost is $120.

21. $S = C + rC$

$82.60 = C + (0.40)C$

$82.60 = 1.40C$

$59 = C$

The cost is $59.

23. True. You can find the store's cost by dividing the selling price by 1.35.

25. $S = C + rC$

$229 = 149.18 + (r)149.18$

$79.82 = 149.18r$

$0.54 \approx r$

$54\% = r$

The markup rate is 54%.

27. Discount is the amount by which a retailer reduces the regular price of a product for sale. Sale price is the regular price of the product minus the amount of the discount.

29. In the formula $S = R - rR$, $S =$ <u>$318.75</u>, $R =$ <u>$375</u>, and $r =$ <u>unknown</u>.

31. $S = R - rR$

$S = 195 - (0.25)195$

$S = 195 - 48.75$

$S = 146.25$

The sale price is $146.25.

33. $S = R - rR$

$S = 230 - (0.05)230$

$S = 230 - 11.50$

$S = 218.50$

The discount price is $218.50.

35. $S = R - rR$

$318.75 = 425 - (r)425$

$-106.25 = -425r$

$0.25 = r$

$25\% = r$

The discount rate is 25%.

37. $S = R - rR$

$201.50 = 325 - (r)325$

$-123.50 = -325r$

$0.38 = r$

$38\% = r$

The markdown rate is 38%.

39. $S = R - rR$
$568 = R - (0.2)R$
$568 = 0.8R$
$710 = R$
The regular price is \$710.

41. $S = R - rR$
$180 = R - (0.40)R$
$180 = 0.60R$
$300 = R$
The regular price is \$300.

43. True. You can find the sale price by multiplying the regular price by 0.85 when the discount rate is 15%.

45. $S = C + rC$
$63 = C + (0.40)C$
$63 = 1.40C$
$45 = C$
The markup is \$63 – 45 = \$18.

47. $S - M = C$
$1087.50 - 217.50 = 870 = C$
$S = C + rC$
$1087.50 = 870 + (r)870$
$217.50 = 870r$
$0.25 \approx r$
$25\% = r$
The markup rate is 25%.

49. $S = C + rC$
$299 = C + (0.30)C$
$299 = 1.30C$
$230 = C$
The cost is \$230.

51. $608 = 3x + 0.80x$
$608 = 3.80x$
$160 = x$
The regular price for one tire is \$160.

53. No. The result is not a 30% discount because the 20% was off the regular price and the additional 10% was off the sale price. The same thing could be accomplished with a discount of 28%.
Original discount = 0.20(8500) = \$1700
Sale price = \$8500 – 1700 = \$6800
Additional discount = 0.10(6800) = \$680
Final price = \$6800 – 680 = \$6120
$S = R - rR$
$6120 = 8500 - (r)8500$
$-2300 = -8500r$
$0.28 = r$
$28\% = r$

55. Retailers who purchase products from a wholesaler may receive various discounts on the items purchased. The discounts are usually related to the number of items purchased. Two or more discounts given in the same item constitute a series discount. For instance, a series discount of 25/10 means that the retailer receives a 25% discount off the list price and then a 10% discount off the discounted price. If the recommended retail price of an item is $40, then the cost to the retailer receiving a series discount of 25/10 is calculated as follows: $40 - 40(0.25) = 30$. Then $30 - 30(0.10) = 27$. The retailer would pay $27 for the item. The wholesaler may also offer more than two discounts, such as the series discount 25/10/5. To convert a series discount into a single-discount equivalent, subtract the complements of the discounts from 1 (100%). The cost to the retailer can be calculated directly by multiplying the complements of the discounts by the retail price.

Section 4.4 Investment Problems

1. P = principal, the amount invested; r = simple interest rate; I = simple interest earned on the investment

3. a. principal = $1250

b. interest rate = 5%

c. interest earned is 0.05 x 1250 = $62.50

5. True. Since $I = Pr$, when $P = x$ dollars and $r = 7\% = 0.07, I = 0.07x$.

7. False. If the two accounts earn the same amount of interest, then the equation is formed by setting the expressions equal to each other.

$0.05x = 0.08(9000 - x)$.

9.

	P	·	r	=	I
Amount at 5.2%	x	·	0.052	=	$0.052x$
Amount at 7.2%	$1000 + x$	·	0.072	=	$0.072(1000+x)$

11. He invested $9000 at 7% and $6000 at 6.5%.

	P	·	r	=	I
Amount at 7%	x	·	0.07	=	$0.07x$
Amount at 6.5%	$15000 - x$	·	0.065	=	$0.065(15000 - x)$

$$0.07x + 0.065(15{,}000 - x) = 1020$$
$$0.07x + 975 - 0.065x = 1020$$
$$0.005x + 975 = 1020$$
$$0.005x = 45$$
$$x = 9000$$
$$15{,}000 - x = 6000$$

13. He invested $1500 in the mutual fund.

	P	·	r	=	I
Amount at 13%	x	·	0.13	=	$0.13x$
Amount at 7%	$x + 2500$	·	0.07	=	$0.07(x + 2500)$

$$0.13x + 0.07(x + 2500) = 475$$
$$0.13x + 0.07x + 175 = 475$$
$$0.2x + 175 = 475$$
$$0.2x = 300$$
$$x = 1500$$

15. They invested $200,000 at 10% and $100,000 at 8.5%.

	P	·	r	=	, I
Amount at 10%	x	·	0.10	=	$0.10x$
Amount at 8.5%	300,000 - x	·	0.08 5	=	0.085(300,0 00 - x)

$$0.1x + 0.085(300{,}000 - x) = 28{,}500$$
$$0.1x + 25{,}500 - 0.085x = 28{,}500$$
$$0.015x + 25{,}500 = 28{,}500$$
$$0.015x = 3000$$
$$x = 200{,}000$$
$$300{,}000 - x = 100{,}000$$

17. She invested $3000 at 8%.

	P	·	r	=	I
Amount at 7.5%	5000	·	0.075	=	0.075(5000)
Amount at 8%	x	·	0.08	=	$0.08x$

$$0.075(5000) + 0.08x = 615$$
$$375 + 0.08x = 615$$
$$0.08x = 240$$
$$x = 3000$$

19. She invested $2500 at 11%.

	P	·	r	=	I
Amount at 7%	2500	·	0.07	=	0.07(2500)
Amount at 11%	x	·	0.11	=	$0.11x$
Total	$x + 2500$	·	0.09	=	$0.09(x + 2500)$

$$0.07(2500) + 0.11x = 0.09(x + 2500)$$
$$175 + 0.11x = 0.09x + 225$$
$$0.02x = 50$$
$$x = 2500$$

21. He invested a total of $650,000.

	P	·	r	=	I
Amount at 8.25%	$0.55x$	·	0.0825	=	$0.00825(0.55x)$
Amount at 10%	$0.45x$	·	0.10	=	$0.10(0.45x)$

$$0.0825(0.55x) + 0.10(0.45x) = 58{,}743.75$$
$$0.045375x + 0.045x = 58{,}743.75$$
$$0.090375x = 58{,}743.75$$
$$x = 650{,}000$$

23. Sarah invested a total of $500,000.

	P	·	, r	=	I
Amount at 6%	$0.3x$	·	0.06	=	$0.06(0.3x)$
Amount at 8%	$0.25x$	·	0.08	=	$0.08(0.25x)$
Amount at 7.5%	$0.45x$	·	0.075	=	$0.075(0.45x)$

$$0.06(0.3x) + 0.08(0.25x) + 0.075(0.45x) = 35{,}875$$
$$0.018x + 0.02x + 0.03375x = 35{,}875$$
$$0.07175x = 35{,}875$$
$$x = 500{,}000$$

25. **a.** The interest rates were $0.03 = 3\%$ and $0.05 = 5\%$.

b. Beth invested a total of $x + (5000 - x) = \$5000$.

27. The research consultant invested $45,000.

	P	·	r	=	I
Amount at 9%	x	·	0.09	=	$0.09x$
Amount at 8%	$x + 5000$	·	0.08	=	$0.08(x + 5000)$

$$0.09x = 0.08(x+5000)$$
$$0.09x = 0.08x + 400$$
$$0.01x = 400$$
$$x = 40,000$$
$$x + 5000 = 45,000$$

29. She invested a total of $16,000.

Interest earned :

$$0.08(16,000 + 3000) + 0.095(16,000) = \$3040.$$

	P	·	r	=	I
Amount at 8%	$x + 3000$	·	0.08	=	$0.08(x + 3000)$
Amount at 9.5%	x	·	0.095	=	$0.095x$

$$0.08(x+3000) = 0.095x$$
$$0.08x + 240 = 0.095x$$
$$240 = 0.015x$$
$$16,000 = x$$

31. Year 1: $P = 3000$;

$I = \text{Pr} = 3000(0.065) = \$195$

Year 2: $P = 3195$;

$I = \text{Pr} = 3195(0.065) = 207.675$

Year 3: $P = 3402.675$;

$I = \text{Pr} = 3402.675(0.065) = 221.174$

Year 4: $P = 3623.849$;

$I = \text{Pr} = 3623.849(0.065) = 235.55$

At the end of year 4, the value of the investment is $3859.40.

33. **a.** From the table, savings should be $298,000.

b. From the table, savings should be between $524,000 and $1,144,000.

c. From the table, savings should be between $28,000 and $75,000.

d. Student explanations should include the fact that money that is invested earns interest; the amount of interest the money earns depends on (1) the interest rate and (2) the number of years it is invested. The closer one is to retirement, the less time any retirement savings have to earn interest, and therefore the more one should have saved. At the same time that retirement savings are earning interest, inflation is eroding the purchasing power of the money saved. If a loaf of bread costs $1.00 today and the inflation rate is 7% per year, then the cost of a loaf of bread in 10 years will be $1.97. Therefore, if your retirement savings are not earning at least 7% annual interest during that 10 years, you will not be able to purchase in 10 years what you are able to purchase today with that amount.

Section 4.5 Mixture Problems

1. **True.** Both mixture and percent mixture problems involve combining two or more ingredients into a single substance.

3. **True.** When mixing two salt solutions, the variable Q in the percent mixture equation $Q = Ar$ represents the amount of salt in a solution.

5. **False.** If we combine a 9% acid solution with a solution that is 4% acid, the resulting solution will between 4% and 9% acid.

7. If sugar is added to a solution of sugar and water, the percent sugar in the mixture **increases**.

9. If 8 L of a solvent costs \$75 per liter, then the value of the 8 L solvent is 8(75) = \$600.

11. **a.**

	Amount, A	·	Unit cost, C	=	Value, V
Whole-wheat flour	x	·	1.70	=	$1.70x$
Unbleached flour	10 - x	·	0.50	=	$0.50(10-x)$
Mixture	10	·	1.30	=	$10(1.30)$

b. $1.70x + 0.50(10 - x) = 10(1.30)$

13. The value of the amount of peanuts used plus the value of the amount of raisins used is equal to the value of the mixture.

15. The mixture contains 56 oz of Alloy #1 and 144 oz of Alloy #2.

	Amount, A	·	Unit cost, C	=	Value, V
Alloy #1	x	·	4.30	=	$4.30x$
Alloy #2	200 - x	·	1.80	=	$1.80(200-x)$
Mixture	200	·	2.50	=	$2.50(200)$

$$4.30x + 1.80(200 - x) = 2.50(200)$$
$$4.30x + 360 - 1.80x = 500$$
$$2.5x + 360 = 500$$
$$2.5x = 140$$
$$x = 56$$
$$200 - x = 144$$

17. The mixture contains 8 lb of chamomile tea.

	Amount, A	$\cdot$	Unit cost, C	=	Value, V
Chamomile tea	x	$\cdot$	18.20	=	$18.20x$
Orange tea	12	$\cdot$	12.25	=	12.25(12)
Mixture	$x + 12$	$\cdot$	14.63	=	$14.63(x + 12)$

$$18.20x + 12.25(12) = 14.63(x + 12)$$
$$18.20x + 147 = 14.63x + 175.56$$
$$3.57x = 28.56$$
$$x = 8$$

19. The mixture costs \$5 per pound.

	Amount, A	$\cdot$	Unit cost, C	=	Value, V
Cologne #1	200	$\cdot$	7.50	=	7.50(200)
Cologne #2	500	$\cdot$	4.00	=	4.00(500)
Mixture	700	$\cdot$	x	=	$x(700)$

$$7.50(200) + 4.00(500) = x(700)$$
$$1500 + 2000 = 700x$$
$$3500 = 700x$$
$$5 = x$$

21. The mixture contains 3 lb of caramel.

	Amount, A	$\cdot$	Unit cost, C	=	Value, V
Popcorn	5	$\cdot$	0.80	=	0.80(5)
Caramel	x	$\cdot$	2.40	=	$2.40x$
Mixture	$x + 5$	$\cdot$	1.40	=	$1.40(x + 5)$

$$0.80(5) + 2.40x = 1.40(x + 5)$$
$$4 + 2.40x = 1.40x + 7$$
$$x = 3$$

23. The mixture contains 20 lb of oak chips and 60 lb of pine chips.

	Amount, A	$\cdot$	Unit cost, C	=	Value, V
Oak chips	x	$\cdot$	3.10	=	$3.10x$
Pine chips	$80 - x$	$\cdot$	2.50	=	$2.50(80-x)$
Mixture	80	$\cdot$	2.65	=	2.65(80)

$$3.10x + 2.50(80 - x) = 2.65(80)$$
$$3.10x + 200 - 2.50x = 212$$
$$0.60x = 12$$
$$x = 20$$
$$80 - x = 60$$

25. The mixture contains 75 gal of fruit juice and 25 gal of ice cream.

	Amount, A	$\cdot$	Unit cost, C	=	Value, V
Fruit juice	x	$\cdot$	4.50	=	$4.50x$
Ice cream	$100 - x$	$\cdot$	8.50	=	$8.50(100-x)$
Mixture	100	$\cdot$	5.50	=	5.50(100)

$$4.50x + 8.50(100 - x) = 5.50(100)$$
$$4.50x + 850 - 8.50x = 550$$
$$-4x = -300$$
$$x = 75$$
$$100 - x = 25$$

27. The house blend costs \$5.50 per pound.

	Amount, A	$\cdot$	Unit cost, C	=	Value, V
Central American	12	$\cdot$	8.00	=	8.00(12)
South American	30	$\cdot$	4.50	=	4.50(30)
House blend	42	$\cdot$	x	=	$x(42)$

$$8.00(12) + 4.50(30) = x(42)$$
$$96 + 135 = 42x$$
$$231 = 42x$$
$$5.5 = x$$

29. The department bought 8 bundles of seedlings and 6 bundles of container-grown plants..

	Amount, A	·	Unit cost, C	=	Value, V
Seedlings	x	·	17	=	$17x$
Container-grown	14 - x	·	45	=	$45(14-x)$
Bundles	14	·		=	406

$$17x+45(14-x)=406$$
$$17x+630-45x=406$$
$$-28x=-224$$
$$x=8$$
$$14-x=6$$

31. The mixture costs about $1042.92 per ounce.

	Amount, A	·	Unit cost, C	=	Value, V
Pure gold	25	·	1282	=	25(1282)
Alloy	40	·	900	=	900(40)
Mixture	65	·	x	=	$x(65)$

$$25(1282)+900(40)=x(65)$$
$$32{,}050+36{,}000=65x$$
$$68{,}050=65x$$
$$1046.92\approx x$$

33. If the mixture is made up of peanuts at $3 per pound and almonds at $7 per pound, the mixture will cost somewhere between $3 and $7 per pound, so statements **iii**, **v**, and **vi** could be true.

35. A 250-mL bottle contains a solution that is 90% isopropyl alcohol. The amount of isopropyl alcohol in the solution is (250)<u>0.90</u> = <u>225</u> mL.

37. A 10-pound box of chocolate-covered cherries is 15% chocolate. There are <u>0.15(10) = 1.5</u> lb of chocolate in the box. There are <u>10 – 1.5 = 8.5</u> lb of cherries in the box.

39. In the equation $Q=Ar$, Q is the quantity of a substance in the solution, r is the percent of concentration, and A is the amount of solution. The equation would be used in a problem such as the following: A juice drink is to be made using 6 oz of a juice mix that is 30% real fruit juice and 10 oz of a fruit juice that is 45% real fruit juice. How much real fruit juice is in the juice drink?

41. The resulting gold alloy has a concentration of 24%.

	Amount, A	·	Percent of concentration, r	=	Quantity, Q
30% gold alloy	40	·	0.30	=	0.30(40)
20% gold alloy	60	·	0.20	=	0.20(60)
Mixture	100	·	x	=	x(100)

$$0.30(40)+0.20(60)=x(100)$$
$$12+12=100x$$
$$24=100x$$
$$0.24=x$$

43. 20 gallons of the 15% acid solution are required.

	Amount, A	·	Percent of concentration, r	=	Quantity, Q
15% acid solution	x	·	0.15	=	0.15(x)
20% acid solution	5	·	0.20	=	0.20(5)
Mixture	$x+5$	·	0.16	=	0.16(x +5)

$$0.15x+0.20(5)=0.16(x+5)$$
$$0.15x+1=0.16x+0.8$$
$$-0.01x=-0.2$$
$$x=20$$

45. 30 pounds of the 25% wool yarn must be used.

	Amount, A	·	Percent of concentration, r	=	Quantity, Q
50% wool yarn	20	·	0.50	=	0.50(20)
25% wool yarn	x	·	0.25	=	0.25(x)
Finished rug	$x+20$	·	0.35	=	0.35(x +20)

$$0.50(20)+0.25x=0.35(x+20)$$
$$10+0.25x=0.35x+7$$
$$-0.10x=-3$$
$$x=30$$

47. The mixture requires 6.25 gallons of the 9% nitrogen plant food.

	Amount, A	·	Percent of concentration, r	=	Quantity, Q
9% nitrogen food	x	·	0.09	=	$0.09(x)$
25% nitrogen food	$10 - x$	·	0.25	=	$0.25(10 - x)$
Mixture	10	·	0.15	=	$0.15(10)$

$$0.09x + 0.25(10 - x) = 0.15(10)$$
$$0.09x + 2.5 - 0.25x = 1.5$$
$$-0.16x + 2.5 = 1.5$$
$$-0.16x = -1$$
$$x = 6.25$$

49. The mixture is 19% sugar.

	Amount, A	·	Percent of concentration, r	=	Quantity, Q
Sugar	5	·	1.00	=	$1(5)$
10% sugar cereal	45	·	0.10	=	$0.10(45)$
Mixture	50	·	x	=	$(x)50$

$$1(5) + 0.10(45) = 50x$$
$$5 + 4.5 = 50x$$
$$9.5 = 50x$$
$$0.19 = x$$

51. The mixture will have 30 oz of the 60% lavender potpourri.

	Amount, A	·	Percent of concentration, r	=	Quantity, Q
80% lavender	70	·	0.80	=	$0.80(70)$
60% lavender	x	·	0.60	=	$0.60(x)$
Mixture	$x + 70$	·	0.74	=	$0.74(x + 70)$

$$0.80(70) + 0.60x = 0.74(x + 70)$$
$$56 + 0.60x = 0.74x + 51.8$$
$$-0.14x = -4.2$$
$$x = 30$$

53. The mixture will have 100 mL of the 7% solution and 200 mL of the 4% solution.

	Amount, A	·	Percent of concentration, r	=	Quantity, Q
7% hydrogen peroxide	x	·	0.07	=	$0.07(x)$
4% hydrogen peroxide	$300 - x$	·	0.04	=	$0.04(300-x)$
Mixture	300	·	0.05	=	$0.05(300)$

$$\begin{aligned} 0.07x + 0.04(300-x) &= 0.05(300) \\ 0.07x + 12 - 0.04x &= 15 \\ 0.03x &= 3 \\ x &= 100 \\ 300 - x &= 200 \end{aligned}$$

55. The mixture is 80% aloe.

	Amount, A	·	Percent of concentration, r	=	Quantity, Q
Pure aloe cream	40	·	1.00	=	$1.00(40)$
64% aloe	50	·	0.64	=	$0.64(50)$
Mixture	90	·	x	=	$x(90)$

$$\begin{aligned} 1.00(40) + 0.64(50) &= 90x \\ 40 + 32 &= 90x \\ 72 &= 90x \\ 0.8 &= x \end{aligned}$$

57. The mixture contains 12.5 gal of pure ethanol.

	Amount, A	·	Percent of concentration, r	=	Quantity, Q
E10	100	·	0.10	=	$0.10(100)$
Ethanol	x	·	1.00	=	$1.00(x)$
E20	$x + 100$	·	0.20	=	$0.20(x + 100)$

$$\begin{aligned} 0.10(100) + 1.00x &= 0.20(x+100) \\ 10 + x &= 0.20x + 20 \\ 0.8x &= 10 \\ x &= 12.5 \end{aligned}$$

59. The mixture will be 52% wild rice.

	Amount, A	$\cdot$	Percent of concentration, r	=	Quantity, Q
20% wild rice	12	$\cdot$	0.20	=	0.20(12)
Pure wild rice	8	$\cdot$	1.00	=	1.00(8)
Mixture	20	$\cdot$	x	=	$x(20)$

$$0.20(12)+1.00(8)=20x$$
$$2.4+8=20x$$
$$10.4=20x$$
$$0.52=x$$

61. False. The resulting solution cannot be 50% acid because there are two parts of water for only one part acid.

63. The mixture has 10 pounds of walnuts and 20 pounds of cashews.

	Amount, A	$\cdot$	Unit cost, C	=	Quantity, Q
Walnuts	x	$\cdot$	5.60	=	$5.60(x)$
Cashews	$30 - x$	$\cdot$	7.50	=	$7.50(30-x)$
Peanuts	20	$\cdot$	4.00	=	4.00(20)
Mixture	50	$\cdot$	5.72	=	5.72(50)

$$5.60x+7.50(30-x)+4.00(20)=5.72(50)$$
$$5.60x+225-7.50x+80=286$$
$$-1.9x+305=286$$
$$-1.9x=-19$$
$$x=10$$
$$30-x=20$$

65. The chemist used 3 L of pure acid and 7 L of water.

	Amount, A	$\cdot$	Percent of concentration, r	=	Quantity, Q
Pure acid	x	$\cdot$	1.00	=	$1.00x$
Water	$10 - x$	$\cdot$	0	=	$0(10-x)$
Mixture	10	$\cdot$	0.30	=	0.30(10)

$$1.00x=0.30(10)$$
$$x=3$$
$$10-x=7$$

67. There were 85 adult tickets and 35 children's tickets sold.

	A	·	C	=	V
Adults	x	·	5.50	=	$5.50x$
Tickets	$120 - x$	·	2.75	=	$2.75(120-x)$
Total	120	·		=	563.75

$$\begin{aligned} 5.50x+2.75(120-x) &= 563.75 \\ 5.50x+330-2.75x &= 563.75 \\ 2.75x+330 &= 563.75 \\ 2.75x &= 233.75 \\ x &= 85 \\ 120-x &= 35 \end{aligned}$$

69. When equal amounts of pure water and pure acid are added to a mixture, the resulting mixture has a concentration equal to the concentration of the original solution. So if the resulting mixture is 50% acid, then the original mixture was also 50% acid.

Section 4.6 Uniform Motion Problems

1. In the formula $d = rt$, d is distance, r is rate, and t is time.

3. If t is the time that the Boeing 757 is in the air, then the Boeing 767 has been in the air $(t - 1)$ hours because it left an hour later.

5. When they meet, they have traveled a total distance of 50 feet, which is the initial distance between them.

7. The rate for plane #1 is 105 mph and for plane #2 the rate is 130 mph.

	r	·	t	=	d
Plane #1	$r - 25$	·	2	=	$2(r-25)$
Plane #2	r	·	2	=	$2r$
Total		·		=	470

$$\begin{aligned} 2(r-25)+2r &= 470 \\ 2r-50+2r &= 470 \\ 4r-50 &= 470 \\ 4r &= 520 \\ r &= 130 \\ r-25 &= 105 \end{aligned}$$

9. The second skater overtakes the first skater 40 seconds after he starts.

	r	·	t	=	d
Skater #1	8	·	$t + 10$	=	$8(t+10)$
Skater #2	10	·	t	=	$10t$

When the second skater catches up, they have traveled the same distance.

$$\begin{aligned} 8(t+10) &= 10t \\ 8t+80 &= 10t \\ 80 &= 2t \\ 40 &= t \end{aligned}$$

11. The tour boat is alongside Michael after 2 hours.

	r	$\cdot$	t	$=$	d
Michael's boat	9	$\cdot$	$t+2$	$=$	$9(t+2)$
Tour boat	18	$\cdot$	t	$=$	$18t$

When the tour boat is alongside, they have traveled the same distance.

$$9(t+2)=18t$$
$$9t+18=18t$$
$$18=9t$$
$$2=t$$

13. The airport is 120 miles from the corporate offices.

	r	$\cdot$	t	$=$	d
Car	30	$\cdot$	$\frac{150-x}{30}$	$=$	$150-x$
Helicopter	60	$\cdot$	$\frac{x}{60}$	$=$	x
Total			3		150

The sum of the time in the car and the time in the helicopter is 3 hours.

$$\frac{150-x}{30}+\frac{x}{60}=3$$
$$60\left(\frac{150-x}{30}\right)+60\left(\frac{x}{60}\right)=60(3)$$
$$2(150-x)+x=180$$
$$300-2x+x=180$$
$$300-x=180$$
$$-x=-120$$
$$x=120$$

15. The sailboat travels 36 miles in the first 3 hours.

	r	$\cdot$	t	$=$	d
Part #1	r	$\cdot$	3	$=$	$3r$
Part #2	$r-5$	$\cdot$	3	$=$	$3(r-5)$

The sum of the distances from the two parts of the trip is 57 miles.

$$3r+3(r-5)=57$$
$$3r+3r-15=57$$
$$6r-15=57$$
$$6r=72$$
$$r=12$$
$$3r=36$$

17. The speed of the passenger train is 50 mph and the speed of the freight train is 30 mph.

	r	$\cdot$	t	$=$	d
Passenger train	r	$\cdot$	3	$=$	$3r$
Freight train	$r-20$	$\cdot$	5	$=$	$5(r-20)$

After the passenger train has been moving 3 hours, the trains have traveled the same distance.

$$3r=5(r-20)$$
$$3r=5r-100$$
$$-2r=-100$$
$$r=50$$
$$r-20=30$$

19. The second ship caught up in 1 hour.

	Rate, r	$\cdot$	Time, t	$=$	Distance, d
Ship #1	25	$\cdot$	$\frac{d-10}{25}$	$=$	$d - 10$
Ship #2	35	$\cdot$	$\frac{d}{35}$	$=$	d

$$\frac{d-10}{25} = \frac{d}{35}$$
$$35(d-10) = 25d$$
$$35d - 350 = 25d$$
$$-350 = -10d$$
$$35 = d$$
$$time = \frac{d}{35} = \frac{35}{35}$$
$$time = 1hour$$

21. The trains meet 4 hours after the train leaves Charleston.

	Rate, r	$\cdot$	Time, t	$=$	Distance, d
DC to Charleston	60	$\cdot$	t	$=$	$60t$
Charleston to DC	50	$\cdot$	$t - 1$	$=$	$50(t-1)$

When the trains meet, the sum of their distances is 500 miles.

$$60t + 50(t-1) = 500$$
$$60t + 50t - 50 = 500$$
$$110t - 50 = 500$$
$$110t = 550$$
$$t = 5$$
$$t - 1 = 4$$

23.

	Rate, r	$\cdot$	Time, t	$=$	Distance, d
Driver #1	90	$\cdot$	$50/90 = 33.3$ min	$=$	50
Driver #2	120	$\cdot$	$50/120 = 25$ min	$=$	50

Driver #1 completes the course in 33.3 minutes. When driver #2 starts the course, driver #1 will finish the course in a little more than 18 minutes. Since it will take driver #2 25 minutes to complete the course, he will not catch up before driver #1 completes the course.

25. The bus overtakes the car 180 miles from the starting point.

	Rate, r	$\cdot$	Time, t	$=$	Distance, d
Bus	60	$\cdot$	t	$=$	$60t$
Car	45	$\cdot$	$t+1$	$=$	$45(t+1)$

When the bus overtakes the car they have traveled the same distance.

$$60t = 45(t+1)$$
$$60t = 45t + 45$$
$$15t = 45$$
$$t = 3$$
$$60t = 180$$

27. The plane traveled for 2 hours at 115 mph and for 3 hours at 125 mph.

	r	·	t	=	d
Part #1	115	·	t	=	$115t$
Part #2	125	·	5 - t	=	125(5 - t)

The sum of the distances for the two parts of the trip is 605 miles.

$$115t + 125(5-t) = 605$$
$$115t + 625 - 125t = 605$$
$$-10t + 625 = 605$$
$$-10t = -20$$
$$t = 2$$
$$5 - t = 3$$

29. The rate of the cyclist is 14 mph.

	r	·	t	=	d
Cyclist	$2r$	·	1	=	$2r$
Jogger	r	·	1	=	r

After 1 hour the distance for the cyclist is equal to the distance of the jogger plus 7 miles.

$$2r = r + 7$$
$$r = 7$$
$$2r = 14$$

31. The campers turned around at 10:15 A.M.

	r	·	t	=	d
Downstream	12	·	t	=	$12t$
Upstream	4	·	1 - t	=	4(1 - t)

The distance downstream is equal to the distance upstream.

$$12t = 4(1-t)$$
$$12t = 4 - 4t$$
$$16t = 4$$
$$t = \frac{1}{4} hr = 15 \min$$

33. The van overtakes the truck after 2.25 hours, or 2 hours 15 minutes. The time then is 2:15 P.M.

	r	·	t	=	d
Truck	45	·	$t+1$	=	$45(t+1)$
Van	65	·	t	=	$65t$

When the van overtakes the truck, their distances are equal.

$$45(t+1) = 65t$$
$$45t + 45 = 65t$$
$$45 = 20t$$
$$2.25 = t$$

35. $$\frac{60\,\cancel{mi}}{1\,\cancel{hr}} \cdot \frac{5280\,ft}{1\,\cancel{mi}} \cdot \frac{1\,\cancel{hr}}{60\,\cancel{min}} \cdot \frac{1\,\cancel{min}}{60\sec}$$
$$= \frac{60 \cdot 5280}{60 \cdot 60} ft/\sec$$
$$= 88\,ft/\sec$$

37. $$\frac{total\ miles}{total\ time} = 60mph$$
$$\frac{2}{\frac{1}{30} + \frac{1}{r}} = 60$$
$$2 = 60\left(\frac{1}{30} + \frac{1}{r}\right)$$
$$2 = 2 + \frac{60}{r}$$
$$0 = \frac{60}{r}$$

This equation has no solution so it is not possible to have an average speed of 60 mph in this situation.

39. It takes Pat $\frac{80}{5} = 16$ seconds to run 80 meters, so now find how far Chris can skate in 16 seconds.

$$(8m/\sec)(16\sec) = 128m$$

Chris can skate 128 meters in 16 seconds.

Section 4.7 Inequalities

1. False. Greater than is represented by the symbol >. Greater than or equal to is represented by the symbol ≥.

3. True.

5. False. The expression 45 + 0.20*m* represents the cost to drive the car for one day.

7. $n \geq 102$

9. $n + 0.45n \leq 200$

11.
$$\begin{aligned} 4(n+5) &< 6n \\ 4n + 20 &< 6n \\ 20 &< 2n \\ 10 &< n \end{aligned}$$

The smallest integer that satisfies the inequality is 11.

13.
$$\begin{aligned} 2050 &\leq 0.38x \\ 5385 &\leq x \end{aligned}$$

The monthly household income is $5395 or more.

15.
$$\begin{aligned} 505 + 493 + 412 + x &> 1850 \\ 1410 + x &> 1850 \\ x &> 440 \end{aligned}$$

They must collect more than 440 pounds.

17.
$$\begin{aligned} \frac{75 + 83 + 86 + 78 + x}{5} &\geq 80 \\ 75 + 83 + 86 + 78 + x &\geq 400 \\ 322 + x &\geq 400 \\ x &\geq 78 \end{aligned}$$

The student must make at least 78 on the fifth test to earn a B.

19.
$$\begin{aligned} 0.35x &> 2000 \\ x &> 5714 \end{aligned}$$

Sales of more than $5714 will result in a commission higher than the monthly salary.

21.
$$\begin{aligned} 1.5x &< 68 \\ x &< 45 \end{aligned}$$

The turbine can be 45 feet tall or less.

23.
$$\begin{aligned} 1000 + 0.11x &< 3200 \\ 0.11x &< 2200 \\ x &< 20{,}000 \end{aligned}$$

He expects sales of less than $20,000.

25.
$$\begin{aligned} 0.80x &> 32 \\ x &> 40 \end{aligned}$$

To have 80% fruit juice, 8 oz of artificial flavoring may be added to 32 oz of pure fruit juice.

27.
$$\begin{aligned} 4(120) + 1.50x &< 3(120 + x) \\ 480 + 1.50x &< 360 + 3x \\ -1.50x &< -120 \\ x &> 80 \end{aligned}$$

More than 80 ounces are needed.

29.
$$\begin{aligned} 45 + 0.25x &< 64 \\ 0.25x &< 19 \\ x &< 76 \end{aligned}$$

The car is cheaper if the round trip is less than 76 miles so the ski area needs to be less than 38 miles away for the car to be less expensive. If they choose to take the shuttle, then the ski area must be more than 38 miles away.

31. $35 + 0.14x < 45 + 0.08x$

$0.06x < 10$

$x < 166.7$

Company B costs less for 166 miles or less.

33. $3(n + n + 2) < 4(n + 4)$

$3(2n + 2) < 4(n + 4)$

$6n + 6 < 4n + 16$

$2n < 10$

$n < 5$

If $n < 5$, then n must be 1 or 3. Then the three consecutive odd integers are either 3, 5, and 7 or 1, 3, and 5.

35. $30x \le 360$ or $45x \le 360$

$x \le 12$ $\quad x \le 8$

The crew can prepare between 8 and 12 aircraft In 6 hours.

37. More than 50 oz of pretzels should be used.

	A	∘	C	=	Q
Raisins	100	∘	0.24	=	100(0.24)
Nuts	100	∘	0.50	=	100(0.50)
Pretzels	x	∘	0.12	=	$0.12x$
Mixture	$200 + x$	∘	0.32	=	$0.32(200 + x)$

$0.24(100) + 0.50(100) + 0.12x < 0.32(200 + x)$

$24 + 50 + 0.12x < 64 + 0.32x$

$74 + 0.12x < 64 + 0.32x$

$-0.20x < -10$

$x > 50$

39. He should use between 3.3 pounds and 30 pounds of Black tea.

	A	∘	C	=	Q
Black tea	x	∘	3	=	$3x$
Green tea	10	∘	4	=	4(10)
Mixture	$10 + x$	∘		=	$?(10 + x)$

$3x + 4(10) > 3.25(10 + x)$

$3x + 40 > 32.5 + 3.25x$

$-0.25x > -7.5$

$x < 30$

$3x + 4(10) < 3.75(10 + x)$

$3x + 40 < 37.5 + 3.75x$

$-0.75x < -2.5$

$x > 3.3$

Chapter 4 Review

1. $2x - 7 = x$

$-7 = -x$

$7 = x$

The number is 7.

2. $3x = 2(35 - x)$

$3x = 70 - 2x$

$5x = 70$

$x = 14$

The shorter piece is 14 inches long.

3. $3x = 2(21 - x) - 2$

$3x = 42 - 2x - 2$

$3x = 40 - 2x$

$5x = 40$

$x = 8$

$21 - x = 13$

The numbers are 8 and 13.

4. $2n+6=4(n-2)$

$2n+6=4n-8$

$-2n=-14$

$n=7$

5. $S=R-Rr$

$60=90-90r$

$-30=-90r$

$\frac{1}{3}=r$

$33\frac{1}{3}\%=r$

The discount rate is $33\frac{1}{3}\%$.

6. \$8000 was invested at 6% and \$7000 was invested at 7%.

	P	·	R	=	I
Amount at 6%	x	·	0.06	=	$0.06x$
Amount at 7%	$15{,}000-x$	·	0.07	=	$0.07(15{,}000-x)$

$0.06x+0.07(15{,}000-x)=970$

$0.06x+1050-0.07x=970$

$-0.01x+1050=970$

$-0.01x=-80$

$x=8000$

$15{,}000-x=7000$

7. $x+15+x+15+x=180$

$3x+30=180$

$3x=150$

$x=50$

$x+15=65$

The angles measure $50°$, $65°$ and $65°$.

8. The rate of the motorcyclist is 45 mph.

	Rate, r	·	Time, t	=	Distance, d
Motorcyclist	$3r$	·	2	=	$2(3r)$
Bicyclist	r	·	2	=	$2r$

After 2 hours the motorcyclist's distance is the bicyclist's distance plus 60 miles.

$2(3r)=2r+60$

$6r=2r+60$

$4r=60$

$r=15$

$3r=45$

9. $x+x+\frac{x}{2}-25=180$

$2\left(x+x+\frac{x}{2}-25\right)=2(180)$

$2x+2x+x-50=360$

$5x-50=360$

$5x=410$

$x=82°$

$\frac{x}{2}-25=16°$

The angles measure $82°$, $82°$, and $16°$.

10. $S=C+rC$

$18.45=10.25+r(10.25)$

$8.20=10.25r$

$0.8=r$

$80\%=r$

The markup rate is 80%.

11. The mixture is 14% butterfat.

	A	·	r	=	Q
Cream	5	·	0.30	=	0.30(5)
Milk	8	·	0.04	=	0.04(8)
Mixture	13	·	x	=	x(13)

$$
\begin{aligned}
0.30(5)+0.04(8) &= 13x \\
1.5+0.32 &= 13x \\
1.82 &= 13x \\
0.14 &= x \\
14\% &= x
\end{aligned}
$$

12.

$$
\begin{aligned}
L &= 4W \\
2(4W)+2W &= 200 \\
8W+2W &= 200 \\
10W &= 200 \\
W &= 20 \\
4W &= 80
\end{aligned}
$$

The width is 20 feet and the length is 80 feet.

13.

$$
\begin{aligned}
50+0.03x &< 27+0.05x \\
-0.02x &< -23 \\
x &> 1150
\end{aligned}
$$

Copy Center is less expensive for 1151 copies or more.

14.

$$
\begin{aligned}
S &= R-rR \\
51.46 &= R-0.17R \\
51.46 &= 0.83R \\
62 &= R
\end{aligned}
$$

The regular price is $62.

15. The mixture will contain 7 qt of cranberry juice and 3 qt of apple juice.

	A	·	c	=	V
Cranberry Juice	x	·	7.79	=	7.79x
Apple juice	10 - x	·	7.19	=	7.19(10 − x)
Mixture	10	·	7.61	=	7.61(10)

$$
\begin{aligned}
7.79x+7.19(10-x) &= 7.61(10) \\
7.79x+71.9-7.19x &= 76.1 \\
0.60x+71.9 &= 76.1 \\
0.60x &= 4.2 \\
x &= 7 \\
10-x &= 3
\end{aligned}
$$

16. Let x = the larger integer.

$$
\begin{aligned}
x-8 &= 4+3(36-x) \\
x-8 &= 4+108-3x \\
x-8 &= 112-3x \\
4x &= 120 \\
x &= 30 \\
36-x &= 6
\end{aligned}
$$

The numbers are 30 and 6.

17. $5600 was invested at 12%.

	P	·	r	=	I
Amount at 8.15%	14,000	·	0.0815	=	0.0815(14,000)
Amount at 12%	x	·	0.12	=	0.12(x)
Total	x +14,000	·	0.0925	=	0.0925(x + 14,000)

$$
\begin{aligned}
0.0815(14,000)+0.12x &= 0.0925(x+14,000) \\
1141+0.12x &= 0.0925x+1295 \\
0.0275x &= 154 \\
x &= 5600
\end{aligned}
$$

18. He should add 1 L of pure water to the solution.

	A	$\cdot$	r	$=$	Q
80% alcohol	15	$\cdot$	0.80	$=$	0.80(15)
Pure water	x	$\cdot$	0	$=$	$0(x)$
Mixture	$x + 15$	$\cdot$	0.75	$=$	$0.75(x+15)$

$$0.80(15) = 0.75(x+15)$$
$$12 = 0.75x + 11.25$$
$$0.75 = 0.75x$$
$$1 = x$$

19. $68+82+90+73+95+x \geq 480$

$$408 + x \geq 480$$
$$x \geq 72$$

The student needs at least 72 on the sixth test.

20.

$$-7 = \frac{1}{2}n - 10$$
$$-7 + 10 = \frac{1}{2}n - 10 + 10$$
$$3 = \frac{1}{2}n$$
$$2(3) = 2\left(\frac{1}{2}n\right)$$
$$6 = n$$

21. Let E = the height of the Eiffel Tower.

$$1472 = 2E - 654$$
$$2126 = 2E$$
$$1063 = E$$

The Eiffel Tower is 1063 feet tall.

22. Let x = the measure of the second angle.

Then the other angles are $x + 15$ and $x - 15$.

$$x+15+x+x-15 = 180$$
$$3x = 180$$
$$x = 60°$$
$$x + 15 = 75°$$
$$x - 15 = 45°$$

The angles measure 75°, 60°, and 45°.

23. Let x = the length of the shorter piece.

Then the longer piece = $10 - x$.

$$4x = 2(10-x) - 2$$
$$4x = 20 - 2x - 2$$
$$4x = 18 - 2x$$
$$6x = 18$$
$$x = 3$$
$$10 - x = 7$$

The longer piece if 7 feet long.

24. Let h = the number of hours worked.

$$80 + 65h = 600$$
$$65h = 520$$
$$h = 8$$

The consultation took 8 hours.

25.

$$S = C + rC$$
$$1074 = C + 0.60C$$
$$1074 = 1.6C$$
$$671.25 = C$$

The cost is $671.25.

26.

$$d = rt$$
$$480 = (65+55)t$$
$$480 = 120t$$
$$4 = t$$

The swimmers meet after 4 minutes.

27. Let x = the length of the first side. The other sides are $x + 4$ and $2x - 1$.

$$35 = x + x + 4 + 2x - 1$$
$$35 = 4x + 3$$
$$32 = 4x$$
$$8 = x$$
$$x + 4 = 12$$
$$2x - 1 = 15$$

The sides measure 8 inches, 12 inches, and 15 inches.

28. Let the three consecutive odd integers be n, $n + 2$, and $n + 4$.

$$n + n + 2 + n + 4 = -45$$
$$3n + 6 = -45$$
$$3n = -51$$
$$n = -17$$
$$n + 2 = -15$$
$$n + 4 = -13$$

The integers are -17, -15, and -13.

29. $2(15) + 2(2x - 4) < 52$

$$30 + 4x - 8 < 52$$
$$22 + 4x < 52$$
$$4x < 30$$
$$x < 7.5$$
$$2x - 4 < 11$$

The maximum width is 10 feet.

30. $\angle a$ must equal 138° because they are alternate interior angles. $\angle a$ and $\angle b$ are supplementary angles, so $\angle b = 42°$.

Chapter 4 Test

1. $6n + 13 = 3n - 5$

$$3n = -18$$
$$n = -6$$

2. $3n - 15 = 27$

$$3n = 42$$
$$n = 14$$

3. Let x = the smaller number. Then the larger number is $18 - x$.

$$4x - 7 = 2(18 - x) + 5$$
$$4x - 7 = 36 - 2x + 5$$
$$4x - 7 = 41 - 2x$$
$$6x = 48$$
$$x = 8$$
$$18 - x = 10$$

The numbers are 8 and 10.

4. Let x = the length of the smaller piece. Then $18 - x$ = the length of the longer piece.

$$5x - 2 = 3(18 - x) - 8$$
$$5x - 2 = 54 - 3x - 8$$
$$5x - 2 = 46 - 3x$$
$$8x = 48$$
$$x = 6$$
$$18 - x = 12$$

The pieces are 6 feet and 12 feet long.

5. $S = C + rC$

$$300 = C + 0.50C$$
$$300 = 1.5C$$
$$200 = C$$

The golf clubs cost \$200.

6. $S = R - rR$

$$80 = 100 - r(100)$$
$$-20 = -100r$$
$$0.2 = r$$
$$20\% = r$$

The discount rate is 20%.

7. 20 gallons of the 15% acid solution are required.

	A	$\cdot$	r	$=$	Q
15% acid solution	x	$\cdot$	0.15	$=$	$0.15(x)$
20% acid solution	5	$\cdot$	0.20	$=$	$0.20(5)$
Mixture	$x+5$	$\cdot$	0.16	$=$	$0.16(x+5)$

$$0.15x+0.20(5)=0.16(x+5)$$
$$0.15x+1=0.16x+0.8$$
$$-0.01x=-0.2$$
$$x=20$$

8. Let W = the width of the rectangle. Then the length is $3W-1$.

$$38=2(3W-1)+2W$$
$$38=6W-2+2W$$
$$38=8W-2$$
$$40=8W$$
$$5=W$$
$$3W-1=14$$

The length is 14 meters and the width is 5 meters.

9. Let the integers be n, $n+2$, and $n+4$.

$$3n=n+2+n+4-1$$
$$3n=2n+5$$
$$n=5$$
$$n+2=7$$
$$n+4=9$$

The integers are 5, 7, and 9.

10. $3+21x>15+18x$

$$3x>12$$
$$x>4$$

Florist B is more economical if 5 or more residents are in the nursing home.

11. \$5000 was invested at 10% and \$2000 was invested at 15%.

	P	$\cdot$	R	$=$	I
Amount at 10%	x	$\cdot$	0.10	$=$	$0.10x$
Amount at 15%	$7000-x$	$\cdot$	0.15	$=$	$0.15(7000-x)$
Total	7000				800

$$0.10x+0.15(7000-x)=800$$
$$0.10x+1050-0.15x=800$$
$$-0.05x=-250$$
$$x=5000$$
$$7000-x=2000$$

12. The mixture has 8 pounds if \$7 coffee and 4 pounds of \$4 coffee.

	A	$\cdot$	C	$=$	Q
\$7 grade	x	$\cdot$	7	$=$	$7x$
\$4 grade	$12-x$	$\cdot$	4	$=$	$4(12-x)$
Mixture	12	$\cdot$	6	$=$	$6(12)$

$$7x+4(12-x)=6(12)$$
$$7x+48-4x=72$$
$$3x+48=72$$
$$3x=24$$
$$x=8$$
$$12-x=4$$

13. The speed of the first plane is 225 mph and the speed of the second plane is 125 mph.

	r	$\cdot$	t	$=$	d
Plane #1	$r+100$	$\cdot$	3	$=$	$3(r+100)$
Plane #2	r	$\cdot$	3	$=$	$3r$

In three hours the sum of the distances is 1050 miles.

$$3(r+100)+3r=1050$$
$$3r+300+3r=1050$$
$$6r+300=1050$$
$$6r=750$$
$$r=125$$
$$r+100=225$$

14. Let x = the measure of the second angle. Then the other angles are $x+15$ and $3x$.

$$x+x+15+3x=180$$
$$5x+15=180$$
$$5x=165$$
$$x=33$$
$$x+15=48$$
$$3x=99$$

The angles measure 33°, 48°, and 99°.

15. \$1400 was invested at 6.75% and \$1000 was invested at 9.45%.

	P	$\cdot$	r	$=$	I
Amount at 6.75%	x	$\cdot$	0.0675	$=$	$0.0675x$
Amount at 9.45%	$2400-x$	$\cdot$	0.0945	$=$	$0.0945(2400-x)$
Total	2400				

$$0.0675x=0.0945(2400-x)$$
$$0.0675x=226.8-0.0945x$$
$$0.162x=226.8$$
$$x=1400$$
$$2400-x=1000$$

16.

$$LW=A$$
$$(3x+5)(12)>276$$
$$36x+60>276$$
$$36x>216$$
$$x>6$$
$$3x+5>23$$

The length needs to be at least 24 feet.

17.

$$S=R-rR$$
$$S=99-0.20(99)$$
$$S=99-19.8$$
$$S=79.20$$

The sale price is \$79.20.

Chapter 4 Cumulative Review

1. Elements of B less than -4: -12, -6

2. $-2+(-8)-(-16)$
$-2+(-8)+16$
$-10+16$
6

3. $\left(-\frac{2}{3}\right)^3\left(-\frac{3}{4}\right)^2=\left(-\frac{8}{27}\right)\left(\frac{9}{16}\right)$
$=-\frac{1}{6}$

4. $\frac{5}{6}-\left(\frac{2}{3}\right)^2\div\left(\frac{1}{2}-\frac{1}{3}\right)$
$=\frac{5}{6}-\left(\frac{2}{3}\right)^2\div\left(\frac{3}{6}-\frac{2}{6}\right)$
$=\frac{5}{6}-\left(\frac{2}{3}\right)^2\div\frac{1}{6}$
$=\frac{5}{6}-\frac{4}{9}\div\frac{1}{6}$
$=\frac{5}{6}-\frac{4}{9}\cdot\frac{6}{1}=\frac{5}{6}-\frac{8}{3}$
$=\frac{5}{6}-\frac{16}{6}=-\frac{11}{6}$

5. $-|-18|=-18$

6. $b^2-(a-b)^2=(-1)^2-[4-(-1)]^2$
$=(-1)^2-5^2=1-25=-24$

7. $5x-3y-(-4x)+7y=5x-3y+4x+7y$
$=9x+4y$

8. $-4(3-2x-5x^3)=-12+8x+20x^3$

9. $-2[x-3(x-1)-5]=-2[x-3x+3-5]$
$=-2[-2x-2]=4x+4$

10. $-3x^2-(-5x^2)+4x^2=-3x^2+5x^2+4x^2=6x^2$

11. $4-2x-x^2=2-4x$
$4-2(2)-2^2=2-4(2)$
$4-4-4=2-8$
$-4\neq-6$

No, 2 is not a solution.

12. $9-x=12$
$9-x-9=12-9$
$-x=3$
$x=-3$

13. $-\frac{4}{5}x=12$
$\left(-\frac{5}{4}\right)\left(-\frac{4}{5}x\right)=12\left(-\frac{5}{4}\right)$
$x=-15$

14. $8-5x=-7$
$8-5x-8=-7-8$
$-5x=-15$
$x=3$

15. $-6x-4(3-2x)=4x+8$
$-6x-12+8x=4x+8$
$2x-12=4x+8$
$-2x=20$
$x=-10$

16. $40\%=40\left(\frac{1}{100}\right)=\frac{40}{100}=\frac{2}{5}$

17. $4x\geq16$
$x\geq4$

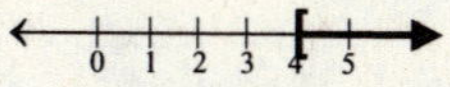

18. $-15x \le 45$

$\dfrac{-15x}{-15} \ge \dfrac{45}{-15}$

$x \ge -3$

19. $2x-3 > x+15$

$2x-3+3 > x+15+3$

$2x-x > x+18-x$

$x > 18$

20. $12-4(x-1) \le 5(x-4)$

$12-4x+4 \le 5x-20$

$16-4x \le 5x-20$

$-4x \le 5x-36$

$-9x \le -36$

$x \ge 4$

21. $0.025 = 0.025(100\%) = 2.5\%$

22. $\dfrac{3}{25} = \dfrac{3}{\cancel{25}_{1}}\left(\dfrac{\cancel{100}^{4}}{1}\%\right) = 12\%$

23. $A = PB$

$A = 16\frac{2}{3}\%(18)$

$A = \frac{1}{6}(18) = 3$

24. $A = PB$

$18 = 0.40B$

$\dfrac{18}{0.40} = \dfrac{0.40B}{0.40}$

$45 = B$

25. $8n+12 = 4n$

$12 = -4n$

$-3 = n$

26. Let G = area of the garage.

$2000 = 200 + 3G$

$1800 = 3G$

$600 = G$

The area of the garage is 600 square feet.

27. Let h = the hours of labor.

$563 = 188 + 75h$

$375 = 75h$

$5 = h$

The bill included 5 hours of labor.

28. $A = PB$

$50 = P(250)$

$\dfrac{50}{250} = P$

$0.20 = P$

$20\% = P$

29. \$2000 was invested at 14%.

	P	·	R	=	I
Amount at 11%	4000	·	0.11	=	0.11(4000)
Amount at 14%	x	·	0.14	=	$0.14x$
Total	$x + 4000$		0.12	=	$0.12(x+4000)$

$0.11(4000) + 0.14x = 0.12(x+4000)$

$440 + 0.14x = 0.12x + 480$

$0.02x = 40$

$x = 2000$

30. $S = C + rC$

$140 = 80 + r(80)$

$60 = 80r$

$\dfrac{60}{80} = r$

$0.75 = r$

$75\% = r$

The markup rate is 75%.

31. The mixture will have 60 grams of the $4 alloy.

	A	°	C	=	Q
Alloy #1	x	°	4	=	$4x$
Alloy #2	30	°	7	=	7(30)
Mixture	$x+30$	°	5	=	$5(x+30)$

$$4x+7(30)=5(x+30)$$
$$4x+210=5x+150$$
$$-x=-60$$
$$x=60$$

32. 30 ounces of water need to be added to the solution.

	A	·	R	=	Q
Pure water	x	·	0	=	0
10% solution	70	·	0.10	=	0.10(70)
Mixture	$x+70$	·	0.07	=	$0.07(x+70)$

$$0.10(70)=0.07(x+70)$$
$$7=0.07x+4.9$$
$$2.1=0.07x$$
$$30=x$$

33. Let x = the measure of each of the equal angles. Then the other angle measures $2x-8$.

$$x+x+2x-8=180$$
$$4x-8=180$$
$$4x=188$$
$$x=47°$$

The equal angles each measure 47°.

34. Let the three consecutive even integers be n, $n+2$, and $n+4$.

$$3(n+2)=14+n+n+4$$
$$3n+6=2n+18$$
$$n=12$$
$$n+2=14$$

The second integer is 14.

CHAPTER 5 Linear Equations and Inequalities

Chapter 5 Prep Test

1. $\dfrac{5-(-7)}{4-8}=\dfrac{5+7}{4+(-8)}=\dfrac{12}{-4}=-3$

2. $\dfrac{a-b}{c-d};\ a=3, b=-2, c=-3, d=2$

$\dfrac{3-(-2)}{-3-2}=\dfrac{3+2}{-3+(-2)}=\dfrac{5}{-5}=-1$

3. $-3(x-4)=-3x+12$

4.
$$\begin{aligned} 3x+6&=0\\ 3x+6-6&=0-6\\ 3x&=-6\\ \frac{3x}{3}&=\frac{-6}{3}\\ x&=-2 \end{aligned}$$

5.
$$\begin{aligned} 4x+5y&=20;\ y=0\\ 4x+5(0)&=20\\ 4x&=20\\ \frac{4x}{4}&=\frac{20}{4}\\ x&=5 \end{aligned}$$

6.
$$\begin{aligned} 3x-7y&=11;\ x=-1\\ 3(-1)-7y&=11\\ -3-7y&=11\\ -3+3-7y&=11+3\\ -7y&=14\\ \frac{-7y}{-7}&=\frac{14}{-7}\\ y&=-2 \end{aligned}$$

7.
$$\begin{aligned} y&=-4x+5;\ x=-2\\ y&=-4(-2)+5\\ y&=8+5\\ y&=13 \end{aligned}$$

8. $\dfrac{1}{4}(3x-16)=\dfrac{3}{4}x-4$

9. **i.** 0 is a solution because $-4<3$.

ii. -3 is a solution because $-4<0$.

iii. 5 is a solution because $-4<8$.

iv. -7 is not a solution because $-4\not<-4$.

v. -10 is not a solution because $-4\not<-7$.

Section 5.1 The Rectangular Coordinate System

1. The point (-3, 4) is in Quadrant II because the *x*-coordinate is negative and the *y*-coordinate is positive.

3. The point (0, -4) lies on the *y*-axis because the *x*-coordinate is 0.

5. The value of the *y*-coordinate at any point on the *x*-axis is 0.

7. Answers will vary. Any point with a *y*-coordinate of 2 will lie 2 units above the *x*-axis. Possible answers: (-5, 2) and (4, 2).

9. To graph the point (5, -4), start at the origin and move 5 units <u>right</u> and 4 units <u>down</u>.

11.

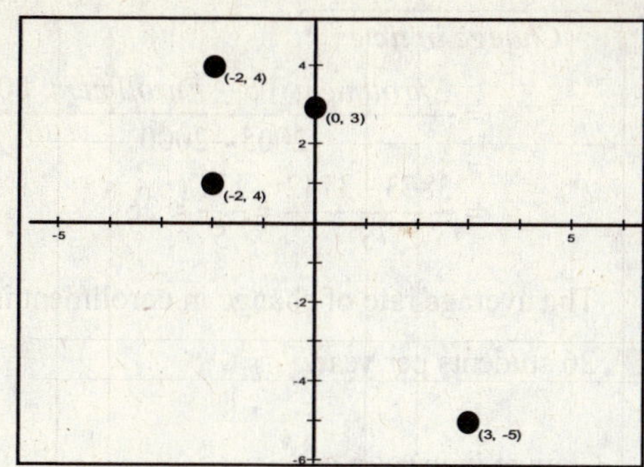

13.

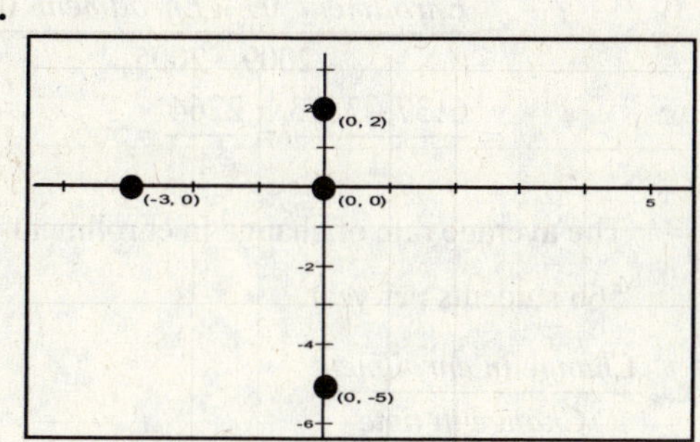

15.

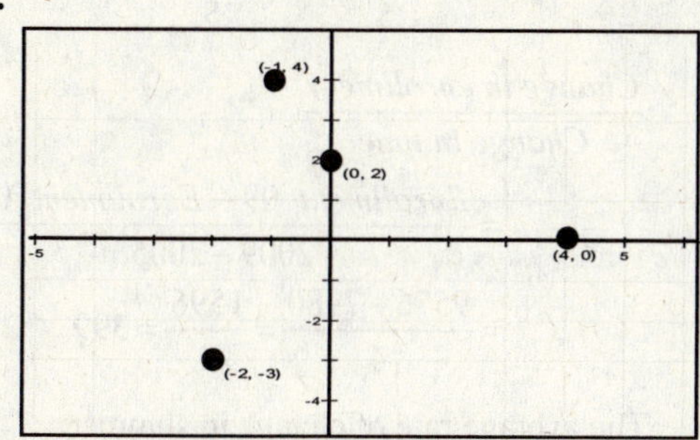

17. A: (2, 3); B: (4, 0); C: (-4, 1); D; (-2, -2)

19. A: (-2, 5); B: (3, 4); C: (0, 0); D; (-3, -2)

21. **a.** Since a and b are poth positive, the point (a, b) will be in Quadrant I.

b. Since a is positive, $-a$ is negative, so the point (-a, b) will be in Quadrant II.

c. Since $a < b$, $b - a$ is positive and since b is positive, $-b$ will be negative, the point $(b - a, -b)$ will be in Quadrant IV.

d. Since $a < b$, $a - b$ is negative and since $-b - a = -(b+a)$, this coordinate is negative, so the point $(a - b, -b - a)$ will be in Quadrant III.

23. From the data table in Exercise 25: The lowest x value is <u>18</u> and the highest x value is <u>51</u>. The lowest y value is <u>22</u> and the highest y value is <u>48</u>. In a scatter diagram of the data, the x-axis must show values from at least <u>18</u> to <u>51</u>, and the y-axis must show values from at least <u>22</u> to <u>48</u>.

25.

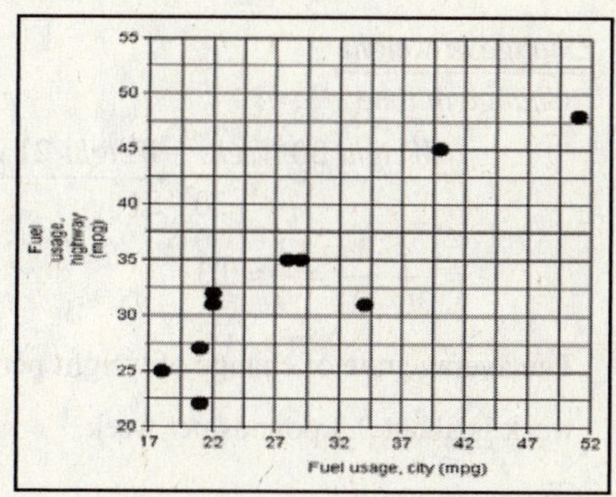

27.

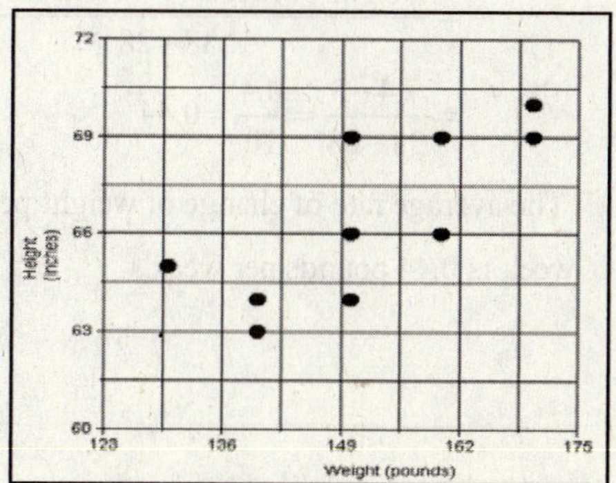

29.

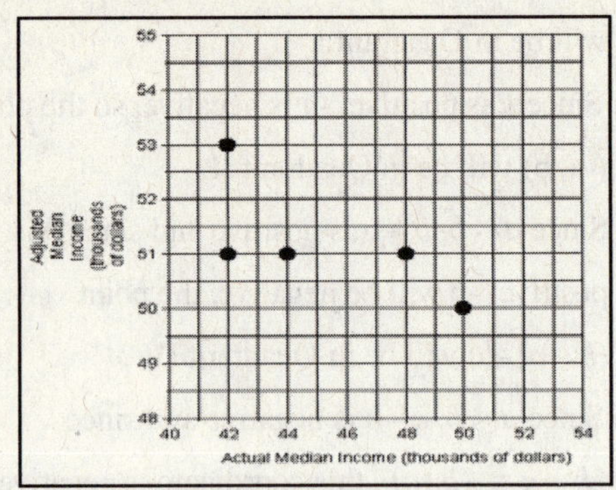

31. In calculating average rate of change, the values that represent time are in the denominator.

33. $\dfrac{\textit{Change in temperature}}{\textit{Change in time}}$

$$= \frac{\textit{Temp. at 2PM} - \textit{Temp. at 6AM}}{2PM - 6AM}$$

$$= \frac{77-45}{8} = \frac{32}{8} = 4$$

The average rate of change is 4 degrees/hour.

35. a. $\dfrac{\textit{Change in weight}}{\textit{Change in time}}$

$$= \frac{\textit{Weight 30 weeks} - \textit{Weight 21 weeks}}{30-21}$$

$$= \frac{4-1}{30-21} = \frac{3}{9} \approx 0.3$$

The average rate of change of weight per week is about 0.3 pounds per week.

b. $\dfrac{\textit{Change in weight}}{\textit{Change in time}}$

$$= \frac{\textit{Weight 38 weeks} - \textit{Weight 28 weeks}}{38-28}$$

$$= \frac{7.4-3}{38-28} = \frac{4.4}{10} = 0.44$$

The average rate of change of weight per week is 0.44 pounds per week.

37. a. $\dfrac{\textit{Change in enrollment}}{\textit{Change in time}}$

$$= \frac{\textit{Enrollment `05} - \textit{Enrollment `00}}{2005-2000}$$

$$= \frac{3873-3742}{5} = \frac{131}{5} \approx 26$$

The average rate of change in enrollment is 26 students per year.

b. $\dfrac{\textit{Change in enrollment}}{\textit{Change in time}}$

$$= \frac{\textit{Enrollment `09} - \textit{Enrollment `05}}{2009-2005}$$

$$= \frac{6137-3873}{4} = \frac{2264}{4} = 566$$

The average rate of change in enrollment is 566 students per year.

c. $\dfrac{\textit{Change in enrollment}}{\textit{Change in time}}$

$$= \frac{\textit{Enrollment `05} - \textit{Enrollment `00}}{2005-2000}$$

$$= \frac{7980-6254}{5} = \frac{1726}{5} \approx 345$$

$\dfrac{\textit{Change in enrollment}}{\textit{Change in time}}$

$$= \frac{\textit{Enrollment `09} - \textit{Enrollment `05}}{2009-2005}$$

$$= \frac{9576-7980}{4} = \frac{1596}{4} = 399$$

The average rate of change in summer enrollment between 2005 and 2009 is greater.

39. **a.** $\frac{\textit{Change in wage}}{\textit{Change in time}}$

$$= \frac{Wage\ `90 - Wage\ `70}{1990 - 1970}$$

$$= \frac{6.33 - 8.12}{20} = \frac{-1.79}{20} = -\$0.0895$$

The average annual rate of change was -8.95 cents/year.

b. The greatest change was from 1975 – 1980 and averaged 2 cents/year.

41. During the 10-minute interval from 30 to 40 minutes the average rate of change was closest to -1 degree per minute.

43. The point (-5, 1) is **1 unit** from the horizontal axis.

45. The point (-6, 0) is **0 units** from the horizontal axis.

47. The point (1, -3) is **1 unit** from the vertical axis.

49. The point at the origin of the rectangular coordinate system has coordinates (0,0).

51. Descriptions should include the idea that plotting points requires starting at the origin, then moving right or left the number of units indicated by the *x*-coordinate; the movement is right if the *x*-coordinate is positive and left if the *x*-coordinate is negative. From there, move up or down the number of units indicated by the *y*-coordinate; move up if the *y*-coordinate is positive and down if the *y*-coordinate is negative. At this point, draw a dot.

53. Longitude indicates the distance east or west of the zero meridian (the prime meridian in Greenwich, England) and is expressed as a number of degrees between $0°$ and $180°$. Latitude indicates the distance north or south of the equator and is expressed as a number of degrees between $0°$ and $90°$. The degree measurement is further broken down into minutes and seconds. Most information almanacs provide the latitudes and longitudes of the major cities in the world.

Section 5.2 Graphs of Straight Lines

1. Both **i** and **ii** are linear equations in two variables. **iii** and **iv** are not because they both have a term with an exponent of 2.

3. **a.** The graph is not a straight line.
b. The graph is a straight line.
c. The graph is not a straight line.
d. The graph is a straight line.
e. The graph is not a straight line.

5. **a.** This equation is in the form $Ax + By = C$.
b. This equation is in the form $y = mx + b$.
c. This equation is in neither form.
d. This equation is in the form $Ax + By = C$.

7. The *x*-coordinate of a *y*-intercept is 0.

9. A line that has a *y*-intercept but no *x*-intercept is a horizontal line.

11. To decide whether the ordered pair (1, 7) is a solution of the equation $y = 2x + 5$, substitute 1 for $\underline{x}$ and 7 for $\underline{y}$ to see whether the ordered pair (1, 7) makes the equation $y = 2x + 5$ a true statement.

13. $y = -x + 7$

$4 \stackrel{?}{=} -3 + 7$

$4 = 4$

Yes, (3, 4) is a solution.

15. $y = \frac{1}{2}x - 1$

$2 \stackrel{?}{=} \frac{1}{2}(-1) - 2$

$2 \stackrel{?}{=} -\frac{1}{2} - 2$

$2 \neq -\frac{5}{2}$

No, (-1, 2) is not a solution.

17. $y = \frac{1}{4}x + 1$

$1 \stackrel{?}{=} \frac{1}{4} \cdot 4 + 1$

$1 \neq 2$

No, (4, 1) is not a solution.

19. $y = \frac{3}{4}x + 4$

$4 \stackrel{?}{=} \frac{3}{4} \cdot 0 + 4$

$4 = 4$

Yes, (0, 4) is a solution.

21. $y = 3x + 2$

$0 \stackrel{?}{=} 3 \cdot 0 + 2$

$0 \neq 2$

No, (0, 0) is not a solution.

23. $y = 3x - 2$

$y = 3(3) - 2$

$y = 7$

The ordered pair is (3, 7) .

25. $y = \frac{2}{3}x - 1$

$y = \frac{2}{3} \cdot 6 - 1$

$y = 3$

The solution is (6, 3).

27. $y = -3x + 1$

$y = -3(0) + 1$

$y = 1$

The solution is (0, 1).

29. $y = \frac{2}{5}x + 2$

$y = \frac{2}{5}(-5) + 2$

$y = 0$

The solution is (-5,0).

31. a. $y = -3x + 6$

$0 = -3x + 6$

$-6 = -3x$

$2 = x$

The ordered pair is (2, 0).

b. If $x > 2$, then y will be negative.

33. a. When $x = -1$, $y = 3(-1) - 4 = -7$. A point on the graph is (-1, -7).

b. When $x = 0$, $y = 3(0) - 4 = -4$. A point on the graph is (0, -4).

c. When $x = 1$, $y = 3(1) - 4 = -1$. A point on the graph is (1, -1).

35. $y = 2x - 3$

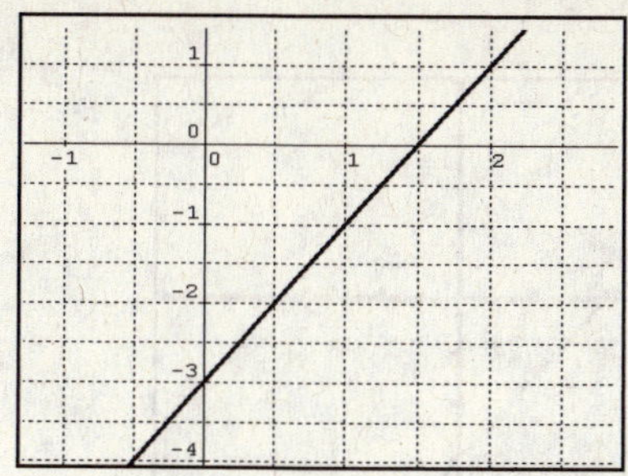

37. $y = \frac{1}{3}x$

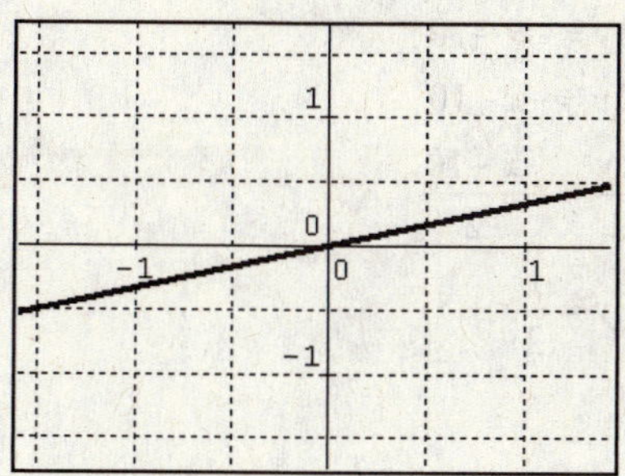

39. $y = \frac{2}{3}x - 1$

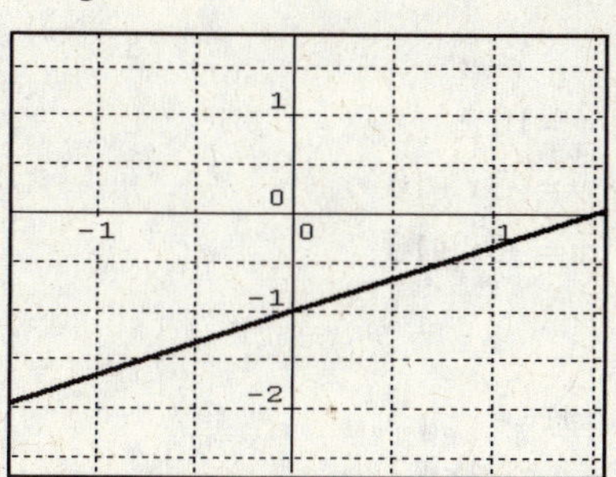

41. $y = -\frac{1}{4}x + 2$

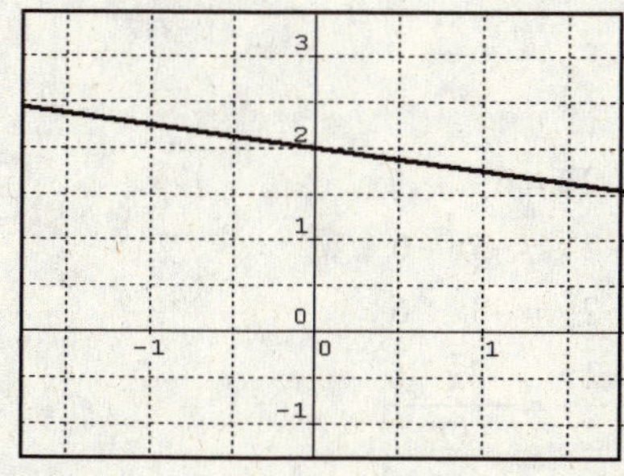

43. $y = -\frac{2}{5}x + 1$

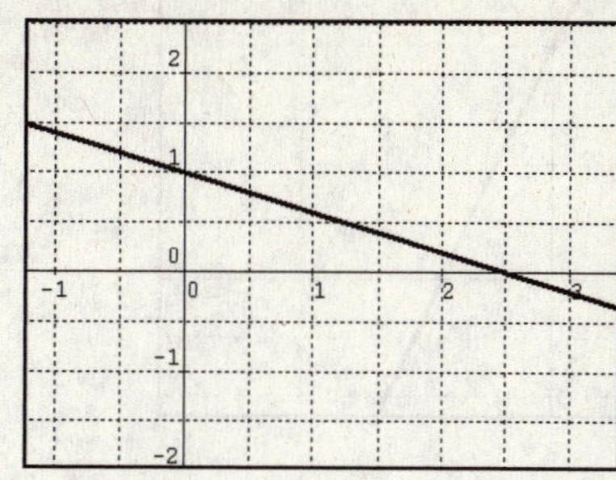

45. $y = 2x - 4$

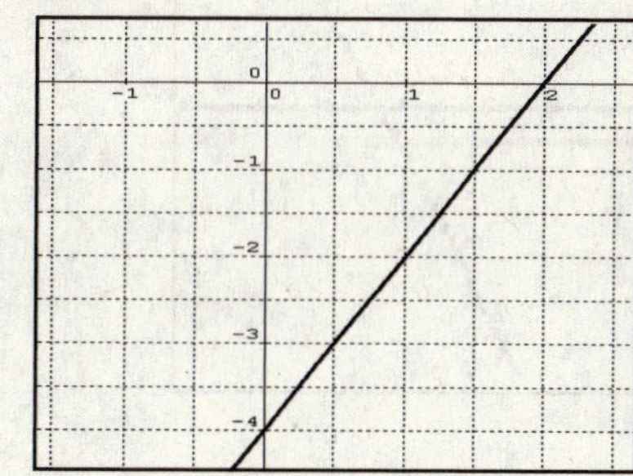

47. $y = -x + 2$

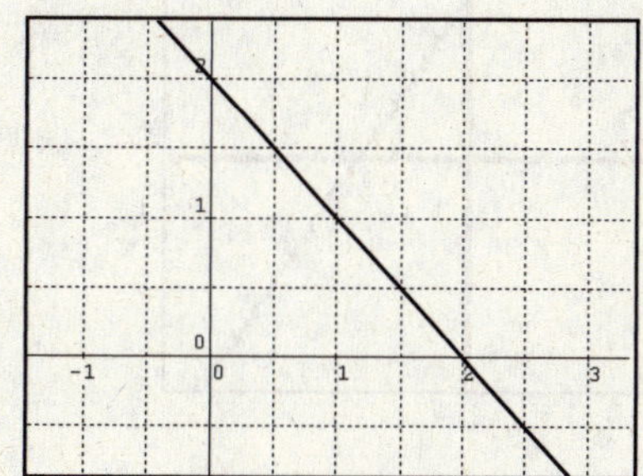

49. $y = -\frac{2}{3}x + 1$

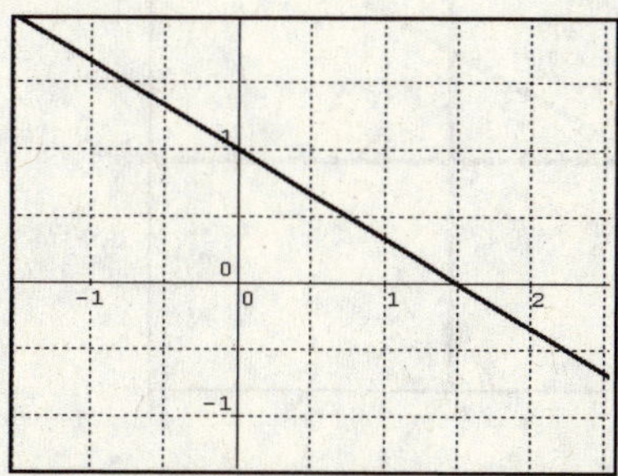

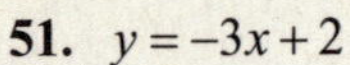

51. $y = -3x + 2$

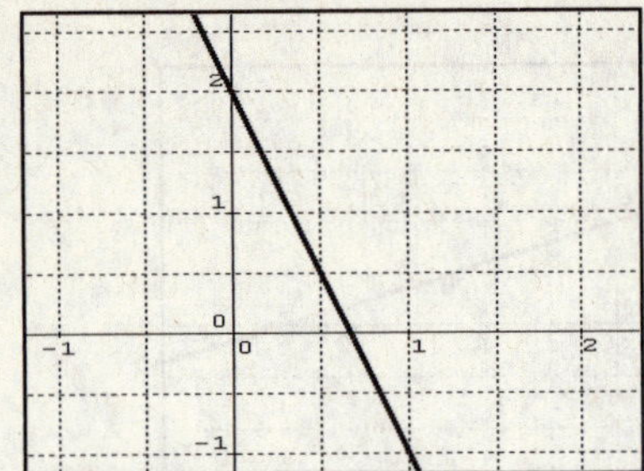

53. $y = 3x - 4$

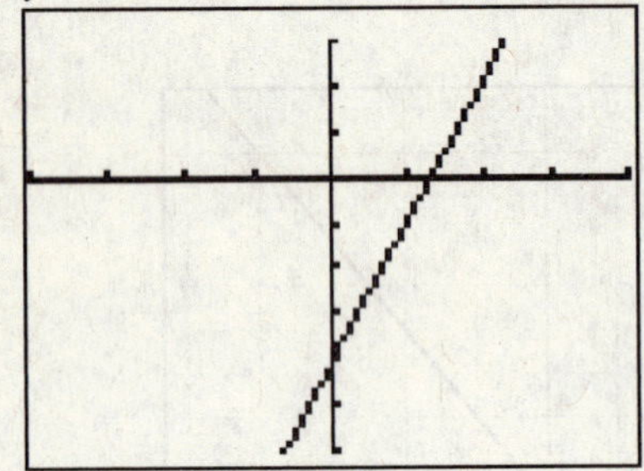

55. $y = -2x + 3$

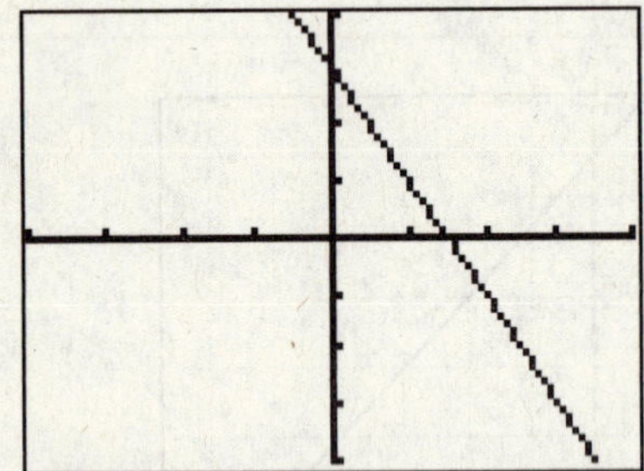

57. $y = \frac{3}{4}x + 2$

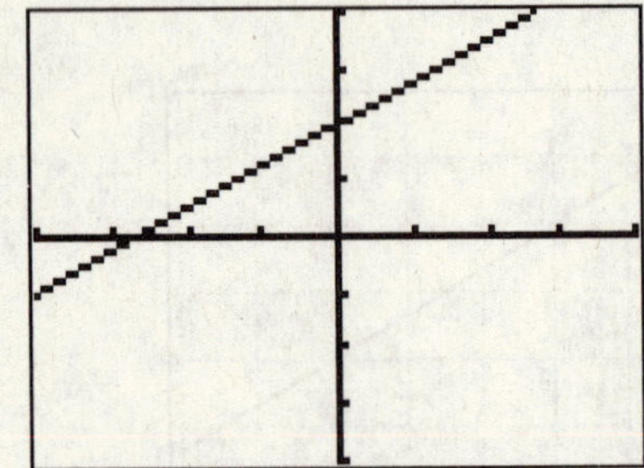

59. $y = -\frac{3}{2}x - 3$

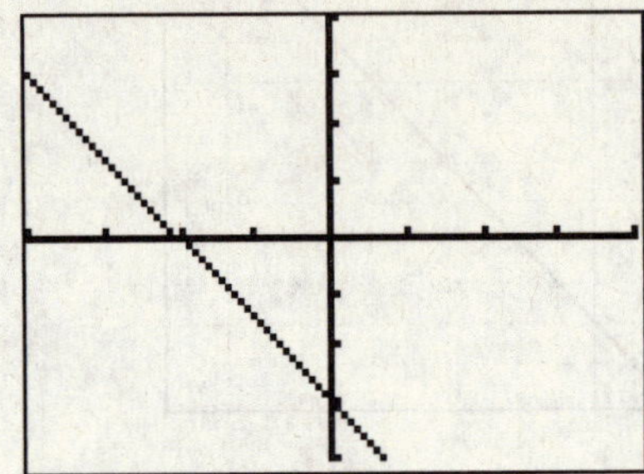

61.
$$\begin{aligned} x - 2y &= -10 \\ x - x - 2y &= -x - 10 \\ -2y &= -x - 10 \\ \frac{-2y}{-2} &= \frac{-x - 10}{-2} \\ y &= \frac{1}{2}x + 5 \end{aligned}$$

63. The graph of the equation $y = 7$ is a <u>horizontal</u> line passing through the point whose coordinates are (0, 7).

65.
$$\begin{aligned} 3x + y &= 10 \\ 3x - 3x + y &= -3x + 10 \\ y &= -3x + 10 \end{aligned}$$

67.
$$\begin{aligned} 4x - y &= 3 \\ 4x - 4x - y &= -4x + 3 \\ -y &= -4x + 3 \\ \frac{-y}{-1} &= \frac{-4x + 3}{-1} \\ y &= 4x - 3 \end{aligned}$$

69.
$$\begin{aligned} 3x + 2y &= 6 \\ 3x - 3x + 2y &= -3x + 6 \\ 2y &= -3x + 6 \\ \frac{2y}{2} &= \frac{-3x + 6}{2} \\ y &= -\frac{3}{2}x + 3 \end{aligned}$$

71.
$$2x-5y=10$$
$$2x-2x-5y=-2x+10$$
$$-5y=-2x+10$$
$$\frac{-5y}{-5}=\frac{-2x+10}{-5}$$
$$y=\frac{2}{5}x-2$$

73.
$$2x+7y=14$$
$$2x-2x+7y=-2x+14$$
$$7y=-2x+14$$
$$\frac{7y}{7}=\frac{-2x+14}{7}$$
$$y=-\frac{2}{7}x+2$$

75.
$$x+3y=6$$
$$x-x+3y=-x+6$$
$$3y=-x+6$$
$$\frac{3y}{3}=\frac{-x+6}{3}$$
$$y=-\frac{1}{3}x+2$$

77. $3x+y=3$

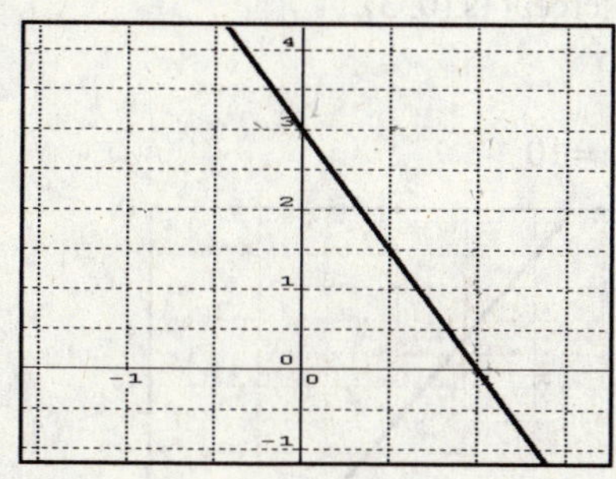

79. $2x+3y=6$

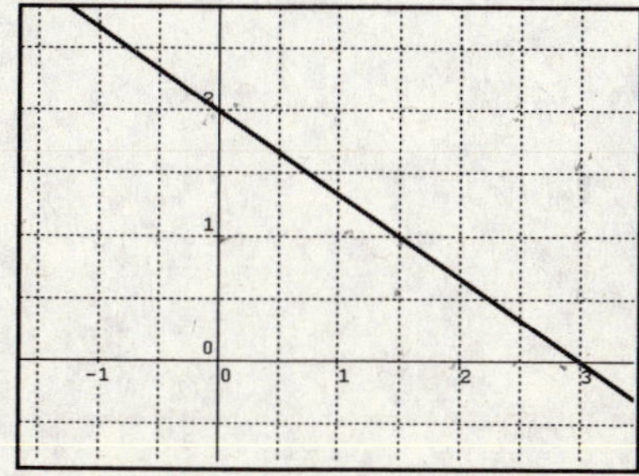

81. $x-2y=4$

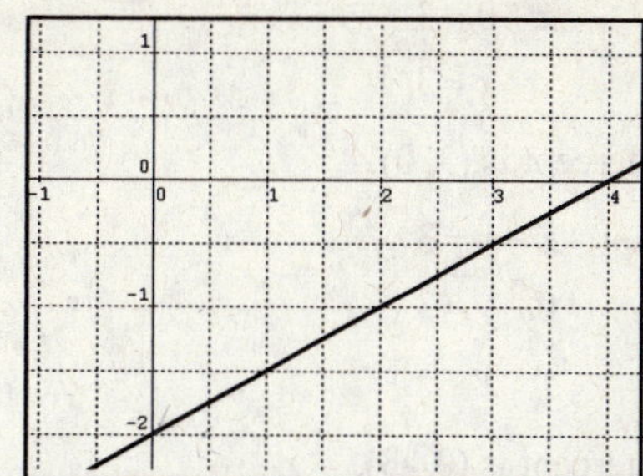

83. $2x-3y=6$

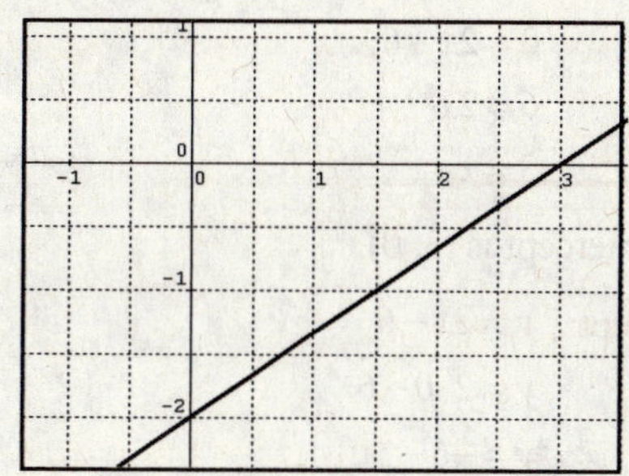

85. $y=4$

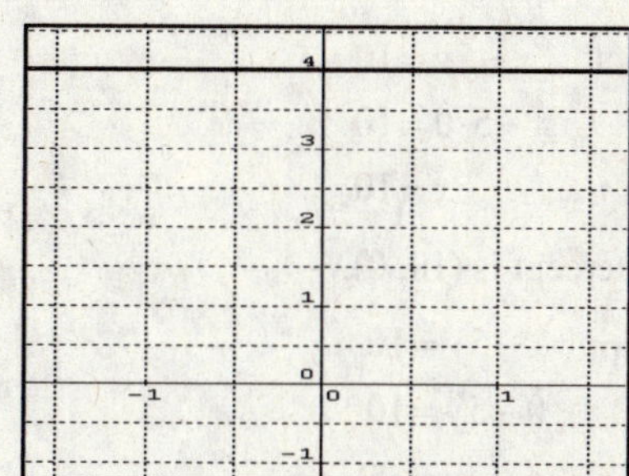

87. $x=-2$

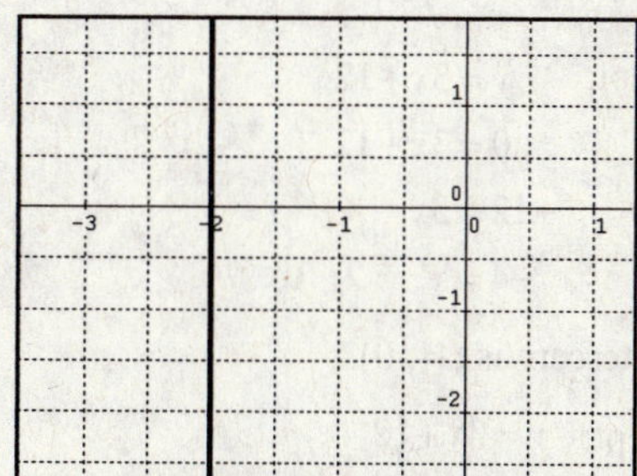

89. x-intercept: $x-y=3$

$x-0=3$

$x=3$

The x-intercept is (3, 0).

y-intercept: $x-y=3$

$0-y=3$

$y=-3$

The y-intercept is (0, -3).

91. x-intercept: $y=2x-6$

$0=2x-6$

$6=2x$

$3=x$

The x-intercept is (3, 0).

y-intercept: $y=2x-6$

$y=2\cdot 0-6$

$y=-6$

The y-intercept is (0, -6).

93. x-intercept: $x-5y=10$

$x-5\cdot 0=10$

$x=10$

The x-intercept is (10, 0).

y-intercept: $x-5y=10$

$0-5y=10$

$y=-2$

The y-intercept is (0, -2).

95. x-intercept: $y=3x+12$

$0=3x+12$

$-12=3x$

$-4=x$

The x-intercept is (-4, 0).

y-intercept: $y=3x+12$

$y=3\cdot 0+12$

$y=12$

The y-intercept is (0, 12).

97. x-intercept: $2x-3y=0$

$2x-3\cdot 0=0$

$2x=0$

$x=0$

The x-intercept is (0, 0).

y-intercept: $2x-3y=0$

$2\cdot 0-3y=0$

$-3y=0$

$y=0$

The y-intercept is (0, 0).

99. x-intercept: $y=\frac{1}{2}x+3$

$0=\frac{1}{2}x+3$

$-3=\frac{1}{2}x$

$-6=x$

The x-intercept is (-6, 0).

y-intercept: $y=\frac{1}{2}x+3$

$y=\frac{1}{2}\cdot 0+3$

$y=3$

The y-intercept is (0, 3).

101. $5x+2y=10$

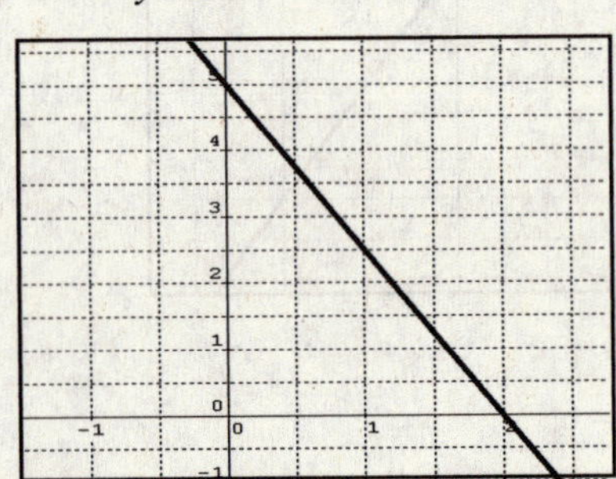

103. $3x-4y=12$

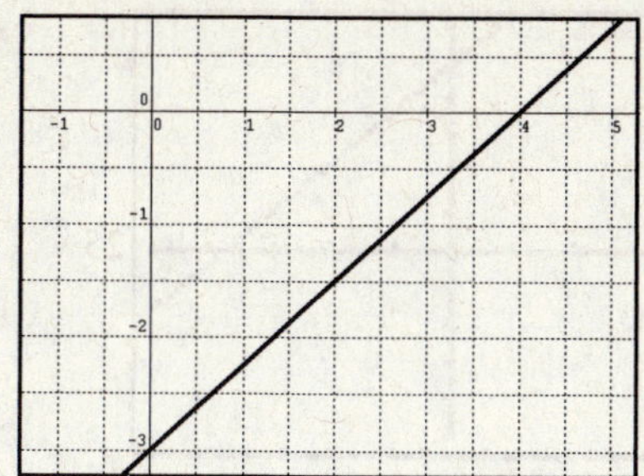

105. $2x+3y=6$

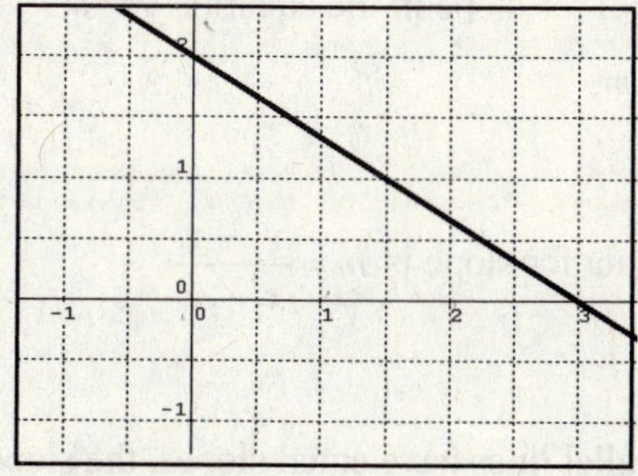

107. $2x+5y=10$

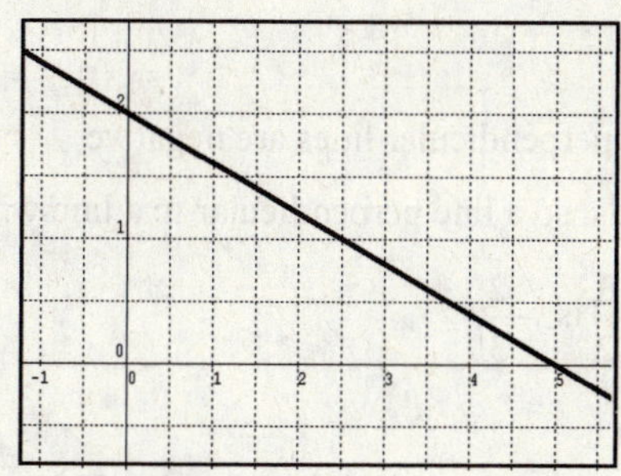

109. $x-3y=6$

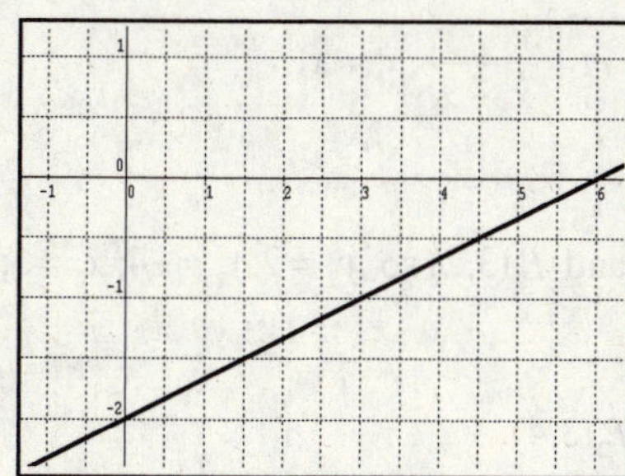

111. $x-4y=8$

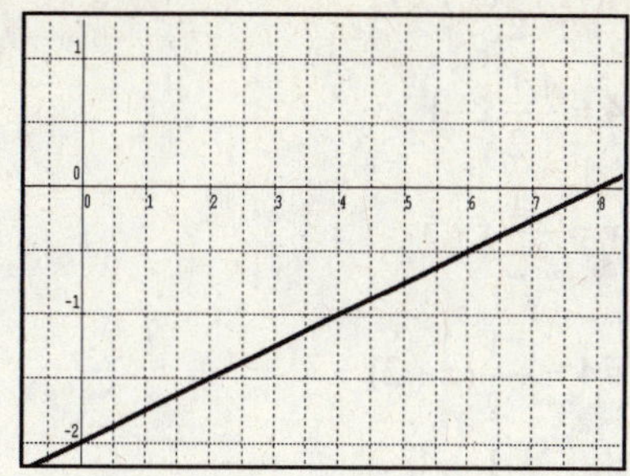

113. $3x+4y=-12$

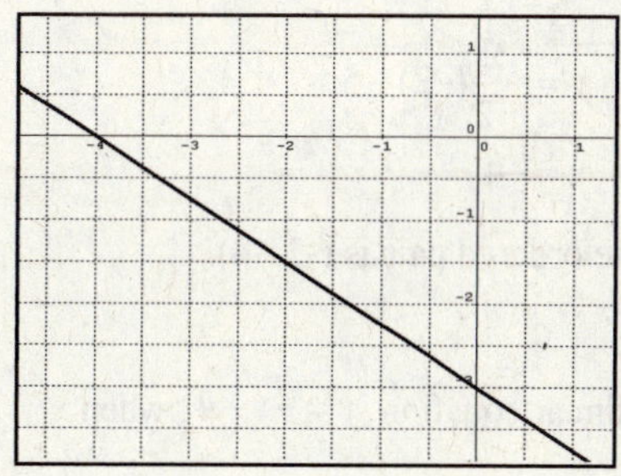

115. $2x-y=4$

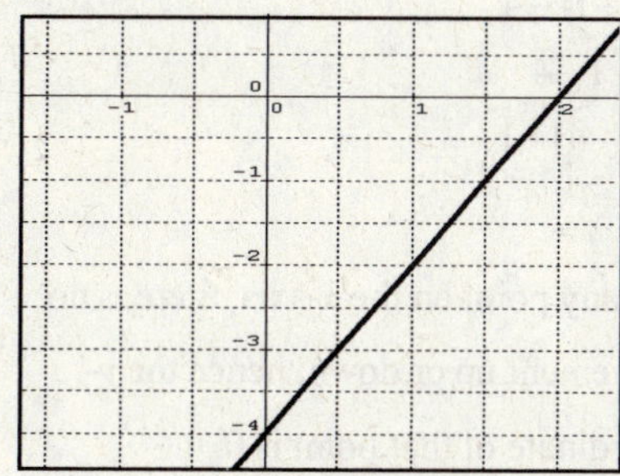

117. If A and C are negative and B is positive, then the x-intercept of $Ax + By = C$ has a positive x-coordinate so the x-intercept is **to the right** of the y-axis.

119. a. $y+4=-\frac{1}{2}(x+2)$

$y+4=-\frac{1}{2}x-1$

$y=-\frac{1}{2}x-5$

b. $y+4=-\frac{1}{2}(x+2)$

$y+4=-\frac{1}{2}x-1$

$y=-\frac{1}{2}x-5$

$y=-\frac{1}{2}(-2)-5$

$y=-4$

The ordered pair is (-2, -4).

121. In the linear equation $y=-x-4$, when x increases by 1, y decreases by 1.

$y=-x-4$

$y=-(x+1)-4$

$y=-x-1-4$

$y=(-x-4)-1$

123. a. For any point on the x-axis, there is no movement up or down; hence the y-coordinate of that point is 0.

b. For any point on the y-axis, there is no

125.

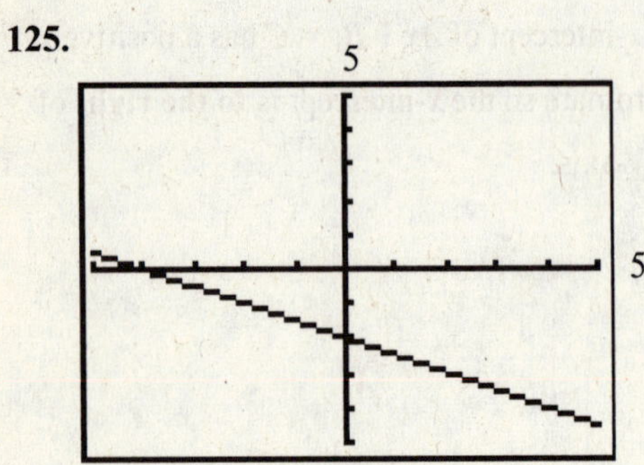

127.

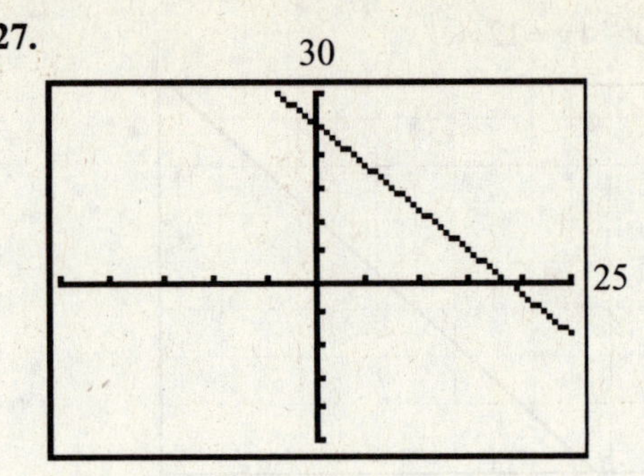

Section 5.3 Slopes of Straight Lines

1. The symbol for slope in the equation $y = mx + b$ is the letter m.

3. The formula for slope is $m=\frac{y_2-y_1}{x_2-x_1}$.

5. Since parallel lines have equal slopes, the slope of any line parallel to a line with a slope of -4 is also -4.

7. Slopes of perpendicular lines are negative reciprocals, so a line perpendicular to a line with a slope of $\frac{3}{2}$ is $-\frac{2}{3}$.

9. Find the slope using the slope formula. If the given points are (x_1, y_1) and (x_2, y_2), the slope of the line is $m=\frac{y_2-y_1}{x_2-x_1}, x_1 \neq x_2$.

11. $P_1(1,-4)$ and $P_2(3,2)$ so $y_2=2; y_1=-4; x_2=3; x_1=1$.

13. $P_1(4,2)$, $P_2(3,4)$

$$m=\frac{y_2-y_1}{x_2-x_1}=\frac{4-2}{3-4}=\frac{2}{-1}=-2$$

15. $P_1(-1,3),\ P_2(2,4)$

$$m=\frac{y_2-y_1}{x_2-x_1}=\frac{4-3}{2-(-1)}=\frac{1}{3}$$

17. $P_1(2,4),\ P_2(4,-1)$

$$m=\frac{y_2-y_1}{x_2-x_1}=\frac{-1-4}{4-2}=\frac{-5}{2}$$

19. $P_1(-2,3),\ P_2(2,1)$

$$m=\frac{y_2-y_1}{x_2-x_1}=\frac{1-3}{2-(-2)}=\frac{-2}{4}=-\frac{1}{2}$$

21. $P_1(8,-3),\ P_2(4,1)$

$$m=\frac{y_2-y_1}{x_2-x_1}=\frac{1-(-3)}{4-8}=\frac{4}{-4}=-1$$

23. $P_1(3,-4),\ P_2(3,5)$

$$m=\frac{y_2-y_1}{x_2-x_1}=\frac{5-(-4)}{3-3}=\frac{9}{0}\ \textit{undefined}$$

25. $P_1(4,-2),\ P_2(3,-2)$

$$m=\frac{y_2-y_1}{x_2-x_1}=\frac{-2-(-2)}{3-4}=\frac{0}{-1}=0$$

27. $P_1(0,-1),\ P_2(3,-2)$

$$m=\frac{y_2-y_1}{x_2-x_1}=\frac{-2-(-1)}{3-0}=\frac{-1}{3}$$

29. $P_1(-2,3),\ P_2(1,3)$

$$m=\frac{y_2-y_1}{x_2-x_1}=\frac{3-3}{1-(-2)}=\frac{0}{3}=0$$

31. $P_1(-2,4),\ P_2(-1,-1)$

$$m=\frac{y_2-y_1}{x_2-x_1}=\frac{-1-4}{-1-(-2)}=\frac{-5}{1}=-5$$

33. $P_1(-2,-3),\ P_2(-2,1)$

$$m=\frac{y_2-y_1}{x_2-x_1}=\frac{1-(-3)}{-2-(-2)}=\frac{4}{0}\ \textit{undefined}$$

35. $P_1(-1,5),\ P_2(5,1)$

$$m=\frac{y_2-y_1}{x_2-x_1}=\frac{1-5}{5-(-1)}=\frac{-4}{6}=-\frac{2}{3}$$

37. $y=4x+6;\ m=4\ \textit{so}\ m_{parallel}=4.$

39. $y=-\frac{1}{3}x-2;\ m=-\frac{1}{3}\ \textit{so}\ m_{perpendicular}=3.$

41. $y=\frac{3}{8}x-5;\ m=\frac{3}{8}$

$y=\frac{3}{8}x+2;\ m=\frac{3}{8}$

The lines are parallel.

43. $y=\frac{7}{2}x;\ m=\frac{7}{2}$

$y=-\frac{2}{7}x+2;\ m=-\frac{2}{7}$

The lines are perpendicular.

45. Slope of the line through (-4, 2) and (1, 6):

$$m=\frac{y_2-y_1}{x_2-x_1}=\frac{6-2}{1-(-4)}=\frac{4}{5}$$

Slope of the line through (2, -4) and (7, 0):

$$m=\frac{y_2-y_1}{x_2-x_1}=\frac{0-(-4)}{7-2}=\frac{4}{5}$$

The lines are parallel.

47. Slope of the line through (-2, -3) and (7, 1):

$$m=\frac{y_2-y_1}{x_2-x_1}=\frac{1-(-3)}{7-(-2)}=\frac{4}{9}$$

Slope of the line through (6, -5) and (4, 1):

$$m=\frac{y_2-y_1}{x_2-x_1}=\frac{1-(-5)}{4-6}=\frac{6}{-2}=-3$$

The lines are not parallel.

49. Slope of the line through (1, -1) and (3, -2):

$$m=\frac{y_2-y_1}{x_2-x_1}=\frac{-2-(-1)}{3-1}=\frac{-1}{2}$$

Slope of the line through (-4, 1) and (2, -5):

$$m=\frac{y_2-y_1}{x_2-x_1}=\frac{-5-1}{2-(-4)}=\frac{-6}{6}=-1$$

The lines are not perpendicular.

51. Slope of the line through (5, 1) and (3, -2):

$$m=\frac{y_2-y_1}{x_2-x_1}=\frac{-2-1}{3-5}=\frac{-3}{-2}=\frac{3}{2}$$

Slope of the line through (0, -2) and (3, -4):

$$m=\frac{y_2-y_1}{x_2-x_1}=\frac{-4-(-2)}{3-0}=\frac{-2}{3}$$

The lines are perpendicular.

53. $P_1(15,750)$ and $P_2(5,250)$

$$m=\frac{y_2-y_1}{x_2-x_1}=\frac{250-750}{5-15}=\frac{-500}{-10}=50$$

The electronic mail sorter sorts 50 pieces of mail per minute.

55. $P_1(0,20)$ and $P_2(200,16)$

$$m=\frac{y_2-y_1}{x_2-x_1}=\frac{16-20}{200-0}=\frac{-4}{200}=-0.02$$

The car uses 0.02 gallons of gasoline per mile driven.

57. $(x_1,y_1)=(2003,79);\ (x_2,y_2)=(2008,83)$

$$m=\frac{y_2-y_1}{x_2-x_1}=\frac{83-79}{2008-2003}=\frac{4}{5}=0.8$$

The slope is 0.8. This means that the percent of people using seat belts has increased by 0.8% per year.

59. a. For the slope to be undefined, the denominator in the slope formula must be 0 and the numerator must not be 0. This will happen when $a=c$ and $b\neq d$.

b. For the slope to be 0, the numerator must be 0 and the denominator must not be 0. This will happen when $b=d$ and $a\neq c$.

c. If $a>c$ then $a-c>0$. If $d>b$ then $b-d<0$. The ratio will be negative so the line slants **downward to the right**.

61. The slope of the line with equation $y=5x-3$ is <u>5</u> and its y-intercept is <u>(0, -3)</u>.

63. $y=3x+1$

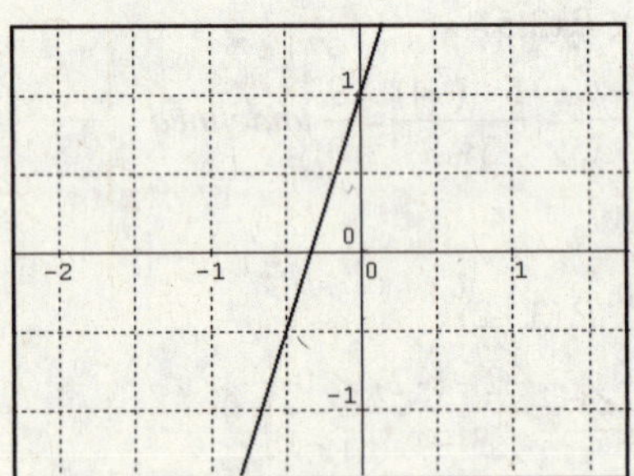

65. $y=\frac{2}{5}x-2$

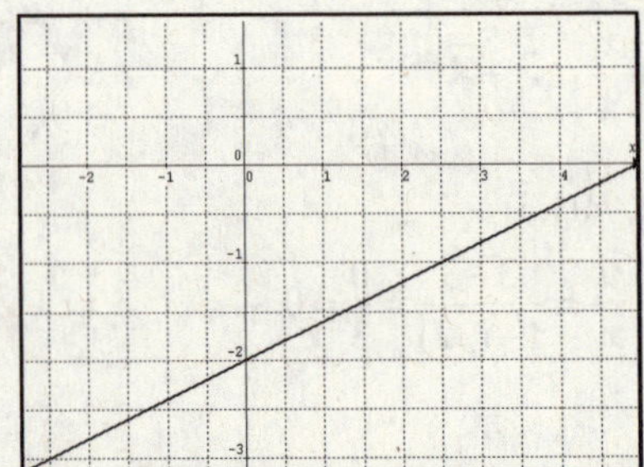

67. $2x+y=3$

$y=-2x+3$

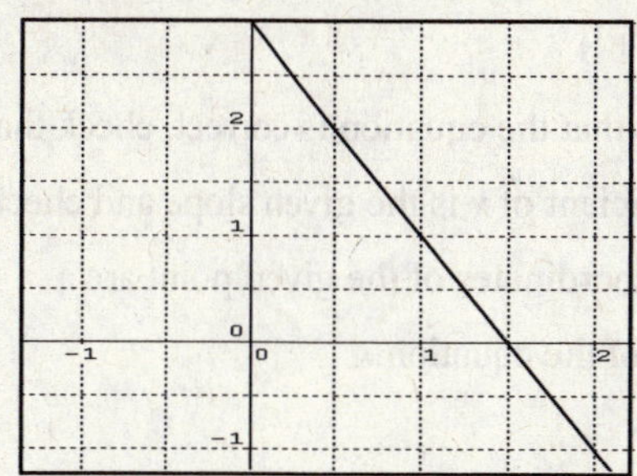

69. $x-2y=4$

$y=\frac{1}{2}x-2$

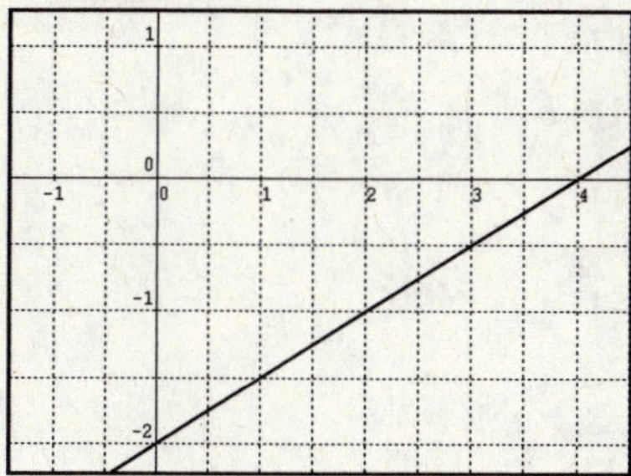

71. $y=\frac{2}{3}x$

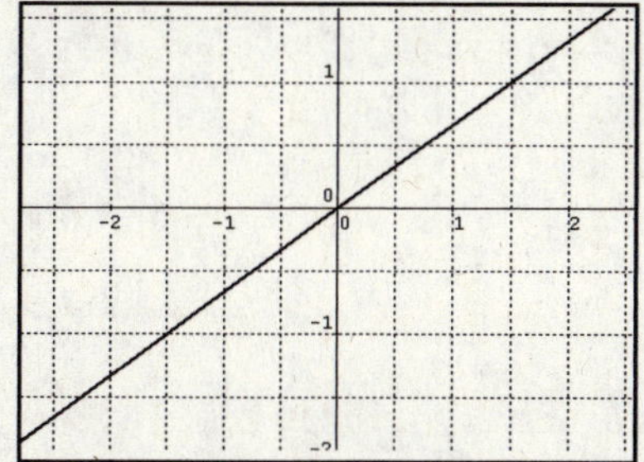

73. $y=-x+1$

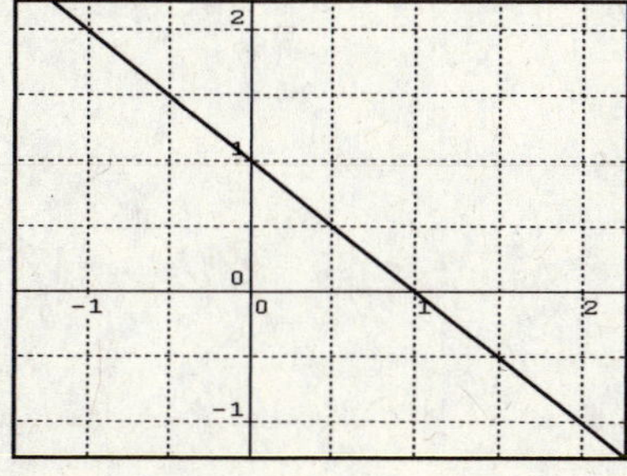

75. $3x-4y=12$

$y=\frac{3}{4}x-3$

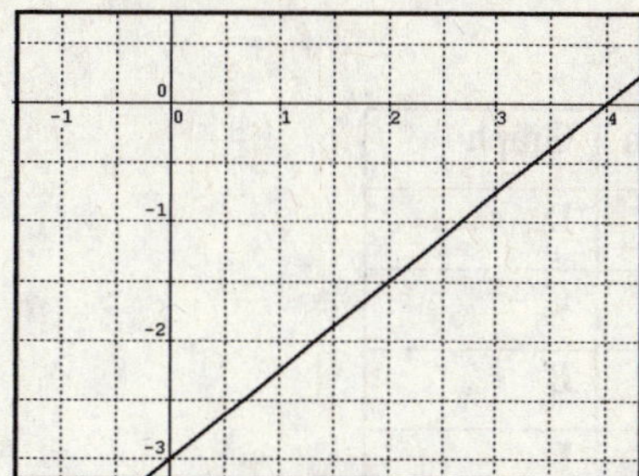

77. $y=-4x+2$

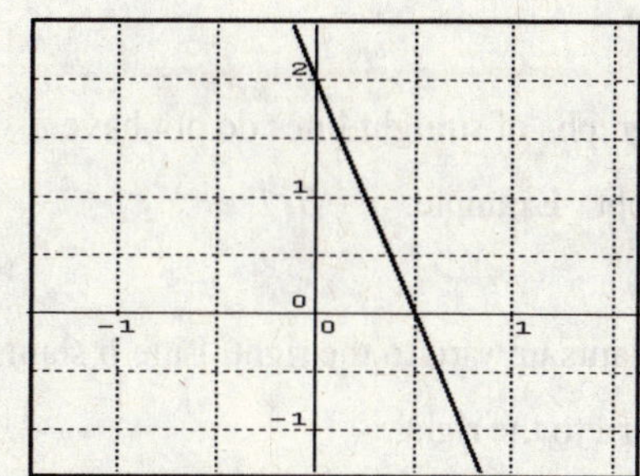

79. $4x-5y=20$

$y=\frac{4}{5}x-4$

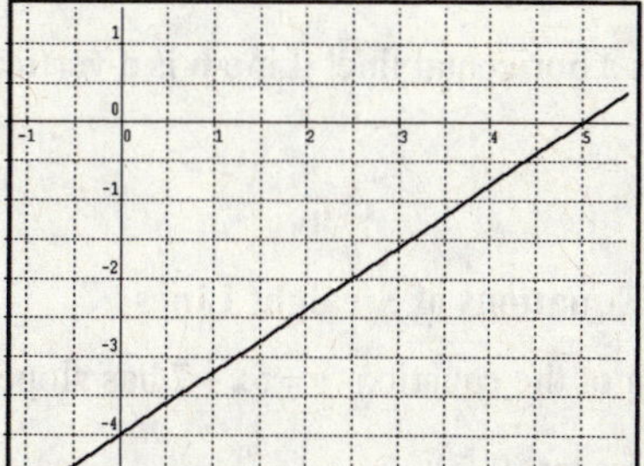

81. If A, B, and C are all positive, then the y-intercept of the graph of $Ax + By = C$ will lie **above** the x-axis. The slope will be negative so the graph slants **downward to the right**.

83. Increasing the coefficient of x **increases the slope** of the line.

85. Increasing the constant term **increases the *y*-intercept**.

87.

Equation	Graph
i	**D**
ii	**C**
iii	**B**
iv	**F**
v	**E**
vi	**A**

89. No, all graphs of straight lines do not have a *y*-intercept. Example: $x = 5$.

91. Line **a** slants upward to the right. Line **b** slants downward to the right.

93. Line **a** has a *y*-intercept of (0, 5). Line **b** has a *y*-intercept of (0, -5).

95. Line **a** is a horizontal line. Line **b** is a vertical line.

Section 5.4 Equations of Straight Lines

1. The graph of the equation $y = 5x + 7$ has slope $\underline{5}$ and *y*-intercept $\underline{(0, 7)}$.

3. If it is stated that the *y*-intercept is 2, then the *y*-intercept is the point $\underline{(0, 2)}$.

5. The point-slope formula is $y - y_1 = m(x - x_1)$.

7. In the equation of the line that has slope 3 and *y*-intercept (0, 1), $m = \underline{3}$ and $b = \underline{1}$. The equation is $y = 3x + 1$.

9. To verify that the equation is correct, check that the coefficient of *x* is the given slope and check that the coordinates of the given point are a solution of the equation.

11.
$$y = mx + b$$
$$y = 2x + 2$$

13.
$$y = mx + b$$
$$y = -3x + b$$
$$2 = -3(-1) + b$$
$$2 = 3 + b$$
$$-1 = b$$
$$y = -3x - 1$$

15.
$$y = mx + b$$
$$y = \frac{1}{3}x + b$$
$$1 = \frac{1}{3} \cdot 3 + b$$
$$1 = 1 + b$$
$$0 = b$$
$$y = \frac{1}{3}x$$

17.
$$y = mx + b$$
$$y = \frac{3}{4}x + b$$
$$-2 = \frac{3}{4}(4) + b$$
$$-2 = 3 + b$$
$$-5 = b$$
$$y = \frac{3}{4}x - 5$$

19. $y = mx + b$

$y = -\frac{3}{5}x + b$

$-3 = -\frac{3}{5} \cdot 5 + b$

$-3 = -3 + b$

$0 = b$

$y = -\frac{3}{5}x$

21. $y = mx + b$

$y = \frac{1}{4}x + b$

$3 = \frac{1}{4} \cdot 2 + b$

$3 = \frac{1}{2} + b$

$\frac{5}{2} = b$

$y = \frac{1}{4}x + \frac{5}{2}$

23. $y = mx + b$

$y = -\frac{2}{3}x + b$

$-5 = -\frac{2}{3}(-3) + b$

$-5 = 2 + b$

$-7 = b$

$y = -\frac{2}{3}x - 7$

25. The point-slope formula is used when the slope is known or can be determined and a point on the line is known.

27. In the equation of the line that has slope $-\frac{4}{5}$ and y-intercept (0, 3), m is $-\frac{4}{5}$ and b is 3. The equation is $y = -\frac{4}{5}x + 3$.

29. $y - y_1 = m(x - x_1)$

$y - (-1) = 2(x - 1)$

$y + 1 = 2x - 2$

$y = 2x - 3$

31. $y - y_1 = m(x - x_1)$

$y - 1 = -2(x - (-2))$

$y - 1 = -2(x + 2)$

$y - 1 = -2x - 4$

$y = -2x - 3$

33. $y - y_1 = m(x - x_1)$

$y - 0 = \frac{2}{3}(x - 0)$

$y = \frac{2}{3}x$

35. $y - y_1 = m(x - x_1)$

$y - 3 = \frac{1}{2}(x - 2)$

$y - 3 = \frac{1}{2}x - 1$

$y = \frac{1}{2}x + 2$

37. $y - y_1 = m(x - x_1)$

$y - 1 = -\frac{3}{4}(x - (-4))$

$y - 1 = -\frac{3}{4}(x + 4)$

$y - 1 = -\frac{3}{4}x - 3$

$y = -\frac{3}{4}x - 2$

39. $y - y_1 = m(x - x_1)$

$y - 1 = \frac{3}{4}(x - (-2))$

$y - 1 = \frac{3}{4}(x + 2)$

$y - 1 = \frac{3}{4}x + \frac{3}{2}$

$y = \frac{3}{4}x + \frac{5}{2}$

41.
$$y - y_1 = m(x - x_1)$$
$$y - (-5) = -\frac{4}{3}(x - (-3))$$
$$y + 5 = -\frac{4}{3}(x + 3)$$
$$y + 5 = -\frac{4}{3}x - 4$$
$$y = -\frac{4}{3}x - 9$$

43. a.
$$y - y_1 = m(x - x_1)$$
$$y - b = m(x - 0)$$
$$y - b = mx$$
$$y = mx + b$$

b. Yes. A line with slope m and y-intercept (0, b) also has the equation $y = mx + b$.

45. Before using the point-slope formula, you must find the slope of the line between the two points.

47. $(x_1, y_1) = (-2, -2)$ *and* $(x_2, y_2) = (1, 7)$
$$m = \frac{y_2 - y_1}{x_2 - x_1} = \frac{7 - (-2)}{1 - (-2)} = \frac{9}{3} = 3$$
$$y - y_1 = m(x - x_1)$$
$$y - (-2) = 3[x - (-2)]$$
$$y + 2 = 3(x + 2)$$
$$y + 2 = 3x + 6$$
$$y = 3x + 4$$

49. $(x_1, y_1) = (-5, 1)$ *and* $(x_2, y_2) = (2, -6)$
$$m = \frac{y_2 - y_1}{x_2 - x_1} = \frac{-6 - 1}{2 - (-5)} = \frac{-7}{7} = -1$$
$$y - y_1 = m(x - x_1)$$
$$y - 1 = -1[x - (-5)]$$
$$y - 1 = -1(x + 5)$$
$$y - 1 = -x - 5$$
$$y = -x - 4$$

51. $(x_1, y_1) = (5, -1)$ *and* $(x_2, y_2) = (-5, 11)$
$$m = \frac{y_2 - y_1}{x_2 - x_1} = \frac{11 - (-1)}{-5 - 5} = \frac{12}{-10} = -\frac{6}{5}$$
$$y - y_1 = m(x - x_1)$$
$$y - (-1) = -\frac{6}{5}(x - 5)$$
$$y + 1 = -\frac{6}{5}(x - 5)$$
$$y + 1 = -\frac{6}{5}x + 6$$
$$y = -\frac{6}{5}x + 5$$

53. $(x_1, y_1) = (-10, -3)$ *and* $(x_2, y_2) = (5, -9)$
$$m = \frac{y_2 - y_1}{x_2 - x_1} = \frac{-9 - (-3)}{5 - (-10)} = \frac{-6}{15} = -\frac{2}{5}$$
$$y - y_1 = m(x - x_1)$$
$$y - (-3) = -\frac{2}{5}[x - (-10)]$$
$$y + 3 = -\frac{2}{5}(x + 10)$$
$$y + 3 = -\frac{2}{5}x - 4$$
$$y = -\frac{2}{5}x - 7$$

55. $(x_1, y_1) = (1, 5)$ *and* $(x_2, y_2) = (-6, 5)$
$$m = \frac{y_2 - y_1}{x_2 - x_1} = \frac{5 - 5}{-6 - 1} = \frac{0}{-7} = 0$$
$$y - y_1 = m(x - x_1)$$
$$y - 5 = 0(x - 1)$$
$$y - 5 = 0$$
$$y = 5$$

57. $(x_1, y_1) = (5, -1)$ *and* $(x_2, y_2) = (5, -7)$
$$m = \frac{y_2 - y_1}{x_2 - x_1} = \frac{-7 - (-1)}{5 - 5} = \frac{-6}{0} \text{ undefined}$$
$$x = 5$$

59. $(x_1, y_1) = (-20, -8)$ *and* $(x_2, y_2) = (5, 12)$

$$m = \frac{y_2 - y_1}{x_2 - x_1} = \frac{12-(-8)}{5-(-20)} = \frac{20}{25} = \frac{4}{5}$$

$$y - y_1 = m(x - x_1)$$
$$y-(-8) = \frac{4}{5}[x-(-20)]$$
$$y+8 = \frac{4}{5}(x+20)$$
$$y+8 = \frac{4}{5}x+16$$
$$y = \frac{4}{5}x+8$$

61. $(x_1, y_1) = (0, -2)$ *and* $(x_2, y_2) = (-6, 1)$

$$m = \frac{y_2 - y_1}{x_2 - x_1} = \frac{1-(-2)}{-6-0} = \frac{3}{-6} = -\frac{1}{2}$$

$$y - y_1 = m(x - x_1)$$
$$y-(-2) = -\frac{1}{2}(x-0)$$
$$y+2 = -\frac{1}{2}x$$
$$y = -\frac{1}{2}x-2$$

63. $(x_1, y_1) = (6, -11)$ *and* $(x_2, y_2) = (-3, 1)$

$$m = \frac{y_2 - y_1}{x_2 - x_1} = \frac{1-(-11)}{-3-6} = \frac{12}{-9} = -\frac{4}{3}$$

$$y - y_1 = m(x - x_1)$$
$$y-(-11) = -\frac{4}{3}(x-6)$$
$$y+11 = -\frac{4}{3}x+8$$
$$y = -\frac{4}{3}x-3$$

65. $(x_1, y_1) = (3, 6)$ *and* $(x_2, y_2) = (0, -3)$

$$m = \frac{y_2 - y_1}{x_2 - x_1} = \frac{-3-6}{0-3} = \frac{-9}{-3} = 3$$

$$y - y_1 = m(x - x_1)$$
$$y-6 = 3(x-3)$$
$$y-6 = 3x-9$$
$$y = 3x-3$$

67. $(x_1, y_1) = (-1, -3)$ *and* $(x_2, y_2) = (2, 6)$

$$m = \frac{y_2 - y_1}{x_2 - x_1} = \frac{6-(-3)}{2-(-1)} = \frac{9}{3} = 3$$

$$y - y_1 = m(x - x_1)$$
$$y-(-3) = 3[x-(-1)]$$
$$y+3 = 3(x+1)$$
$$y+3 = 3x+3$$
$$y = 3x$$

69. $(x_1, y_1) = (3, -5)$ *and* $(x_2, y_2) = (3, 1)$

$$m = \frac{y_2 - y_1}{x_2 - x_1} = \frac{1-(-5)}{3-3} = \frac{6}{0} \text{ undefined}$$

$$x = 3$$

71. Yes; however, all of the points must be on the same line.

73. For a linear equation to contain all of the given ordered pairs, they must all lie on the same line. So, the equation formed from any pair of points should be satisfied by the third point.

$(x_1, y_1) = (5, 1)$ *and* $(x_2, y_2) = (4, 2)$

$$m = \frac{y_2 - y_1}{x_2 - x_1} = \frac{2-1}{4-5} = \frac{1}{-1} = -1$$

$$y - y_1 = m(x - x_1)$$
$$y-1 = -1(x-5)$$
$$y-1 = -x+5$$
$$y = -x+6$$

The third point also satisfies this equation.

75. For a linear equation to contain all of the given ordered pairs, they must all lie on the same line. So, the equation formed from any pair of points should be satisfied by the third point.

$(x_1, y_1) = (-1, -5)$ *and* $(x_2, y_2) = (2, 4)$

$$m = \frac{y_2 - y_1}{x_2 - x_1} = \frac{4-(-5)}{2-(-1)} = \frac{9}{3} = 3$$

$$\begin{aligned} y - y_1 &= m(x - x_1) \\ y - (-5) &= 3[x - (-1)] \\ y + 5 &= 3(x+1) \\ y + 5 &= 3x + 3 \\ y &= 3x - 2 \end{aligned}$$

The third point does not this equation.

77. $(x_1, y_1) = (0,1)$ *and* $(x_2, y_2) = (4,9)$

$$m = \frac{y_2 - y_1}{x_2 - x_1} = \frac{9-1}{4-0} = \frac{8}{4} = 2$$

$(x_1, y_1) = (0,1)$ *and* $(x_2, y_2) = (3,n)$

$$m = \frac{y_2 - y_1}{x_2 - x_1} = \frac{n-1}{3-0} = \frac{n-1}{3}$$

$$\frac{n-1}{3} = 2$$

$$n - 1 = 6$$

$$n = 7$$

79. $(x_1, y_1) = (2,-2)$ *and* $(x_2, y_2) = (-2,-4)$

$$m = \frac{y_2 - y_1}{x_2 - x_1} = \frac{-4-(-2)}{-2-2} = \frac{-2}{-4} = \frac{1}{2}$$

$(x_1, y_1) = (2,-2)$ *and* $(x_2, y_2) = (4,n)$

$$m = \frac{y_2 - y_1}{x_2 - x_1} = \frac{n-(-2)}{4-2} = \frac{n+2}{2}$$

$$\frac{n+2}{2} = \frac{1}{2}$$

$$n + 2 = 1$$

$$n = -1$$

81. $(x_1, y_1) = (2,3)$ *and* $(x_2, y_2) = (-1,-3)$

$$m = \frac{y_2 - y_1}{x_2 - x_1} = \frac{-3-3}{-1-2} = \frac{-6}{-3} = 2$$

$(x_1, y_1) = (2,3)$ *and* $(x_2, y_2) = (1, y)$ **82.** *x*-

$$m = \frac{y_2 - y_1}{x_2 - x_1} = \frac{y-3}{1-2} = \frac{y-3}{-1} = 3 - y$$

$$3 - y = 2$$

$$y = 1$$

intercept: (-3, 0); *y*-intercept: (0, 4)

$$m = \frac{4}{3};\ \ y = \frac{4}{3}x + 4$$

83. *x*-intercept: (3, 0); *y*-intercept: (0, 2)

$$m = -\frac{2}{3};\ \ y = -\frac{2}{3}x + 2$$

85. *x*-intercept: (-1, 0); *y*-intercept: (0, -2)

$$m = -2;\ \ y = -2x - 2$$

Section 5.5 Functions

1. a. Yes; $f(x) = -\frac{3}{5}x - 2$

b. Yes; $f(x) = x + 1$

c. No

d. No

e. Yes; $f(x) = 2x - 4$

f. Yes; $f(x) = 6$

3. For the relation graphed at the right, the domain is {-2, -1, 2, 3}.

5. A relation is a set if ordered pairs. The set of first coordinates of the ordered pairs is called the domain of the relation. The set of second coordinates is called the range of the relation. function *f* pairs with *x*.

7. {(35, 7.50), (45, 7.58), (38, 7.63), (24, 7.78), (47, 7.80), (51, 7.86), (35, 7.89), (48, 7.92)}
No, this relation is not a function.

9. No, this relation is not a function because there are more people in the world than there are days in the year so some of the dates must be repeated with different people.

11. {(4, 411), (5, 514), (6, 618), (7, 720), (8, 823)}
Yes, the relation is a function.

13. For the function $f(x) = 3x - 4$, $f(-1) = -7$. The number -1 is in the <u>domain</u> of the function, and the number -7 is in the <u>range</u> of the function.

15. Domain: {0, 2, 4, 6}; Range:{0}; Function

17. Domain: {2}; Range:{2, 4, 6, 8}; Not a function

19. Domain: {0, 1, 2, 3}; Range:{0, 1, 2, 3}; Function

21. Domain: {-3, -2, -1, 1, 2, 3}; Range:{-3, 2, 3, 4}; Function

23. $f(x) = 4x$
$f(10) = 40$
$(10, 40)$

25. $f(x) = x - 5$
$f(-6) = -6 - 5 = -11$
$(-6, -11)$

27. $f(x) = 3x^2$
$f(-2) = 3(-2)^2$
$f(-2) = 3(4) = 12$
$(-2, 12)$

29. $f(x) = 5x + 1$
$f\left(\frac{1}{2}\right) = 5 \cdot \frac{1}{2} + 1$
$f\left(\frac{1}{2}\right) = \frac{5}{2} + \frac{2}{2} = \frac{7}{2}$
$\left(\frac{1}{2}, \frac{7}{2}\right)$

31. $f(x) = \frac{2}{5}x + 4$
$f(-5) = \frac{2}{5}(-5) + 4$
$f(-5) = -2 + 4 = 2$
$(-5, 2)$

33. $f(x) = 2x^2$
$f(-4) = 2(-4)^2$
$f(-4) = 2(16) = 32$
$(-4, 32)$

35. Range: {-19, -13, -7, -1, 5}
(-5, -19), (-3, -13), (-1, -7), (1, -1), (3, 5)

37. Range: {1, 2, 3, 4, 5}
(-4, 1), (-2, 2), (0, 3), (2, 4), (4, 5)

39. Range: {6, 7, 15}
(-3, 15), (-1, 7), (0, 6), (1, 7), (3, 15)

41. In the linear equation $y = 5x - 9$, the independent variable is <u>x</u> and the dependent variable is <u>y</u>. To write this equation in function notation, replace y with the symbol <u>$f(x)$</u>.

43. $f(x) = 5x$

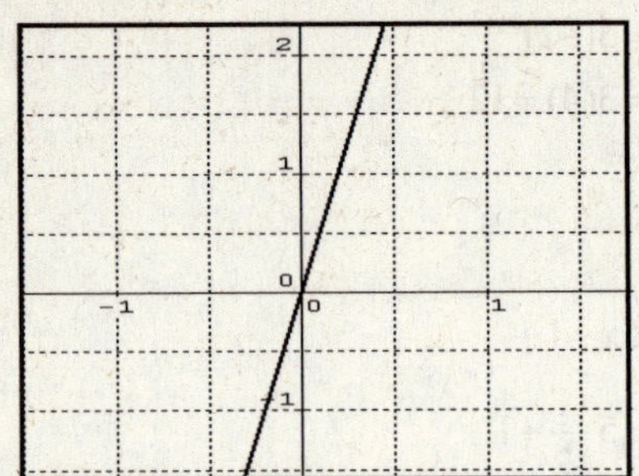

45. $f(x) = x + 2$

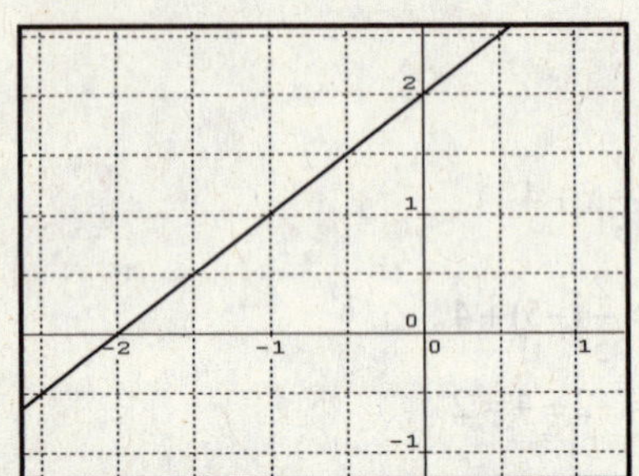

47. $f(x) = 6x - 1$

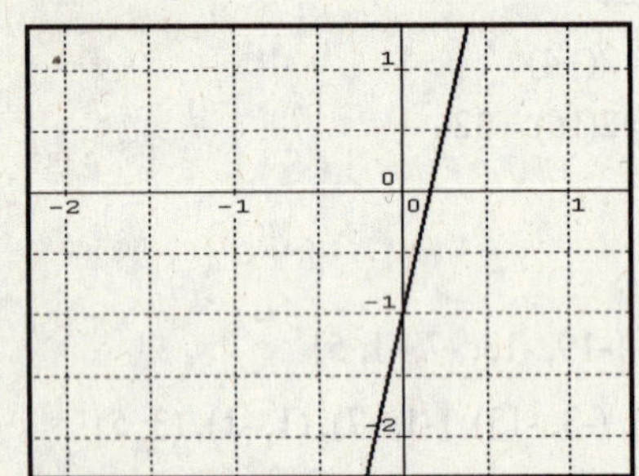

49. $f(x) = -2x + 3$

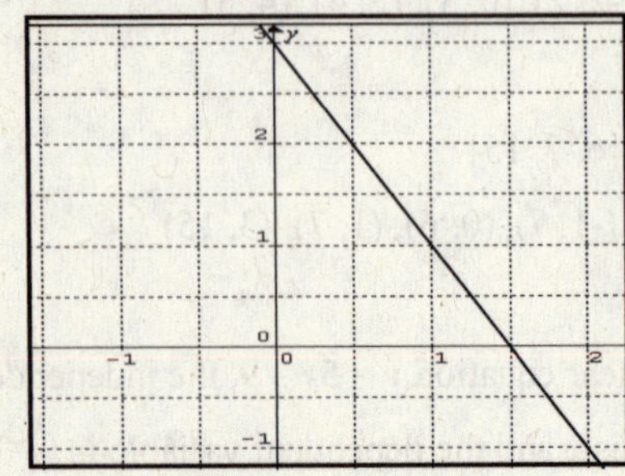

51. $f(x) = \frac{1}{3}x - 4$

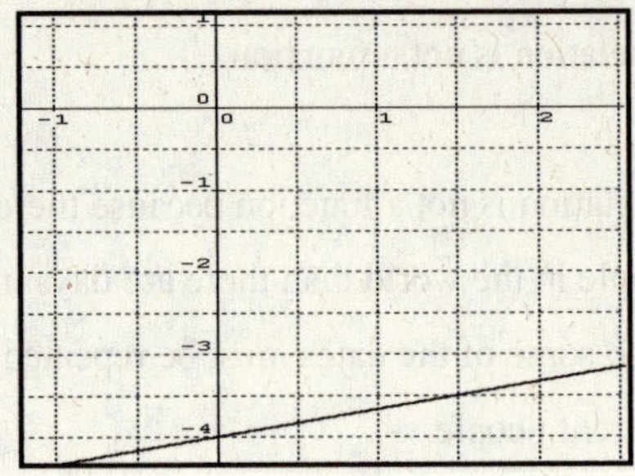

53. $f(x) = 4$

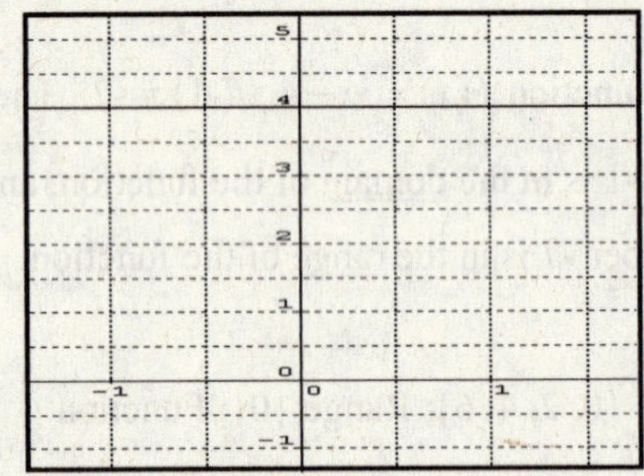

55. $f(x) = 2x - 1$

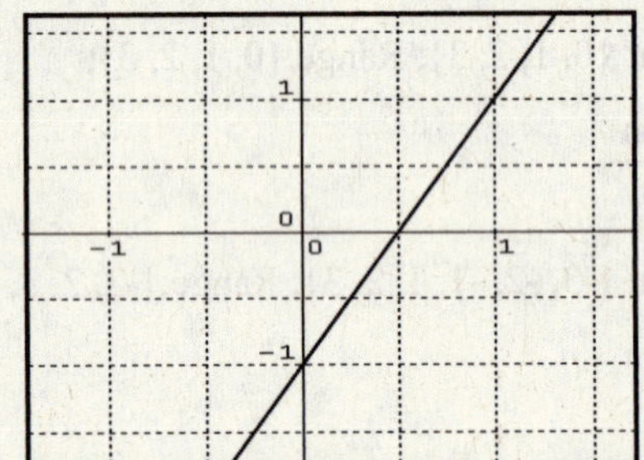

57. $f(x) = -\frac{1}{2}x + 1$

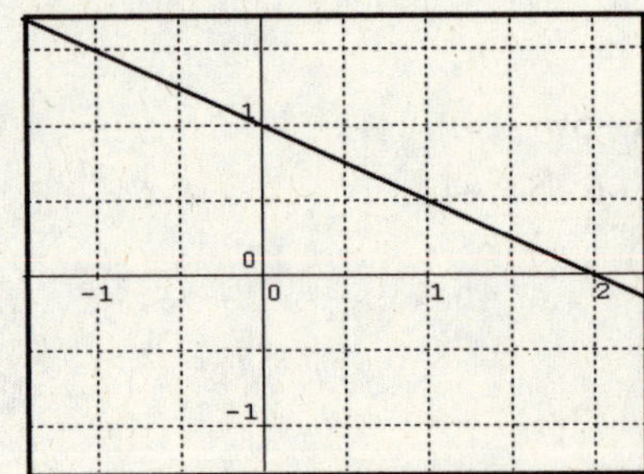

59. $f(x)=\frac{5}{2}x$

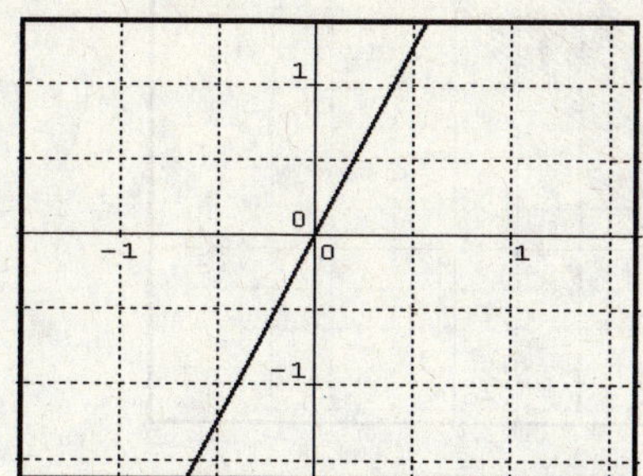

61. a. $f(M)=2.25M+2.65$

b.

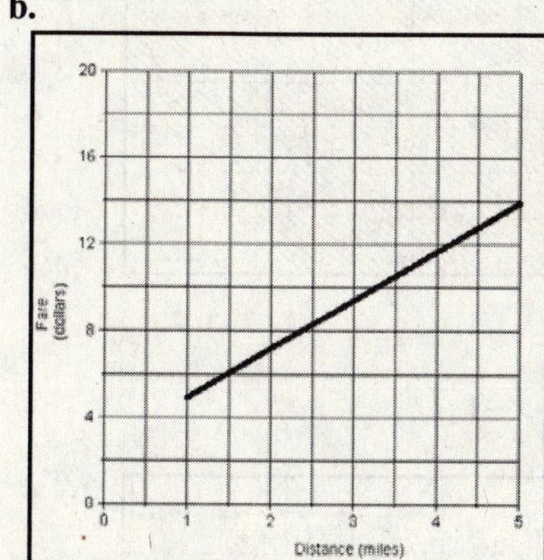

c. A 3-mile taxi ride costs $9.40.

63. a. $f(x)=30{,}000-5000x$

b.

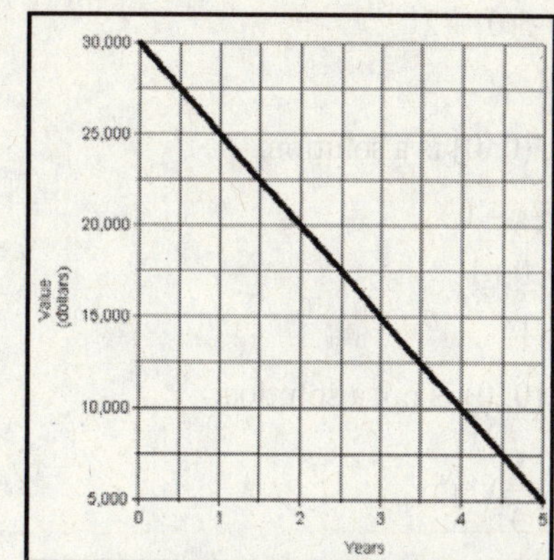

c. The value of the computer after 1 year is $25,000.

65. a. Since the ordered pair (2, 6) is part of the function, $f(2) = 6$.

b. Since the ordered pair (1, 8) is part of the function, $f(1) = 8$.

67. a. This situation corresponds with graph **iii.**

b. This situation corresponds with graph **iv.**

c. This situation corresponds with graph **ii.**

d. This situation corresponds with graph **i.**

69. Answers will vary.

71.
$$f(x)=x^2+4x+6$$
$$f(-5)=(-5)^2+4(-5)+6$$
$$=25-20+6=11$$
$$f(1)=1^2+4(1)+6=11$$

It does not follow that if $f(a) = f(b)$ that $a = b$.

Section 5.6 Graphing Linear Inequalities

1. a. $y < -5x + 2$

$0 < -5(0) + 2$

$0 < 2$

Yes, (0, 0) is a solution.

b. $y > x + 1$

$0 > 0 + 1$

$0 > 1$

No, (0, 0) is not a solution.

c. $y \le \frac{1}{4}x - 5$

$0 \le \frac{1}{4} \cdot 0 - 5$

$0 \le -5$

No, (0, 0) is not a solution.

d. $y \ge -\frac{2}{3}x - 6$

$0 \ge -\frac{2}{3}(0) - 6$

$0 \ge -6$

Yes, (0, 0) is a solution.

3. When graphing the solution set of an inequality, draw a dashed line when the inequality symbol is < or >.

5. When graphing the solution set of an inequality of the form $y \le mx + b$, you should shade the **lower** half plane.

7. In the graph of a linear inequality, points on a solid line <u>are</u> elements of the solution set of the inequality. Points on a dashed line <u>are not</u> elements of the solution set of the inequality.

9. $y > 2x + 3$

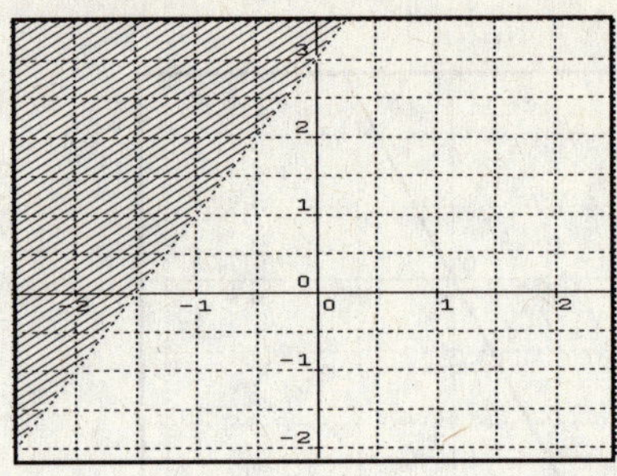

11. $y > \frac{3}{2}x - 4$

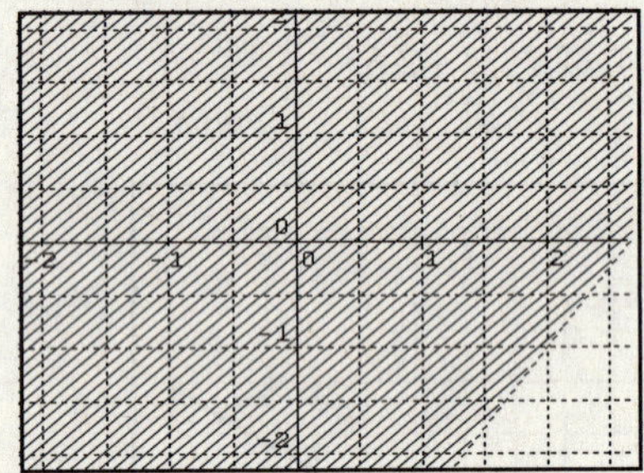

13. $y \le -\frac{3}{4}x - 1$

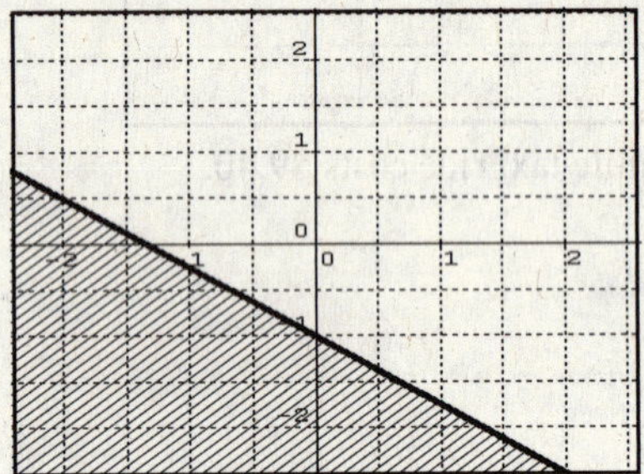

15. $y \le -\frac{6}{5}x - 2$

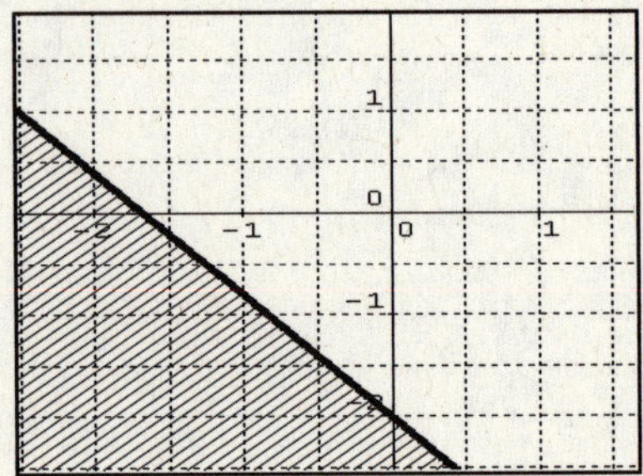

17. $x + y > 4$

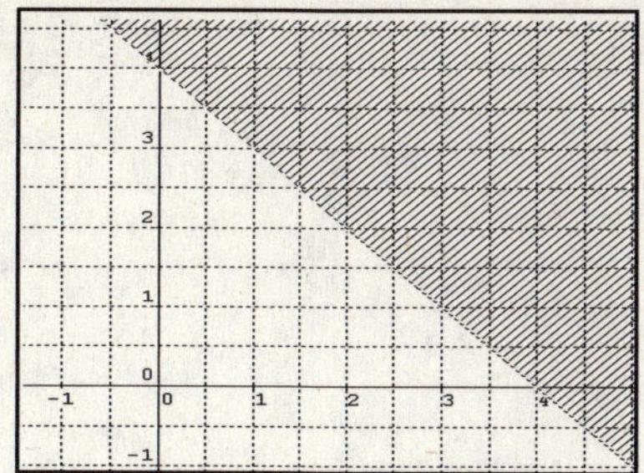

19. $2x + y \geq 4$

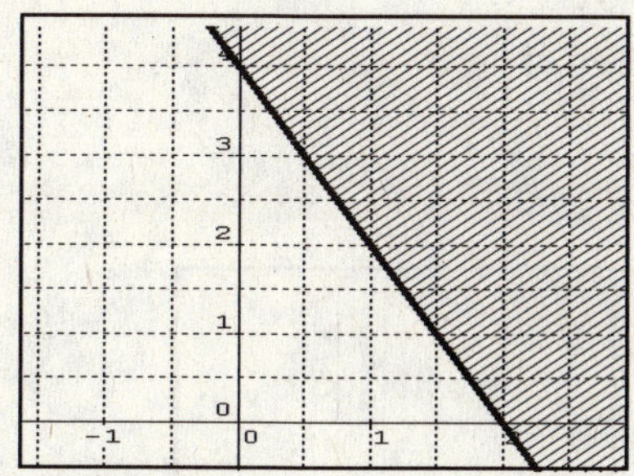

21. $y \leq -2$

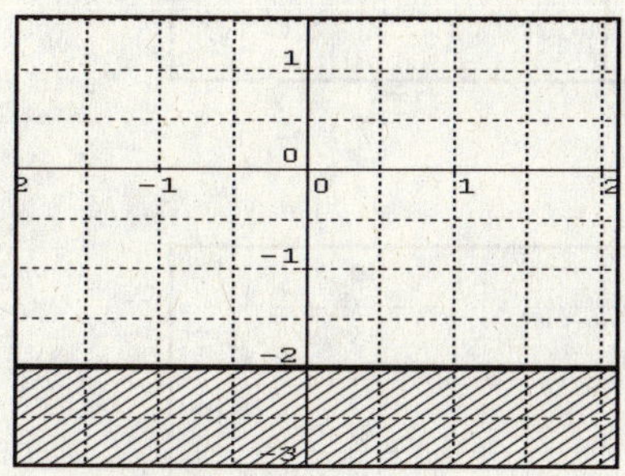

23. $2x + 3y \leq -6$

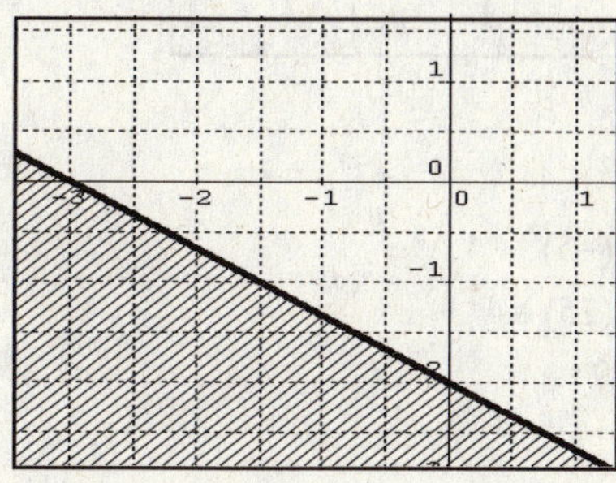

25. $5x - 2y > 10$

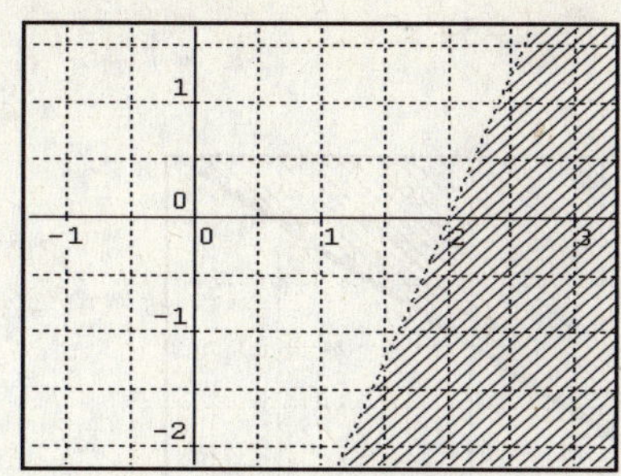

27. If (0, 0) is a point on the graph of the linear inequality $Ax + By > C$, where C is not zero, then C must be negative. After substituting 0 for x and 0 for y, the resulting inequality would be $0 > C$, so C must be negative.

29. From the graph, $m = 2$ and $b = 2$, the line is solid and the half plane above the line is shaded. The inequality is $y \geq 2x + 2$.

31. The graph is a horizontal, dashed line with all y-coordinates of 2. The half plane above the line is shaded. The inequality is $y > 2$.

33. $y - 5 < 4(x - 2)$
$y < 4x - 3$

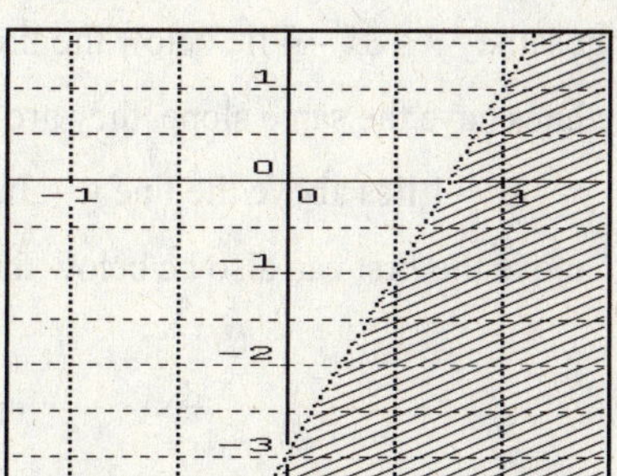

35. $3x-2(y+1)\le y-(5-x)$

$3x-2y-2\le y-5+x$

$2x-3y\le -3$

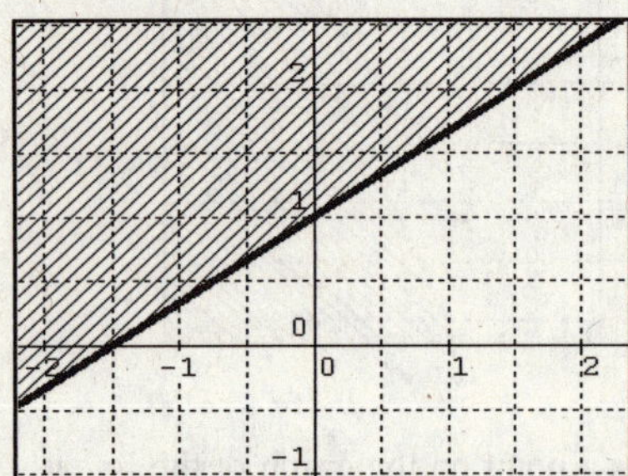

37. An inequality in two variables defines a relation because an inequality in two variables is a set of ordered pairs that satisfy the inequality. An inequality in two variables does not define a function because, given a value of *x*, there is more than one *y* that satisfies the inequality. For example, for the inequality $y > 2x + 1$, both (2, 6) and (2, 7) are ordered pairs that satisfy the inequality. This contradicts the definition of function because there are two ordered pairs with the same *x*-coordinate but different *y*-coordinates.

39. There are no points that satisfy both inequalities. Points that satisfy $y > 3x + 1$ lie above the line. Points that satisfy $y < 3x - 4$ lie below the line. Since the lines have the same slope, they are parallel. $y = 3x + 1$ lies above the line $y = 3x - 4$. No point above the first can also be below the second.

Chapter 5 Review Exercises

1. $y=-\dfrac{2}{3}x+2$

$y=-\dfrac{2}{3}(3)+2$

$y=0$

$(3,0)$

2. $y-y_1=m(x-x_1)$

$y-(-1)=3(x-0)$

$y+1=3x$

$y=3x-1$

3. $x = -3$

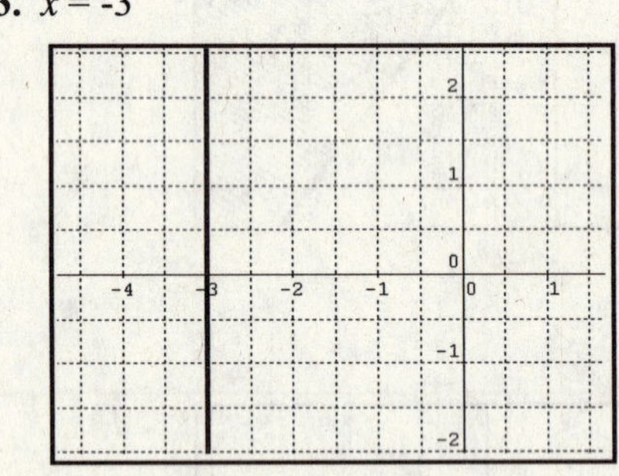

4.

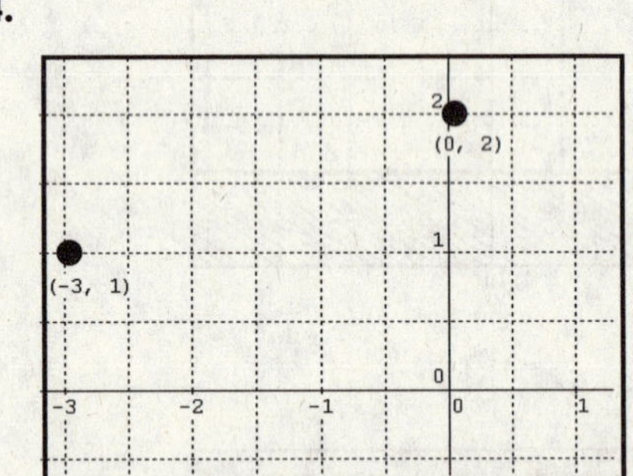

5. $f(x)=3x^2+4$

$f(-5)=3(-5)^2+4$

$f(-5)=3(25)+4$

$f(-5)=79$

6. $(x_1,y_1)=(3,-4)$ *and* $(x_2,y_2)=(1,-4)$

$$m=\frac{y_2-y_1}{x_2-x_1}=\frac{-4-(-4)}{1-3}=\frac{0}{-2}=0$$

7. $y = 3x + 1$

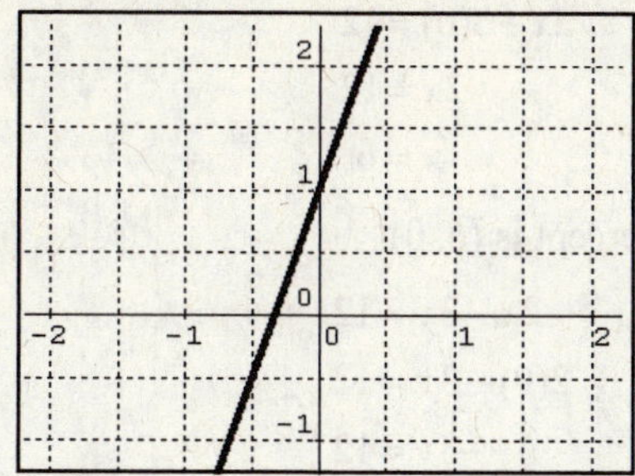

8. $3x - 2y = 6$

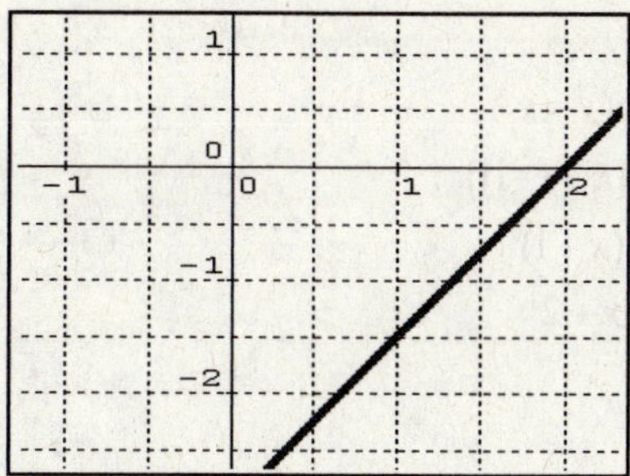

9. $y - y_1 = m(x - x_1)$

$$y - 2 = -\frac{2}{3}[x - (-1)]$$
$$y - 2 = -\frac{2}{3}(x + 1)$$
$$y - 2 = -\frac{2}{3}x - \frac{2}{3}$$
$$y = -\frac{2}{3}x + \frac{4}{3}$$

10. x-intercept: $6x - 4y = 12$

$$6x - 4(0) = 12$$
$$6x = 12$$
$$x = 2$$

The x-intercept is (2, 0).

y-intercept: $6x - 4y = 12$

$$6(0) - 4y = 12$$
$$-4y = 12$$
$$y = -3$$

The y-intercept is (0, -3).

11.

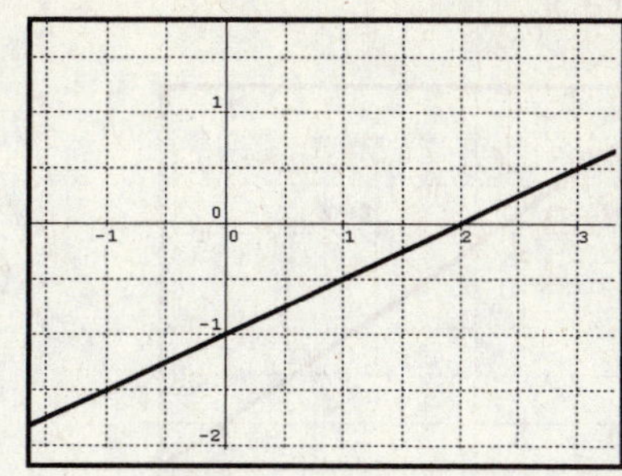

12. $y = -\frac{2}{3}x + 4$

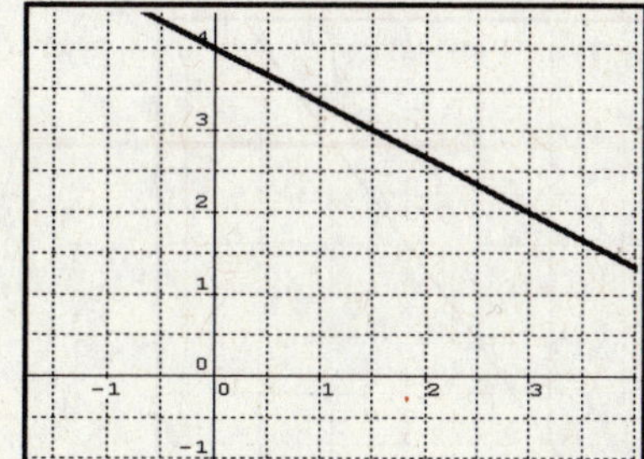

13. $y - y_1 = m(x - x_1)$

$$y - 1 = \frac{2}{3}[x - (-3)]$$
$$y - 1 = \frac{2}{3}(x + 3)$$
$$y - 1 = \frac{2}{3}x + 2$$
$$y = \frac{2}{3}x + 3$$

14. $(x_1, y_1) = (2, -3)$ *and* $(x_2, y_2) = (4, 1)$

$$m = \frac{y_2 - y_1}{x_2 - x_1} = \frac{1 - (-3)}{4 - 2} = \frac{4}{2} = 2$$

15. $f(x) = \frac{3}{5}x + 2$

$$f(-10) = \frac{3}{5}(-10) + 2$$
$$f(-10) = -6 + 2 = -4$$

16. Domain: {-20, -10, 0, 10};

Range: {-10, -5, 0, 5}: Function

17. $y=-\dfrac{3}{4}x+3$

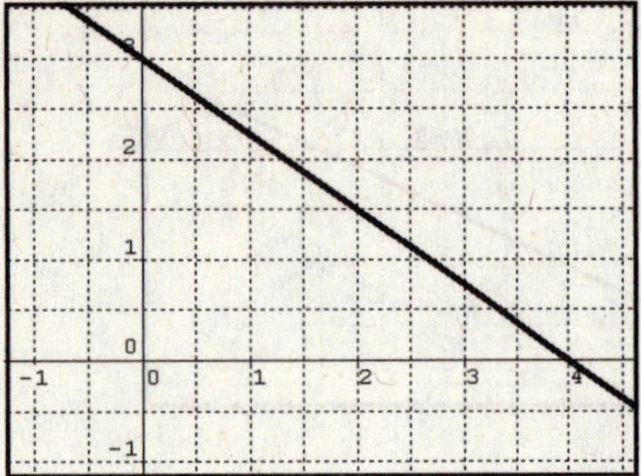

18.

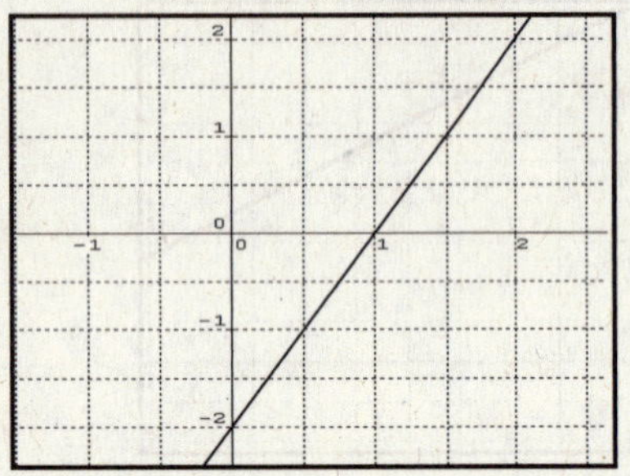

19.

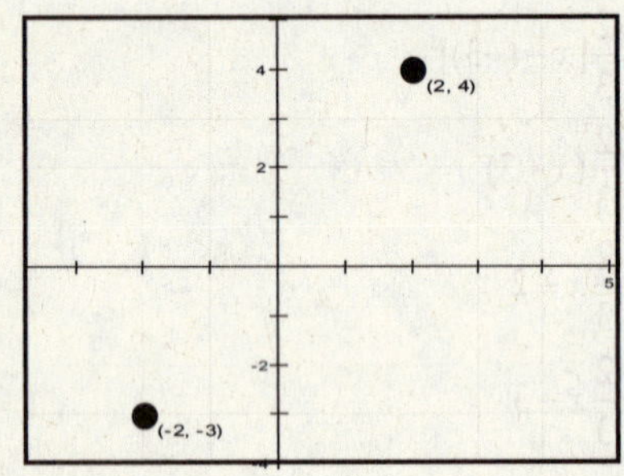

20. $f(x)=5x+1$

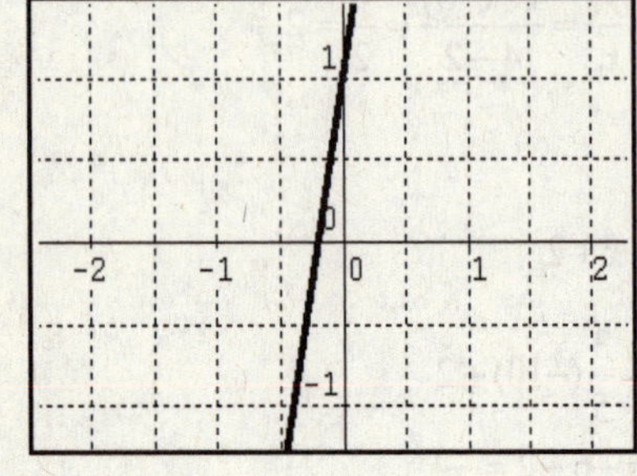

21. x-intercept:

$$\begin{aligned}2x+3y&=12\\2x+3(0)&=12\\2x&=12\\x&=6\end{aligned}$$

The x-intercept is (6, 0).

y-intercept:

$$\begin{aligned}2x-3y&=12\\2(0)-3y&=12\\-3y&=12\\y&=-4\end{aligned}$$

The y-intercept is (0, -4).

22.

$$\begin{aligned}y-y_1&=m(x-x_1)\\y-0&=2[x-(-1)]\\y&=2(x+1)\\y&=2x+2\end{aligned}$$

23. $y=3$

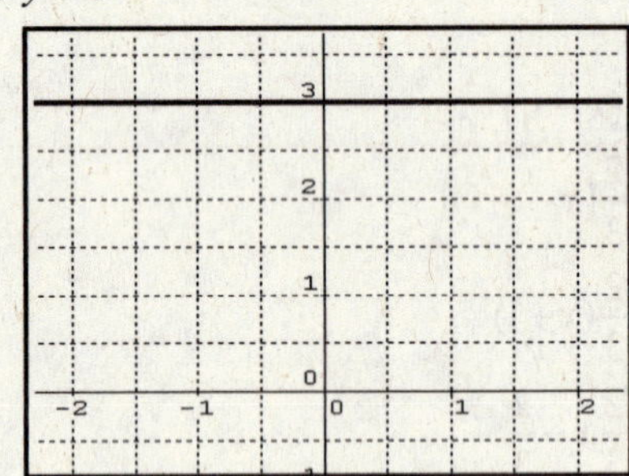

24. $3x+2y\le 12$

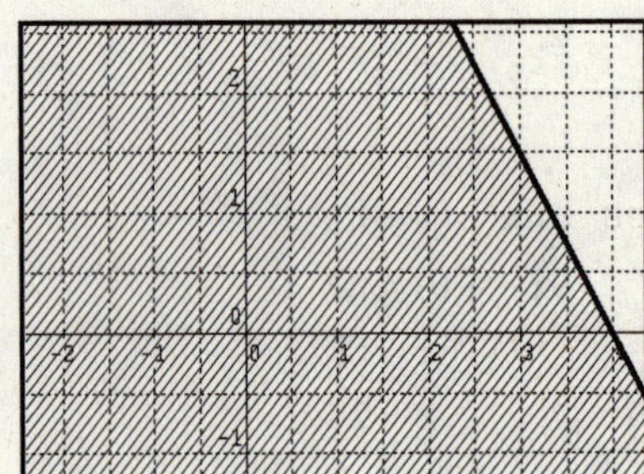

25. $(x_1, y_1) = (-2,1)$ *and* $(x_2, y_2) = (3,5)$

$$m = \frac{y_2 - y_1}{x_2 - x_1} = \frac{5-1}{3-(-2)} = \frac{4}{5}$$

$(x_1, y_1) = (4,-5)$ *and* $(x_2, y_2) = (9,-1)$

$$m = \frac{y_2 - y_1}{x_2 - x_1} = \frac{-1-(-5)}{9-4} = \frac{4}{5}$$

The lines are parallel.

26. $(x_1, y_1) = (4,14)$ *and* $(x_2, y_2) = (-8,-1)$

$$m = \frac{y_2 - y_1}{x_2 - x_1} = \frac{-1-14}{-8-4} = \frac{-15}{-12} = \frac{5}{4}$$

$$y - y_1 = m(x - x_1)$$

$$y - 14 = \frac{5}{4}(x-4)$$

$$y - 14 = \frac{5}{4}x - 5$$

$$y = \frac{5}{4}x + 9$$

27. $y = -x + 2$

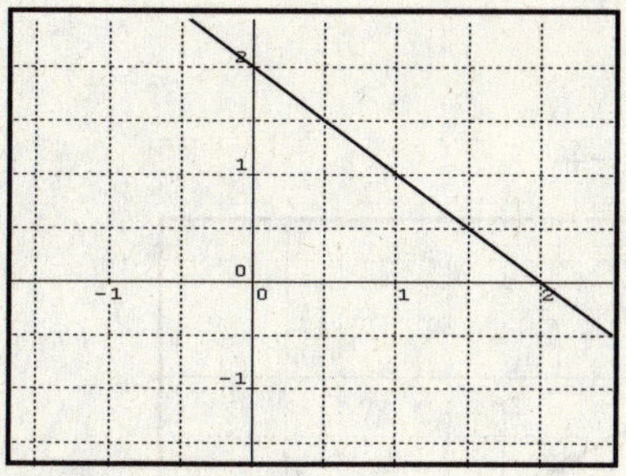

28. $2x - 3y = 6$

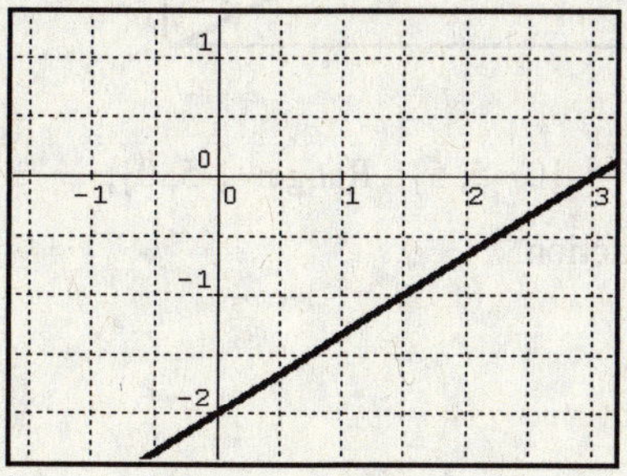

29. $y = 2x - 1$

$$y = 2(-2) - 1$$

$$y = -4 - 1$$

$$y = -5$$

The ordered-pair solution is (-2, -5).

30. $y = mx + b$

$$y = -3x + 2$$

31. $f(x) = 2x$

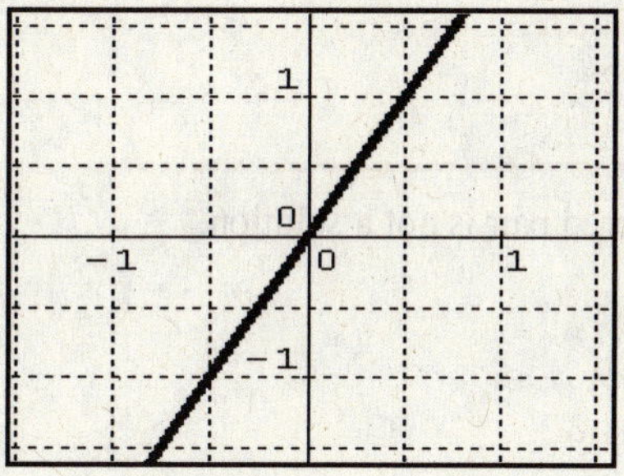

32. $f(x) = 3x - 5$

$$f\left(\frac{5}{3}\right) = 3\left(\frac{5}{3}\right) - 5$$

$$f\left(\frac{5}{3}\right) = 5 - 5 = 0$$

33. $(x_1, y_1) = (-6,0)$ *and* $(x_2, y_2) = (3,-3)$

$$m = \frac{y_2 - y_1}{x_2 - x_1} = \frac{-3-0}{3-(-6)} = \frac{-3}{9} = -\frac{1}{3}$$

$$y - y_1 = m(x - x_1)$$

$$y - 0 = -\frac{1}{3}[x - (-6)]$$

$$y = -\frac{1}{3}(x+6)$$

$$y = -\frac{1}{3}x - 2$$

34. $y - y_1 = m(x - x_1)$

$y - (-1) = \frac{1}{2}(x - 2)$

$y + 1 = \frac{1}{2}x - 1$

$y = \frac{1}{2}x - 2$

35. $y = \frac{1}{5}x + 2$

$0 \stackrel{?}{=} \frac{1}{5}(-10) + 2$

$0 \stackrel{?}{=} -2 + 2$

$0 \neq 0$

The ordered pair is not a solution.

36. $y = 4x - 9$

$y = 4(2) - 9$

$y = 8 - 9$

$y = -1$

The ordered-pair solution is (2, -1).

37. x-intercept: $4x - 3y = 0$

$4x - 3(0) = 0$

$4x = 0$

$x = 0$

The x-intercept is (0, 0).

y-intercept: $4x - 3y = 0$

$4(0) - 3y = 0$

$-3y = 0$

$y = 0$

The y-intercept is (0, 0).

38. $y - y_1 = m(x - x_1)$

$y - 3 = 0[x - (-2)]$

$y - 3 = 0$

$y = 3$

39. $6x - y > 6$

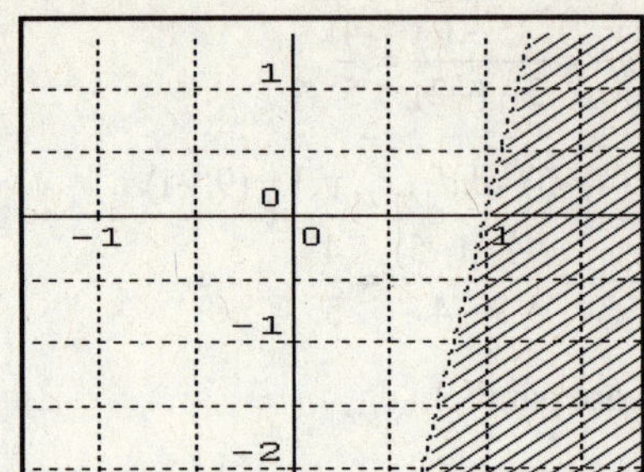

40. $y = -3x + 1$

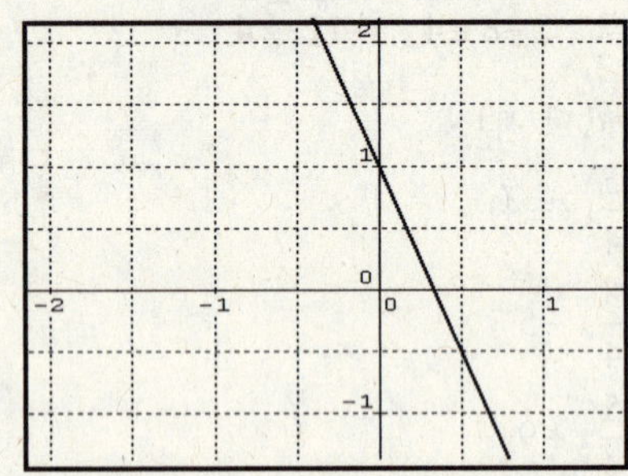

41. $y = mx + b$

$y = 3x - 4$

42. $x + 2y = -4$

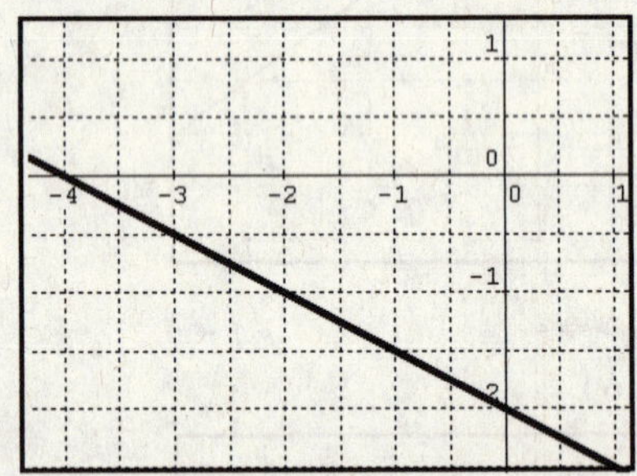

43. Domain: {-10, -5, 5}; Range: {-5, 0};

Not a function

44. $f(x) = 3x + 7$

$f(-20) = -60 + 7 = -53$

$f(-10) = -30 + 7 = -23$

$f(0) = 0 + 7 = 7$

$f(10) = 30 + 7 = 37$

$f(20) = 60 + 7 = 67$

Range: {-53, -23, 7, 37, 67}

45. $f(x) = \frac{1}{3}x + 4$

$f(-6) = \frac{1}{3}(-6) + 4 = 2$

$f(-3) = \frac{1}{3}(-3) + 4 = 3$

$f(0) = \frac{1}{3}(0) + 4 = 4$

$f(3) = \frac{1}{3}(3) + 4 = 5$

$f(6) = \frac{1}{3}(6) + 4 = 6$

Range: {2, 3, 4, 5, 6}

46. $(x_1, y_1) = (-3, 6)$ *and* $(x_2, y_2) = (2, -4)$

$m = \frac{y_2 - y_1}{x_2 - x_1} = \frac{-4 - 6}{2 - (-3)} = \frac{-10}{5} = -2$

$(x_1, y_1) = (-5, -1)$ *and* $(x_2, y_2) = (7, 5)$

$m = \frac{y_2 - y_1}{x_2 - x_1} = \frac{5 - (-1)}{7 - (-5)} = \frac{6}{12} = \frac{1}{2}$

The lines are perpendicular.

47.

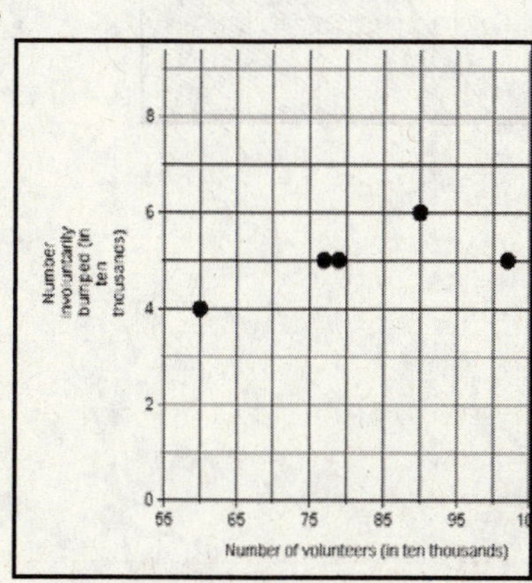

48. {(25, 4.8), (35, 3.5), (40, 2.1), (20, 5.5), (45,1.0)}

The relation is a function.

49. a. $C(s) = 70s + 40{,}000$

b.

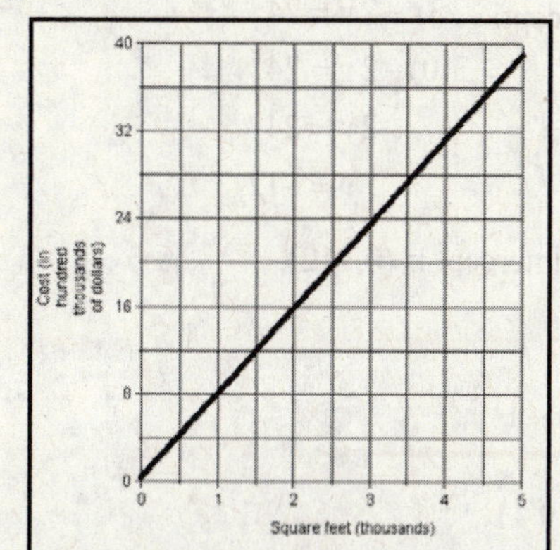

c. The contractor's cost to build a house of 1500 square feet is $145,000.

50. *average annual rate of change*

$= \frac{\textit{change in wages}}{\textit{change in time}}$

$= \frac{18.97 - 6.85}{2010 - 1980} = \frac{12.12}{30} = 0.404$

The average annual rate of change in wage was 40.4 cents per year.

Chapter 5 Test

1. $y - y_1 = m(x - x_1)$

$y - (-3) = -\frac{1}{3}(x - 9)$

$y + 3 = -\frac{1}{3}x + 3$

$y = -\frac{1}{3}x$

2. $(x_1, y_1) = (9, 8)$ *and* $(x_2, y_2) = (-2, 1)$

$m = \frac{y_2 - y_1}{x_2 - x_1} = \frac{1 - 8}{-2 - 9} = \frac{-7}{-11} = \frac{7}{11}$

3. x-intercept:
$$3x-2y=24$$
$$3x-2(0)=24$$
$$3x=24$$
$$x=8$$

The x-intercept is (8, 0).

y-intercept:
$$3x-2y=24$$
$$3(0)-2y=24$$
$$-2y=24$$
$$y=-12$$

The y-intercept is (0, -12).

4. $y=-\frac{4}{3}x-1$

$$y=-\frac{4}{3}\cdot 9-1$$
$$y=-12-1$$
$$y=-13$$

The ordered pair solution is (9, -13).

5. $5x+3y=15$

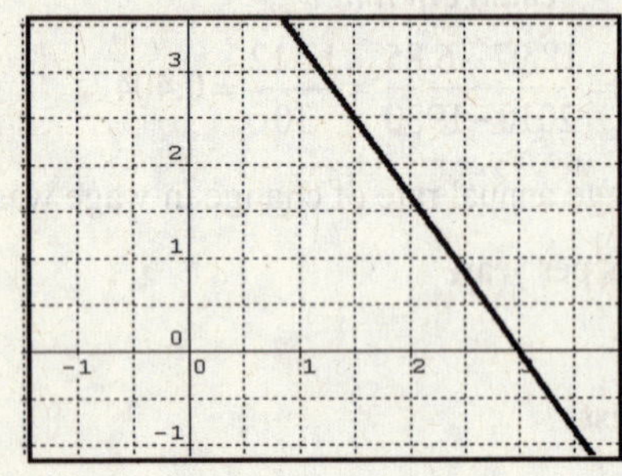

6. $y=\frac{1}{4}x+3$

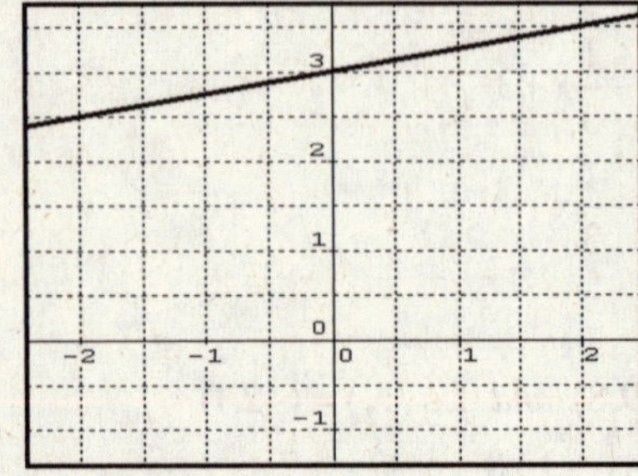

7. $f(x)=4x+7$

$$f\left(\frac{3}{4}\right)=4\left(\frac{3}{4}\right)+7$$
$$f\left(\frac{3}{4}\right)=3+7$$
$$f\left(\frac{3}{4}\right)=10$$

8. $y=\frac{2}{3}x+1$

$$3\overset{?}{=}\frac{2}{3}(6)+1$$
$$3\overset{?}{=}4+1$$
$$3\neq 5$$

9.

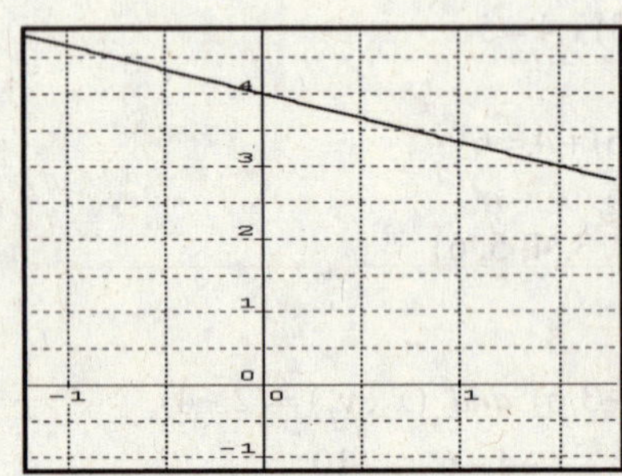

10. $2x-y\geq 2$

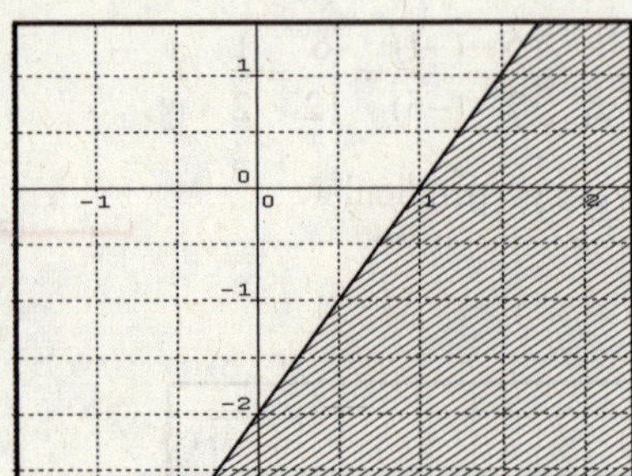

11.

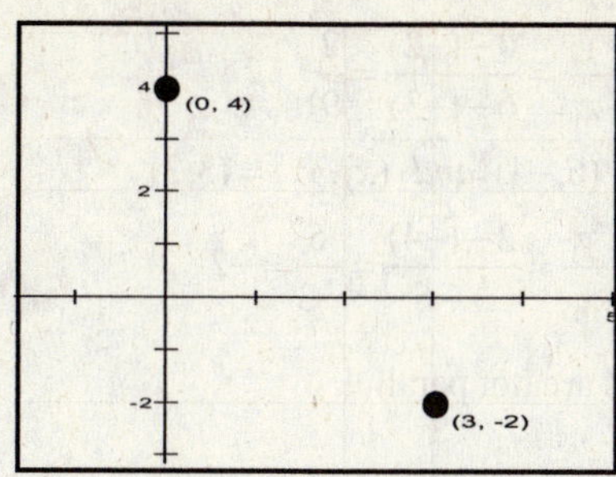

12.

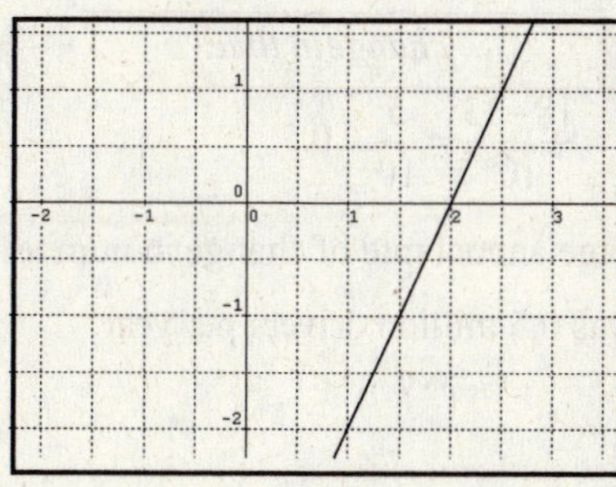

13. $(x_1, y_1) = (-5, 5)$ *and* $(x_2, y_2) = (10, 14)$

$$m = \frac{y_2 - y_1}{x_2 - x_1} = \frac{14-5}{10-(-5)} = \frac{9}{15} = \frac{3}{5}$$

$$y - y_1 = m(x - x_1)$$

$$y - 5 = \frac{3}{5}[x - (-5)]$$

$$y - 5 = \frac{3}{5}(x + 5)$$

$$y - 5 = \frac{3}{5}x + 3$$

$$y = \frac{3}{5}x + 8$$

14. $y = mx + b$

$$y = -\frac{2}{5}x + 7$$

15. $y - y_1 = m(x - x_1)$

$$y - 1 = 4(x - 2)$$

$$y - 1 = 4x - 8$$

$$y = 4x - 7$$

16. $f(x) = 4x^2 - 3$

$$f(-2) = 4(-2)^2 - 3$$

$$f(-2) = 16 - 3$$

$$f(-2) = 13$$

17. $y = -2x - 1$

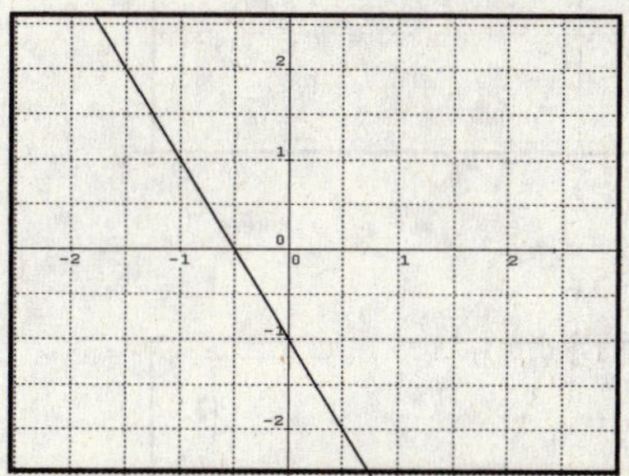

18. $y > 2$

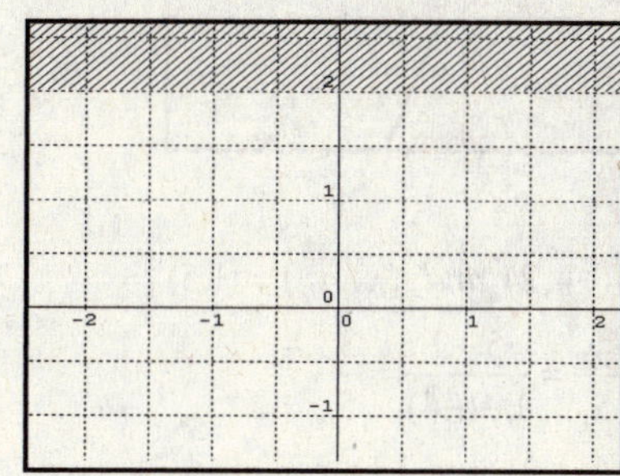

19. $x = 4$

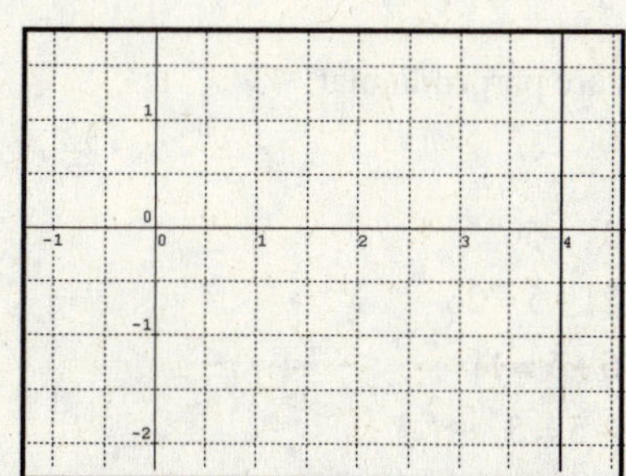

20. $y = \frac{1}{2}x - 3$

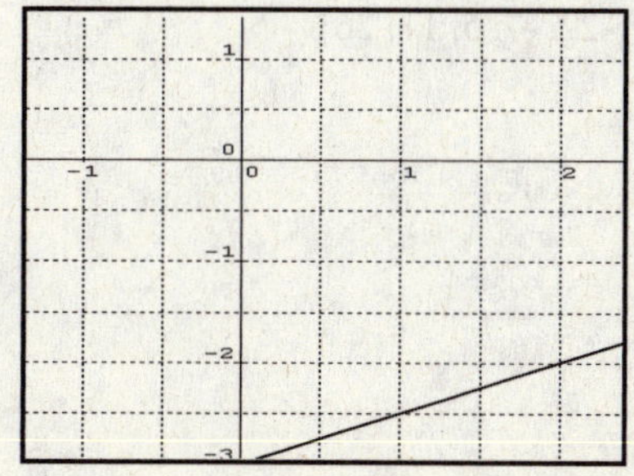

21. $f(x) = -\frac{2}{3}x + 2$

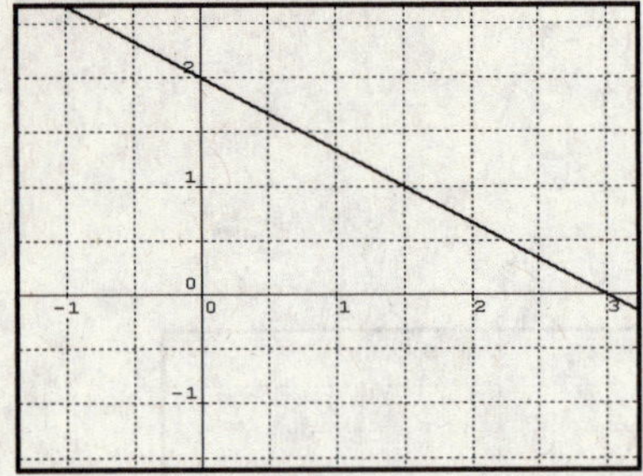

22. $f(x) = -5x$

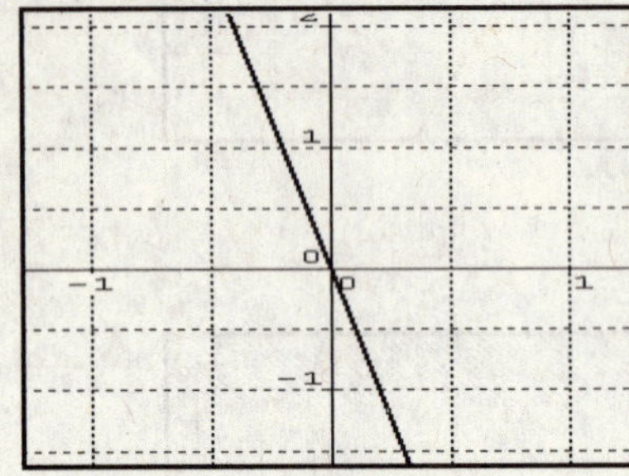

23. $(x_1, y_1) = (-4, 2)$ *and* $(x_2, y_2) = (1, 5)$

$$m = \frac{y_2 - y_1}{x_2 - x_1} = \frac{5 - 2}{1 - (-4)} = \frac{3}{5}$$

$(x_1, y_1) = (-1, 2)$ *and* $(x_2, y_2) = (-4, 7)$

$$m = \frac{y_2 - y_1}{x_2 - x_1} = \frac{7 - 2}{-4 - (-1)} = \frac{5}{-3}$$

The lines are perpendicular.

24. $f(x) = -3x + 5$

$f(-7) = 21 + 5 = 26$

$f(-2) = 6 + 5 = 11$

$f(0) = 0 + 5 = 5$

$f(4) = -12 + 5 = -7$

$f(9) = -27 + 5 = -22$

Range: {-22, -7, 5, 11, 26}

25. $(x_1, y_1) = (-3, -2)$ *and* $(x_2, y_2) = (6, 0)$

$$m = \frac{y_2 - y_1}{x_2 - x_1} = \frac{0 - (-2)}{6 - (-3)} = \frac{2}{9}$$

$(x_1, y_1) = (6, -4)$ *and* $(x_2, y_2) = (3, 2)$

$$m = \frac{y_2 - y_1}{x_2 - x_1} = \frac{2 - (-4)}{3 - 6} = \frac{6}{-3} = -2$$

The lines are not parallel.

26. *average annual rate of change*

$$= \frac{\textit{change in number of drivers}}{\textit{change in time}}$$

$$= \frac{18 - 13}{10} = \frac{5}{10} = 0.5$$

The average annual rate of change in number of drivers was 0.5 million drivers per year.

27.

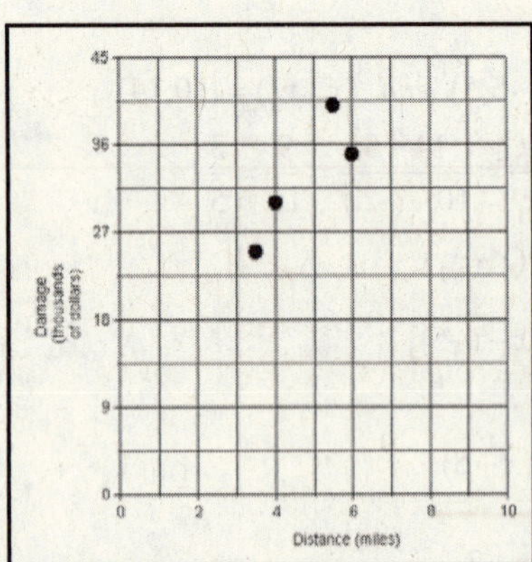

28. a. $C(t) = 8t + 1000$

b.

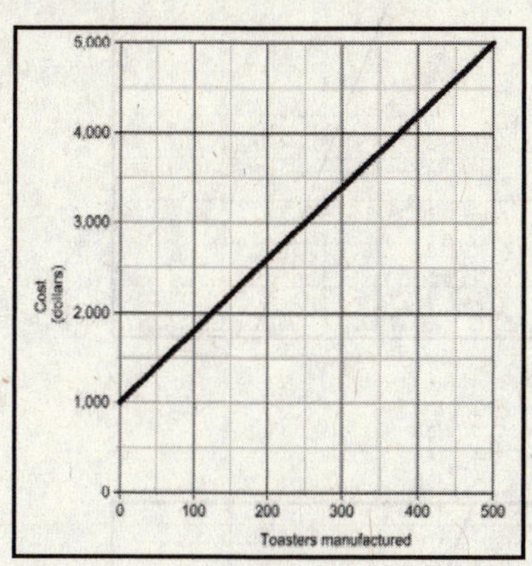

c. The cost to manufacture 340 toasters is $3720.

29. {(8.5, 64), (9.4, 68), (10.1, 76), (11.4, 87), (12.0, 92)}

Function.

30. $(x_1, y_1) = (1,300)$ *and* $(x_2, y_2) = (6,480)$

$$m = \frac{y_2 - y_1}{x_2 - x_1} = \frac{480-300}{6-1} = \frac{180}{5} = 36$$

The slope is 36. The cost of 1000 board feet of lumber has been increasing by $36 per month.

Cumulative Review Exercises

1. $12 - 18 \div 3 \cdot (-2)^2 = 12 - 18 \div 3 \cdot 4$

$= 12 - 6 \cdot 4$

$= 12 - 24$

$= -12$

2. $\frac{a-b}{a^2-c} = \frac{-2-3}{(-2)^2 - (-4)}$

$= \frac{-5}{4+4} = \frac{-5}{8}$

3. $4(2-3x) - 5(x-4) = 8 - 12x - 5x + 20$

$= -17x + 28$

4. $2x - \frac{2}{3} = \frac{7}{3}$

$2x - \frac{2}{3} + \frac{2}{3} = \frac{7}{3} + \frac{2}{3}$

$2x = 3$

$x = \frac{3}{2}$

5. $3x - 2[x - 3(2-3x)] = x - 6$

$3x - 2[x - 6 + 9x] = x - 6$

$3x - 2[10x - 6] = x - 6$

$3x - 20x + 12 = x - 6$

$-17x + 12 = x - 6$

$-18x = -18$

$x = 1$

6. $6\frac{2}{3}\% = \frac{\overset{1}{\cancel{20}}}{3} \cdot \frac{1}{\underset{5}{\cancel{100}}} = \frac{1}{15}$

7. Natural numbers less than 9:

{1, 2, 3, 4, 5, 6, 7, 8}

8. Elements of *D* greater than -16: -4, 0, 5

9. $8a - 3 \geq 5a - 6$

$3a \geq -3$

$a \geq -1$

10. $4x - 5y = 15$

$4x - 4x - 5y = -4x + 15$

$-5y = -4x + 15$

$\frac{-5y}{-5} = \frac{-4x+15}{-5}$

$y = \frac{4}{5}x - 3$

11. $y = 3x - 1$

$y = 3(-2) - 1$

$y = -6 - 1$

$y = -7$

The ordered pair solution is (-2, -7).

12. $(x_1, y_1) = (2,3)$ *and* $(x_2, y_2) = (-2,3)$

$$m = \frac{y_2 - y_1}{x_2 - x_1} = \frac{3-3}{-2-2} = \frac{0}{-4} = 0$$

13. x-intercept: $5x + 2y = 20$

$5x + 2(0) = 20$

$5x = 20$

$x = 4$

The x-intercept is (4, 0).

y-intercept: $5x + 2y = 20$

$5(0) + 2y = 20$

$2y = 20$

$y = 10$

The y-intercept is (0, 10).

14. $y - y_1 = m(x - x_1)$

$y - 2 = -1(x - 3)$

$y - 2 = -x + 3$

$y = -x + 5$

15. $y = \frac{1}{2}x + 2$

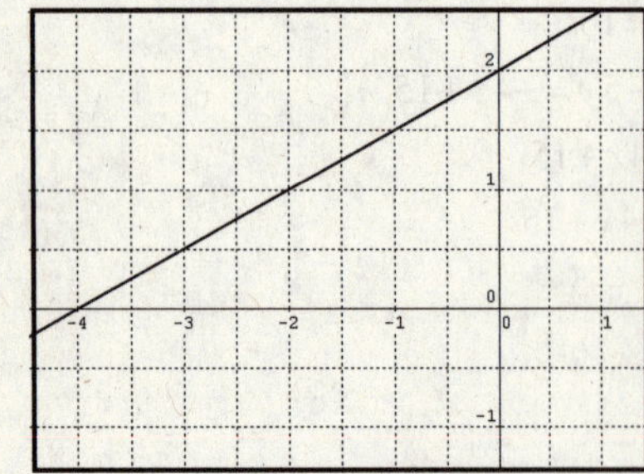

16. $3x + y = 2$

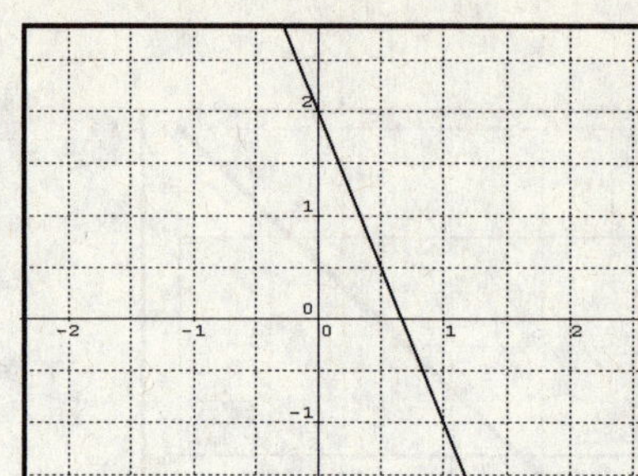

17. $f(x) = -4x - 1$

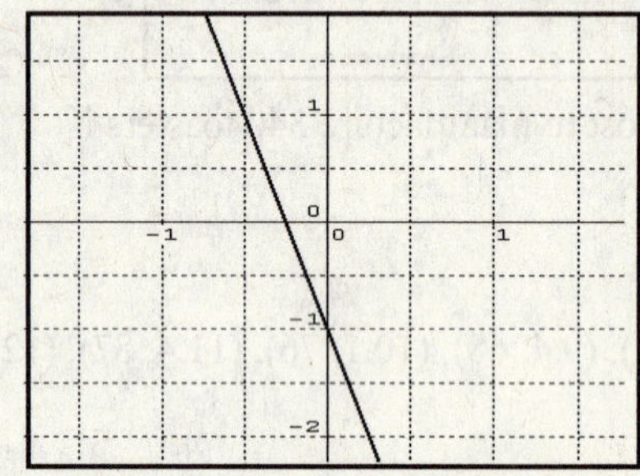

18. $x - y \le 5$

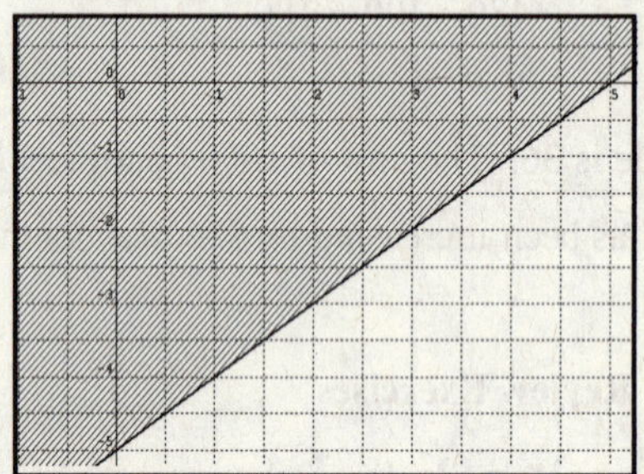

19. Domain: $\{0, 1, 2, 3, 4\}$

Range: $\{0, 1, 2, 3, 4\}$

The relation is a function.

20. $f(x) = -4x + 9$

$f(5) = -4(5) + 9$

$f(5) = -20 + 9$

$f(5) = -11$

21. $f(x) = -\frac{5}{3}x + 3$

$f(-9) = -\frac{5}{3}(-9) + 3 = 18$

$f(-6) = -\frac{5}{3}(-6) + 3 = 13$

$f(-3) = -\frac{5}{3}(-3) + 3 = 8$

$f(0) = 0 + 3 = 3$

$f(3) = -\frac{5}{3}(3) + 3 = -2$

$f(6) = -\frac{5}{3}(6) + 3 = -7$

Range: {-7, -2, 3, 8, 13, 18}

22.

	P	·	R	=	I
Amount at 4.5%	x	·	0.045	=	$0.045x$
Amount at 3.2%	15,000 -x	·	0.032	=	$0.032(15{,}000 - x)$
Total	15,000	·		=	584

$0.045x + 0.032(15{,}000 - x) = 584$

$0.045x + 480 - 0.032x = 584$

$0.013x + 480 = 584$

$0.013x = 104$

$x = 8000$

$8000 was invested at 4.5%.

23. $F_1 x = F_2(d - x)$

$80x = 560(8 - x)$

$80x = 4480 - 560x$

$640x = 4480$

$x = 7$

The fulcrum is located 7 feet from the 80 pound force.

24. Let x = the length of the third side. Then the other two sides are $2x$ and $x + 5$.

$P = \textit{sum of lengths of sides}$

$49 = 2x + x + x + 5$

$49 = 4x + 5$

$44 = 4x$

$11 = x$

$2x = 22$

The first side is 22 feet long.

25. $S = R - rR$

$S = 89 - (0.3)(89)$

$S = 89 - 26.7$

$S = 62.30$

The sale price is $62.30.

CHAPTER 6 Systems of Linear Equations

Chapter 6 Prep Test

1. $3x-4y=24$

$3x-4y-3x=-3x+24$

$-4y=-3x+24$

$-\frac{1}{4}(-4y)=\left(-\frac{1}{4}\right)(-3x+24)$

$y=\frac{3}{4}x-6$

2. $50+0.07x=0.05(x+1400)$

$50+0.07x=0.05x+70$

$0.07x=0.05x+20$

$0.02x=20$

$\frac{0.02x}{0.02}=\frac{20}{0.02}$

$x=1000$

3. $-3(2x-7y)+3(2x+4y)$

$-6x+21y+6x+12y$

$-6x+6x+21y+12y=33y$

4. $4x+2(3x-5)=4x+6x-10=10x-10$

5. $3x-5y=-22;\ x=-4,\ y=2$

$3(-4)-5(2)\stackrel{?}{=}-22$

$-12-10\stackrel{?}{=}-22$

$-22=-22$

Yes, the ordered pair is a solution.

6. x-intercept: $3x-4y=12$

$3x-4(0)=12$

$3x=12$

$x=4$

$(4,0)$

y-intercept: $3x-4y=12$

$3(0)-4y=12$

$-4y=12$

$y=-3$

$(0,-3)$

7. The graphs are parallel because both lines have a slope of -3.

8. $y=\frac{5}{4}x-2$

9. The hikers are side by side when their distances are equal.

Let t = the time for the second hiker.

First hiker: $d=3(t+0.5)$

Second hiker: $d=4t$

$3(t+0.5)=4t$

$3t+1.5=4t$

$1.5=t$

The hikers are side by side 1.5 hours after the second hiker starts.

Section 6.1 Solving System of Linear Equations by Graphing

1. It is **always true** that a solution of a system of linear equations in two variables is an ordered pair (x, y).

3. It is **sometimes true** that a system of equations has no solution if an ordered pair is a solution of one equation in the system but not the other.

5. It is **never true** that the system shown has no solution. Every ordered pair solution of one equation is also an ordered pair solution of the other equation.

7. $5x-2y=14 \qquad x+y=8$
$5(4)-2(3)\stackrel{?}{=}14 \qquad 4+3\stackrel{?}{=}8$
$20-6\stackrel{?}{=}14 \qquad 7\neq 8$
$14=14$
(4, 3) is not a solution.

9. $4x-y=-5$
$4(-1)-3\stackrel{?}{=}-5$
$-4-3\stackrel{?}{=}-5$
$-7\neq -5$
(-1, 3) is not a solution.

11. $4x+3y=0 \qquad 2x-y=1$
$4(0)+3(0)\stackrel{?}{=}0 \qquad 2(0)-0\stackrel{?}{=}1$
$0+0\stackrel{?}{=}0 \qquad 0-0\stackrel{?}{=}1$
$0=0 \qquad 0\neq 1$
(0, 0) is not a solution.

13. $y=2x-7 \qquad 3x-y=9$
$-3\stackrel{?}{=}2(2)-7 \qquad 3(2)-(-3)\stackrel{?}{=}9$
$-3\stackrel{?}{=}4-7 \qquad 6+3\stackrel{?}{=}9$
$-3=-3 \qquad 9=9$
(2, -3) is a solution.

15. $y=2x-8 \qquad y=3x-13$
$2\stackrel{?}{=}2(5)-8 \qquad 2\stackrel{?}{=}3(5)-13$
$2\stackrel{?}{=}10-8 \qquad 2\stackrel{?}{=}15-13$
$2=2 \qquad 2=2$
(5, 2) is a solution.

17. To solve a linear system of two equations in two variables by graphing, graph the equations on the same set of coordinate axes. The point at which the graphs intersect is the ordered pair that is the solution of the system.

19. Since there is one point of intersection, the system is **independent**.

21. Since the equations graph as the same line, the system is **dependent**.

23. Since the lines are parallel, the system is **inconsistent**.

25. The solution point is (2, -1).

27. There is no solution. The lines are parallel.

29. The solution point is (-2, 4).

31. The solution point is (4, 1).

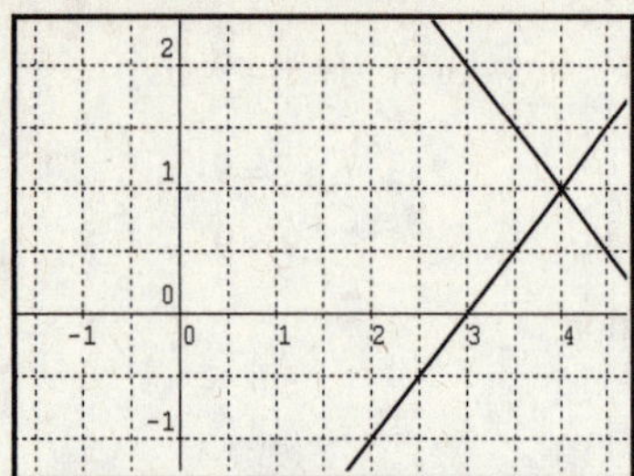

33. The solution point is (4, 1).

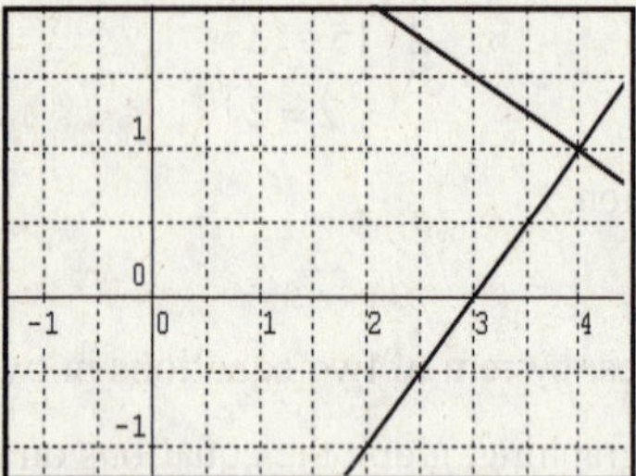

35. The solution point is (4, 3).

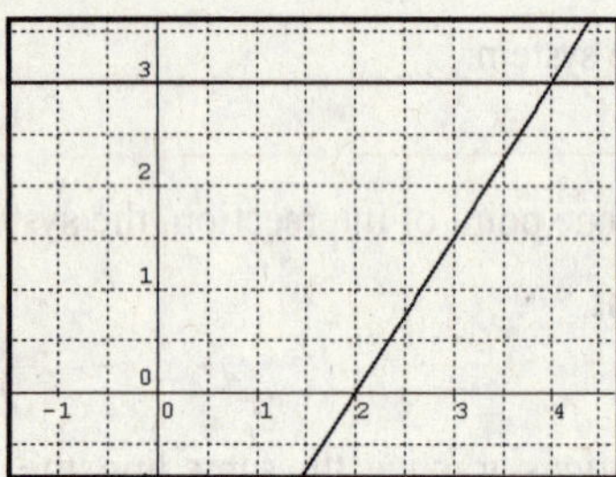

37. The solution point is (3, -2).

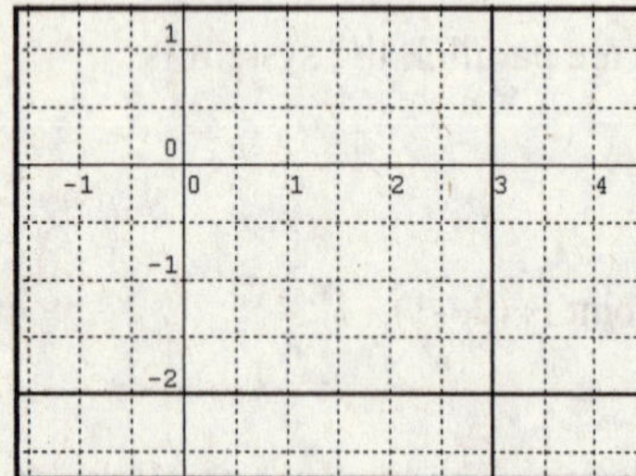

39. The solution point is (2, -2).

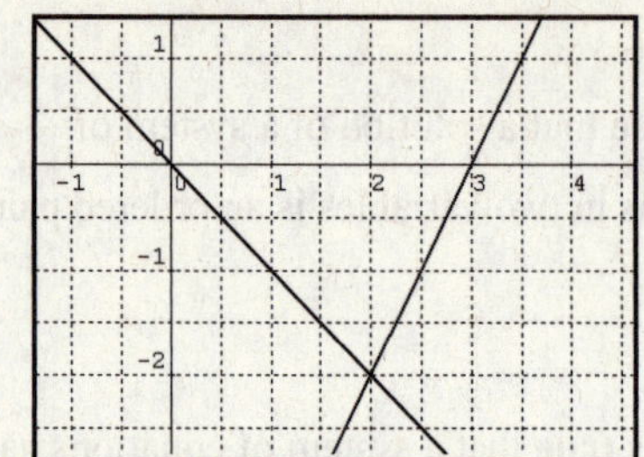

41. The system is inconsistent

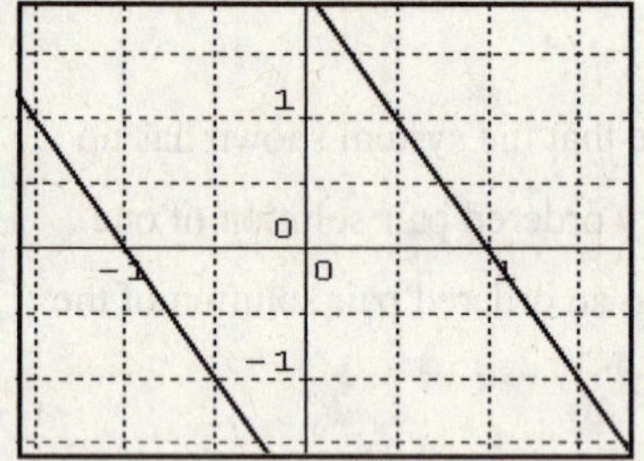

43. The system is dependent.

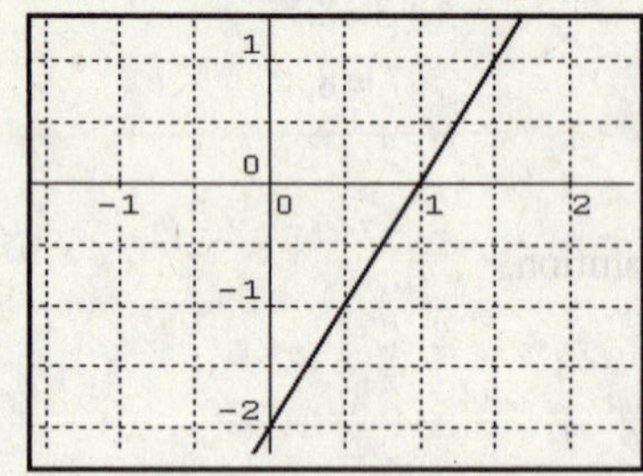

45. The solution point is (1, -4).

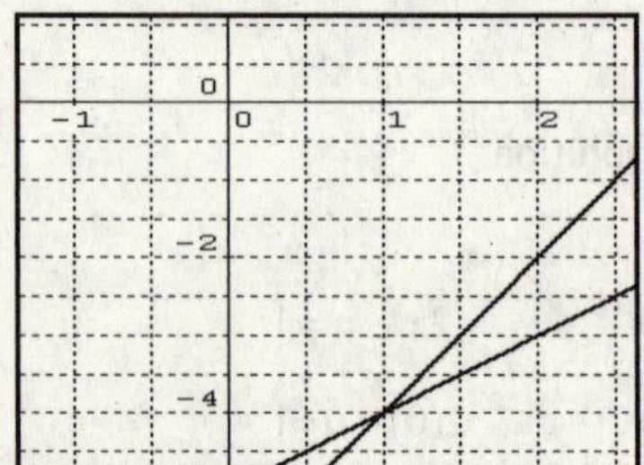

47. The solution point is (0,0).

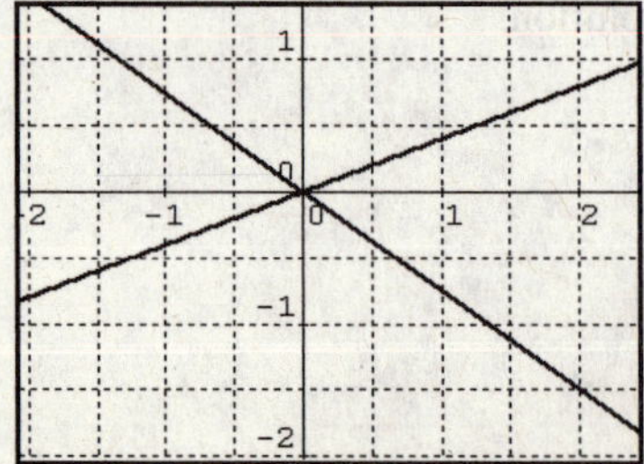

49. The system is inconsistent.

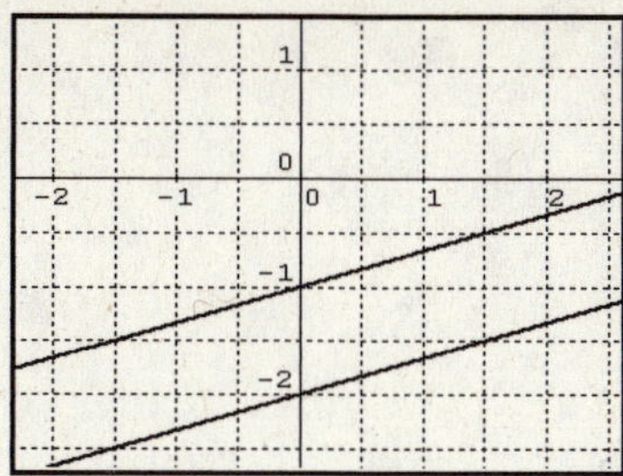

51. The solution point is (0, -2).

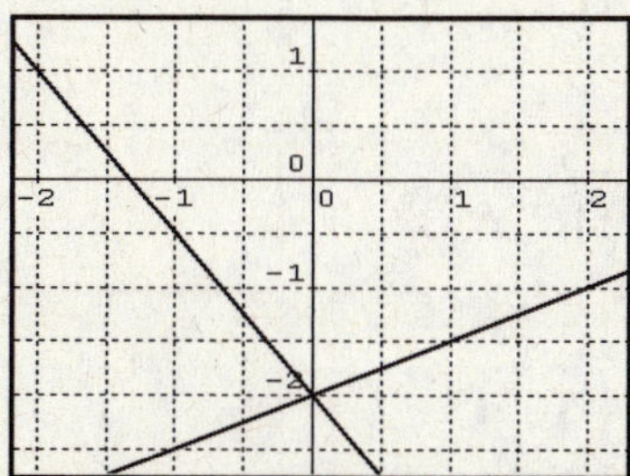

53. The solution point is (1, -1).

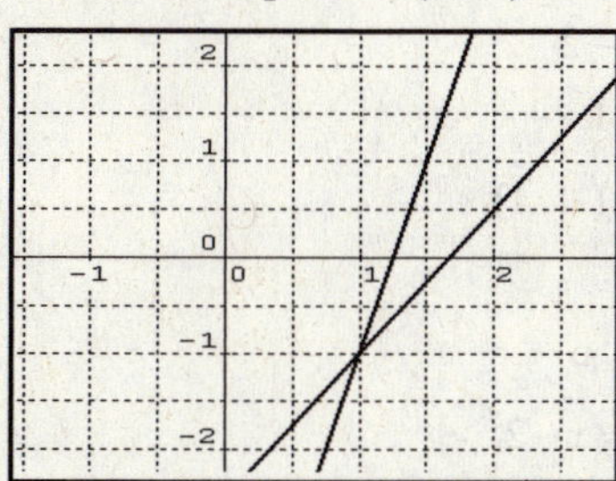

55. The solution point is (2, 0).

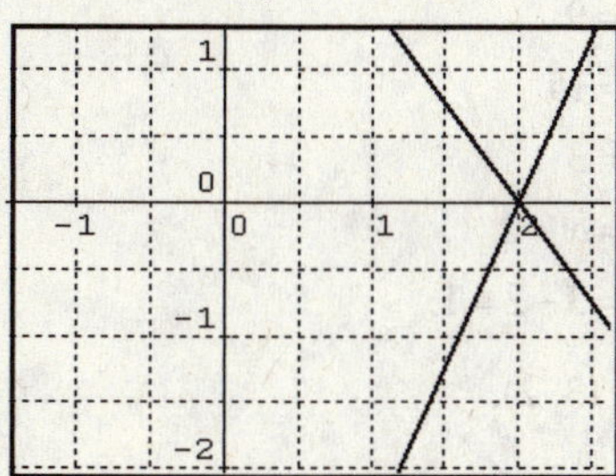

57. The solution point is (-3, -2).

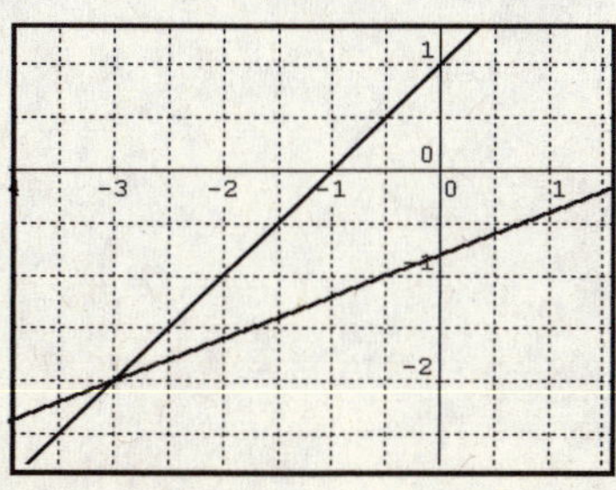

59. The system is inconsistent.

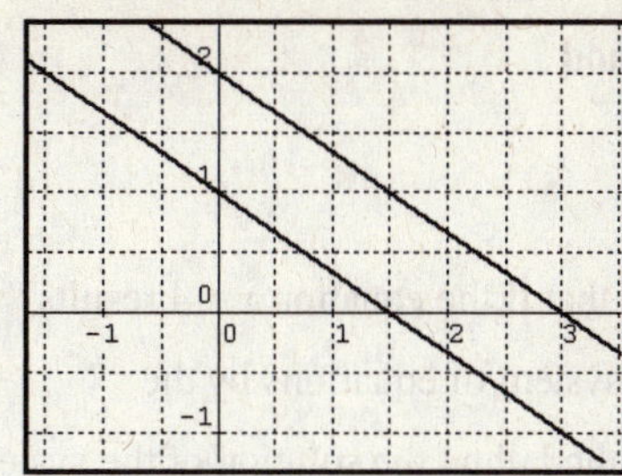

61. The system is inconsistent. The lines have the same slopes but different *y*-intercepts so they are parallel.

63. The horizontal line has the equation $y = 2$. The other line has a slope of 1 and a *y*-intercept of (0, 4).

$$y = 2$$
$$y = x + 4$$

65. One line has a slope of 1 and a *y*-intercept of (0, 0). The other line has a slope of -1 and a *y*-intercept of (0, 2).

$$y = x$$
$$y = -x + 2$$

67. **a.** It is **always true** that two parallel lines always have the same slope.

b. It is **never true** that two different lines with the same *y*-intercepts are parallel.

c. It is **always true** that two different lines with the same slope are parallel.

69.

System	Graph
a	**iii**
b	**iv**
c	**i**
d	**ii**

Section 6.2 Solving Systems of Equations by the Substitution Method

1. True

3. It is **never true** that if the equation $x = 4$ results from solving a system of equations by the substitution method, then the solution of the system of equations is 4. The solution of the system is an ordered pair.

5. A false statement like $0 = 7$ means that the system is inconsistent.

7. To solve the system by substitution, substitute 2 for x in equation (1): $y = 3(2) - 5 = 1$. The solution of the system of equations is (2, 1).

9.
$$\begin{aligned} 2x + 3y &= 7 \\ x &= 2 \\ 2(2) + 3y &= 7 \\ 4 + 3y &= 7 \\ 3y &= 3 \\ y &= 1 \end{aligned}$$

The solution is (2, 1).

11.
$$\begin{aligned} y &= x - 3 \\ x + y &= 5 \\ x + x - 3 &= 5 \\ 2x - 3 &= 5 \\ 2x &= 8 \\ x &= 4 \\ y &= x - 3 \\ y &= 4 - 3 \\ y &= 1 \end{aligned}$$

The solution is (4, 1).

13.
$$\begin{aligned} x &= y - 2 \\ x + 3y &= 2 \\ y - 2 + 3y &= 2 \\ 4y - 2 &= 2 \\ 4y &= 4 \\ y &= 1 \\ x &= y - 2 \\ x &= 1 - 2 \\ x &= -1 \end{aligned}$$

The solution is (-1, 1).

15.
$$\begin{aligned} 2x + 3y &= 9 \\ y &= x - 2 \\ 2x + 3(x - 2) &= 9 \\ 2x + 3x - 6 &= 9 \\ 5x - 6 &= 9 \\ 5x &= 15 \\ x &= 3 \\ y &= x - 2 \\ y &= 3 - 2 = 1 \end{aligned}$$

The solution is (3, 2).

17.
$$\begin{aligned} 3x - y &= 2 \\ y &= 2x - 1 \\ 3x - (2x - 1) &= 2 \\ 3x - 2x + 1 &= 2 \\ x &= 1 \\ y &= 2x - 1 \\ y &= 2(1) - 1 \\ y &= 1 \end{aligned}$$

The solution is (1, 1).

19.
$$\begin{aligned} x &= 2y - 3 \\ 2x - 3y &= -5 \\ 2(2y - 3) - 3y &= -5 \\ 4y - 6 - 3y &= -5 \\ y - 6 &= -5 \\ y &= 1 \\ x &= 2y - 3 \\ x &= 2(1) - 3 \\ x &= -1 \end{aligned}$$

The solution is (-1, 1).

21.
$$\begin{aligned} y &= 4 - 3x \\ 3x + y &= 5 \\ 3x + 4 - 3x &= 5 \\ 4 &= 5 \textit{ false} \end{aligned}$$

The system is inconsistent.

23.
$$\begin{aligned} x &= 3y + 3 \\ 2x - 6y &= 12 \\ 2(3y + 3) - 6y &= 12 \\ 6y + 6 - 6y &= 12 \\ 6 &= 12 \textit{ false} \end{aligned}$$

The system is inconsistent.

25.
$$\begin{aligned} 3x + 5y &= -6 \\ x &= 5y + 3 \\ 3(5y + 3) + 5y &= -6 \\ 15y + 9 + 5y &= -6 \\ 20y &= -15 \\ y &= -\frac{15}{20} = -\frac{3}{4} \\ x &= 5y + 3 \\ x &= 5\left(-\frac{3}{4}\right) + \frac{12}{4} \\ x &= -\frac{15}{4} + \frac{12}{4} \\ x &= -\frac{3}{4} \end{aligned}$$

The solution is $\left(-\frac{3}{4}, -\frac{3}{4}\right)$.

27.
$$\begin{aligned} x &= 4y - 3 \\ 2x - 3y &= 0 \\ 2(4y - 3) - 3y &= 0 \\ 8y - 6 - 3y &= 0 \\ 5y - 6 &= 0 \\ 5y &= 6 \\ y &= \frac{6}{5} \\ x &= 4y - 3 \\ 2x - 3y &= 0 \\ x &= 4\left(\frac{6}{5}\right) - 3 \\ x &= \frac{24}{5} - \frac{15}{5} \\ x &= \frac{9}{5} \end{aligned}$$

The solution is $\left(\frac{9}{5}, \frac{6}{5}\right)$.

29.

$$y = 2x - 9$$
$$3x - y = 2$$
$$3x - (2x - 9) = 2$$
$$3x - 2x + 9 = 2$$
$$x + 9 = 2$$
$$x = -7$$
$$y = 2x - 9$$
$$y = 2(-7) - 9$$
$$y = -14 - 9$$
$$y = -23$$

The solution is (-7, -23).

31.

$$2x - y = 4$$
$$-y = -2x + 4$$
$$y = 2x - 4$$
$$3x + 2y = 6$$
$$3x + 2(2x - 4) = 6$$
$$3x + 4x - 8 = 6$$
$$7x - 8 = 6$$
$$7x = 14$$
$$x = 2$$
$$y = 2x - 4$$
$$y = 2(2) - 4$$
$$y = 0$$

The solution is (2, 0).

33.

$$x + 2y = 4$$
$$x = 4 - 2y$$
$$4x - 3y = 5$$
$$4(4 - 2y) - 3y = 5$$
$$16 - 8y - 3y = 5$$
$$16 - 11y = 5$$
$$-11y = -11$$
$$y = 1$$
$$x = 4 - 2y$$
$$x = 4 - 2(1)$$
$$x = 2$$

The solution is (2, 1).

35.

$$7x - y = 4$$
$$-y = -7x + 4$$
$$y = 7x - 4$$
$$5x + 2y = 1$$
$$5x + 2(7x - 4) = 1$$
$$5x + 14x - 8 = 1$$
$$19x - 8 = 1$$
$$19x = 9$$
$$x = \frac{9}{19}$$
$$y = 7x - 4$$
$$y = 7\left(\frac{9}{19}\right) - 4$$
$$y = \frac{63}{19} - \frac{76}{19}$$
$$y = -\frac{13}{19}$$

The solution is $\left(\frac{9}{19}, -\frac{13}{19}\right)$.

37.

$$7x + y = 14$$
$$y = -7x + 14$$
$$2x - 5y = -33$$
$$2x - 5(-7x + 14) = -33$$
$$2x + 35x - 70 = -33$$
$$37x = 37$$
$$x = 1$$
$$y = -7x + 14$$
$$y = -7(1) + 14$$
$$y = 7$$

The solution is (1, 7).

39.
$$\begin{aligned} x-4y&=9\\ x&=4y+9\\ 2x-3y&=11\\ 2(4y+9)-3y&=11\\ 8y+18-3y&=11\\ 5y&=-7\\ y&=-\frac{7}{5}\\ x&=4y+9\\ x&=4\left(-\frac{7}{5}\right)+9\\ x&=-\frac{28}{5}+\frac{45}{5}\\ x&=\frac{17}{5} \end{aligned}$$

The solution is $\left(\frac{17}{5},-\frac{7}{5}\right)$.

41.
$$\begin{aligned} 3x-y&=6\\ -y&=-3x+6\\ y&=3x-6\\ x+3y&=2\\ x+3(3x-6)&=2\\ x+9x-18&=2\\ 10x&=20\\ x&=2\\ 2+3y&=2\\ 3y&=0\\ y&=0 \end{aligned}$$

The solution is (2, 0).

43.
$$\begin{aligned} 2x-y&=0\\ 2x&=y\\ 4x+3y&=0\\ 4x+3(2x)&=0\\ 4x+6x&=0\\ 10x&=0\\ x&=0\\ y&=2x\\ y&=2(0)=0 \end{aligned}$$

The solution is (0, 0).

45.
$$\begin{aligned} 2x-y&=2\\ -y&=-2x+2\\ y&=2x-2\\ 6x-3y&=6\\ 6x-3(2x-2)&=6\\ 6x-6x+6&=6\\ 6&=6 \end{aligned}$$

The system is dependent.

47.
$$\begin{aligned} y&=2x+11\\ y&=5x-1\\ 2x+11&=5x-1\\ -3x&=-12\\ x&=4\\ y&=2x+11\\ y&=2(4)+11\\ y&=8+11\\ y&=19 \end{aligned}$$

The solution is (4, 19).

49.
$$\begin{aligned} y&=-4x+2\\ y&=-3x-1\\ -4x+2&=-3x-1\\ -x&=-3\\ x&=3\\ y&=-4x+2\\ y&=-4(3)+2\\ y&=-12+2\\ y&=-10 \end{aligned}$$

The solution is (3, -10).

51.
$$\begin{aligned} x&=4y-2\\ x&=6y+8\\ 4y-2&=6y+8\\ -2y&=10\\ y&=-5\\ x&=4y-2\\ x&=4(-5)-2\\ x&=-20-2\\ x&=-22 \end{aligned}$$

The solution is (-22, -5).

53. $y=2x-7$

$y=4x+5$

$2x-7=4x+5$

$-2x=12$

$x=-6$

$y=2x-7$

$y=2(-6)-7$

$y=-12-7$

$y=-19$

The solution is (-6, -19).

55. $y=C-x$

$x+y=B$

$x+C-x=B$

$C=B$ *false*

The system is inconsistent.

57. $0.8x-0.1y=0.3 \quad \Rightarrow 8x-y=3 \Rightarrow y=8x-3$

$0.5x-0.2y=-0.5 \quad \Rightarrow 5x-2y=-5$

$5x-2(8x-3)=-5$

$5x-16x+6=-5$

$-11x=-11$

$x=1$

$y=8x-3$

$y=8(1)-3$

$y=5$

The solution is $(1,5)$.

59. $-0.1x+0.3y=1.1$

$0.4x-0.1y=-2.2$

$-x+3y=11 \Rightarrow x=3y-11$

$4x-y=-22$

$4(3y-11)-y=-22$

$12y-44-y=-22$

$11y-44=-22$

$11y=22$

$y=2$

$x=3y-11$

$x=3(2)-11$

$x=-5$

The solution is (-5, 2).

61. $1.25x-0.01y=1.5$

$0.24x-0.02y=-1.52$

$125x-y=150 \Rightarrow y=125x-150$

$24x-2y=-152$

$24x-2(125x-150)=-152$

$24x-250x+300=-152$

$-226x=-452$

$x=2$

$y=125x-150$

$y=125(2)-150$

$y=100$

The solution is (2, 100).

63.

$2x-3y=7$	$kx-3y=4$
$-3y=-2x+7$	$-3y=-kx+4$
$y=\frac{2}{3}x-\frac{7}{3}$	$y=\frac{k}{3}x-\frac{4}{3}$
$m=\frac{2}{3}$	$m=\frac{k}{3}$

The system has no solution when the lines are parallel –that is, when they have the same slope. So, setting the slopes equal, $k = 2$.

65.
$$x = 4y + 4$$
$$x - 4 = 4y$$
$$\frac{1}{4}x - 1 = y$$
$$m = \frac{1}{4}$$
$$kx - 8y = 4$$
$$-8y = -kx + 4$$
$$y = \frac{k}{8}x - \frac{1}{2}$$
$$m = \frac{k}{8}$$

The system has no solution when the lines are parallel –that is, when they have the same slope. So, setting the slopes equal, $k = 2$.

Section 6.3 Solving Systems of Linear Equations by the Addition Method

1. a. This system is **inconsistent**. The lines have the same slopes and different *y*-intercepts so they are parallel.

b. This system is **independent**. The lines have different slopes so they do intersect.

c. This system is **dependent**. The lines have the same slopes and the same *y*-intercepts so they are the same line.

d. This system is **independent**. The lines have different slopes so they do intersect.

3. True

5. a. To eliminate *x* from the system of equations by using the addition method, multiply each side of equation (2) by 3.

b. To eliminate the *y* from the system of equations by using the addition method, multiply each side of equation (1) by -4.

7.
$$\begin{aligned} x + y &= 4 \\ \underline{x - y} &\underline{= 6} \\ 2x \quad &= 10 \\ x &= 5 \\ x + y &= 4 \\ 5 + y &= 4 \\ y &= -1 \end{aligned}$$

The solution is (5, -1).

9.
$$\begin{aligned} -1(x + y) &= -1(4) \\ \underline{2x + y} &\underline{= 5} \\ -x - y &= -4 \\ \underline{2x + y} &\underline{= 5} \\ x \quad &= 1 \\ x + y &= 4 \\ 1 + y &= 4 \\ y &= 3 \end{aligned}$$

The solution is (1, 3).

11.
$$\begin{aligned} 3(2x - y) &= 3 \cdot 1 \\ \underline{x + 3y} &\underline{= 4} \\ 6x - 3y &= 3 \\ \underline{x + 3y} &\underline{= 4} \\ 7x \quad &= 7 \\ x &= 1 \\ x + 3y &= 4 \\ 1 + 3y &= 4 \\ 3y &= 3 \\ y &= 1 \end{aligned}$$

The solution is (1, 1).

13.
$$\begin{aligned} 4x-5y&=22\\ -4(x+2y)&=-1\cdot(-4) \end{aligned}$$
$$\begin{aligned} 4x-5y&=22\\ -4x-8y&=4\\ \hline -13y&=26\\ y&=-2\\ x+2y&=-1\\ x+2(-2)&=-1\\ x-4&=-1\\ x&=3 \end{aligned}$$
The solution is (3, -2).

15.
$$\begin{aligned} -2(2x-y)&=-2(1)\\ 4x-2y&=2\\ \hline -4x+2y&=-2\\ 4x-2y&=2\\ \hline 0&=0 \end{aligned}$$
The system is dependent.

17.
$$\begin{aligned} 4x+3y&=15\\ -2(2x-5y)&=-2\cdot1\\ \hline 4x+3y&=15\\ -4x+10y&=-2\\ \hline 13y&=13\\ y&=1\\ 2x-5y&=1\\ 2x-5(1)&=1\\ 2x&=6\\ x&=3 \end{aligned}$$
The solution is (3, 1).

19.
$$\begin{aligned} -2(2x-3y)&=-2(5)\\ 4x-6y&=3\\ \hline -4x+6y&=-10\\ 4x-6y&=3\\ \hline 0&=-7 \end{aligned}$$
The system is inconsistent.

21.
$$\begin{aligned} 5x-2y&=-1\\ 2(x+y)&=2(4)\\ \hline 5x-2y&=-1\\ 2x+2y&=8\\ \hline 7x&=7\\ x&=1\\ x+y&=4\\ 1+y&=4\\ y&=3 \end{aligned}$$
The solution is (1, 3).

23.
$$\begin{aligned} 2(5x+7y)&=2\cdot10\\ 3x-14y&=6\\ \hline 10x+14y&=20\\ 3x-14y&=6\\ \hline 13x&=26\\ x&=2\\ 5x+7y&=10\\ 5(2)+7y&=10\\ 10+7y&=10\\ 7y&=0\\ y&=0 \end{aligned}$$
The solution is (2, 0).

25.
$$\begin{aligned} -2(3x-2y)&=-2(0)\\ 6x+5y&=0\\ \hline -6x+4y&=0\\ 6x+5y&=0\\ \hline 9y&=0\\ y&=0\\ 3x-2y&=0\\ 3x-2(0)&=0\\ 3x&=0\\ x&=0 \end{aligned}$$
The solution is (0,0).

27.
$$\begin{aligned} 4(2x-3y) &= 4(16) \\ 3(3x+4y) &= 3(7) \\ \hline 8x-12y &= 64 \\ 9x+12y &= 21 \\ \hline 17x &= 85 \\ x &= 5 \\ 3x+4y &= 7 \\ 3(5)+4y &= 7 \\ 15+4y &= 7 \\ 4y &= -8 \\ y &= -2 \end{aligned}$$

The solution is (5, -2).

29.
$$\begin{aligned} -2(x+3y) &= -2(4) \\ 2x+5y &= 1 \\ \hline -2x-6y &= -8 \\ 2x+5y &= 1 \\ \hline -y &= -7 \\ y &= 7 \\ x+3y &= 4 \\ x+3(7) &= 4 \\ x+21 &= 4 \\ x &= -17 \end{aligned}$$

The solution is (-17, 7).

31.
$$\begin{aligned} 3(7x-2y) &= 3(13) \\ 2(5x+3y) &= 2(27) \\ \hline 21x-6y &= 39 \\ 10x+6y &= 54 \\ \hline 31x &= 93 \\ x &= 3 \\ 7x-2y &= 13 \\ 7(3)-2y &= 13 \\ 21-2y &= 13 \\ -2y &= -8 \\ y &= 4 \end{aligned}$$

The solution is (3,4).

33.
$$\begin{aligned} 5(8x-3y) &= 5(11) \\ -3(6x-5y) &= -3(11) \\ \hline 40x-15y &= 55 \\ -18x+15y &= -33 \\ \hline 22x &= 22 \\ x &= 1 \\ 8x-3y &= 11 \\ 8(1)-3y &= 11 \\ 8-3y &= 11 \\ -3y &= 3 \\ y &= -1 \end{aligned}$$

The solution is (1, -1).

35.
$$\begin{aligned} -2(5x+15y) &= -2(20) \\ 5(2x+6y) &= 5(8) \\ \hline -10x-30y &= -40 \\ 10x+30y &= 40 \\ \hline 0 &= 0 \end{aligned}$$

The system is dependent.

37.
$$\begin{aligned} 3x &= 2y+7 \\ 5x-2y &= 13 \\ \hline -1(3x-2y) &= -1(7) \\ 5x-2y &= 13 \\ \hline -3x+2y &= -7 \\ 5x-2y &= 13 \\ \hline 2x &= 6 \\ x &= 3 \\ 3x &= 2y+7 \\ 9 &= 2y+7 \\ 2 &= 2y \\ 1 &= y \end{aligned}$$

The solution is (3, 1).

39.

$$\begin{aligned} 2x+9y&=5 \\ 5x&=6-3y \end{aligned}$$

$$\begin{aligned} 2x+9y&=5 \\ -3(5x+3y)&=-3(6) \end{aligned}$$

$$\begin{aligned} 2x+9y&=5 \\ -15x-9y&=-18 \end{aligned}$$

$$\begin{aligned} -13x&=-13 \\ x&=1 \\ 5x&=6-3y \\ 5(1)&=6-3y \\ 5&=6-3y \\ -1&=-3y \\ \frac{1}{3}&=y \end{aligned}$$

The solution is $\left(1, \frac{1}{3}\right)$.

41.

$$\begin{aligned} 2x+3y&=7-2x \\ 7x+2y&=9 \end{aligned}$$

$$\begin{aligned} -2(4x+3y)&=-2(7) \\ 3(7x+2y)&=3(9) \end{aligned}$$

$$\begin{aligned} -8x-6y&=-14 \\ 21x+6y&=27 \end{aligned}$$

$$\begin{aligned} 13x&=13 \\ x&=1 \\ 7x+2y&=9 \\ 7(1)+2y&=9 \\ 7+2y&=9 \\ 2y&=2 \\ y&=1 \end{aligned}$$

The solution is (1, 1).

43.

$$\begin{aligned} 3x+y&=1 \\ 5x+y&=2 \end{aligned}$$

$$\begin{aligned} -(3x+y)&=-1 \\ 5x+y&=2 \end{aligned}$$

$$\begin{aligned} -3x-y&=-1 \\ 5x+y&=2 \end{aligned}$$

$$\begin{aligned} 2x&=1 \\ x&=\frac{1}{2} \\ 3x+y&=1 \\ 3\left(\frac{1}{2}\right)+y&=1 \\ \frac{3}{2}+y&=1 \\ y&=-\frac{1}{2} \end{aligned}$$

The solution is $\left(\frac{1}{2}, -\frac{1}{2}\right)$.

45.

$$\begin{aligned} 4x+3y&=\ \ 3 \\ -4(x+3y)&=-4(1) \end{aligned}$$

$$\begin{aligned} 4x+3y&=3 \\ -4x-12y&=-4 \end{aligned}$$

$$\begin{aligned} -9y&=-1 \\ y&=\frac{1}{9} \\ x+3y&=1 \\ x+3\left(\frac{1}{9}\right)&=1 \\ x+\frac{1}{3}&=1 \\ x&=\frac{2}{3} \end{aligned}$$

The solution is $\left(\frac{2}{3}, \frac{1}{9}\right)$.

47.
$$\begin{aligned} x-0.2y&=0.2\\ 0.2x+0.5y&=2.2\end{aligned}$$
$$\begin{aligned} 10x-2y&=2\\ -5(2x+5y)&=-5(22)\end{aligned}$$
$$\begin{aligned} 10x-2y&=2\\ -10x-25y&=-110\\ -27y&=-108\\ y&=4\\ x-0.2y&=0.2\\ x-0.2(4)&=0.2\\ x-0.8&=0.2\\ x&=1\end{aligned}$$
The solution is (1, 4).

49.
$$\begin{aligned} 1.25x-1.5y&=-1.75\\ 2.5x-1.75y&=-1\end{aligned}$$
$$\begin{aligned} -2(125x-150y)&=-2(-175)\\ 250x-175y&=-100\end{aligned}$$
$$\begin{aligned} -250x+300y&=350\\ 250x-175y&=-100\\ 125y&=250\\ y&=2\\ 1.25x-1.5(2)&=-1.75\\ 1.25x-3&=-1.75\\ 1.25x&=1.25\\ x&=1\end{aligned}$$
The solution is (1, 2).

51.
$$\begin{aligned} Ax-4y=9\Rightarrow A(-1)-4(-3)&=9\\ -A+12&=9\\ -A&=-3\\ A&=3\end{aligned}$$
$$\begin{aligned} 4x+By=-1\Rightarrow 4(-1)+B(-3)&=-1\\ -4-3B&=-1\\ -3B&=3\\ B&=-1\end{aligned}$$

53.
$$\begin{aligned} 3x-2y&=-2\\ 2x-y&=0\\ 2x&=y\end{aligned}$$
$$\begin{aligned} 3x-2(2x)&=-2\\ 3x-4x&=-2\\ -x&=-2\\ x&=2\\ 2x-y&=0\\ 2(2)-y&=0\\ y&=4\\ Ax+y&=8\\ A(2)+4&=8\\ 2A&=4\\ A&=2\end{aligned}$$

55. The systems are dependent when the equations represent the same line.

a.
$$\begin{aligned} 2x+3y&=7\\ 3y&=-2x+7\\ y&=-\frac{2}{3}x+\frac{7}{3}\end{aligned}\qquad \begin{aligned} 4x+6y&=k\\ 6y&=-4x+k\\ y&=-\frac{2}{3}x+\frac{k}{6}\end{aligned}\qquad \begin{aligned} \frac{7}{3}&=\frac{k}{6}\\ 3k&=42\\ k&=14\end{aligned}$$

b.
$$\begin{aligned} y&=\frac{2}{3}x-3\\ y&=kx-3\\ k&=\frac{2}{3}\end{aligned}$$

c.
$$\begin{aligned} x&=ky-1\\ x+1&=ky\\ y&=\frac{1}{k}x+\frac{1}{k}\end{aligned}\qquad \begin{aligned} y&=2x+2\\ y&=\frac{1}{k}x+\frac{1}{k}\end{aligned}\qquad \begin{aligned} \frac{1}{k}&=2\\ 1&=2k\\ k&=\frac{1}{2}\end{aligned}$$

57. The graph of the sum of the equations is a line whose graph contains the point of intersection of the graph of the system of equations.

a.

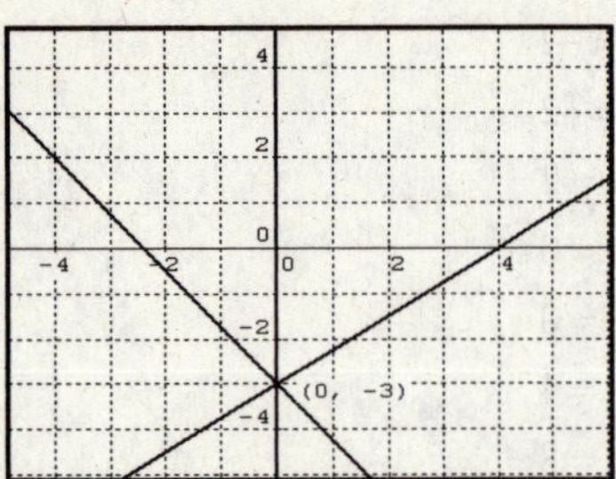

b.

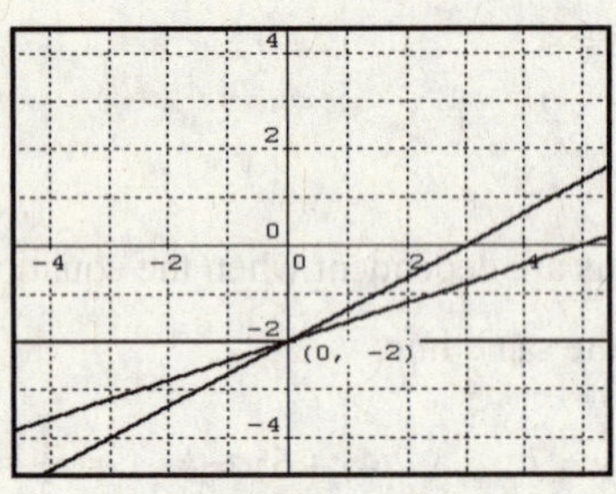

c.

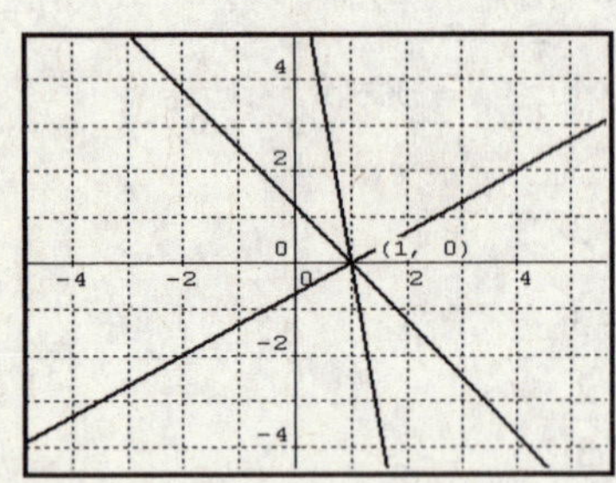

d.

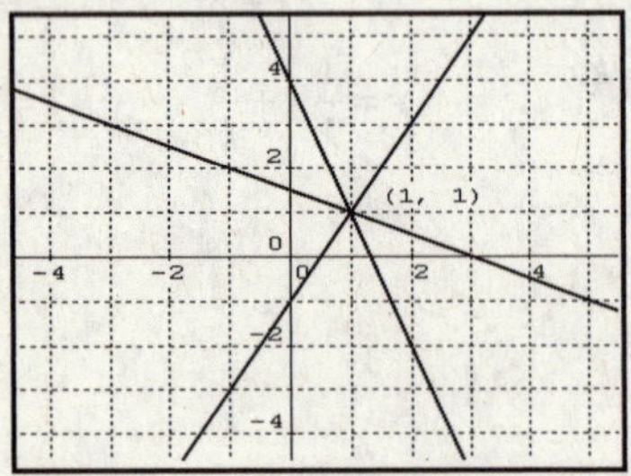

Section 6.4 Application Problems in Two Variables

1. True.

3. False. If b represents the rate of a boat in calm water and c represents that rate of the water current, then $b + c$ represents the rate of the boat while the boat is traveling with the current and $b - c$ represents the rate of the boat while the boat is traveling against the current.

5. True.

7.

	r	$\cdot$	t	$=$	d
With current	$b + c$	$\cdot$	2	$=$	$2(b + c)$
Against current	$b - c$	$\cdot$	3	$=$	$3(b - c)$

9. The plane travels a greater distance in the same time when flying with the wind. So m is **greater than** n.

11.

	r	·	t	=	d
With current	$w+c$	·	1.5	=	$1.5(w+c)=60$
Against current	$w-c$	·	2	=	$2(w-c)=60$

$$1.5(w+c)=60$$
$$\underline{2(w-c)=60}$$
$$w+c=40$$
$$\underline{w-c=30}$$
$$2w \quad =70$$
$$w=35$$
$$w+c=40$$
$$35+c=40$$
$$c=5$$

The speed of the whale is 35 mph and the speed of the current is 5mph.

13.

	r	·	t	=	d
With current	$b+c$	·	2	=	$2(b+c)=40$
Against current	$b-c$	·	2	=	$2(b-c)=16$

$$2(b+c)=40$$
$$\underline{2(b-c)=16}$$
$$b+c=20$$
$$\underline{b-c=\ 8}$$
$$2b \quad =28$$
$$b=14$$
$$b+c=20$$
$$14+c=20$$
$$c=6$$

The speed of the rowing team is 14 km/hr and the speed of the current is 6 km/hr.

15.

	r	·	t	=	d
With current	$b+c$	·	3.5	=	$3.5(b+c)=35$
Against current	$b-c$	·	3	=	$3(b-c)=12$

$$3.5(b+c)=35$$
$$\underline{3(b-c)=12}$$
$$b+c=10$$
$$\underline{b-c=4}$$
$$2b=14$$
$$b=7$$
$$b+c=10$$
$$7+c=10$$
$$c=3$$

The speed of the motorboat is7mph and the speed of the current is 3 mph.

17.

	r	·	t	=	d
With wind	$p+w$	·	2	=	$2(p+w)=1120$
Against wind	$p-w$	·	2	=	$2(p-w)=980$

$$2(p+w)=1120$$
$$\underline{2(p-w)=980}$$
$$p+w=560$$
$$\underline{p-w=490}$$
$$2p=1050$$
$$p=525$$
$$p-w=490$$
$$525-w=490$$
$$-w=-35$$
$$w=35$$

The speed of the plane is 525 mph and the speed of the wind is 35 mph.

19.

	r	$\cdot$	t	$=$	d
With wind	$p+w$	$\cdot$	2	$=$	$2(p+w)=240$
Against wind	$p-w$	$\cdot$	3	$=$	$3(p-w)=240$

$$
\begin{aligned}
2(p+w)&=240\\
3(p-w)&=240\\
p+w&=120\\
p-w&=80\\
2p&=200\\
p&=100\\
p+w&=120\\
100+w&=120\\
w&=20
\end{aligned}
$$

The speed of the seaplane was 100 mph and the speed of the wind was 20 mph.

21.

	r	$\cdot$	t	$=$	d
With wind	$h+w$	$\cdot$	$\frac{5}{3}$	$=$	$\frac{5}{3}(h+w)=450$
Against wind	$h-w$	$\cdot$	2.5	$=$	$2.5(h-w)=450$

$$
\begin{aligned}
\frac{5}{3}(h+w)&=450\\
2.5(h-w)&=450\\
h+w&=270\\
h-w&=180\\
2h&=450\\
h&=225\\
h+w&=270\\
225+w&=270\\
w&=45
\end{aligned}
$$

The speed of the helicopter is 225 mph and the speed of the wind is 45 mph.

23.

Your purchase	Amount	·	Unit cost	=	Value
Adult's ticket	4	·	A	=	4A
Child's ticket	2	·	C	=	2C

Neighbor's purchase	Amount	·	Unit cost	=	Value
Adult's ticket	2	·	A	=	2A
Child's ticket	3	·	C	=	3C

25. Since the mixture with more cinnamon tea costs less, the cost per pound of the cinnamon tea must be **less than** the cost per pound of the spice tea.

27.

First shipment	Amount	·	Unit Cost	=	Value
Antivirus program	12	·	A	=	12A
Design program	10	·	D	=	10D

Second shipment	Amount	·	Unit Cost	=	Value
Antivirus program	5	·	A	=	5A
Design program	8	·	D	=	8D

$$4(12A+10D)=4(1780)$$
$$-5(5A+8D)=-5(1125)$$
$$48A+40D=7120$$
$$-25A-40D=-5625$$
$$23A=1495$$
$$A=65$$

The antivirus program cost \$65.

29.

First purchase	Amount	·	Unit Cost	=	Value
Wheat flour	12	·	W	=	12W
Rye flour	15	·	R	=	15R

Second purchase	Amount	·	Unit Cost	=	Value
Wheat flour	15	·	W	=	15W
Rye flour	10	·	R	=	10R

$$-2(12W+15R)=-2(39.87)$$
$$3(15W+10R)=3(33.30)$$
$$-24W-30R=-79.74$$
$$45W+30R=99.90$$
$$21W=20.16$$
$$W=0.96$$
$$15W+10R=33.30$$
$$15(0.96)+10R=33.30$$
$$14.40+10R=33.30$$
$$10R=18.90$$
$$R=1.89$$

A pound of wheat flour costs \$0.96 and a pound of rye flour costs \$1.89.

31.

	Base	·	Percentage	=	Amount
20% solution	x	·	0.20	=	$0.2x$
40% solution	y		0.40	=	$0.4y$
Total	50	·	0.34	=	0.34(50)

$$x+y=50$$
$$0.2x+0.4y=0.34(50)$$
$$x+y=50$$
$$2x+4y=170$$
$$-2x-2y=-100$$
$$2x+4y=170$$
$$2y=70$$
$$y=35$$
$$x+35=50$$
$$x=15$$

The mixture has 15 ounces of the 20% solution and 35 ounces of the 40% solution.

33. Let t = the original number of two point baskets and let b = the original number of three point baskets.

$$-3(2t+3b)=-3(87)$$
$$2(3t+2b)=2(93)$$
$$-6t-9b=-261$$
$$6t+4b=186$$
$$-5b=-75$$
$$b=15$$
$$2t+3b=87$$
$$2t+45=87$$
$$2t=42$$
$$t=21$$

The team scored 21 two-point baskets and 15 three-point baskets.

35.

$$W = 52 + 1.9x$$
$$W = 56.2 + 1.41x$$
$$52 + 1.9x = 56.2 + 1.41x$$
$$0.49x = 4.2$$
$$x \approx 9$$

Since x is the height above 60 inches, the height is 69 inches.

37.

	Amount	·	Unit rate	=	Total
City miles	g	·	18	=	$18g$
Highway miles	h	·	25	=	$25h$
	10				208

$$g + h = 10$$
$$18g + 25h = 208$$

$$-18g - 18h = -180$$
$$18g + 25h = 208$$
$$7h = 28$$
$$h = 4$$
$$g + h = 10$$
$$g + 4 = 10$$
$$g = 6$$
$$18g = 18 \cdot 6 = 108$$
$$208 - 108 = 100$$

The car was driven 108 miles in the city and 100 miles on the highway.

39.

	Principle	·	Rate	=	Interest
Amount at 5%	x	·	0.05	=	$0.05x$
Amount at 4.5%	y		0.045		$0.045y$
Total					240

	Principle	·	Rate	=	Interest
Amount at 4.5%	x	·	0.045	=	$0.045x$
Amount at 5%	y		0.05		$0.05y$
Total					235

$$1000(0.05x + 0.045y) = 1000(240)$$
$$1000(0.045x + 0.05y) = 1000(235)$$

$$-10(50x + 45y) = -10(240{,}000)$$
$$9(45x + 50y) = 9(235{,}000)$$

$$-500x - 450y = -2{,}400{,}000$$
$$405x + 450y = 2{,}115{,}000$$

$$-95x = -285{,}000$$
$$x = 3000$$

\$3000 was invested at 5%.

41. Let x = the smaller angle.

Then y = the larger angle.

$$x + y = 90$$
$$y = 4x$$
$$x + 4x = 90$$
$$5x = 90$$
$$x = 18$$
$$4x = 72$$

The angles measure 18° and 72°.

43. Let W = the width and L = the length.

$$2L+2W=68$$
$$L=3W-2$$
$$2(3W-2)+2W=68$$
$$6W-4+2W=68$$
$$8W=72$$
$$W=9$$
$$L=3W-2$$
$$L=25$$

The length is 25 inches.

Chapter 6 Review Exercises

1. $4x+7y=3$

$$x=y-2$$

$$4(y-2)+7y=3$$
$$4y-8+7y=3$$
$$11y-8=3$$
$$11y=11$$
$$y=1$$
$$x=y-2$$
$$x=1-2$$
$$x=-1$$

The solution is (-1, 1).

2. $3x+y=3$

$$x=2$$

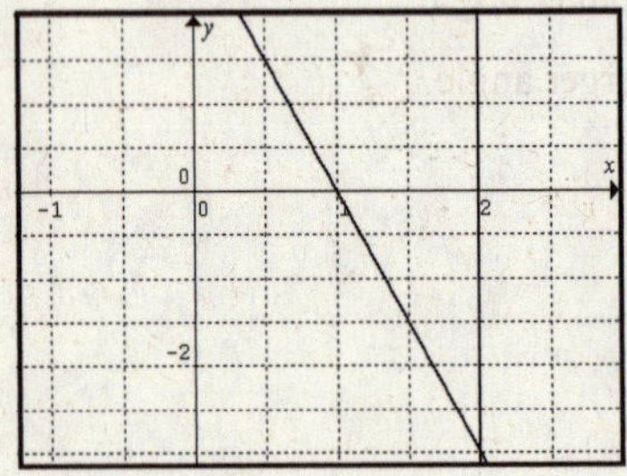

The solution is (2, -3).

3. $3x+8y=\ \ -1$

$$\underline{4(x-2y)=4(-5)}$$

$$3x+8y=-1$$
$$\underline{4x-8y=-20}$$
$$7x\quad=-21$$
$$x=-3$$
$$x-2y=-5$$
$$-3-2y=-5$$
$$-2y=-2$$
$$y=1$$

The solution is (-3, 1).

4. $8x-y=2$

$$y=5x+1$$
$$8x-(5x+1)=2$$
$$8x-5x-1=2$$
$$3x=3$$
$$x=1$$
$$y=5x+1$$
$$y=6$$

The solution is (1, 6).

5.

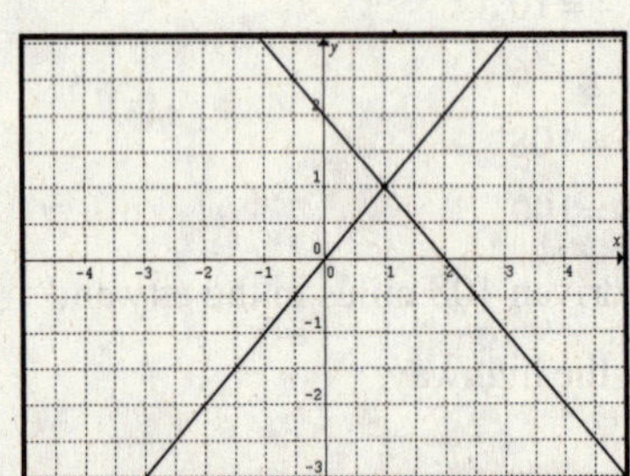

The solution is (1, 1).

6. The solution is (1,-5).

$$3(4x-y)=3(9)$$
$$\underline{2x+3y=-13}$$
$$12x-3y=27$$
$$\underline{2x+3y=-13}$$
$$14x=14$$
$$x=1$$
$$4x-y=9$$
$$4-y=9$$
$$y=-5$$

7.
$$-2x+y=-4$$
$$x=y+1$$
$$-2(y+1)+y=-4$$
$$-2y-2+y=-4$$
$$-y=-2$$
$$y=2$$
$$x=y+1$$
$$x=2+1$$
$$x=3$$

The solution is (3, 2).

8.
$$-4(8x-y)=-4(25)$$
$$\underline{32x-4y=100}$$
$$-32x+4y=-100$$
$$\underline{32x-4y=100}$$
$$0=0$$

The system is dependent.

9.
$$2(5x-15y)=2(30)$$
$$\underline{5(2x+6y)=5(0)}$$
$$10x-30y=60$$
$$\underline{10x+30y=0}$$
$$20x=60$$
$$x=3$$
$$2x+6y=0$$
$$2(3)+6y=0$$
$$6y=-6$$
$$y=-1$$

The solution is (3, -1).

10.

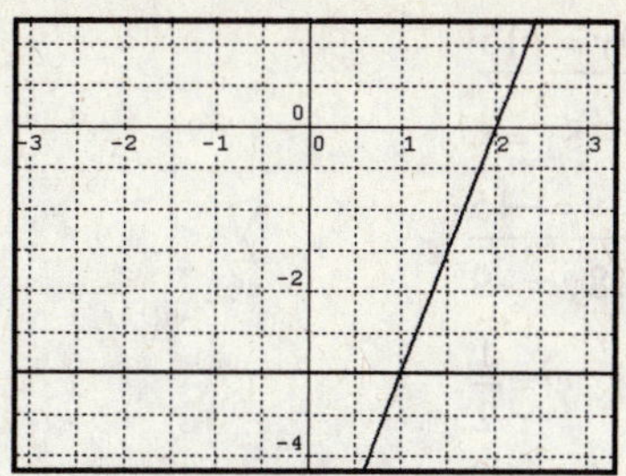

The solution is (1, -3).

11. The solution is (-2, -7).

$$7x-2y=0$$
$$\underline{2(2x+y)=2(-11)}$$
$$7x-2y=0$$
$$\underline{4x+2y=-22}$$
$$11x=-22$$
$$x=-2$$
$$7x-2y=0$$
$$7(-2)-2y=0$$
$$-2y=14$$
$$y=-7$$

12.
$$x-5y=4$$
$$y=x-4$$
$$x-5(x-4)=4$$
$$x-5x+20=4$$
$$-4x=-16$$
$$x=4$$
$$y=x-4$$
$$y=4-4$$
$$y=0$$

The solution is (4, 0).

13.
$$5x+4y=-17 \qquad 2x-y=1$$
$$5(-1)+4(-3)\overset{?}{=}-17 \qquad 2(-1)-(-3)\overset{?}{=}1$$
$$-5+(-12)\overset{?}{=}-17 \qquad -2+3\overset{?}{=}1$$
$$-17=-17 \qquad 1=1$$

(-1, -3) is a solution.

14. $-2(6x+4y)=-2(-3)$

$$12x-10y=-15$$

$$-12x-8y=6$$
$$12x-10y=-15$$
$$-18y=-9$$
$$y=\frac{1}{2}$$
$$6x+4y=-3$$
$$6x+4\left(\frac{1}{2}\right)=-3$$
$$6x+2=-3$$
$$6x=-5$$
$$x=-\frac{5}{6}$$

The solution is $\left(-\frac{5}{6},\frac{1}{2}\right)$.

15. $7(5x+2y)=7(-9)$

$$2(12x-7y)=7(2)$$

$$35x+14y=-63$$
$$24x-14y=14$$
$$49x=-49$$
$$x=-1$$
$$5x+2y=-9$$
$$-5+2y=-9$$
$$2y=-4$$
$$y=-2$$

The solution is (-1, -2).

16.

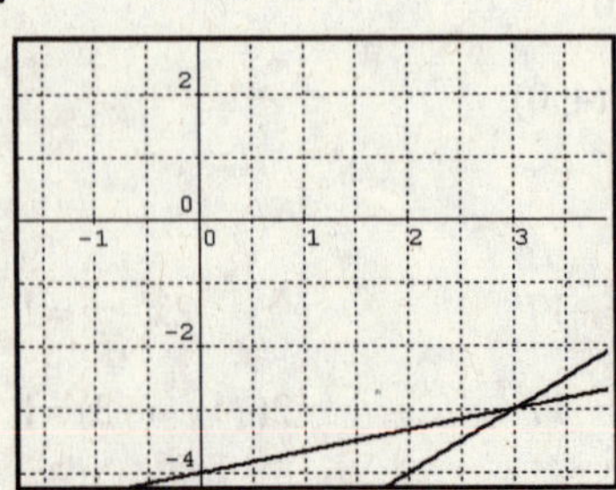

The solution is (3, -3).

17. $-4(5x+7y)=-4(21)$

$$20x+28y=63$$

$$-20x-28y=-84$$
$$20x+28y=63$$
$$0=-21$$

The system is inconsistent.

18. $9x+12y=-1$

$$x-4y=-1$$
$$x=4y-1$$

$$9(4y-1)+12y=-1$$
$$36y-9+12y=-1$$
$$48y-9=-1$$
$$48y=8$$
$$y=\frac{8}{48}$$
$$y=\frac{1}{6}$$
$$9x+12y=-1$$
$$9x+12\left(\frac{1}{6}\right)=-1$$
$$9x+2=-1$$
$$9x=-3$$
$$x=-\frac{3}{9}$$
$$x=-\frac{1}{3}$$

The solution is $\left(-\frac{1}{3},\frac{1}{6}\right)$.

19.

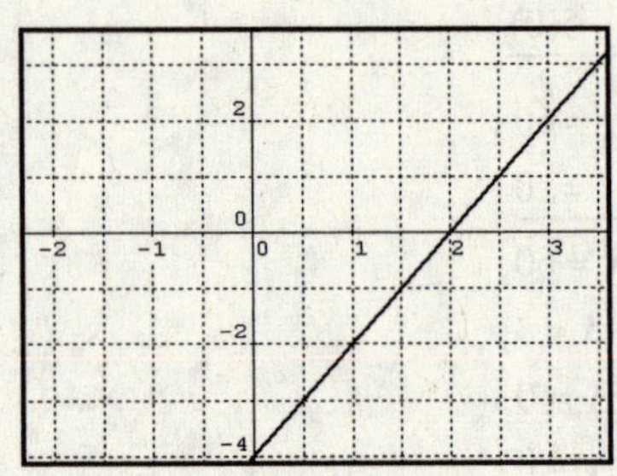

The system is dependent.

20. $3(3x+y)=3(-2)$

$$\begin{aligned} -9x-3y&=6 \\ \hline 9x+3y&=-6 \\ -9x-3y&=6 \\ \hline 0&=0 \end{aligned}$$

The system is dependent.

21. $-2(11x-2y)=-2(4)$

$$\begin{aligned} 25x-4y&=\ \ 2 \\ \hline -22x+4y&=-8 \\ 25x-4y&=2 \\ \hline 3x&=-6 \\ x&=-2 \\ 11x-2y&=4 \\ 11(-2)-2y&=4 \\ -22-2y&=4 \\ -2y&=26 \\ y&=-13 \end{aligned}$$

The solution is (-2, -13).

22.

$$\begin{aligned} 4x+3y&=12 \\ y&=-\frac{4}{3}x+4 \\ 4x+3\left(-\frac{4}{3}x+4\right)&=12 \\ 4x-4x+12&=12 \\ 12&=12 \end{aligned}$$

The system is dependent.

23.

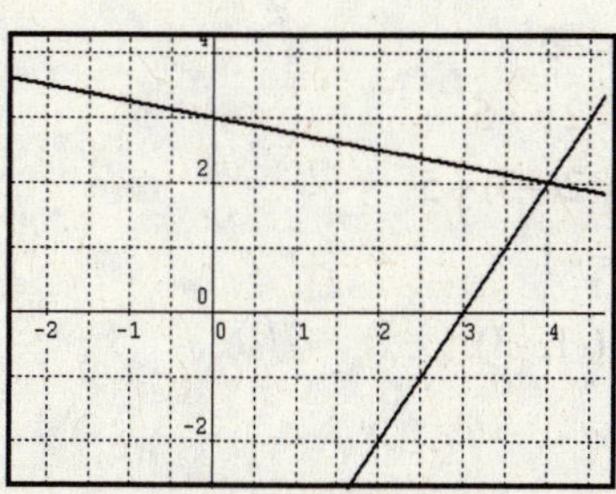

The solution is (4,2).

24. $-5(2x-y)=-5(5)$

$$\begin{aligned} 10x-5y&=20 \\ \hline -10x+5y&=-25 \\ 10x-5y&=20 \\ \hline 0&=-5 \end{aligned}$$

The system is inconsistent.

25. $6x+5y=-2$

$$\begin{aligned} y&=2x-2 \\ 6x+5(2x-2)&=-2 \\ 6x+10x-10&=-2 \\ 16x-10&=-2 \\ 16x&=8 \\ x&=\frac{1}{2} \\ y&=2x-2 \\ y&=2\left(\frac{1}{2}\right)-2 \\ y&=1-2 \\ y&=-1 \end{aligned}$$

The solution is $\left(\frac{1}{2},-1\right)$.

26.

$$\begin{aligned} -x+9y&=2 & 6x-4y&=12 \\ -(-2)+9(0)&\overset{?}{=}2 & 6(-2)-4(0)&\overset{?}{=}12 \\ 2+0&\overset{?}{=}2 & -12-0&\overset{?}{=}12 \\ 2&=2 & -12&\neq 12 \end{aligned}$$

The ordered pair is not a solution.

27.
$$3(6x-18y)=3(7)$$
$$-2(9x+24y)=-2(2)$$
$$18x-54y=21$$
$$-18x-48y=-4$$
$$-102y=17$$
$$y=-\frac{17}{102}=-\frac{1}{6}$$
$$6x-18y=7$$
$$6x-18\left(-\frac{1}{6}\right)=7$$
$$6x+3=7$$
$$6x=4$$
$$x=\frac{4}{6}=\frac{2}{3}$$

The solution is $\left(\frac{2}{3},-\frac{1}{6}\right)$.

28.
$$12x-9y=18$$
$$y=\frac{4}{3}x-3$$
$$12x-9\left(\frac{4}{3}x-3\right)=18$$
$$12x-12x+27=18$$
$$27=18$$

The system is inconsistent.

29.
$$9x-y=-3$$
$$18x-y=0$$
$$18x=y$$
$$9x-18x=-3$$
$$-9x=-3$$
$$x=\frac{1}{3}$$
$$18x-y=0$$
$$18\left(\frac{1}{3}\right)-y=0$$
$$6-y=0$$
$$y=6$$

The solution is $\left(\frac{1}{3},6\right)$.

30.
$$x+2y=3$$
$$y=-\frac{1}{2}x+1$$

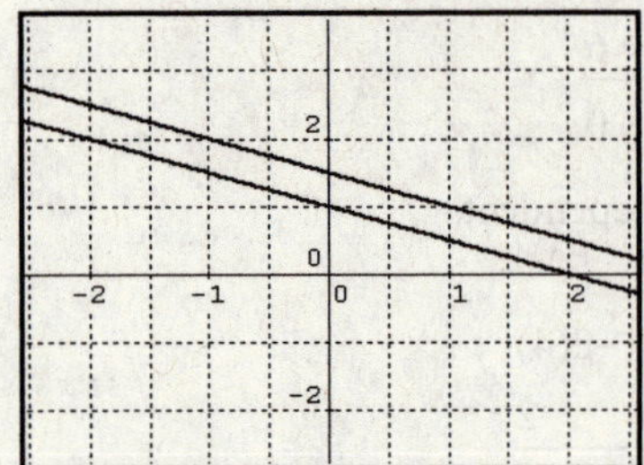

The system is inconsistent.

31.
$$7x-9y=9$$
$$-9(3x-y)=-9(1)$$
$$7x-9y=9$$
$$-27x+9y=-9$$
$$-20x=0$$
$$x=0$$
$$3x-y=1$$
$$3(0)-y=1$$
$$-y=1$$
$$y=-1$$

The solution is (0,-1).

32.
$$7x+3y=-16$$
$$x-2y=5$$
$$x=2y+5$$
$$7(2y+5)+3y=-16$$
$$14y+35+3y=-16$$
$$17y=-51$$
$$y=-3$$
$$x=2y+5$$
$$x=2(-3)+5$$
$$x=-1$$

The solution is (-1, -3).

33. $5x-3y=6$

$x-y=2$

$x=y+2$

$5(y+2)-y=2$

$5y+10-y=2$

$4y=-8$

$y=-2$

$x-y=2$

$x-(-2)=2$

$x+2=2$

$x=0$

The solution is (0, -2).

34. $-2(6x+y)=-2(12)$

$9x+2y=18$

$-12x-2y=-24$

$9x+2y=18$

$-3x=-6$

$x=2$

$6x+y=12$

$6(2)+y=12$

$12+y=12$

$y=0$

The solution is (2, 0).

35. $5x+12y=4$

$-2(x+6y)=-2(8)$

$5x+12y=4$

$-2x-12y=-16$

$3x=-12$

$x=-4$

$x+6y=8$

$-4+6y=8$

$6y=12$

$y=2$

The solution is (-4, 2).

36. $6x-y=0$

$7x-y=1$

$y=6x$

$7x-6x=1$

$x=1$

$6(1)-y=0$

$6-y=0$

$y=6$

The solution is (1, 6).

37.

	r	·	t	=	d
With wind	$p+w$	·	4	=	$4(p+w)=$ 800
Against wind	p - w	·	5	=	$5(p-w)=$ 800

$4(p+w)=800$

$5(p-w)=800$

$p+w=200$

$p-w=160$

$2p=360$

$p=180$

$p+w=200$

$p+180=200$

$p=20$

The speed of the plane is 180 mph and the speed of the wind is 20 mph.

38.

	Amount	·	Unit cost	=	Value
Adult's ticket	x	·	11	=	$11x$
Child's ticket	200 - x	·	8	=	$8(200-x)$

$$11x+8(200-x)=1780$$
$$11x+1600-8x=1780$$
$$3x=180$$
$$x=60$$
$$200-x=140$$

140 children's tickets and 60 adult tickets where sold.

39.

	r	·	t	=	d
With current	$b+c$	·	3	=	$3(b+c)=30$
Against current	b - c	·	5	=	$5(b-c)=30$

$$3(b+c)=30$$
$$\underline{5(b-c)=30}$$
$$b+c=10$$
$$\underline{b-c=6}$$
$$2b=16$$
$$b=8$$
$$b+c=10$$
$$8+c=10$$
$$c=2$$

The speed of the canoe is 8 mph and the speed of

the

current is 2 mph.

40.

	Amount	·	Unit Cost	=	Value
46 cent mailings	x	·	0.46	=	$0.46x$
64 cent mailings	y	·	0.64	=	$0.64y$

$$x+y=190$$
$$\underline{0.46x+0.64y=98.2}$$
$$-64(x+y=190)$$
$$100(0.46x+0.64y=98.2)$$
$$-64x-64y=-12160$$
$$\underline{46x+64y=9820}$$
$$-18x=-2340$$
$$x=130$$

There were 130 mailings that cost 46 cents.

41.

	r	·	t	=	d
With current	$b+c$	·	3	=	$3(b+c)=48$
Against current	b - c	·	2	=	$2(b-c)=24$

$$3(b+c)=48$$
$$\underline{2(b-c)=24}$$
$$b+c=16$$
$$\underline{b-c=12}$$
$$2b=28$$
$$b=14$$
$$b+c=16$$
$$14+c=16$$
$$c=2$$

The speed of the boat is 14 km/hr and the speed of the current is 2 km/hr.

42.

	Amount	·	Unit Cost	=	Value
$15 disks	x	·	15	=	$15x$
$10 disks	10 - x	·	10	=	$10(10-x)$

$$15x+10(10-x)=120$$
$$15x+100-10x=120$$
$$5x=20$$
$$x=4$$
$$10-x=6$$

There were 4 disks at $15 and 6 disks at $10.

43.

	r	·	t	=	d
With wind	$p+w$	·	3	=	$3(p+w)=420$
Against wind	p - w	·	4	=	$3(p-w)=440$

$$3(p+w)=420$$
$$\underline{4(p-w)=440}$$
$$p+w=140$$
$$\underline{p-w=110}$$
$$2p=250$$
$$p=125$$
$$p+w=140$$
$$125+w=140$$
$$w=15$$

The speed of the plane is 125 km/hr and the speed of the wind is 15 km/hr.

44.

	r	·	t	=	d
With current	$b+$ c	·	1	=	$1(b+c)=4$
Against current	b - c	·	1	=	$1(b-c)=2$

$$b+c=4$$
$$\underline{b-c=2}$$
$$2b=6$$
$$b=3$$
$$b+c=4$$
$$3+c=4$$
$$c=1$$

The speed of the boat is 3mph and the speed of the current is 1 mph.

45. Let L = bushels of lentils and C = bushels of corn.

$$L+C+50=2C+C\Rightarrow L-2C=\ -50$$
$$L+C+150=L+L\Rightarrow \underline{-L+C=-150}$$
$$-C=-200$$
$$C=200$$
$$L-2C=-50$$
$$L-400=-50$$
$$L=350$$

The silo contains 200 bushels of corn and 350 bushels of lentils.

46.

	r	·	t	=	d
With wind	$p+w$	·	3	=	$3(p+w)=360$
Against wind	$p-w$	·	4	=	$4(p-w)=360$

$$3(p+w)=360$$
$$4(p-w)=360$$
$$p+w=120$$
$$p-w=90$$
$$2p=210$$
$$p=105$$
$$p+w=120$$
$$105+w=120$$
$$w=15$$

The speed of the plane is 105 mph and the speed of the wind is 15 mph.

47.

	Amount	·	Unit cost	=	Value
\$6 shares	x	·	6	=	$6x$
\$25 share	$1500-x$	·	25	=	$25(1500-x)$

$$6x+25(1500-x)=12{,}800$$
$$6x+37{,}500-25x=12{,}800$$
$$-19x=-24{,}700$$
$$x=1300$$
$$1500-x=200$$

He bought 1300 shares of \$6 stock and 200 shares of \$25 stocks.

48.

	r	·	t	=	d
With current	$b+c$	·	2	=	$2(b+c)=24$
Against current	$b-c$	·	3	=	$2(b-c)=18$

$$2(b+c)=24$$
$$3(b-c)=18$$
$$b+c=12$$
$$b-c=6$$
$$2b=18$$
$$b=9$$
$$b+c=12$$
$$9+c=12$$
$$c=3$$

The speed of the sculling team is 9 mph and the speed of the current was 3 mph.

Chapter 6 Test

1. $4x-y=11$

$y=2x-5$

$$4x-(2x-5)=11$$
$$4x-2x+5=11$$
$$2x=6$$
$$x=3$$
$$y=2x-5$$
$$y=2(3)-5$$
$$y=6-5$$
$$y=1$$

The solution is (3, 1).

2. $4x+3y=11$

$\underline{5x-3y=7}$

$9x=18$

$x=2$

$4x+3y=11$

$4(2)+3y=11$

$8+3y=11$

$3y=3$

$y=1$

The solution is (2, 1).

3. $2x+5y=11$ $\quad x+3y=7$

$2(-2)+5(3)\stackrel{?}{=}11$ $\quad -2+3(3)\stackrel{?}{=}7$

$-4+15\stackrel{?}{=}11$ $\quad -2+9\stackrel{?}{=}7$

$11=11$ $\quad 7=7$

(-2, 3) is a solution.

4. $x=2y+3$

$3x-2y=5$

$3(2y+3)-2y=5$

$6y+9-2y=5$

$4y=-4$

$y=-1$

$x=2y+3$

$x=2(-1)+3$

$x=-2+3$

$x=1$

The solution is (1, -1).

5. $-2(2x-5y)=-2(6)$

$\underline{4x+3y=-1}$

$-4x+10y=-12$

$\underline{4x+3y=-1}$

$13y=-13$

$y=-1$

$2x-5y=6$

$2x-5(-1)=6$

$2x+5=6$

$2x=1$

$x=\frac{1}{2}$

The solution is $\left(\frac{1}{2},-1\right)$.

6.

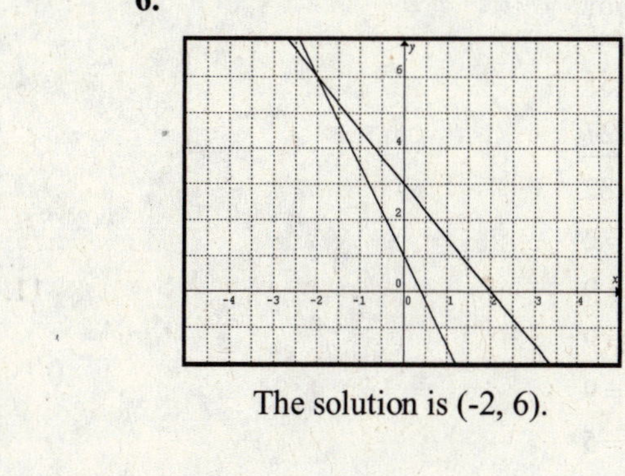

The solution is (-2, 6).

7. $4x+2y=3$

$y=-2x+1$

$4x+2(-2x+1)=3$

$4x-4x+2=3$

$2=3$

The system is inconsistent.

8.
$$3x+5y=1$$
$$2x-y=5$$
$$y=2x-5$$
$$3x+5(2x-5)=1$$
$$3x+10x-25=1$$
$$13x=26$$
$$x=2$$
$$2x-y=5$$
$$2(2)-y=5$$
$$4-y=5$$
$$-y=1$$
$$y=-1$$

The solution is (2, -1).

9.
$$5(7x+3y)=5(11)$$
$$3(2x-5y)=3(9)$$
$$35x+15y=55$$
$$6x-15y=27$$
$$41x=82$$
$$x=2$$
$$2x-5y=9$$
$$2(2)-5y=9$$
$$4-5y=9$$
$$-5y=5$$
$$y=-1$$

The solution is (2, -1).

10.
$$3x-5y=13$$
$$x+3y=1$$
$$x=1-3y$$
$$3(1-3y)-5y=13$$
$$3-9y-5y=13$$
$$3-14y=13$$
$$-14y=10$$
$$y=-\frac{10}{14}$$
$$y=-\frac{5}{7}$$
$$x=1-3y$$
$$x=1-3\left(-\frac{5}{7}\right)$$
$$x=\frac{7}{7}+\frac{15}{7}$$
$$x=\frac{22}{7}$$

The solution is $\left(\frac{22}{7},-\frac{5}{7}\right)$.

11.
$$2(5x+6y)=2(-7)$$
$$-3(3x+4y)=-3(-5)$$
$$10x+12y=-14$$
$$-9x-12y=15$$
$$x=1$$
$$5x+6y=-7$$
$$5(1)+6y=-7$$
$$5+6y=-7$$
$$6y=-12$$
$$y=-2$$

The solution is (1, -2).

12.
$$3x-2y=8$$
$$3(2)-2(1)\stackrel{?}{=}8$$
$$6-2\stackrel{?}{=}8$$
$$4\neq 8$$

The point is not a solution.

13.
$$3x - y = 5$$
$$y = 2x - 3$$
$$3x - (2x - 3) = 5$$
$$3x - 2x + 3 = 5$$
$$x = 2$$
$$y = 2x - 3$$
$$y = 2(2) - 3$$
$$y = 4 - 3$$
$$y = 1$$

The solution is (2, 1).

14.

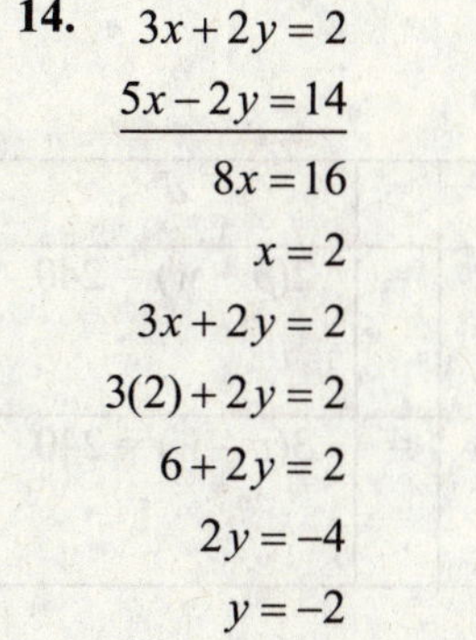

$$\begin{array}{r} 3x + 2y = 2 \\ \underline{5x - 2y = 14} \\ 8x = 16 \end{array}$$
$$x = 2$$
$$3x + 2y = 2$$
$$3(2) + 2y = 2$$
$$6 + 2y = 2$$
$$2y = -4$$
$$y = -2$$

The solution is (2, -2).

15.

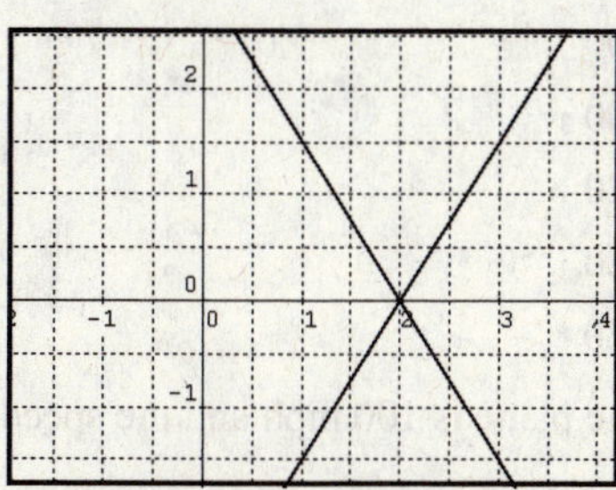

The solution is (2, 0).

16.
$$x = 3y + 1$$
$$2x + 5y = 13$$
$$2(3y + 1) + 5y = 13$$
$$6y + 2 + 5y = 13$$
$$11y + 2 = 13$$
$$11y = 11$$
$$y = 1$$
$$x = 3y + 1$$
$$x = 3(1) + 1$$
$$x = 4$$

The solution is (4, 1).

17.
$$\begin{array}{r} 5x + 4y = 7 \\ \underline{2(3x - 2y) = 2(13)} \end{array}$$
$$\begin{array}{r} 5x + 4y = 7 \\ \underline{6x - 4y = 26} \\ 11x = 33 \end{array}$$
$$x = 3$$
$$5x + 4y = 7$$
$$5(3) + 4y = 7$$
$$15 + 4y = 7$$
$$4y = -8$$
$$y = -2$$

The solution is (3, -2).

18.

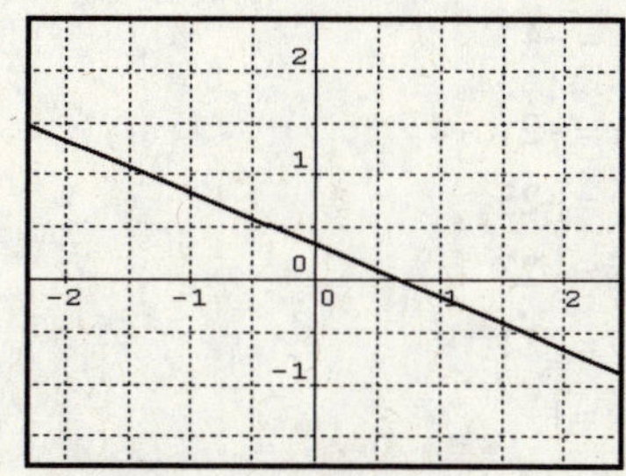

The system is dependent.

19.
$$4x-3y=1$$
$$2x+y=3$$

$$y=3-2x$$
$$4x-3(3-2x)=1$$
$$4x-9+6x=1$$
$$10x=10$$
$$x=1$$
$$y=3-2x$$
$$y=3-2(1)$$
$$y=1$$

The solution is (1, 1).

20.
$$7(5x-3y)=7(29)$$
$$3(4x+7y)=3(-5)$$

$$35x-21y=203$$
$$12x+21y=-15$$
$$47x=188$$
$$x=4$$
$$4x+7y=-5$$
$$4(4)+7y=-5$$
$$16+7y=-5$$
$$7y=-21$$
$$y=-3$$

The solution is (4, -3).

21.
$$3x-5y=-23$$
$$x+2y=-4$$

$$x=-2y-4$$
$$3(-2y-4)-5y=-23$$
$$-6y-12-5y=-23$$
$$-11y=-11$$
$$y=1$$
$$x=-4-2y$$
$$x=-4-2(1)$$
$$x=-6$$

The solution is (-6, 1).

22.
$$3(9x-2y)=3(17)$$
$$2(5x+3y)=2(-7)$$

$$27x-6y=51$$
$$10x+6y=-14$$
$$37x=37$$
$$x=1$$
$$5x+3y=-7$$
$$5(1)+3y=-7$$
$$3y=-12$$
$$y=-4$$

The solution is (1, -4).

23.

	r	$\cdot$	t	=	d
With wind	$p+w$	$\cdot$	2	=	$2(p+w)=240$
Against wind	$p-w$	$\cdot$	3	=	$3(p-w)=240$

$$2(p+w)=240$$
$$3(p-w)=240$$
$$p+w=120$$
$$p-w=80$$
$$2p=200$$
$$p=100$$
$$p+w=120$$
$$100+w=120$$
$$w=20$$

The speed of the plane is 100 mph and the speed of the wind is 20 mph.

24.

	r	·	t	=	d
With current	$b+c$	·	3	=	$3(b+c)=48$
Against current	$b-c$	·	4	=	$4(b-c)=48$

$$3(b+c)=48$$
$$\underline{4(b-c)=48}$$
$$b+c=16$$
$$\underline{b-c=12}$$
$$2b=28$$
$$b=14$$
$$b+c=16$$
$$14+c=16$$
$$c=2$$

The speed of the boat is 14 mph and the speed of the current is 2 mph.

Cumulative Review Exercises

1. Elements of A less than or equal to -4: -8, -4

2. Positive integers less than or equal to 10:
{1, 2, 3, 4, 5, 6, 7, 8, 9, 10}

3. $12-2(7-5)^2 \div 4 = 12-2(2)^2 \div 4$
$= 12-2(4) \div 4$
$= 12-8 \div 4$
$= 12-2$
$= 10$

4. $2[5a-3(2-5a)-8] = 2[5a-6+15a-8]$
$= 2[20a-14]$
$= 40a-28$

5. $\dfrac{a^2-b^2}{2a} = \dfrac{4^2-(-2)^2}{2(4)}$
$= \dfrac{16-4}{8}$
$= \dfrac{12}{8}$
$= \dfrac{3}{2}$

6. $-\dfrac{3}{4}x = \dfrac{9}{8}$
$8\left(-\dfrac{3}{4}x\right) = 8\left(\dfrac{9}{8}\right)$
$-6x = 9$
$x = -\dfrac{9}{6}$
$x = -\dfrac{3}{2}$

7. $4-3(2-3x) = 7x-9$
$4-6+9x = 7x-9$
$9x-2 = 7x-9$
$2x = -7$
$x = -\dfrac{7}{2}$

8. $3[2-4(x+1)] = 6x-2$
$3[2-4x-4] = 6x-2$
$3[-4x-2] = 6x-2$
$-12x-6 = 6x-2$
$-18x = 4$
$x = -\dfrac{4}{18}$
$x = -\dfrac{2}{9}$

9. $-7x-5 > 4x+50$
$-11x > 55$
$\dfrac{-11x}{-11} < \dfrac{55}{-11}$
$x < -5$

10.
$$\begin{aligned}5+2(x+1)&\le 13\\ 5+2x+2&\le 13\\ 2x+7&\le 13\\ 2x&\le 6\\ x&\le 3\end{aligned}$$

11.
$$\begin{aligned}A&=PB\\ 12&=P(50)\\ \frac{12}{50}&=P\\ 0.24&=P\\ 24\%&=P\end{aligned}$$

12.
$$\begin{aligned}3x-6y&=12\\ 3x-6(0)&=12\\ 3x&=12\\ x&=4\end{aligned}\qquad\begin{aligned}3x-6y&=12\\ 3(0)-6y&=12\\ -6y&=12\\ y&=-2\end{aligned}$$

x-intercept: (4, 0)

y-intercept: (0, -2)

13. $m=\dfrac{y_2-y_1}{x_2-x_1}=\dfrac{4-(-3)}{-3-2}$

$=\dfrac{4+3}{-5}=-\dfrac{7}{5}$

14.
$$\begin{aligned}y-y_1&=m(x-x_1)\\ y-3&=-\frac{3}{2}[x-(-2)]\\ y-3&=-\frac{3}{2}(x+2)\\ y-3&=-\frac{3}{2}x-3\\ y&=-\frac{3}{2}x\end{aligned}$$

15. $3x-2y=6$

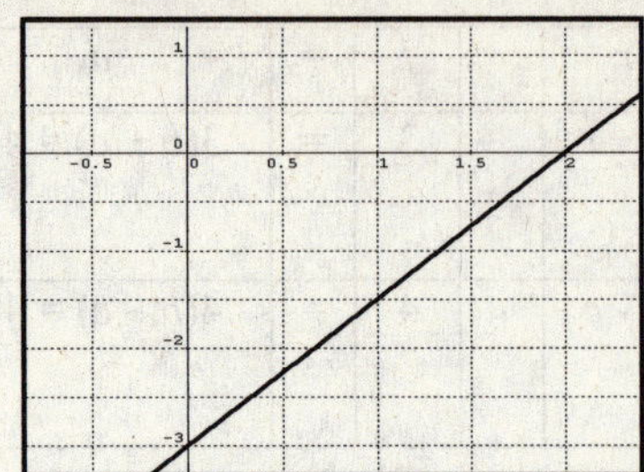

16. $y=-\dfrac{1}{3}x+3$

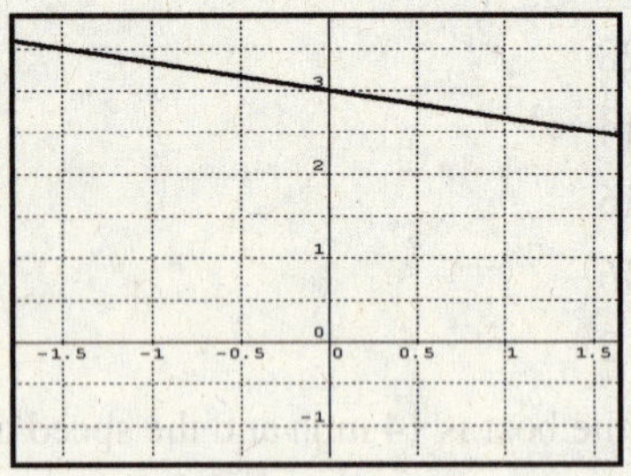

17. $y>3x+4$

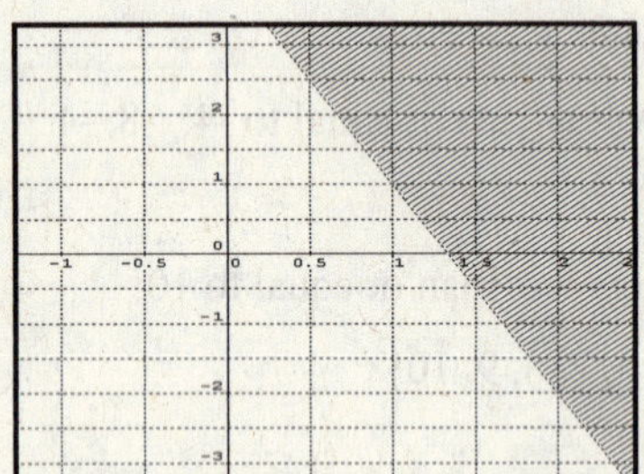

18. $f(x)=\dfrac{3}{4}x+2$

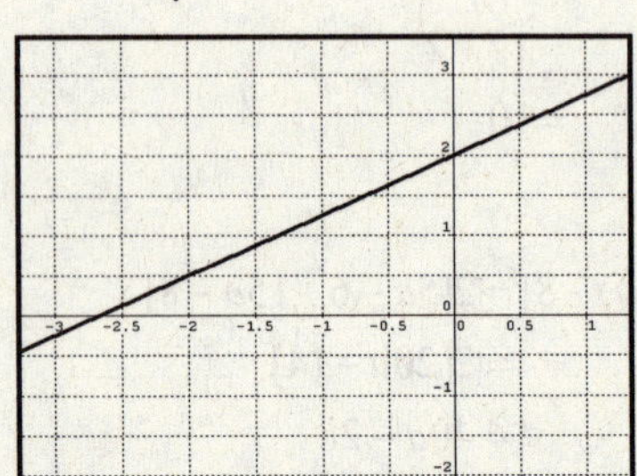

19. Domain: {-5, 0, 1, 5}; Range: {5}; Function

20. $f(x) = -2x - 5$

$$f(-4) = -2(-4) - 5$$
$$= 8 - 5$$
$$= 3$$

21. $5x - 3y = 10 \qquad 4x + 7y = 8$

$$5(2) - 3(0) \stackrel{?}{=} 10 \qquad 4(2) + 7(0) \stackrel{?}{=} 8$$
$$10 - 0 \stackrel{?}{=} 10 \qquad 8 + 0 \stackrel{?}{=} 8$$
$$10 = 10 \qquad 8 = 8$$

The point is a solution.

22. $2x - 3y = -7$

$$x + 4y = 2$$
$$x = 2 - 4y$$
$$2(2 - 4y) - 3y = -7$$
$$4 - 8y - 3y = -7$$
$$4 - 11y = -7$$
$$-11y = -11$$
$$y = 1$$
$$x = 2 - 4y$$
$$x = 2 - 4(1)$$
$$x = -2$$

The solution is (-2, 1).

23. The solution is (0, 2).

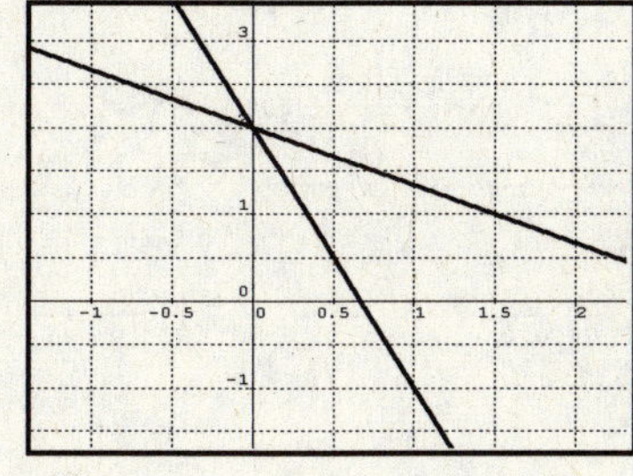

24. $3(5x - 2y) = 3(8)$

$$\underline{2(4x + 3y) = 2(11)}$$
$$15x - 6y = 24$$
$$\underline{8x + 6y = 22}$$
$$23x = 46$$
$$x = 2$$
$$5x - 2y = 8$$
$$5(2) - 2y = 8$$
$$10 - 2y = 8$$
$$-2y = -2$$
$$y = 1$$

The solution is (2, 1).

25. $f(x) = 5x - 2$

$$f(-4) = -20 - 2 = -22$$
$$f(-2) = -10 - 2 = -12$$
$$f(0) = 0 - 2 = -2$$
$$f(2) = 10 - 2 = 8$$
$$f(4) = 20 - 2 = 18$$

Domain: {-22, -12, -2, 8, 18}

26. $T = UN + F$

$$21{,}500 = 90N + 3500$$
$$18{,}000 = 90N$$
$$200 = N$$

200 cameras were produced.

27.

	P	·	R	=	I
A at 9.6%	x	·	0.096	=	$0.096x$
A at 7.2%	8750 - x		0.072	=	$0.072(8750-x)$
Total	8750				

$$0.096x = 0.072(8750 - x)$$
$$0.096x = 630 - 0.072x$$
$$0.168x = 630$$
$$x = 3750$$
$$8750 - x = 5000$$

\$3750 was invested at 9.6% and \$5000 was invested at 7.2%.

28.

	r	·	t	=	d
With wind	$p+w$	·	3	=	$3(p+w) = 570$
Against wind	p - w	·	3	=	$3(p-w) = 390$

$$3(p+w) = 570$$
$$\underline{3(p-w) = 390}$$
$$p+w = 190$$
$$\underline{p-w = 130}$$
$$2p = 320$$
$$p = 160$$
$$p+w = 190$$
$$160 + w = 190$$
$$w = 30$$

The speed of the plane is 160 mph and the speed of the wind is 30 mph.

29.

	r	·	t	=	d
With current	$b+c$	·	2	=	$2(b+c)=24$
Against current	$b - c$	·	3	=	$3(b-c)=24$

$$\begin{aligned} 2(b+c) &= 24 \\ 3(b-c) &= 24 \\ b+c &= 12 \\ b-c &= 8 \\ 2b &= 20 \\ b &= 10 \end{aligned}$$

The speed of the boat is 10 mph.

30.

	Amount	·	Unit cost	=	Value
Milk chocolate	10	·	m	=	$10m$
Dark chocolate	8	·	d	=	$8d$
					98

	Amount	·	Unit cost	=	Value
Milk chocolate	5	·	m	=	$5m$
Dark chocolate	12	·	d	=	$12d$
					97

$$\begin{aligned} 10m+8d &= 98 \\ -2(5m+12d) &= -2(97) \\ 10m+8d &= 98 \\ -10m-24d &= -194 \\ -16d &= -96 \\ d &= 6 \\ 10m+8d &= 98 \\ 10m+48 &= 98 \\ 10m &= 50 \\ m &= 5 \end{aligned}$$

Milk chocolate costs \$5 per pound and dark chocolate costs \$6 per pound.

CHAPTER 7 Polynomials

Chapter 7 Prep Test

1. $-2-(-3)=-2+3=1$

2. $-3(6)=-18$

3. $-\dfrac{24}{-36}=\dfrac{2\cdot 12}{3\cdot 12}=\dfrac{2}{3}$

4. $3n^4;\ n=-2$
   $3(-2)^4=3(16)=48$

5. The denominator b cannot be 0.

6. They are not like terms because the variable parts are not exactly the same.

7. $3x^2-4x+1+2x^2-5x-7$
   $(3x^2+2x^2)+(-4x-5x)+(1-7)$
   $5x^2-9x-6$

8. $-4y+4y=(-4+4)y=0y=0$

9. $-3(2x-8)=-6x+24$

10. $3xy-4y-2(5xy-7y)$
    $3xy-4y-10xy+14y$
    $-7xy+10y$

Section 7.1 Addition and Subtraction of Polynomials

1. Binomial (two terms)

3. Monomial (one term)

5. Not a monomial (variable in the denominator)

7. Polynomial

9. Polynomial

11. To use vertical format to add polynomials, arrange the terms of each polynomial is descending order with like terms in the same column.

13.
$$\begin{array}{r} x^2+7x \\ -3x^2-4x \\ \hline -2x^2+3x \end{array}$$

15.
$$\begin{array}{rr} y^2+4y & \\ -4y & -8 \\ \hline y^2 & -8 \end{array}$$

17.
$$\begin{array}{r} 2x^2+6x+12 \\ 3x^2+\ \ x+\ 8 \\ \hline 5x^2+7x+20 \end{array}$$

19.
$$\begin{array}{rrrr} x^3 & & -7x & +4 \\ & 2x^2 & +x & -10 \\ \hline x^3 & +2x^2 & -6x & -6 \end{array}$$

21.
$$\begin{array}{rrrr} 2a^3 & & -7a & +1 \\ & -3a^2 & -4a & +1 \\ \hline 2a^3 & -3a^2 & -11a & +2 \end{array}$$

23.
$$\begin{array}{r} 4x^2+2x \\ x^2+6x \\ \hline 5x^2+8x \end{array}$$

25.
$$\begin{array}{r} 4x^2 - 5xy \\ 3x^2 + 6xy - 4y^2 \\ \hline 7x^2 + xy - 4y^2 \end{array}$$

27.
$$\begin{array}{r} 2a^2 - 7a + 10 \\ a^2 + 4a + 7 \\ \hline 3a^2 - 3a + 17 \end{array}$$

29.
$$\begin{array}{r} 5x^3 + 7x - 7 \\ 10x^2 - 8x + 3 \\ \hline 5x^3 + 10x^2 - x - 4 \end{array}$$

31.
$$\begin{array}{r} 2r^2 - 5r + 7 \\ 3r^3 - 6r \\ \hline 3r^3 + 2r^2 - 11r + 7 \end{array}$$

33.
$$\begin{array}{r} 3x^2 + 7x + 10 \\ -2x^3 + 3x + 1 \\ \hline -2x^3 + 3x^2 + 10x + 11 \end{array}$$

35. The sum $Q + R$ will be a polynomial of degree 3. In the other combinations, the coefficients of the x^3 term are opposites.

37. The opposite of $7x^2 + 5x - 3$ is

$-(7x^2 + 5x - 3) = -7x^2 - 5x + 3.$

39. $(x^2 - 6x) - (x^2 - 10x) = (x^2 - 6x) + (-x^2 + 10x)$
$$\begin{array}{r} x^2 - 6x \\ -x^2 + 10x \\ \hline 4x \end{array}$$

41. $(2y^2 - 4y) - (-y^2 + 2) = (2y^2 - 4y) + (y^2 - 2)$
$$\begin{array}{r} 2y^2 - 4y \\ -y^2 + 2 \\ \hline y^2 - 4y + 2 \end{array}$$

43. $(x^2 - 2x + 1) - (x^2 + 5x + 8)$

$= (x^2 - 2x + 1) + (-x^2 - 5x - 8)$
$$\begin{array}{r} x^2 - 2x + 1 \\ -x^2 - 5x - 8 \\ \hline -7x - 7 \end{array}$$

45. $(4x^3 + 5x + 2) - (-3x^2 + 2x + 1)$

$= (4x^3 + 5x + 2) + (3x^2 - 2x - 1)$
$$\begin{array}{r} 4x^3 + 5x + 2 \\ 3x^2 - 2x - 1 \\ \hline 4x^3 + 3x^2 + 3x + 1 \end{array}$$

47. $(2y^3 + 6y - 2) - (y^3 + y^2 + 4)$

$= (2y^3 + 6y - 2) + (-y^3 - y^2 - 4)$
$$\begin{array}{r} 2y^3 + 6y - 2 \\ -y^3 - y^2 - 4 \\ \hline y^3 - y^2 + 6y - 6 \end{array}$$

49. $(y^2 - 10xy) - (2y^2 + 3xy)$

$= (y^2 - 10xy) + (-2y^2 - 3xy)$

$= (y^2 - 2y^2) + (-10xy - 3xy)$

$= -y^2 - 13xy$

51. $(3x^2 + x - 3) - (x^2 + 4x - 2)$

$= (3x^2 + x - 3) + (-x^2 - 4x + 2)$

$= (3x^2 - x^2) + (x - 4x) + (-3 + 2)$

$= 2x^2 - 3x - 1$

53. $(-2x^3+x-1)-(-x^2+x-3)$
$=(-2x^3+x-1)+(x^2-x+3)$
$=-2x^3+x^2+(x-x)+(-1+3)$
$=-2x^3+x^2+2$

55. $(4a^3-2a+1)-(a^3-2a+3)$
$=(4a^3-2a+1)+(-a^3+2a-3)$
$=(4a^3-a^3)+(-2a+2a)+(1-3)$
$=3a^3-2$

57. $(4y^3-y-1)-(2y^2-3y+3)$
$=(4y^3-y-1)+(-2y^2+3y-3)$
$=4y^3-2y^2+(-y+3y)+(-1-3)$
$=4y^3-2y^2+2y-4$

59. **a.** Coefficient of the x^2 term: $a-(-d)=a+d$ which is positive.

b. Coefficient of the x term: $-b-(-e)=-b+e=e-b$ which is negative since $b>e$.

c. Coefficient of the constant term: $c-f$ which is positive.

61. $\left(\frac{2}{3}a^2+\frac{1}{2}a-\frac{3}{4}\right)-\left(\frac{5}{3}a^2+\frac{1}{2}a+\frac{1}{4}\right)$
$=\left(\frac{2}{3}a^2+\frac{1}{2}a-\frac{3}{4}\right)+\left(-\frac{5}{3}a^2-\frac{1}{2}a-\frac{1}{4}\right)$
$=\left(\frac{2}{3}a^2-\frac{5}{3}a^2\right)+\left(\frac{1}{2}a-\frac{1}{2}a\right)+\left(-\frac{3}{4}-\frac{1}{4}\right)$
$=-a^2-1$

63. $(3x^2-4x-2)+(ax^2+bx+c)=-x^2+2x+1$
$(3+a)x^2+(-4+b)x+(-2+c)=-x^2+2x+1$

$3+a=-1$ $\quad -4+b=2$ $\quad -2+c=1$
$a=-4$ $\quad b=6$ $\quad c=3$

Polynomial to be added: $-4x^2+6x+3$

65. $(6x^2-4x-2)-(ax^2+bx+c)=2x^2+2x-5$
$(6-a)x^2+(-4-b)x+(-2-c)=2x^2+2x-5$

$6-a=2$ $\quad -4-b=2$ $\quad -2-c=-5$
$a=4$ $\quad b=-6$ $\quad c=3$

Polynomial to be subtracted: $4x^2-6x+3$

67. Yes, it is possible if the x^3 coefficients are equal.
Example:
$(3x^3-2x^2+3x-4)-(3x^3+4x^2-6x+5)$
$=-6x^2+9x-9$

69. Answers vary.
One example: $5x^2+6x-9$ and $3x^2+8x+1$

71. Answers vary.
One example: $5x^2+6x-9$ and $-5x^2-6x+1$

Section 7.2 Multiplication of Monomials

1. **a.** No. The rule does not apply to terms.
b. Yes
c. No. The bases are not the same.
d. No. The rule does not apply to terms.

3. **a.** Product
b. Power
c. Power
d. Product

5. **a.** Yes
b. No. The expression inside the parentheses is a sum instead of a power.
c. No. The expression inside the parentheses is a sum instead of a power.
d. Yes

7. $(x^7)(x^2)=x^{7+2}=x^9$

9. $(x)(2x) = 2(x \cdot x) = 2x^2$

11. $(3x)(4x) = (3 \cdot 4)(x \cdot x) = 12x^2$

13. $(-2a^3)(-3a^4) = [(-2)(-3)](a^3 \cdot a^4) = 6a^7$

15. $(x^2y)(xy^4) = (x^2 \cdot x)(y \cdot y^4) = x^3y^5$

17. $(-2x^4)(5x^5y) = (-2 \cdot 5)(x^4 \cdot x^5)y = -10x^9y$

19. $(x^2y^4)(x^5y^4) = (x^2 \cdot x^5)(y^4 \cdot y^4) = x^7y^8$

21. $(2xy)(-3x^2y^4) = [(2)(-3)](x \cdot x^2)(y \cdot y^4)$
$= -6x^3y^5$

23. $(x^2yz)(x^2y^4) = (x^2 \cdot x^2)(y \cdot y^4)z = x^4y^5z$

25. $(a^2b^3)(ab^2c^4) = (a^2 \cdot a)(b^3 \cdot b^2)c^4 = a^3b^5c^4$

27. $(-a^2b^2)(a^3b^6) = (-a^2 \cdot a^3)(b^2 \cdot b^6) = -a^5b^8$

29. $(-6a^3)(a^2b) = -6(a^3 \cdot a^2)b = -6a^5b$

31. $(-5y^4z)(-8y^6z^5) = [(-5)(-8)](y^4 \cdot y^6)(z \cdot z^5)$
$= 40y^{10}z^6$

33. $(10ab^2)(-2ab) = [(10)(-2)](a \cdot a)(b^2 \cdot b)$
$= -20a^2b^3$

35. $(xy^2z)(x^2y)(z^2y^2) = (x^2 \cdot x^2)(y^2 \cdot y \cdot y^2)(z \cdot z^2)$
$= x^4y^5z^3$

37. $(4a^2b)(-3a^3b^4)(a^5b^2)$
$= [(4)(-3)](a^2 \cdot a^3 \cdot a^5)(b \cdot b^4 \cdot b^2)$
$= -12a^{10}b^7$

39. To simplify a power of an exponential expression, multiply each exponent inside the parentheses by the exponent outside the parentheses. Example: $(x^2y^3)^4 = x^8y^{12}$

41. $(x^4y^2)^6 = x^{(4)(2)}y^{(2)(6)} = x^8y^{12}$

43. $(x^3)^3 = x^9$

45. $(x^7)^2 = x^{14}$

47. $(2^2)^3 = 2^6 = 64$

49. $(-2)^2 = 4$

51. $(-2^2)^3 = -2^6 = -64$

53. $(-x^2)^2 = x^4$

55. $(2x)^2 = 4x^2$

57. $(-2x^2)^3 = -8x^6$

59. $(x^2y^3)^2 = x^4y^6$

61. $(3x^2y)^2 = 9x^4y^2$

63. $(a^2)(3a^2)^3 = a^2 \cdot 27a^6 = 27a^8$

65. $(-2x)(2x^3)^2 = -2x \cdot 4x^6 = -8x^7$

67. $(x^2y)(x^2y)^3 = x^2y \cdot x^6y^3 = x^8y^4$

69. $(ab^2)^2(ab)^2 = a^2b^4 \cdot a^2b^2 = a^4b^6$

71. $(-2x)(-2x^3y)^3 = -2x(-8)x^9y^3 = 16x^{10}y^3$

73. $(-2x)(-3xy^2)^2 = -2x \cdot 9x^2y^4 = -18x^3y^4$

75. $(ab^2)(-2a^2b)^3 = ab^2(-8)a^6b^3 = -8a^7b^5$

77. $(-2a^3)(3a^2b)^3 = -2a^3(27)a^6b^3 = -54a^9b^3$

79. $(-3ab)^2(-2ab)^3 = 9a^2b^2(-8)a^3b^3 = -72a^5b^5$

81. $(6x)(2x^2) + (4x^2)(5x) = 12x^3 + 20x^3 = 32x^3$

83. $(3a^2b^2)(2ab) - (9ab^2)(a^2b) = 6a^3b^3 - 9a^3b^3$
$= -3a^3b^3$

85. $(5xy^3)(3x^4y^2) - (2x^3y)(x^2y^4)$
$= 15x^5y^5 - 2x^5y^5 = 13x^5y^5$

87. $4a^2(2ab)^3 - 5b^2(a^5b) = 4a^2 \cdot 8a^3b^3 - 5a^5b^3$
$= 32a^5b^3 - 5a^5b^3 = 27a^5b^3$

89. $-2xy(x^2y)^3 - 3x^5(xy^2)^2 = -2xy \cdot x^6y^3 - 3x^5 \cdot x^2y^4$
$= -2x^7y^4 - 3x^7y^4$
$= -5x^7y^4$

91. $a^n \cdot a^n = a^{n+n} = a^{2n}$

93. $(a^2)^n = a^{2n}$

95. $P = 2(length) + 2(width)$
$P = 2(4ab) + 2(2ab)$
$P = 8ab + 4ab = 12ab$

97. **a.** No, they are not the same. $(2^3)^2 = 2^6$ and $2^{(3^2)} = 2^9$. $2^{(3^2)}$ is larger.

b. First raise m to the n power. Then raise x to the m^n power.

Section 7.3 Multiplication of Polynomials

1. It is **always true** that to multiply a monomial times a polynomial, use the Distributive Property.

3. It is **sometimes true** that a binomial is polynomial of degree 2.

5. It is **always true** that to square a binomial means to multiply it times itself.

7. It is **sometimes true** that the FOIL method is used to multiply two polynomials.

9. It is **sometimes true** that the product of two binomials is a trinomial.

11. $-3y(y+7) = -3y(y) + (-3y)(7) = -3y^2 - 21y$

13. $x(x-2) = x^2 - 2x$

15. $-x(x+7) = -x^2 - 7x$

17. $3a^2(a-2) = 3a^3 - 6a^2$

19. $-5x^2(x^2 - x) = -5x^4 + 5x^3$

21. $-x^3(3x^2-7)=-3x^5+7x^3$

25. $(2x-4)(3x)=6x^2-12x$

27. $-xy(x^2-y^2)=-x^3y+xy^3$

29. $x(2x^3-3x+2)=2x^4-3x^2+2x$

31. $-a(-2a^2-3a-2)=2a^3+3a^2+2a$

33. $x^2(3x^4-3x^2-2)=3x^6-3x^4-2x^2$

35. $2y^2(-3y^2-6y+7)=-6y^4-12y^3+14y^2$

37. $(a^2+3a-4)(-2a)=-2a^3-6a^2+8a$

39. $-3y^2(-2y^2+y-2)=6y^4-3y^3+6y^2$

41. $xy(x^2-3xy+y^2)=x^3y-3x^2y^2+xy^3$

43. $(x-3)(x^2-4x+5)$

$$\begin{aligned}&=(x-3)x^2-(x-3)(4x)+(x-3)(5)\\&=x^3-3x^2-4x^2+12x+5x-15\\&=x^3-7x^2+17x-15\end{aligned}$$

45.

$$\begin{array}{r}x^2+3x+2\\ x+1\\ \hline x^2+3x+2\\ x^3+3x^2+2x\quad\\ \hline x^3+4x^2+5x-15\end{array}$$

23. $2x(6x^2-3x)=12x^3-6x^2$

47.

$$\begin{array}{r}a^2-3a+4\\ a-3\\ \hline -3a^2+9a-12\\ a^3-3a^2+4a\quad\\ \hline a^3-6a^2+13a-12\end{array}$$

49.

$$\begin{array}{r}-2b^2-3b+4\\ b-5\\ \hline 10b^2+15b-20\\ -2b^3-3b^2+4b\quad\\ \hline -2b^3+7b^2+19b-20\end{array}$$

51.

$$\begin{array}{r}-2x^2+7x-2\\ 3x-5\\ \hline 10x^2-35x+10\\ -6x^3+21x^2-6x\quad\\ \hline -6x^3+31x^2-41x+10\end{array}$$

53.

$$\begin{array}{r}x^3\quad-3x+2\\ x-4\\ \hline -4x^3\quad+12x-8\\ x^4\quad-3x^2+2x\quad\\ \hline x^4-4x^3-3x^2+14x-8\end{array}$$

55.

$$\begin{array}{r}5y^2+8y-2\\ 3y-8\\ \hline -40y^2-64y+16\\ 15y^3+24y^2-6y\quad\\ \hline 15y^3-16y^2-70y+16\end{array}$$

57.
$$\begin{array}{r} 5a^3 \qquad -15a+2 \\ a-4 \\ \hline -20a^3 \qquad +60a-8 \\ 5a^4 \qquad -15a^2+2a \qquad \\ \hline 5a^4-20a^3-15a^2+62a-8 \end{array}$$

59.
$$\begin{array}{r} y^3+2y^2-3y+1 \\ y+2 \\ \hline 2y^3+4y^2-6y+2 \\ y^4+2y^3-3y^2+y \qquad \\ \hline y^4+4y^3+y^2 \quad -5y+2 \end{array}$$

61. **a.** The simplified product will have a degree of 4.

b. The simplified product will have at most 5 terms.

63. The **F**irst terms are 4x and x.

The **O**uter terms are 4x and 5.

The **I**nner terms are -3 and x.

The **L**ast terms are -3 and 5.

65. **a.** 3d and d

b. 3d and -1

c. 4 and d

d. 4 and -1

67. $(x+1)(x+3)=x^2+4x+3$

69. $(a-3)(a+4)=a^2+a-12$

71. $(y+3)(y-8)=y^2-5y-24$

73. $(y-7)(y-3)=y^2-10y+21$

75. $(2x+1)(x+7)=2x^2+15x+7$

77. $(3x-1)(x+4)=3x^2+11x-4$

79. $(4x-3)(x-7)=4x^2-31x+21$

81. $(3y-8)(y+2)=3y^2-2y-16$

83. $(3x+7)(3x+11)=9x^2+54x+77$

85. $(7a-16)(3a-5)=21a^2-83a+80$

87. $(3b+13)(5b-6)=15b^2+47a-78$

89. $(2a+b)(a+3b)=2a^2+7ab+3b^2$

91. $(2a-b)(3a+2b)=6a^2+ab-2b^2$

93. $(2x+y)(x-2y)=2x^2-3xy-2y^2$

95. $(2x+3y)(5x+7y)=10x^2+29xy+21y^2$

97. $(3a-2b)(2a-7b)=6a^2-25ab+14b^2$

99. $(a-9b)(2a+7b)=2a^2-11ab-63b^2$

101. $(5x+2y)(2x-5y)=10x^2-21xy-10y^2$

103. The coefficient of x will be $a-b$, which is positive since $a>b$.

105. $(5x+1)(5x-1)=(5x)^2-1=25x^2-1$

107. $(y-5)(y+5)=y^2-25$

109. $(2x+3)(2x-3)=4x^2-9$

111. $(x+1)^2=x^2+2x+1$

113. $(3a-5)^2=9a^2-30a+25$

115. $(3x-7)(3x+7)=9x^2-49$

117. $(2a+b)^2=4a^2+4ab+b^2$

119. $(x-2y)^2=x^2-4xy+4y^2$

121. $(4-3y)(4+3y)=16-9y^2$

123. $(5x+2y)^2=25x^2+20xy+4y^2$

125. Coefficient of x: $-ab+ab=0$

127. Coefficient of x: $2ab>0$

129. $A=(2x+5)(2x-5)=4x^2-25$

131. $(4x+1)^2=(4x)^2+2(4x)(1)+1^2=16x^2+8x+1$

133. $A=(x-6)(2x+3)=(2x^2-9x-18)m^2$

135. $A=2(4x-3)(4x-3)=2(16x^2-24x+9)$
$=(32x^2-48x+18)cm^2$

137. $A=(2x-3)^2=(4x^2-12x+9)yd^2$

139. $A=\frac{1}{2}bh=\frac{1}{2}(2x+6)(x-8)$
$=(x+3)(x-8)$
$=(x^2-5x-24x)in^2$

141. $A=\pi r^2=\pi(x-3)^2=\pi(x^2-6x+9)$
$=(\pi x^2-6\pi x+9\pi)ft^2$

143. $A=45\cdot 45+2\cdot x\cdot 45=(2025+90x)ft^2$

145. Since both rectangles have the same length, the rectangle with the greater width is larger. The rectangle that measures $(ax+b)ft$ by $(cx+d)ft$ is larger.

147. $(a+b)^2-(a-b)^2$
$=[(a+b)-(a-b)][(a+b)+(a-b)]$
$=(a+b-a+b)(a+b+a-b)$
$=2b(2a)=4ab$

149. $(3a^2-4a+2)^2=[(3a^2-4a)+2]^2$
$=(3a^2-4a)^2+2(2)(3a^2-4a)+2^2$
$=9a^4-24a^3+16a^2+12a^2-16a+4$
$=9a^4-24a^3+28a^2-16a+4$

151. $3x^2(2x^3+4x-1)-6x^3(x^2-2)$
$=6x^5+12x^3-3x^2-6x^5+12x^3$
$=24x^3-3x^2$

153. $2n-3=4n-7$
$-2n=-4$
$n=2$

$(4n^3)^2=16n^6=16(2)^6=16(64)=1024$

155. $(3x-4)(4x+5)=12x^2-x-20$

157. $(2x-5)(3x+1)+(x^2+2x-3)$
$=6x^2-13x-5+x^2+2x-3$
$=7x^2-11x-8$

159. No, it is not possible to multiply a polynomial of degree 2 by a polynomial of degree 2 and have the product be a polynomial of degree 3. A polynomial of degree 2 contains the term ax^2, with a not 0. Two polynomials of degree 2 will have terms ax^2 and bx^2, with neither a nor b equal to 0. Multiplying those will yield abx^4, where ab is not 0. So the product will have an x^4 term and will be of degree 4.

161. $(x+1)(-x^2+x-1)=-x^3+x^2-x-x^2+x-1$
$=-x^3-1$

163. $(x+1)(-x^4+x^3-x^2+x-1)$
$=-x^5+x^4-x^3+x^2-x-x^4+x^3-x^2+x-1$
$=-x^5-1$

165. $(x+1)(-x^6+x^5-x^4+x^3-x^2+x-1)=-x^7-1$

Section 7.4 Integer Exponents and Scientific Notation

1. False. The expression cannot be simplified because the bases are not the same.

3. True. $3^{-2}=\frac{1}{3^2}$

5. False. $4^{-3}=\frac{1}{4^3}=\frac{1}{64}$

7. If a variable has a negative exponent and the variable is in the denominator of a fraction, remove the variable from the denominator, put it in the numerator, and make the exponent positive. If a variable has a negative exponent and it is not in the denominator, write the variable in the denominator of a fraction whose numerator is 1 and make the exponent positive.

9. $\frac{8x^{10}}{2x^3}=\left(\frac{8}{2}\right)(x^{10-3})=4x^7$

11. $\frac{y^7}{y^3}=y^{7-3}=y^4$

13. $\frac{a^8}{a^5}=a^{8-5}=a^3$

15. $\frac{p^5}{p}=p^{5-1}=p^4$

17. $\frac{4x^8}{2x^5}=2x^{8-5}=2x^3$

19. $\frac{22k^5}{11k^4}=2k^{5-4}=2k$

21. $\frac{m^9n^7}{m^4n^5}=m^{9-4}n^{7-5}=m^5n^2$

23. $\frac{6r^4}{4r^2}=\frac{6}{4}\cdot r^{4-2}=\frac{3r^2}{2}$

25. $\frac{-16a^7}{24a^6}=-\frac{2}{3}a^{7-6}=-\frac{2a}{3}$

27. $x^{-2}=\dfrac{1}{x^2}$

29. $\dfrac{1}{a^{-6}}=a^6$

31. $5^{-2}=\dfrac{1}{5^2}=\dfrac{1}{25}$

33. $\dfrac{1}{8^{-2}}=8^2=64$

35. $\dfrac{3^{-2}}{3}=3^{-2-1}=3^{-3}=\dfrac{1}{3^3}=\dfrac{1}{27}$

37. $\dfrac{2^3}{2^3}=2^{3-3}=2^0=1$

39. $4x^{-7}=\dfrac{4}{x^7}$

41. $\dfrac{5}{b^{-8}}=5b^8$

43. $\dfrac{1}{3x^{-2}}=\dfrac{x^2}{3}$

45. $(ab^5)^0=1$

47. $\dfrac{y^3}{y^8}=y^{3-8}=y^{-5}=\dfrac{1}{y^5}$

49. $\dfrac{a^5}{a^{11}}=a^{5-11}=a^{-6}=\dfrac{1}{a^6}$

51. $\dfrac{4x^2}{12x^5}=\dfrac{1}{3}x^{2-5}=\dfrac{1}{3}x^{-3}=\dfrac{1}{3x^3}$

53. $\dfrac{-12x}{-18x^6}=\dfrac{2}{3}x^{1-6}=\dfrac{2}{3}x^{-5}=\dfrac{2}{3x^5}$

55. $\dfrac{x^6y^5}{x^8y}=x^{6-8}y^{5-1}=x^{-2}y^4=\dfrac{y^4}{x^2}$

57. $\dfrac{2m^6n^2}{5m^9n^{10}}=\dfrac{2}{5}m^{6-9}n^{2-10}=\dfrac{2}{5}m^{-3}n^{-8}=\dfrac{2}{5m^3n^8}$

59. $\dfrac{pq^3}{p^4q^4}=p^{1-4}q^{3-4}=p^{-3}q^{-1}=\dfrac{1}{p^3q}$

61. $\dfrac{3x^4y^5}{6x^4y^8}=\dfrac{3}{6}\cdot x^{4-4}\cdot y^{5-8}$

$=\dfrac{1}{2}x^0y^{-3}=\dfrac{1}{2y^3}$

63. $\dfrac{14x^4y^6z^2}{16x^3y^9z}=\dfrac{7}{8}x^{4-3}y^{6-9}z^{2-1}=\dfrac{7}{8}xy^{-3}z=\dfrac{7xz}{8y^3}$

65. $(-2xy^{-2})^3=(-2)^3x^3(y^{-2})^3=-8x^3y^{-6}=\dfrac{-8x^3}{y^6}$

67. $(3x^{-1}y^{-2})^2=3^2(x^{-1})^2(y^{-2})^2=9x^{-2}y^{-4}=\dfrac{9}{x^2y^4}$

69. $(2x^{-1})(x^{-3})=2x^{-1+(-3)}=2x^{-4}=\dfrac{2}{x^4}$

71. $(-5a^2)(a^{-5})^2=-5a^2\cdot a^{-10}=-5a^{2+(-10)}$

$=-5a^{-8}=-\dfrac{5}{a^8}$

73. $(-2ab^{-2})\left(4a^{-2}b\right)^{-2} = -2ab^{-2}\cdot 4^{-2}a^{4}b^{-2}$

$= -\frac{2}{4^2}a^5b^{-4} = -\frac{2a^5}{16b^4}$

$= -\frac{a^5}{8b^4}$

75. $\frac{a^{-3}b^{-4}}{a^2b^2} = a^{-3-2}b^{-4-2} = a^{-5}b^{-6} = \frac{1}{a^5b^6}$

77. $\frac{2x^{-2}y}{8xy} = \frac{1}{4}x^{-2-1}y^{1-1} = \frac{1}{4}x^{-3}y^0 = \frac{1}{4x^3}$

79. $\frac{2x^{-1}y^4}{x^2y^3} = 2x^{-1-2}y^{4-3} = 2x^{-3}y = \frac{2y}{x^3}$

81. $\frac{12a^2b^3}{-27a^2b^2} = -\frac{4}{9}a^{2-2}b^{3-2} = -\frac{4}{9}a^0b = -\frac{4b}{9}$

83. $\frac{-8x^2y^4}{44y^2z^5} = -\frac{2}{11}x^2y^{4-2}\frac{1}{z^5}$

$= -\frac{2}{11}x^2y^2\frac{1}{z^5} = -\frac{2x^2y^2}{11z^5}$

85. a. True. $a^{-n}a^n = a^{-n+n} = a^0 = 1$

b. False. $\frac{a^n}{b^m}$ cannot be simplified.

87. A number is written in scientific notation if it is written as a product of a number between 1 and 10 and a power of **10.**

89. No. 0.8 is not between 1 and 10.

91. Yes.

93. To write the number 0.0000000086 in scientific notation, move the decimal point 9 places to the right. The exponent on 10 is -9.

95. $75{,}000 = 7.5\times10^4$

97. $0.000076 = 7.6\times10^{-5}$

99. $819{,}000{,}000 = 8.19\times10^8$

101. $0.00000000096 = 9.6\times10^{-10}$

103. $2.3\times10^7 = 23{,}000{,}000$

105. $9.21\times10^{-7} = 0.000000921$

107. $5.75\times10^9 = 5{,}750{,}000{,}000$

109. $3.54\times10^{-8} = 0.0000000354$

111. $1{,}000{,}000{,}000{,}000{,}000{,}000{,}000 = 1\times10^{21}$

113. $0.0000000004 = 4\times10^{-10}$

115. $12{,}000{,}000{,}000 = 1.2\times10^{10}$

117. $5{,}880{,}000{,}000{,}000 = 5.88\times10^{12}$

119. $5.1156\times10^{15} = 5{,}115{,}600{,}000{,}000{,}000$

121. $(4.2\times10^7)(1.8\times10^{-5}) = 7.56\times10^2$

123. $(3\times10^{-20})(2.4\times10^9) = 7.2\times10^{-11}$

125. $\dfrac{6\times10^{-8}}{2.5\times10^{-2}}=2.4\times10^{-6}$

127. $\dfrac{9.03\times10^{6}}{4.3\times10^{-5}}=2.1\times10^{11}$

129. The product will be greater than 1 because the power of 10 will be $n-m$ which is positive since $n>m$.

131. $9^{-2}+3^{-3}=\dfrac{1}{9^2}+\dfrac{1}{3^3}=\dfrac{1}{81}+\dfrac{1}{27}=\dfrac{1}{81}+\dfrac{3}{81}=\dfrac{4}{81}$

133.

x	3^x	3^{-x}
-2	$3^{-2}=\dfrac{1}{3^2}=\dfrac{1}{9}$	$3^2=9$
-1	$3^{-1}=\dfrac{1}{3}$	$3^1=3$
0	$3^0=1$	$3^0=1$
1	$3^1=3$	$3^{-1}=\dfrac{1}{3}$
2	$3^2=9$	$3^{-2}=\dfrac{1}{3^2}=\dfrac{1}{9}$

135. $25^{-2}=\dfrac{1}{25^2}=\dfrac{1}{625}=0.0016$

137. If $m=n+1$ and $a\neq0$, then $\dfrac{a^m}{a^n}=a.$

139. $-6.3^x=-1$

$x=0$

141. True.

143. True.

145. The expression x^{-2} is positive for all nonzero real numbers x. $x^{-2}=\dfrac{1}{x^2}$ and x^2 is positive for all nonzero real values of x.

147. a. $Area=2.619\times10^5 mi^2\left(\dfrac{2.8\times10^7 ft^2}{1mi^2}\right)$

$=7.3332\times10^{12} ft^2$

$\dfrac{Area\ of\ Texas}{World\ Population}=\dfrac{7.3332\times10^{12} ft^2}{7\times10^9\ people}$

$=1047.6 ft^2 / person$

b. $Area=1.0\times10^3 mi^2\left(\dfrac{2.8\times10^7 ft^2}{1mi^2}\right)$

$=2.8\times10^{10} ft^2$

$\dfrac{Area\ of\ RI}{U.S.\ Population}=\dfrac{2.8\times10^{10} ft^2}{3.1\times10^8\ people}$

$\approx 90 ft^2 / person$

c. i. $Area\ per\ person=(22\cdot22) ft^2$

$=484 ft^2$

$\dfrac{2.8\times10^7 ft^2}{1mi^2}\cdot\dfrac{1\ person}{484 ft^2}$

$\approx 57{,}900\ people / mi^2$

ii. $\dfrac{1mi^2}{57{,}900\ people}\cdot 7\times10^9\ people$

$\approx 120{,}900 mi^2$

d. $(5.7\times10^7 mi^2)\left(\dfrac{2.8\times10^7 ft^2}{1mi^2}\right)\left(\dfrac{1acre}{43{,}560 ft^2}\right)$

$=3.664\times10^{10} acres$

$\dfrac{3.664\times10^{10} acres}{7\times10^9\ people}=5.2 acres / person$

e. $(5.7\times10^7 mi^2)\left(\dfrac{2.8\times10^7 ft^2}{1mi^2}\right)\left(\dfrac{1\ person}{484 ft^2}\right)$

$\approx 3{,}300\times10^9\ people$

The carrying capacity of Earth would be 3,300 *billion* people.

Section 7.5 Division of Polynomials

1. Related multiplication: $15x^2+12x=3x(5x+4)$

3. **True**

5. **True**

7. $\dfrac{18y^5+3y}{3y}=\dfrac{18y^5}{3y}+\dfrac{3y}{3y}=6y^4+1$

9. $\dfrac{10a-25}{5}=\dfrac{10a}{5}-\dfrac{25}{5}=2a-5$

11. $\dfrac{3a^2+2a}{a}=\dfrac{a(3a+2)}{a}=3a+2$

13. $\dfrac{3x^2-6x}{3x}=\dfrac{3x^2}{3x}-\dfrac{6x}{3x}=x-2$

15. $\dfrac{5x^2-10x}{-5x}=\dfrac{5x^2}{-5x}-\dfrac{10x}{-5x}=-x+2$

17. $\dfrac{x^3+3x^2-5x}{x}=\dfrac{x^3}{x}+\dfrac{3x^2}{x}-\dfrac{5x}{x}=x^2+3x-5$

19. $\dfrac{x^6-3x^4-x^2}{x^2}=\dfrac{x^6}{x^2}-\dfrac{3x^4}{x^2}-\dfrac{x^2}{x^2}=x^4-3x^2-1$

21. $\dfrac{5x^2y^2+10xy}{5xy}=\dfrac{5x^2y^2}{5xy}+\dfrac{10xy}{5xy}=xy+2$

23. $\dfrac{9y^6-15y^3}{-3y^3}=\dfrac{9y^6}{-3y^3}-\dfrac{15y^3}{-3y^3}=-3y^3+5$

25. $\dfrac{3x^2-2x+1}{x}=\dfrac{3x^2}{x}-\dfrac{2x}{x}+\dfrac{1}{x}=3x-2+\dfrac{1}{x}$

27. $\dfrac{-3x^2+7x-6}{x}=\dfrac{-3x^2}{x}+\dfrac{7x}{x}-\dfrac{6}{x}=-3x+7-\dfrac{6}{x}$

29. $\dfrac{16a^2b-20ab+24ab^2}{4ab}=\dfrac{16a^2b}{4ab}-\dfrac{20ab}{4ab}+\dfrac{24ab^2}{4ab}$

$=4a-5+6b$

31. $\dfrac{9x^2y+6xy-3x}{xy}=\dfrac{9x^2y}{xy}+\dfrac{6xy}{xy}-\dfrac{3x}{xy}$

$=9x+6-\dfrac{3}{y}$

33. To check the division, multiply the divisor times the quotient. $4x(2x^2-3x-1)=8x^3-12x^2-4x$

35.
$$\begin{array}{r} x+2 \\ x-3\overline{)x^2-x\;-6} \\ \underline{x^2-3x} \\ 2x-6 \\ \underline{2x-6} \\ 0 \end{array}$$

$(x^2-x-6)\div(x-3)=x+2$

37.
$$\begin{array}{r} 2x+1 \\ x+2\overline{)2x^2+5x+2} \\ \underline{2x^2+4x} \\ x+2 \\ \underline{x+2} \\ 0 \end{array}$$

$(2x^2+5x+2)\div(x+2)=2x+1$

39.
$$\begin{array}{r} x+1 \\ x-1\overline{)x^2+0x+1} \\ \underline{x^2-x} \\ x+1 \\ \underline{x-1} \\ 2 \end{array}$$

$(x^2+1)\div(x-1)=x+1+\dfrac{2}{x-1}$

41.

$$\begin{array}{r} 2x-1 \\ 3x-2\overline{)6x^2-7x+0} \\ \underline{6x^2-4x} \\ -3x+0 \\ \underline{-3x+2} \\ -2 \end{array}$$

$(6x^2-7x)\div(3x-2)=2x-1-\dfrac{2}{3x-2}$

43.

$$\begin{array}{r} b-5 \\ b-3\overline{)b^2-8b-9} \\ \underline{b^2-3b} \\ -5b-9 \\ \underline{-5b+15} \\ -24 \end{array}$$

$(b^2-8b-9)\div(b-3)=b-5-\dfrac{24}{b-3}$

45.

$$\begin{array}{r} 3x+17 \\ x-4\overline{)3x^2+5x-4} \\ \underline{3x^2-12x} \\ 17x-4 \\ \underline{17x-68} \\ 64 \end{array}$$

$(3x^2+5x-4)\div(x-4)=3x+17+\dfrac{64}{x-4}$

47.

$$\begin{array}{r} 5y+3 \\ 2y+3\overline{)10y^2+21y+10} \\ \underline{10y^2+15y} \\ 6y+10 \\ \underline{6y+9} \\ 1 \end{array}$$

$(10y^2+21y+10)\div(2y+3)=5y+3+\dfrac{1}{2y+3}$

49.

$$\begin{array}{r} 3x-5 \\ 4x-1\overline{)12x^2-23x+5} \\ \underline{12x^2-3x} \\ -20x+5 \\ \underline{-20x+5} \\ 0 \end{array}$$

$(12x^2-23x+5)\div(4x-1)=3x-5$

51.

$$\begin{array}{r} x^2+2x+3 \\ x+1\overline{)x^3+3x^2+5x+8} \\ \underline{x^3+x^2} \\ 2x^2+5x \\ \underline{2x^2+2x} \\ 3x+8 \\ \underline{3x+3} \\ 5 \end{array}$$

$(x^3+3x^2+5x+8)\div(x+1)=x^2+2x+3+\dfrac{5}{x+1}$

53.

$$\begin{array}{r} x^2-3 \\ x^2+0x+2\overline{)x^4+0x^3-x^2+0x-6} \\ \underline{x^4+0x^3+2x^2} \\ -3x^2+0x-6 \\ \underline{-3x^2+0x-6} \\ 0 \end{array}$$

$(x^4-x^2-6)\div(x^2+2)=x^2-3$

55. **False** When a sixth-degree polynomial is divided by a third-degree polynomial, the quotient is a third-degree polynomial because the exponents are subtracted.

57. If $\dfrac{x^2+2x-3}{x-2}=x+4+\dfrac{5}{x-2}$, then

$x^2+2x-3=(x+4)(x-2)+5.$

59. $6x(4y^2)=24xy^2$

61. $(x-3)(x^2-x+8)+22$
$= x^3-x^2+8x-3x^2+3x-24+22$
$= x^3-4x^2+11x-2$

63.
$$\begin{array}{r} x^2+2x-3 \\ 4x+1\overline{)4x^3+9x^2-10x-3} \\ \underline{4x^3+x^2} \\ 8x^2-10x \\ \underline{8x^2+2x} \\ -12x-3 \\ \underline{-12x-3} \\ 0 \end{array}$$

$(4x^3+9x^2-10x-3)\div(4x+1)=x^2+2x-3$

Chapter 7 Review Exercises

1. $(12y^2+17y-4)+(9y^2-13y+3)$
$=12y^2+17y-4+9y^2-13y+3$
$=21y^2+4y-1$

2. $(5xy^2)(-4x^2y^3)=-20x^3y^5$

3. $-2x(4x^2+7x-9)=-8x^3-14x^2+18x$

4. $(5a-7)(2a+9)=10a^2+31a-63$

5. $\dfrac{36x^2-42x+60}{6}=\dfrac{36x^2}{6}-\dfrac{42x}{6}+\dfrac{60}{6}$
$=6x^2-7x+10$

6. $(5x^2-2x-1)-(3x^2-5x+7)$
$=5x^2-2x-1-3x^2+5x-7$
$=2x^2+3x-8$

7. $(-3^2)^3=-3^6=-729$

8. $(x^2-5x+2)(x-1)=x^3-5x^2+2x-x^2+5x-2$
$=x^3-6x^2+7x-2$

9. $(a+7)(a-7)=a^2-49$

10. $\dfrac{6^2}{6^{-2}}=6^{2-(-2)}=6^{2+2}=6^4=1296$

11.
$$\begin{array}{r} x-6 \\ x+7\overline{)x^2+x-42} \\ \underline{x^2+7x} \\ -6x-42 \\ \underline{-6x-42} \\ 0 \end{array}$$

$(x^2+x-42)\div(x+7)=x-6$

12. $(2x^3+7x^2+x)+(2x^2-4x-12)$
$=2x^3+7x^2+x+2x^2-4x-12$
$=2x^3+9x^2-3x-12$

13. $(6a^2b^5)(3a^6b)=18a^8b^6$

14. $x^2y(3x^2-2x+12)=3x^4y-2x^3y+12x^2y$

15. $(2b-3)(4b+5)=8b^2-2b-15$

16. $\dfrac{16y^2-32y}{-4y}=\dfrac{16y^2}{-4y}-\dfrac{32y}{-4y}=-4y+8$

17. $(13y^3-7y-2)-(12y^2-2y-1)$
$=13y^3-7y-2-12y^2+2y+1$
$=13y^3-12y^2-5y-1$

18. $(2^3)^2=2^6=64$

19. $(3y^2+4y-7)(2y+3)$
$=6y^3+8y^2-14y+9y^2+12y-21$
$=6y^3+17y^2-2y-21$

20. $(2b-9)(2b+9)=4b^2-81$

21. $(a^{-2}b^3c)^2=a^{-4}b^6c^2=\dfrac{b^6c^2}{a^4}$

22.
$$\begin{array}{r} 2y-9 \\ 3y-4\overline{\smash{\big)}6y^2-35y+36} \\ \underline{6y^2-8y} \\ -27y+36 \\ \underline{-27y+36} \\ 0 \end{array}$$

$(6y^2-35y+36)\div(3y-4)=2y-9$

23. $0.00000397=3.97\times10^{-6}$

24. $(xy^5z^3)(x^3y^3z)=x^4y^8z^4$

25. $(6y^2-2y+9)(-2y^3)=-12y^5+4y^4-18y^3$

26. $(6x-12)(3x-2)=18x^2-48x+24$

27. $6.23\times10^{-5}=0.0000623$

28. $(8a^2-a)-(15a^2-4)=8a^2-a-15a^2+4$
$=-7a^2-a+4$

29. $(-3x^2y^3)^2=9x^4y^6$

30. $(4a^2-3)(3a-2)=12a^3-8a^2-9a+6$

31. $(5y-7)^2=(5y)^2-2(5y)(7)+7^2$
$=25y^2-70y+49$

32. $(-3x^{-2}y^{-3})^{-2}=(-3)^{-2}x^4y^6=\dfrac{x^4y^6}{(-3)^2}=\dfrac{x^4y^6}{9}$

33.
$$\begin{array}{r} x+5 \\ x+12\overline{\smash{\big)}x^2+17x+64} \\ \underline{x^2+12x} \\ 5x+64 \\ \underline{5x+60} \\ 4 \end{array}$$

$(x^2+17x+64)\div(x+12)=x+5+\dfrac{4}{x+12}$

34. $2.4\times10^5=240{,}000$

35. $(a^2b^7c^6)(ab^3c)(a^3bc^2)=a^6b^{11}c^9$

36. $(2ab^3)(4a^2-2ab+3b^2)=8a^3b^3-4a^2b^4+6ab^5$

37. $(3x+4y)(2x-5y)=6x^2-7xy-20y^2$

38. $\dfrac{12b^7+36b^5-3b^3}{3b^3}=\dfrac{12b^7}{3b^3}+\dfrac{36b^5}{3b^3}-\dfrac{3b^3}{3b^3}$
$=4b^4+12b^2-1$

39. $(b^2-11b+19)-(5b^2+2b-9)$
$=b^2-11b+19-5b^2-2b+9$
$=-4b^2-13b+28$

40. $(5a^7b^6)^2(4ab)=25a^{14}b^{12}\cdot4ab=100a^{15}b^{13}$

41. $(6b^3-2b^2-5)(2b^2-1)$
$=12b^5-4b^4-10b^2-6b^3+2b^2+5$
$=12b^5-4b^4-6b^3-8b^2+5$

42. $(6-5x)(6+5x)=36-25x^2$

43. $\dfrac{6x^{-2}y^4}{3xy}=\dfrac{6}{3}\cdot x^{-2-1}y^{4-1}=2x^{-3}y^3=\dfrac{2y^3}{x^3}$

44. $$\begin{array}{r} a^2-2a+6 \\ a+3\overline{\big)a^3+a^2+0a+18} \\ \underline{a^3+3a^2} \\ -2a^2+0a \\ \underline{-2a^2-6a} \\ 6a+18 \\ \underline{6a+18} \\ 0 \end{array}$$

$$(a^3+a^2+18)\div(a+3)=a^2-2a+6$$

45. $$\begin{aligned}&(4b^3-7b^2+10)+(2b^2-9b-3)\\&=4b^3-7b^2+10+2b^2-9b-3\\&=4b^3-5b^2-9b+7\end{aligned}$$

46. $(2a^{12}b^3)(-9b^2c^6)(3ac)=-54a^{13}b^5c^7$

47. $-9x^2(2x^2+3x-7)=-18x^4-27x^3+63x^2$

48. $(10y-3)(3y-10)=30y^2-109y+30$

49. $9{,}176{,}000{,}000{,}000=9.176\times10^{12}$

50. $$\begin{aligned}(-3x^{-4}y)(2xy^{-3})^{-2}&=\frac{-3y}{x^4}\cdot\left(\frac{2x}{y^3}\right)^{-2}\\&=\frac{-3y}{x^4}\cdot\left(\frac{y^3}{2x}\right)^2=\frac{-3y}{x^4}\cdot\frac{y^6}{4x^2}\\&=-\frac{3y^7}{4x^6}\end{aligned}$$

51. $$\begin{aligned}(6x^4y^7z^2)^2(-2x^3y^2z^6)^2&=36x^8y^{14}z^4\cdot4x^6y^4z^{12}\\&=144x^{14}y^{18}z^{16}\end{aligned}$$

52. $$\begin{array}{r} -3x^3-2x^2+x-9 \\ \underline{4x+3} \\ -9x^3-6x^2+3x-27 \\ \underline{-12x^4-8x^3+4x^2-36x} \\ -12x^4-17x^3-2x^2-33x-27 \end{array}$$

53. $(8a+1)^2=64a^2+16a+1$

54. $\dfrac{4a^{-2}b^{-8}}{2a^{-1}b^{-2}}=2a^{-2-(-1)}b^{-8-(-2)}=2a^{-1}b^{-6}=\dfrac{2}{ab^6}$

55. $$\begin{array}{r} b^2+5b+2 \\ b-7\overline{\big)b^3-2b^2-33b-7} \\ \underline{b^3-7b^2} \\ 5b^2-33b \\ \underline{5b^2-35b} \\ 2b-7 \\ \underline{2b-14} \\ 7 \end{array}$$

$$(b^3-2b^2-33b-7)\div(b-7)=b^2+5b+2+\frac{7}{b-7}$$

56. $A=(5x)(4x-7)=(20x^2-35x)m^2$

57. $A=(5x+4)^2=(25x^2+40x+16)in^2$

58. $$\begin{aligned}A&=\frac{1}{2}(3x-2)(6x+4)\\&=\frac{1}{2}(2)(3x+2)(3x-2)\\&=(3x+2)(3x-2)=(9x^2-4)\,ft^2\end{aligned}$$

59. $$\begin{aligned}A&=\pi(x-6)^2=\pi(x^2-12x+36)\\&=(\pi x^2-12\pi x+36\pi)cm^2\end{aligned}$$

60. $A=(5x+4)(3x-8)=(15x^2-28x-32)mi^2$

Chapter 7 Test

1. $(3x^3-2x^2-4)+(8x^2-8x+7)$
$=3x^3-2x^2-4+8x^2-8x+7$
$=3x^3+6x^2-8x+3$

2.
$$\begin{array}{r} -2x^3+x^2+0x-7 \\ 2x-3 \\ \hline 6x^3-3x^2+0x+21 \\ -4x^4+2x^3+0x^2-14x \\ \hline -4x^4+8x^3-3x^2-14x+21 \end{array}$$

3. $2x(2x^2-3x)=4x^3-6x^2$

4. $(-2a^2b)^3=-8a^6b^3$

5. $\dfrac{12x^2}{-3x^{-4}}=-4x^{2-(-4)}=-4x^{2+4}=-4x^6$

6. $(2ab^{-3})(3a^{-2}b^4)=6a^{1+(-2)}b^{-3+4}=6a^{-1}b=\dfrac{6b}{a}$

7. $(3a^2-2a-7)-(5a^3+2a-10)$
$=3a^2-2a-7-5a^3-2a+10$
$=-5a^3+3a^2-4a+3$

8. $(a-2b)(a+5b)=a^2+3ab-10b^2$

9. $\dfrac{16x^5-8x^3+20x}{4x}=\dfrac{16x^5}{4x}-\dfrac{8x^3}{4x}+\dfrac{20x}{4x}$
$=4x^4-2x^2+5$

10.
$$\begin{array}{r} 2x+3 \\ 2x-3\overline{)4x^2+0x-7} \\ \underline{4x^2-6x} \\ 6x-7 \\ \underline{6x-9} \\ 2 \end{array}$$

$(4x^2-7)\div(2x-3)=2x+3+\dfrac{2}{2x-3}$

11. $(-2xy^2)(3x^2y^4)=-6x^{1+2}y^{2+4}=-6x^3y^6$

12. $-3y^2(-2y^2+3y-6)=6y^4-9y^3+18y^2$

13. $\dfrac{27xy^3}{3x^4y^3}=9x^{1-4}y^{3-3}=9x^{-3}y^0=\dfrac{9}{x^3}$

14. $(2x-5)^2=4x^2-20x+25$

15. $(2x-7y)(5x-4y)=10x^2-43xy+28y^2$

16.
$$\begin{array}{r} x^2-4x+5 \\ x-3 \\ \hline 3x^2+12x-15 \\ x^3-4x^2+\quad 5x \\ \hline x^3-x^2+17x-15 \end{array}$$

17. $(a^2b^{-3})^2=a^4b^{-6}=\dfrac{a^4}{b^6}$

18. $0.000029=2.9\times10^{-5}$

19. $(4y-3)(4y+3)=16y^2-9$

20. $(3y^3-5y+8)-(-2y^2+5y+8)$
$=3y^3-5y+8+2y^2-5y-8$
$=3y^3+2y^2-10y$

21. $(-3a^3b^2)^2 = (-3)^2 a^{3\cdot 2}b^{2\cdot 2} = 9a^6b^4$

22.
$$\begin{array}{r} 5a^2 - 2a + 3 \\ 2a - 7 \\ \hline -35a^2 + 14a - 21 \\ 10a^3 - 4a^2 + 6a \quad \\ \hline 10a^3 - 39a^2 + 20a - 21 \end{array}$$

23. $(3b+2)^2 = 9b^2 + 12b + 4$

24. $\frac{-2a^2b^3}{8a^4b^8} = -\frac{2}{8}a^{2-4}b^{3-8} = -\frac{1}{4}a^{-2}b^{-5} = -\frac{1}{4a^2b^5}$

25.
$$\begin{array}{r} 4x+8 \\ 2x-3\overline{)8x^2+4x-3} \\ \underline{8x^2-12x} \\ 16x-3 \\ \underline{16x-24} \\ 21 \end{array}$$

$(8x^2+4x-3)\div(2x-3) = 4x+8+\frac{21}{2x-3}$

26. $(a^2b^5)(ab^2) = a^3b^7$

27. $(a-3b)(a+4b) = a^2 + ab - 12b^2$

28. $3.5\times10^{-8} = 0.000000035$

29. $A = (2x+3)^2 = (4x^2+12x+9)m^2$

30. $A = \pi r^2 = \pi(x-5)^2$
$= \pi(x^2 - 10x + 25)$
$= (\pi x^2 - 10\pi x + 25\pi)in^2$

Cumulative Review Exercises

1. $\frac{3}{16} - \left(-\frac{3}{8}\right) - \frac{5}{9} = \frac{3}{16} + \frac{3}{8} + \left(-\frac{5}{9}\right)$
$= \frac{27+54+(-80)}{144} = \frac{1}{144}$

2. $-5^2\cdot\left(\frac{2}{3}\right)^3\cdot\left(-\frac{3}{8}\right) = -\frac{25}{1}\cdot\frac{\overset{1}{\cancel{8}}}{\underset{9}{\cancel{27}}}\cdot\left(-\frac{\overset{1}{\cancel{3}}}{\underset{1}{\cancel{8}}}\right) = \frac{25}{9}$

3. $\left(-\frac{1}{2}\right)^2 \div \left(\frac{5}{8} - \frac{5}{6}\right) + 2 = \frac{1}{4} \div \left(\frac{5}{8} - \frac{5}{6}\right) + 2$
$= \frac{1}{4} \div \left(\frac{15-20}{24}\right) + 2 = \frac{1}{4} \div \left(-\frac{5}{24}\right) + 2$
$= \frac{1}{\underset{1}{\cancel{4}}}\cdot\left(-\frac{\overset{6}{\cancel{24}}}{5}\right) + 2 = -\frac{6}{5} + 2 = -\frac{6}{5} + \frac{10}{5} = \frac{4}{5}$

4. Opposite of -87: 87

5.
$$\begin{array}{r} 0.775 \\ 40\overline{)31.000} \\ \underline{28\,0} \\ 3\,00 \\ \underline{280} \\ 200 \\ \underline{200} \\ 0 \end{array}$$

$\frac{31}{40} = 0.775$

6. $\frac{b-(a-b)^2}{b^2} = \frac{-2-[3-(-2)]^2}{(-2)^2}$
$= \frac{-2-(3+2)^2}{4} = \frac{-2-5^2}{4}$
$= \frac{-2-25}{4} = -\frac{27}{4}$

7. $-3x-(-xy)+2x-5xy$
$=-3x+xy+2x-5xy$
$=-x-4xy$

8. $(\overset{4}{\cancel{16}}\,x)\left(-\frac{3}{\underset{1}{\cancel{4}}}\right)=-12x$

9. $-2[3x-4(3-2x)+2]=-2[3x-12+8x+2]$
$=-2(11x-10)=-22x+20$

10. Using the Inverse Property of Addition:
$-8+8=0$

11. $12=-\frac{2}{3}x$
$\left(-\frac{3}{\underset{1}{\cancel{2}}}\right)(\overset{6}{\cancel{12}})=\left(-\frac{3}{2}\right)\left(-\frac{2}{3}\right)x$
$-18=x$

12. $3x-7=2x+9$
$3x-7+7=2x+9+7$
$3x=2x+16$
$3x-2x=2x+16-2x$
$x=16$

13. $3-4(2-x)=3x+7$
$3-8+4x=3x+7$
$-5+4x=3x+7$
$-5+4x+5=3x+7+5$
$4x=3x+12$
$4x-3x=3x+12-3x$
$x=12$

14. $-\frac{4}{5}x=16-x$
$5\left(-\frac{4}{5}x\right)=5(16-x)$
$-4x=80-5x$
$-4x+5x=80-5x+5x$
$x=80$

15. $A=PB$
$38.4=P(160)$
$\frac{38.4}{160}=P$
$0.24=P$
$P=24\%$

16. $7x-8\geq-29$
$7x-8+8\geq-29+8$
$7x\geq-21$
$x\geq-3$

17. $m=\frac{y_2-y_1}{x_2-x_1}=\frac{5-(-4)}{-2-3}=\frac{5+4}{-2+(-3)}=-\frac{9}{5}$

18. $y-y_1=m(x-x_1)$
$y-(-3)=-\frac{3}{2}(x-1)$
$y+3=-\frac{3}{2}x+\frac{3}{2}$
$y=-\frac{3}{2}x+\frac{3}{2}-\frac{6}{2}$
$y=-\frac{3}{2}x-\frac{3}{2}$

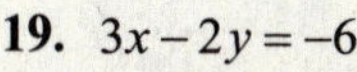

19. $3x-2y=-6$

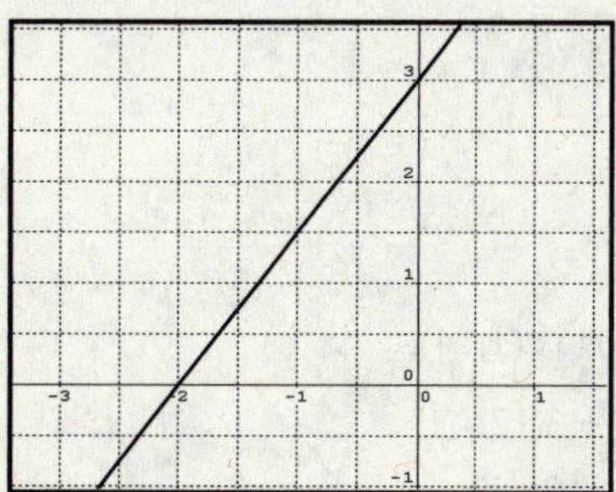

20. $y \le \frac{4}{5}x - 3$

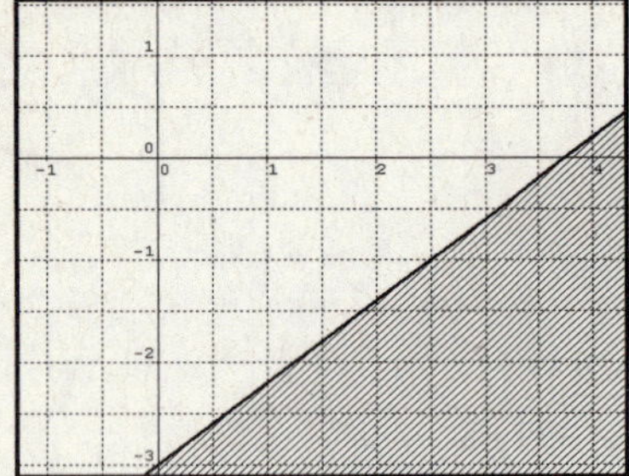

21. Domain: {-8, -6, -4,-2}; Range: {-7, -5, -2, 0); Function

22. $f(x) = -2x + 10$
$f(6) = -2(6) + 10$
$f(6) = -12 + 10 = -2$

23. The solution is (4, 1).
$x = 3y + 1$
$2x + 5y = 13$
$2(3y + 1) + 5y = 13$
$6y + 2 + 5y = 13$
$11y + 2 = 13$
$11y = 11$
$y = 1$
$x = 3y + 1$
$x = 3(1) + 1$
$x = 4$

24. The solution is (1, -4).
$3(9x - 2y) = 3(17)$
$\underline{2(5x + 3y) = 2(-7)}$
$27x - 6y = 51$
$\underline{10x + 6y = -14}$
$37x = 37$
$x = 1$
$5x + 3y = -7$
$5 + 3y = -7$
$3y = -12$
$y = -4$

25. $(5b^3 - 4b^2 - 7) - (3b^2 - 8b + 3)$
$= 5b^3 - 4b^2 - 7 - 3b^2 + 8b - 3$
$= 5b^3 - 7b^2 + 8b - 10$

26.
$$\begin{array}{r} 5x^2 - 2x + 1 \\ 3x - 4 \\ \hline -20x^2 + 8x - 4 \\ 15x^3 - 6x^2 + 3x \quad \\ \hline 15x^3 - 26x^2 + 11x - 4 \end{array}$$

27. $(4b - 3)(5b - 8) = 20b^2 - 47b + 24$

28. $(5b + 3)^2 = 25b^2 + 30b + 9$

29. $\frac{-3a^3b^2}{12a^4b^{-2}} = -\frac{1}{4}a^{3-4}b^{2-(-2)} = -\frac{1}{4}a^{-1}b^{2+2} = -\frac{b^4}{4a}$

30. $\frac{-15y^2 + 12y - 3}{-3y} = \frac{-15y^2}{-3y} + \frac{12y}{-3y} - \frac{3}{-3y}$
$= 5y - 4 + \frac{1}{y}$

31.
$$\begin{array}{r} a - 7 \\ a + 4 \overline{)a^2 - 3a - 28} \\ \underline{a^2 + 4a} \quad\quad \\ -7a - 28 \\ \underline{-7a - 28} \\ 0 \end{array}$$

$(a^2 - 3a - 28) \div (a + 4) = a - 7$

32. $(-3x^{-4}y)(-3x^{-2}y) = 9x^{-4+(-2)}y^{1+1} = 9x^{-6}y^2 = \frac{9y^2}{x^6}$

33. Range: $\{1, 9, 17, 21, 25\}$

$$f(x) = -\frac{4}{3}x + 9$$
$$f(-12) = -\frac{4}{3}(-12) + 9 = 16 + 9 = 25$$
$$f(-9) = -\frac{4}{3}(-9) + 9 = 12 + 9 = 21$$
$$f(-6) = -\frac{4}{3}(-6) + 9 = 8 + 9 = 17$$
$$f(0) = -\frac{4}{3}(0) + 9 = 0 + 9 = 9$$
$$f(6) = -\frac{4}{3}(6) + 9 = -8 + 9 = 1$$

34. $5(n-12) = 5n - 60$

35. $8n - 2n = 18$
$$6n = 18$$
$$n = 3$$

36. $P = 2L + 2W$
$$42 = 2L + 2(0.4L)$$
$$42 = 2L + 0.8L$$
$$42 = 2.8L$$
$$L = 15$$
$$0.4L = 6$$

The length is 15 m and the width is 6 m.

37. $S = C + rC$
$$S = 24 + 0.8(24)$$
$$S = 24 + 19.2$$
$$S = 43.20$$

The selling price is $43.20.

38. Let p = the percent concentration of orange juice in the final mixture

	B	·	P	=	A
Pure juice	50	·	1	=	50
Fruit punch	200	·	0.10	=	0.10(200)
Total	250	·	p	=	$p(250)$

$$50 + 0.10(200) = 250p$$
$$50 + 20 = 250p$$
$$70 = 250p$$
$$\frac{70}{250} = p$$
$$0.28 = p$$
$$28\% = p$$

The mixture is 28% orange juice.

39. They meet when they have each traveled 25 miles.

	r	·	t	=	d
Car	50	·	t	=	$50t$
Cycle	10	·	$t+2$	=	$10(t+2)$

They meet when they have traveled the same distance.

$$50t = 10(t+2)$$
$$50t = 10t + 20$$
$$40t = 20$$
$$t = 0.5$$
$$50t = 50(0.5)$$
$$50t = 25\ mi$$

40. $A = (3x+2)^2 = (9x^2 + 12x + 4)\,ft^2$

CHAPTER 8 Factoring

Chapter 8 Prep Test

1. $30 = 2 \cdot 15 = 2 \cdot 3 \cdot 5$

2. $-3(4y-5) = -12y+15$

3. $-(a-b) = -a+b$

4. $2(a-b)-5(a-b) = 2a-2b-5a+5b$
$= -3a+3b$

5. $4x = 0$
$\frac{4x}{4} = \frac{0}{4}$
$x = 0$

6. $2x+1 = 0$
$2x = -1$
$\frac{2x}{2} = \frac{-1}{2}$
$x = -\frac{1}{2}$

7. $(x+4)(x-6) = x^2-6x+4x-24$
$= x^2-2x-24$

8. $(2x-5)(3x+2) = 6x^2+4x-15x-10$
$= 6x^2-11x-10$

9. $\frac{x^5}{x^2} = x^{5-2} = x^3$

10. $\frac{6x^4y^3}{2xy^2} = \frac{6}{2} \cdot \frac{x^4}{x} \cdot \frac{y^3}{y^2} = 3x^3y$

Section 8.1 Common Factors

1. $4 = 2 \cdot 2$
$12 = 2 \cdot 2 \cdot 3$
$16 = 2 \cdot 2 \cdot 2 \cdot 2$
$GCF = 2 \cdot 2 = 4$

3. **a.** This is a **product.**
b. This is a **sum.**

5. The common binomial factor is $c-6$.

7. A factor of a polynomial is a polynomial itself. For example, a common monomial factor is a monomial that is a factor of each term of a polynomial. To factor a polynomial means to write the polynomial as a product of other polynomials.

9. **a.** z is not a factor of the GCF because it does not appear as a factor in $9x^3y^2$.
b. The GCF of the coefficients 9, 3, and 6 is 3.
c. The GCF of x^3, x, and x^2 is x.
d. The GCF of y^2, y^3, and y^4 is y^2.
e. $GCF = 3xy^2$

11. $x^3 = x \cdot x \cdot x$
$x^7 = x \cdot x \cdot x \cdot x \cdot x \cdot x \cdot x$
$GCF = x \cdot x \cdot x = x^3$

13. $x^2y^4 = x \cdot x \cdot y \cdot y \cdot y \cdot y$
$xy^6 = x \cdot y \cdot y \cdot y \cdot y \cdot y \cdot y$
$GCF = x \cdot y \cdot y \cdot y \cdot y = xy^4$

15. $x^2y^4z^6 = x \cdot x \cdot y \cdot y \cdot y \cdot y \cdot z \cdot z \cdot z \cdot z \cdot z \cdot z$
$xy^8z^2 = x \cdot y \cdot y \cdot y \cdot y \cdot y \cdot y \cdot y \cdot y \cdot z \cdot z$
$GCF = x \cdot y \cdot y \cdot y \cdot y \cdot z \cdot z = xy^4z^2$

17. $14a^3 = 2 \cdot 7 \cdot a \cdot a \cdot a$
$49a^7 = 7 \cdot 7 \cdot a \cdot a \cdot a \cdot a \cdot a \cdot a \cdot a$
$GCF = 7 \cdot a \cdot a \cdot a = 7a^3$

19. $3x^2y^2 = 3 \cdot x \cdot x \cdot y \cdot y$
$5ab^2 = 5 \cdot a \cdot b \cdot b$
$GCF = 1$

21. $9a^2b^4 = 3 \cdot 3 \cdot a \cdot a \cdot b \cdot b \cdot b \cdot b$
$24a^4b^2 = 2 \cdot 2 \cdot 2 \cdot 3 \cdot a \cdot a \cdot a \cdot a \cdot b \cdot b$
$GCF = 3 \cdot a \cdot a \cdot b \cdot b = 3a^2b^2$

23. $ab^3 = a \cdot b \cdot b \cdot b$
$4a^2b = 2 \cdot 2 \cdot a \cdot a \cdot b$
$12a^2b^3 = 2 \cdot 2 \cdot 3 \cdot a \cdot a \cdot b \cdot b \cdot b$
$GCF = a \cdot b = ab$

25. $2x^2y = 2 \cdot x \cdot x \cdot y$
$4xy = 2 \cdot 2 \cdot x \cdot y$
$8x = 2 \cdot 2 \cdot 2 \cdot x$
$GCF = 2 \cdot x = 2x$

27. $3x^2y^2 = 3 \cdot x \cdot x \cdot y \cdot y$
$6x = 2 \cdot 3 \cdot x$
$9x^3y^3 = 3 \cdot 3 \cdot x \cdot x \cdot x \cdot y \cdot y \cdot y$
$GCF = 3 \cdot x = 3x$

29. $5a + 5 = 5(a+1)$

31. $16 - 8a^2 = 8(2 - a^2)$

33. $8x + 12 = 4(2x+3)$

35. $30a - 6 = 6(5a - 1)$

37. $7x^2 - 3x = x(7x - 3)$

39. $3a^2 + 5a^5 = a^2(3 + 5a^3)$

41. $2x^4 - 4x = 2x(x^3 - 2)$

43. $10x^4 - 12x^2 = 2x^2(5x^2 - 6)$

45. $x^2y - xy^3 = xy(x - y^2)$

47. $2a^5b + 3xy^3 = 2a^5b + 3xy^3$

49. $6a^2b^3 - 12b^2 = 6b^2(a^2b - 2)$

51. $6a^2bc + 4ab^2c = 2abc(3a + 2b)$

53. $6x^3y^3 - 12x^6y^6 = 6x^3y^3(1 - 2x^3y^3)$

55. $x^3 - 3x^2 - x = x(x^2 - 3x - 1)$

57. $2x^2 + 8x - 12 = 2(x^2 + 4x - 6)$

59. $b^3 - 5b^2 - 7b = b(b^2 - 5b - 7)$

61. $3x^3 + 6x^2 + 9x = 3x(x^2 + 2x + 3)$

63. $2x^4 - 4x^3 + 6x^2 = 2x^2(x^2 - 2x + 3)$

65. $6a^5 - 3a^3 - 2a^2 = a^2(6a^3 - 3a - 1)$

67. $8x^2y^2 - 4x^2y + x^2 = x^2(8y^2 - 4y + 1)$

69. $4x^5y^5 - 8x^4y^4 + x^3y^3 = x^3y^3(4x^2y^2 - 8xy + 1)$

71. Since $a < b < c$, the GCF is x^a.

73. **a.** The expression is the sum of two terms. One term is $5x(3x-2)$ and the other term is $4(3x-2)$. The terms have the common binomial factor $3x-2$.

b. To write the expression $5x(3x-2)+4(3x-2)$ in factored form, factor out the common binomial factor: $5x(3x-2)+4(3x-2)=(3x-2)(5x+4)$.

75. $x(a+b)+2(a+b)=(a+b)(x+2)$

77. $x(b+2)-y(b+2)=(b+2)(x-y)$

79. $a(y-4)-b(y-4)=(y-4)(a-b)$

81. $a(x-2)-b(2-x)=a(x-2)+b(x-2)$

83. $b(a-7)+3(7-a)=b(a-7)-3(a-7)$

85. $x(a-2b)-y(2b-a)=x(a-2b)+y(a-2b)$

87. $a(x-2)+5(2-x)=a(x-2)-5(x-2)$
$=(x-2)(a-5)$

89. $b(y-3)+3(3-y)=b(y-3)-3(y-3)$
$=(y-3)(b-3)$

91. $a(x-y)-2(y-x)=a(x-y)+2(x-y)$
$=(x-y)(a+2)$

93. $z(c+5)-8(5+c)=z(c+5)-8(c+5)$
$=(c+5)(z-8)$

95. $w(3x-4)-(4-3x)=w(3x-4)+1(3x-4)$
$=(3x-4)(w+1)$

97. $x^3+4x^2+3x+12=(x^3+4x^2)+(3x+12)$
$=x^2(x+4)+3(x+4)$
$=(x+4)(x^2+3)$

99. $3y^3-12y^2+y-4=(3y^3-12y^2)+(y-4)$
$=3y^2(y-4)+1(y-4)$
$=(3y^2+1)(y-4)$

101. $8+2c+4a^2+a^2c=(8+2c)+(4a^2+a^2c)$
$=2(4+c)+a^2(4+c)$
$=(4+c)(2+a^2)$

103. $2y^2-10y+7xy-35x=(2y^2-10y)+(7xy-35x)$
$=2y(y-5)+7x(y-5)$
$=(y-5)(2y+7x)$

105. $ab+3b-2a-6=(ab+3b)-(2a+6)$
$=b(a+3)-2(a+3)$
$=(a+3)(b-2)$

107. $x^2a-2x^2-3a+6=(x^2a-2x^2)-(3a-6)$
$=x^2(a-2)-3(a-2)$
$=(a-2)(x^2-3)$

109. $t^2+4t-st-4s=(t^2+4t)-(st+4s)$
$=t(t+4)-s(t+4)$
$=(t+4)(t-s)$

111. $21x^2+6xy-49x-14y$
$=(21x^2+6xy)-(49x+14y)$
$=3x(7x+2y)-7(7x+2y)$
$=(7x+2y)(3x-7)$

113. $2ra+a^2-2r-a=(2ra+a^2)-(2r+a)$
$=a(2r+a)-1(2r+a)$
$=(2r+a)(a-1)$

115. i. $ab-3a-2b-6=(ab-3a)-(2b+6)$
$=a(b-3)-2(b+3)$

Cannot be factored by grouping.

ii. $2ab+3a-2b-6=(2ab+3a)-(2b+3)$
$=a(2b+3)-1(2b+3)$
$=(2b+3)(a-1)$

Can by factored by grouping.

iii. $ab+3a+2b-6=(ab+3a)+(2b-6)$
$=a(b+3)+2(b-3)$

Cannot be factored by grouping.

117. $P=2L+2W$
$=2(L+W)$
$=2[2(L+W)]=2P$

When $L+W$ doubles, P also doubles.

119. a. $a-3=-1(3-a)$

b. $2-(x-y)=2+(y-x)$

c. $4x+(3a-b)=4x-(b-3a)$

121. a. $2x^2+6x+5x+15=(2x^2+6x)+(5x+15)$
$=2x(x+3)+5(x+3)$
$=(x+3)(2x+5)$

b. $2x^2+5x+6x+15=(2x^2+5x)+(6x+15)$
$=x(2x+5)+3(2x+5)$
$=(2x+5)(x+3)$

123. a. $2a^2-2ab-3ab+3b^2$
$=(2a^2-2ab)-(3ab-3b^2)$
$=2a(a-b)-3b(a-b)$
$=(a-b)(2a-3b)$

b. $2a^2-3ab-2ab+3b^2$
$=(2a^2-3ab)-(2ab-3b^2)$
$=a(2a-3b)-b(2a-3b)$
$=(2a-3b)(a-b)$

Section 8.2 Factoring Polynomials of the Form x^2+bx+c

1. The value of b is the coefficient of x, so b is -8.

3. Two numbers whose sum is 4 and whose product is -12 are -2 and 6.

5. In factoring a trinomial, if the constant term is negative, the signs in both binomials will be different.

7.

Factors of 18	Sum
1, 18	19
2, 9	11
3, 6	9

9.

Factors of -21	Sum
1, -21	-20
-1, 21	20
3, -7	-4
-3, 7	4

11.

Factors of -28	Sum
1, -28	-27
-1, 28	27
2, -14	-12
-2, 14	12
4, -7	-3
-4, 7	3

13. $x^2+3x+2=(x+2)(x+1)$

15. $x^2-x-2=(x-2)(x+1)$

17. $a^2+a-12=(a+4)(a-3)$

19. $a^2-3a+2=(a-2)(a-1)$

21. $a^2+a-2=(a+2)(a-1)$

23. $b^2+7b-8=(b+8)(b-1)$

25. $y^2+6y-55=(y+11)(y-5)$

27. $y^2-5y+6=(y-3)(y-2)$

29. $z^2-14z+45=(z-9)(z-5)$

31. $p^2+2p-35=(p+7)(p-5)$

33. $p^2-6p+8=(p-2)(p-4)$

35. $b^2+13b+40=(b+5)(b+8)$

37. $x^2+9x-70=(x+14)(x-5)$

39. $b^2+3b-40=(b+8)(b-5)$

41. $y^2-y-72=(y-9)(y+8)$

43. $p^2+16p+39=(p+3)(p+13)$

45. $y^2-9y+81$ *nonfactorable*

47. $x^2+21x+38=(x+2)(x+19)$

49. $x^2+5x-36=(x+9)(x-4)$

51. $a^2-15a+36=(a-3)(a-12)$

53. a. If m and n are both positive constants then their product mn is positive. mn corresponds to c, so c is positive.

b. If m and n are both positive constants then mx is positive and nx is positive. $mx + nx$ corresponds to b, so b is positive.

55. a. The first step in factoring the trinomial is to look for a <u>common</u> factor of the three terms.

b. The GCF of the terms of the trinomial $5x^2-10x-40$ is <u>5</u>.

c. Factor the GCF from the terms of the trinomial: $5x^2-10x-40=5(x^2-2x-8)$

d. Factor the trinomial completely:

$$\begin{aligned}5x^2-10x-40&=5(x^2-2x-8)\\&=5(x-4)(x+2)\end{aligned}$$

57. $$\begin{aligned}2x^2+6x+4&=2(x^2+3x+2)\\&=2(x+2)(x+1)\end{aligned}$$

59. $$\begin{aligned}3a^2+3a-18&=3(a^2+a-6)\\&=3(a+3)(a-2)\end{aligned}$$

61. $$\begin{aligned}ab^2+2ab-15a&=a(b^2+2b-15)\\&=a(b+5)(b-3)\end{aligned}$$

63. $$\begin{aligned}xy^2-5xy+6x&=x(y^2-5y+6)\\&=x(y-3)(y-2)\end{aligned}$$

65. $2a^3+6a^2+4a=2a(a^2+3a+2)$
$=2a(a+2)(a+1)$

67. $4y^3+12y^2-72y=4y(y^2+3y-18)$
$=4y(y+6)(y-3)$

69. $5z^2-15z-140=5(z^2-3x-28)$
$=5(z-7)(z+4)$

71. $2a^3+8a^2-64a=2a(a^2+4a-32)$
$=2a(a+8)(a-4)$

73. $x^2-5xy+6y^2=(x-2y)(x-3y)$

75. $a^2-9ab+20b^2=(a-4b)(a-5b)$

77. $x^2-3xy-28y^2=(x-7y)(x+4y)$

79. $y^2-15yz-41z^2$ *nonfactorable*

81. $z^4-12z^3+35z^2=z^2(z^2-12z+35)$
$=z^2(z-5)(z-7)$

83. $b^4-22b^3+120b^2=b^2(b^2-22b+120)$
$=b^2(b-10)(b-12)$

85. $2y^4-26y^3-96y^2=2y^2(y^2-13y-48)$
$=2y^2(y-16)(y+3)$

87. $x^4+7x^3-8x^2=x^2(x^2+7x-8)$
$=x^2(x+8)(x-1)$

89. $3x^3-36x^2+81x=3x(x^2-12x+27)$
$=3x(x-9)(x-3)$

91. $x^2-8xy+15y^2=(x-3y)(x-5y)$

93. $a^2-13ab+42b^2=(a-6b)(a-7b)$

95. $3x^2y+60xy-63y=3y(x^2+20x-21)$
$=3y(x+21)(x-1)$

97. $3x^3+3x^2-36x=3x(x^2+x-12)$
$=3x(x+4)(x-3)$

99. **a.** Yes. $3x^2-3x-36=3(x^2-x-12)$
$=3(x-4)(x+3)$

b. Yes. $x^2y-xy-12y=y(x^2-x-12)$
$=y(x-4)(x+3)$

101. $x^2+2x-3=(x-1)(x+3)$
$=a(x+3)$
So $a=x+3$.

103. $20+c^2+9c=c^2+9c+20$
$=(c+4)(c+5)$

105. $45a^2+a^2b^2-14a^2b=a^2(45+b^2-14b)$
$=a^2(b^2-14b+45)$
$=a^2(b-9)(b-5)$

107. k can be 12, -12, 36, or -36.

Factors of 35	k
5, 7	12
-5, -7	-12
1, 35	36
-1, -35	-36

109. k can be 22, -22, 10, or -10.

Factors of 21	Sum
1, 21	22
-1, -21	-22
3, 7	10
-3, -7	-10

111. Since k is positive, then the possible factorizations of y^2+4y+k are $(y+2)(y+2)$ or $(y+1)(y+3)$. So k must be 3 or 4.

113. Since k is positive, then the possible factorizations of a^2-6a+k are $(a-5)(a-1)$, $(a-3)(a-3)$ or $(a-4)(a-2)$. So k must be 5, 8, or 9.

115. Since k is positive, then the only possible factorization of x^2-3x+k is $(x+2)(x+1)$. So k must be 2.

117. If k is allowed to be negative, then there are infinitely many possible values for k because there are infinitely many combinations of numbers whose difference is the same.

Section 8.3 Factoring Polynomials of the Form ax^2+bx+c

1. $6x^2+11x-10=(3x-2)(2x+5)$

3. $20x^2-31x+12=(5x-4)(4x-3)$

5. To factor $2x^2-5x+2$ by grouping, find two numbers whose product is <u>4</u> and whose sum is <u>-5</u>.

7. To factor $4x^2-8x+3$ by grouping, -8x must be written as <u>-6x</u> + <u>-2x</u>.

9. The **OI** part of **FOIL**, the sum of the Outer and Inner products, determines the middle term.

11. $2x^2+3x+1=(x+1)(2x+1)$

Factors of 2	Positive Factors of 1 (since b is positive)
1, 2	1, 1

Trial Factors	Middle Term
*$(x+1)(2x+1)$	$3x$

13. $2y^2+7y+3=(y+3)(2y+1)$

Factors of 2	Positive Factors of 3
1, 2	1, 3

Trial Factors	Middle Term
$(y+1)(2y+3)$	$5y$
*$(2y+1)(y+3)$	$7y$

15. $2a^2-3a+1=(a-1)(2a-1)$

Factors of 2	Negative Factors of 1 (since b is negative)
1, 2	-1, -1

Trial Factors	Middle Term
*$(a-1)(2a-1)$	$-3a$

17. $2b^2 - 11b + 5 = (b-5)(2b-1)$

Factors of 2	Factors of 5
1, 2	-1, -5

Trial Factors	Middle Term
$(b - 1)(2b - 5)$	$-7b$
*$(2b - 1)(b - 5)$	$11b$

19. $2x^2 + x - 1 = (x+1)(2x-1)$

Factors of 2	Factors of -1
1, 2	1, -1

Trial Factors	Middle Term
*$(x + 1)(2x - 1)$	x
$(2x + 1)(x - 2)$	$-3x$

21. $2x^2 - 5x - 3 = (x-3)(2x+1)$

Factors of 2	Factors of -3
1, 2	1, -3
	-1, 3

Trial Factors	Middle Term
$(x + 1)(2x - 3)$	$-x$
*$(2x + 1)(x - 3)$	$-5x$

23. $6z^2 - 7z + 3$ *nonfactorable*

Factors of 6	Factors of 3
1, 6	-1,- 3
2, 3	

Trial Factors	Middle Term
$(z - 1)(6z - 3)$	Common factor
$(6z - 1)(z - 3)$	$-19z$
$(2z - 1)(3z - 3)$	Common factor
$(2z - 3)(3z - 1)$	$-11z$

25. $6t^2 - 11t + 4 = (2t-1)(3t-4)$

Factors of 6	Factors of 4
1, 6	-1, -4
2, 3	-2, -2

Trial Factors	Middle Term
$(t - 1)(6t - 4)$	Common factor
$(6t - 1)(t - 4)$	$-25t$
*$(2t - 1)(3t - 4)$	$-11t$

27. $8x^2 + 33x + 4 = (x+4)(8x+1)$

Factors of 8	Factors of 4
1, 8	1, 4
2, 4	2, 2

Trial Factors	Middle Term
$(x + 1)(8x + 4)$	Common factor
*$(x + 4)(8x + 1)$	$33x$

29. $6b^2-19b+15=(2b-3)(3b-5)$

Factors of 6	Factors of 15
1, 6	-1, -15
2, 3	-3, -5

Trial Factors	Middle Term
$(b-1)(6b-15)$	Common factor
$(b-15)(6b-1)$	$-91b$
$(b-3)(6b-5)$	$-23b$
$(b-5)(6b-3)$	Common factor
$*(2b-3)(3b-5)$	$-19b$

31. $3p^2+22p-16=(p+8)(3p-2)$

Factors of 3	Factors of -16
1, 3	1, -16
	-1, 16
	8, -2
	-8, 2
	4, -4

Trial Factors	Middle Term
$(p+1)(3p-16)$	$-13p$
$(p-16)(3p+1)$	$-47p$
$(p-1)(3p+16)$	$13p$
$(3p-1)(p+16)$	$47p$
$*(p+8)(3p-2)$	$22p$

33. $6x^2-17x+12=(2x-3)(3x-4)$

Factors of 6	Factors of 12
1, 6	-1, -12
2, 3	-3, -4
	-6, -2

Trial Factors	Middle Term
$(x-1)(6x-12)$	Common factor
$(x-12)(6x-1)$	$-73x$
$*(2x-3)(3x-4)$	$-17x$

35. $5b^2+33b-14=(b+7)(5b-2)$

Factors of 5	Factors of -14
1, 5	1, -14
	-1, 14
	-2, 7
	2, -7

Trial Factors	Middle Term
$(b+1)(5b-14)$	$-9b$
$(b-14)(5b+1)$	$-69b$
$(b-1)(5b+14)$	$9b$
$(b+14)(5b-1)$	$69b$
$(b-2)(5b+7)$	$-3b$
$*(b+7)(5b-2)$	$33b$

37. $6a^2 + 7a - 24 = (2a - 3)(3a + 8)$

Factors of 6	Factors of -24
1, 6	1, -24
2, 3	-1, 24
	2, -12
	-2, 12
	3, -8
	-3, 8
	4, -6
	-4, 6

Trial Factors	Middle Term
$(a + 1)(6a - 24)$	Common factors
$(a - 24)(6a + 1)$	$-143a$
$*(2a - 3)(3a + 8)$	$7a$

39. $18t^2 - 9t - 5 = (3t + 1)(6t - 5)$

Factors of 18	Factors of -5
1, 18	1, -5
2, 9	-1, 5
3, 6	

Trial Factors	Middle Term
$(t + 1)(18t - 5)$	$13t$
$(t - 5)(18t + 1)$	$-89t$
$(2t + 1)(18t - 5)$	$8t$
$(2t - 5)(9t + 1)$	$-43t$
$*(3t + 1)(6t - 5)$	$-9t$

41. $15a^2 + 26a - 21 = (3a + 7)(5a - 3)$

Factors of 15	Factors of -21
1, 15	7, -3
3, 5	-7, 3
	1, -21
	-1, 21

Trial Factors	Middle Term
$(a + 7)(15a - 3)$	Common factors
$(15a + 7)(a - 3)$	$-38a$
$*(3a + 7)(5a - 3)$	$26a$

43. $8y^2 - 26y + 15 = (2y - 5)(4y - 3)$

Factors of 8	Factors of 15
1, 8	-1, -15
2, 4	-3, -5

Trial Factors	Middle Term
$(y - 1)(8y - 15)$	$-23y$
$(y - 15)(8y - 1)$	$-121y$
$(y - 3)(8y - 5)$	$-29y$
$(y - 5)(8y - 3)$	$-43y$
$(2y - 1)(4y - 15)$	$-34y$
$(2y - 15)(4y - 1)$	$-62y$
$(2y - 3)(4y - 5)$	$-22y$
$*(2y - 5)(4y - 3)$	$-26y$

45. $3z^2 + 95z + 10$ *nonfactorable*

Factors of 3	Factors of 10
1, 3	1, 10
	2, 5

Trial Factors	Middle Term
$(z+1)(3z+10)$	$13z$
$(z+10)(3z+1)$	$31z$
$(z+2)(3z+5)$	$11z$
$(z+5)(3z+2)$	$17z$

47. $28 + 3z - z^2 = (4+z)(7-z)$

Factors of 28	Factors of -1
1, 28	-1, 1
2, 14	
4, 7	

Trial Factors	Middle Term
$(1 - z)(28 + z)$	$-27z$
$(1 + z)(28 - z)$	$27z$
$(2 - z)(14 + z)$	$-12z$
$(2 + z)(14 - z)$	$12z$
$(4 - z)(7 + z)$	$-3z$
$*(4 + z)(7 - z)$	$3z$

49. $8 - 7x - x^2 = (1-x)(8+x)$

Factors of 8	Factors of -1
1, 8	-1, 1
2, 4	

Trial Factors	Middle Term
$*(1 - x)(8 + x)$	$-7x$

51. $9x^2 + 33x - 60 = 3(3x^2 + 11x - 20)$
$= 3(x+5)(3x-4)$

Factors of 3	Factors of -20
1, 3	-1, 20
	1, -20
	2, -10
	-2, 10
	-4, 5
	4, -5

Trial Factors	Middle Term
$(x - 1)(3x + 20)$	$17x$
$(x + 20)(3x - 1)$	$59x$
$(x + 1)(3x - 20)$	$-17x$
$(x - 20)(3x + 1)$	$-59x$
$(x + 2)(3x - 10)$	$-4x$
$(x - 10)(3x + 2)$	$-28x$
$(x - 2)(3x + 10)$	$4x$
$(x + 10)(3x - 2)$	$28x$
$(x - 4)(3x + 5)$	$-7x$
$*(x + 5)(3x - 4)$	$11x$

53. $24x^2 - 52x + 24 = 4(6x^2 - 13x + 6)$
$= 4(2x-3)(3x-2)$

Factors of 6	Factors of 6
1, 6	-1, -6
2, 3	-2, -3

Trial Factors	Middle Term
$(x - 1)(6x - 6)$	Common factor
$(x - 6)(6x - 1)$	$-37x$
$(x - 2)(6x - 3)$	Common factor
$(x - 3)(6x - 2)$	Common factor
$(2x - 1)(3x - 6)$	Common factor
$(2x - 6)(3x - 1)$	Common factor
$(2x - 2)(3x - 3)$	Common factor
*$(2x - 3)(3x - 2)$	$-13x$

55. $35a^4 + 9a^3 - 2a^2 = a^2(35a^2 + 9a - 2)$
$= a^2(5a+2)(7a-1)$

Factors of 35	Factors of -2
1, 35	-1, 2
5, 7	1, -2

Trial Factors	Middle Term
$(a - 1)(35a + 2)$	$-33a$
$(a + 2)(35a - 1)$	$69a$
$(5a - 1)(7a + 2)$	$3a$
*$(5a + 2)(7a - 1)$	$9a$

57. $15b^2 - 115b + 70 = 5(3b^2 - 23b + 14)$
$= 5(b-7)(3b-2)$

Factors of 3	Factors of 14
1, 3	-1, -14
2, 9	-2, -7

Trial Factors	Middle Term
$(b - 1)(3b - 14)$	$-17b$
$(3b - 1)(b - 14)$	$-43b$
$(b - 2)(3b - 7)$	$-13b$
*$(b - 7)(3b - 2)$	$-23b$

59. $10x^3 + 12x^2 + 2x = 2x(5x^2 + 6x + 1)$
$= 2(x+1)(5x+1)$

Factors of 5	Factors of 1
1, 5	1, 1

Trial Factors	Middle Term
$(x + 1)(5x + 1)$	$6x$

61. $4yz^3 + 5yz^2 - 6yz = yz(4z^2 + 5z - 6)$
$= yz(z+2)(4z-3)$

Factors of 4	Factors of -6
1, 4	1, -6
2, 2	-1, 6
	2, -3
	3, -2

Trial Factors	Middle Term
$(z + 1)(4z - 6)$	Common factor
$(z - 6)(4z + 1)$	$-23z$
$(z - 1)(4z + 6)$	Common factor
$(z + 6)(4z - 1)$	$23z$
*$(z + 2)(4z - 3)$	$5z$

63. $9x^3y+12x^2y+4xy=xy(9x^2+12x+4)$
$=xy(3x+2)(3x+2)$

Factors of 9	Factors of 4
1, 9	1, 4
3, 3	2, 2

Trial Factors	Middle Term
$(x+1)(9x+4)$	$13x$
$(x+4)(9x+1)$	$37x$
$(x+2)(9x+2)$	$20x$
$(3x+1)(3x+4)$	$15x$
*$(3x+2)(3x+2)$	$12x$

65. $ax^2+bx+c=(nx+p)(mx+q)$
$=(mn)x^2+(pm+qn)x+(pq)$

a is even so *mn* is even; *c* is even so *pq* is even; *b* is odd so *pm* + *qn* is odd. We are given that *n* is even so for *pm* + *qn* to be odd, one addend must be even and one must be odd. No matter what *q* is, *nq* is even. That means that *pm* must be odd, so both *p* and *m* must be odd.

67.

Factors of -18	Sum of the Factors
1, -18	-17
-1, 18	17
2, -9	-7
-2, 9	7
3, -6	-3
-3, 6	3

69. $ac=-20$; Factors of -20 whose sum is -1: 4 and -5.

$$\begin{aligned}2t^2-t-10&=2t^2+4t-5t-20\\&=(2t^2+4t)-(5t+10)\\&=2t(t+2)-5(t+2)\\&=(t+2)(2t-5)\end{aligned}$$

71. $ac=15$; Factors of 15 whose sum is -16: -1 and -15.

$$\begin{aligned}3p^2-16p+5&=3p^2-15p-p+5\\&=(3p^2-15p)-(p-5)\\&=3p(p-5)-1(p-5)\\&=(p-5)(3p-1)\end{aligned}$$

73. $ac=12$; Factors of 12 whose sum is -7: -4 and -3.

$$\begin{aligned}12y^2-7y+1&=12y^2-3y-4y+1\\&=(12y^2-3y)-(4y-1)\\&=3y(4y-1)-1(4y-1)\\&=(4y-1)(3y-1)\end{aligned}$$

75. $ac=-35$; Factors of -35 whose sum is -62: none.

$5x^2-62x-7$ *nonfactorable*

77. $ac=60$; Factors of 60 whose sum is 19: 15 and 4.

$$\begin{aligned}12y^2+19y+5&=12y^2+4y+15y+5\\&=(12y^2+4y)+(15y+5)\\&=4y(3y+1)+5(3y+1)\\&=(3y+1)(4y+5)\end{aligned}$$

79. $ac=98$; Factors of 98 whose sum is 47: 49 and -2.

$$\begin{aligned}7a^2+47a-14&=7a^2+49a-2a-14\\&=(7a^2+49a)-(2a+14)\\&=7a(a+7)-2(a+7)\\&=(a+7)(7a-2)\end{aligned}$$

81. $ac = 24$; Factors of 24 whose sum is 11: 8 and 3.

$$\begin{aligned}4z^2+11z+6 &= 4z^2+8z+3z+6\\ &= (4z^2+8z)+(3z+6)\\ &= 4z(z+2)+3(z+2)\\ &= (z+2)(4z+3)\end{aligned}$$

83. $ac = 72$; Factors of 72 whose sum is 17: 8 and 9.

$$\begin{aligned}8y^2+17y+9 &= 8y^2+8y+9y+9\\ &= (8y^2+8y)+(9y+9)\\ &= 8y(y+1)+9(y+1)\\ &= (y+1)(8y+9)\end{aligned}$$

85. $ac = 36$; Factors of 36 whose sum is -13: -9 and -4.

$$\begin{aligned}6b^2-13b+6 &= 6b^2-9b-4b+6\\ &= (6b^2-9b)-(4b-6)\\ &= 3b(2b-3)-2(2b-3)\\ &= (2b-3)(3b-2)\end{aligned}$$

87. $ac = 360$; Factors of 360 whose sum is -82: none.

$15x^2-82x+24$ *nonfactorable*

89. $ac = 100$; Factors of 100 whose sum is -29: -25 and -4.

$$\begin{aligned}10z^2-29z+10 &= 10z^2-4z-25z+10\\ &= (10z^2-4z)-(25z-10)\\ &= 2z(5z-2)-5(5z-2)\\ &= (5z-2)(2z-5)\end{aligned}$$

91. $ac = 2$; Factors of 2 whose sum is 3: 1 and 2.

$$\begin{aligned}4x^2+6x+2 &= 2(2x^2+3x+1)\\ &= 2(2x^2+2x+x+1)\\ &= 2[(2x^2+2x)+(x+1)]\\ &= 2[2x(x+1)+1(x+1)]\\ &= 2(x+1)(2x+1)\end{aligned}$$

93. $ac = 21$; Factors of 21 whose sum is -10: -7 and -3.

$$\begin{aligned}15y^2-50y+35 &= 5(3y^2-10y+7)\\ &= 5(3y^3-3y-7y+7)\\ &= 5[(3y^2-3y)-(7y-7)]\\ &= 5[3y(y-1)-7(y-1)]\\ &= 5(y-1)(3y-7)\end{aligned}$$

95. $ac = 10$; Factors of 10 whose sum is -11: -10 and -1.

$$\begin{aligned}2x^3-11x^2+5x &= x(2x^2-11x+5)\\ &= x(2x^2-10x-x+5)\\ &= x[(2x^2-10x)-(x-5)]\\ &= x[2x(x-5)-1(x-5)]\\ &= x(x-5)(2x-1)\end{aligned}$$

97. $ac = -6$; Factors of -6 whose sum is 5: 6 and -1.

$$\begin{aligned}3a^2+5ab-2b^2 &= 3a^2+6ab-ab-2b^2\\ &= (3a^2+6ab)-(ab+2b^2)\\ &= 3a(a+2b)-b(a+2b)\\ &= (a+2b)(3a-b)\end{aligned}$$

99. $ac = 24$; Factors of 24 whose sum is -11: -8 and -3.

$$\begin{aligned}4y^2-11yz+6z^2 &= 4y^2-8yz-3yz+6z^2\\ &= (4y^2-8yz)-(3yz-6z^2)\\ &= 4y(y-2z)-3z(y-2z)\\ &= (y-2z)(4y-3z)\end{aligned}$$

101. $ac = -12$; Factors of -12 whose sum is -1: -4 and 3.

$$\begin{aligned}12-x-x^2 &= 12-4x+3x-x^2\\ &= (12-4x)+(3x-x^2)\\ &= 4(3-x)+x(3-x)\\ &= (3-x)(4+x)\end{aligned}$$

103. $ac = -90$; Factors of -90 whose sum is 1: -9 and 10.

$$\begin{aligned}360y^2+4y-4 &= 4(90y^2+y-1)\\ &= 4(90y^2-9y+10y-1)\\ &= 4[(90y^2-9y)+(10y-1)]\\ &= 4[9y(10y-1)+1(10y-1)]\\ &= 4(10y-1)(9y+1)\end{aligned}$$

105. $ac = -24$; Factors of -24 whose sum is 5: 8 and -3.

$$\begin{aligned}16t^2+40t-96 &= 8(2t^2+5t-12)\\ &= 8(2t^2+8t-3t-12)\\ &= 8[(2t^2+8t)-(3t+12)]\\ &= 8[2t(t+4)-3(t+4)]\\ &= 8(t+4)(2t-3)\end{aligned}$$

107. $ac = 6$; Factors of 6 whose sum is 5: 2 and 3.

$$\begin{aligned}6p^3+5p^2+p &= p(6p^2+5p+1)\\ &= p(6p^2+3p+2p+1)\\ &= p[(6p^2+3p)+(2p+1)]\\ &= p[3p(2p+1)+1(2p+1)]\\ &= p(2p+1)(3p+1)\end{aligned}$$

109. $ac = 100$; Factors of 100 whose sum is -29: -25 and -4.

$$\begin{aligned}30z^2-87z+30 &= 3(10z^2-29z+10)\\ &= 3(10z^2-4z-25z+10)\\ &= 3[(10z^2-4z)-(25z-10)]\\ &= 3[2z(5z-2)-5(5z-2)]\\ &= 3(5z-2)(2z-5)\end{aligned}$$

111. $ac = -126$; Factors of -126 whose sum is 15: 21 and -6.

$$\begin{aligned}42a^3+45a^2-27a &= 3a(14a^2+15a-9)\\ &= 3a(14a^2+21a-6a-9)\\ &= 3a[(14a^2+21a)-(6a+9)]\\ &= 3a[7a(2a+3)-3(2a+3)]\\ &= 3a(2a+3)(7a-3)\end{aligned}$$

113. $ac = 225$; Factors of 225 whose sum is -30: -15 and -15.

$$\begin{aligned}9x^2y-30xy^2+25y^3 &= y(9x^2-30xy+25y^2)\\ &= y(9x^2-15xy-15xy+25y^2)\\ &= y[(9x^2-15xy)-(15xy-25y^2)]\\ &= y[3x(3x-5y)-5y(3x-5y)]\\ &= y(3x-5y)(3x-5y)\end{aligned}$$

115. $ac = 144$; Factors of 144 whose sum is -24: -12 and -12.

$$\begin{aligned}9x^3y-24x^2y^2+16xy^3 &= xy(9x^2-24xy+16y^2)\\ &= xy(9x^2-12xy-12xy+16y^2)\\ &= xy[(9x^2-12xy)-(12xy-16y^2)]\\ &= xy[3x(3x-4y)-4y(3x-4y)]\\ &= xy(3x-4y)(3x-4y)\end{aligned}$$

117. If $b > 0$ and $c > 0$, then look for **two positive** factors of ac.

119. If $b < 0$ and $c > 0$, then look for **two negative** factors of ac.

121.
$$\begin{aligned}6y+8y^3-26y^2 &= 8y^3-26y^2+6y\\ &= 2y(4y^2-13y+3)\\ &= 2y(y-3)(4y-1)\end{aligned}$$

123.
$$\begin{aligned}a^3b-24ab-2a^2b &= ab(a^2-24-2a)\\ &= ab(a^2-2a-24)\\ &= ab(a-6)(a+4)\end{aligned}$$

125.
$$\begin{aligned}25t^2+60t-10t^3 &= 5t(5t+12-2t^2)\\ &= 5t(12+5t-2t^2)\\ &= 5t(3+2t)(4-t)\end{aligned}$$

127.
$$\begin{aligned}&2(y+2)^2-(y+2)-3\\ &= [2(y+2)-3][(y+2)+1]\\ &= (2y+4-3)(y+2+1)\\ &= (2y+1)(y+3)\end{aligned}$$

129. $10(x+1)^2-11(x+1)-6$
$=[2(x+1)-3][5(x+1)+2]$
$=(2x+2-3)(5x+5+2)$
$=(2x-1)(5x+7)$

131. $x^3-2x^2-5x+6=(x+2)(x^2-4x+3)$
$=(x+2)(x-3)(x-1)$

133. $ac = 6$; Factors of 6: -1 and -6; 1 and 6; -2 and -3; 2 and 3. k is the sum of the pairs of factors, so k could be 7, -7, -5 or 5.

135. $ac = 6$; Factors of 6. Factors of 6: -1 and -6; 1 and 6; -2 and -3; 2 and 3. k could be 7, -7, -5 or 5.

137. $ac = 10$; Factors of 10: 1 and 10; -1 and -10; 2 and 5; -2 and -5. k could be -11, 11, -7, or 7.

139. The formula for the area of a rectangle is $A=LW$. To find the dimensions of the rectangle, we want to write the given area as a product, LW. The area of the rectangle is $(3x^2+x-2)\,ft^2$. $3x^2+x-2$ factors as $(3x-2)(x+1)$. So, the dimensions of the rectangle are $(3x-2)$ feet and $(x+1)$ feet. The expressions $3x-2$ and $x+1$ are equal when $x=\frac{3}{2}$. When $x<\frac{3}{2}, x+1>3x-2$ in which case the length of the rectangle is $(x+1)$ feet and the width is $(3x-2)$ feet. If $x>\frac{3}{2}, 3x-2>x+1$, in which case the length of the rectangle is $(3x-2)$ feet and the width is $(x+1)$ feet. If $x<0$, the dimension $3x-2$ is negative. Dimensions of geometric figures cannot be negative. So, x cannot be less than 0. If $x=0$, then $3x-2$ is negative. So x cannot be 0.

Section 8.4 Special Factoring

1. Perfect squares: 4, $25x^6$, $100x^4y^4$

3. Difference of two squares: **i.** a^2-36, **iv.** d^2-100

5. It is **sometimes true** that a binomial if factorable.

7. It is **always true** that if a binomial is multiplied by itself, the result is a perfect-square trinomial.

9. **a.** The binomial $9x^2-4$ is in the form a^2-b^2, where $a=\underline{3x}$ and $b=\underline{2}$.

b. $9x^2-4=(3x+2)(3x-2)$

11. Answers will vary.

a. x^2-16

b. $(x+7)(x-7)$

c. $x^2+10x+25$

d. $(x-3)^2$

e. y^2+36

13. $x^2-4=(x-2)(x+2)$

15. $a^2-81=(a-9)(a+9)$

17. $4x^2-9=(2x-3)(2x+3)$

19. $y^2+6y+9=(y+3)^2$

21. $a^2-2a+1=(a-1)^2$

23. $z^2-18z-81$ *nonfactorable*

25. $x^6-9=(x^3-3)(x^3+3)$

27. $25x^2-4=(5x-2)(5x+2)$

29. $1-49x^2=(1-7x)(1+7x)$

31. $x^2+2xy+y^2=(x+y)^2$

33. $4a^2+4a+1=(2a+1)^2$

35. t^2+36 *nonfactorable*

37. $x^4-y^2=(x^2-y)(x^2+y)$

39. $9x^2-16y^2=(3x-4y)(3x+4y)$

41. $16b^2+24b+9=(4b+3)^2$

43. $4b^2+28b+49=(2b+7)^2$

45. $25a^2+30ab+9b^2=(5a+3b)^2$

47. $x^2y^2-4=(xy-2)(xy+2)$

49. $9x^2+13x+4=(x+1)(9x+4)$

51. **i** and **iii** are equivalent expressions.

i. $(x^2+9)(x+3)(x-3)=(x^2+9)(x^2-9)$
$=x^4-81$

ii. $(2x+9)(2x^2-9)=4x^3+18x^2-18x-81$

iii. $(x^2+9)(x^2-9)=x^4-81$

iv. $(x+3)^2(x-3)^2=[(x+3)(x-3)]^2$
$=(x^2-9)^2$
$=x^4-18x^2+81$

53. $A=16x^2+24x+9=(4x+3)^2$

The dimensions of the square are $(4x+3)$ feet by $(4x+3)$ feet. If $x=0$, then the dimensions of the square are 3 feet by 3 feet so 0 is a permissible value for x. Since $4x+3$ must be positive, any values of x that are greater than $-\frac{3}{4}$ are possible.

55. $y^3+27=(y+3)(y^2-3y+9)$

57. $x^3-1=(x-1)(x^2+x+1)$

59. $x^3-125=(x-5)(x^2+5x+25)$

61. $27y^3+1=(3y+1)(9y^2-3y+1)$

63. $25x^2-kx+1=(5x-1)^2$
$=25x^2-10x+1$
or $=25x^2-(-10x)+1$

k = 10, -10

65. $64x^2+kxy+y^2=(8x+y)^2$
$=64x^2+16xy+y^2$
or $=64x^2+(-16xy)y^2$

k = 16, -16

67. $x^2-4x+k=(x-2)^2$
$=x^2-4x+4$

$k=4$

69. $x^2+10x+k=(x+5)^2$
$=x^2+10x+25$

$k=25$

Section 8.5 Factoring Polynomials Completely

1. If the polynomial has two terms and there is no common factor, the next step is to determine whether the polynomial is a difference of two squares. That can be factored.

3. If the polynomial has four terms, try to factor by grouping.

5. When a polynomial is factored completely, each factor is nonfactorable over the integers.

7. $2x^2 - 18 = 2(x^2 - 9) = 2(x-3)(x+3)$

9. $x^4 + 2x^3 - 35x^2 = x^2(x^2 + 2x - 35)$
$= x^2(x+7)(x-5)$

11. $5b^2 + 75b + 180 = 5(b^2 + 15b + 36)$
$= 5(b+12)(b+3)$

13. $3a^2 + 36a + 10$ *nonfactorable*

15. $2x^2y + 16xy - 66y = 2y(x^2 + 8x - 33)$
$= 2y(x+11)(x-3)$

17. $x^3 - 6x^2 - 5x = x(x^2 - 6x - 5)$

19. $3y^2 - 36 = 3(y^2 - 12)$

21. $20a^2 + 12a + 1 = (2a+1)(10a+1)$

23. $x^2y^2 - 7xy^2 - 8y^2 = y^2(x^2 - 7x - 8)$
$= y^2(x-8)(x+1)$

25. $10a^2 - 5ab - 15b^2 = 5(2a^2 - ab - 3b^2)$
$= 5(2a-3b)(a+b)$

27. $50 - 2x^2 = 2(25 - x^2)$
$= 2(5-x)(5+x)$

29. $12a^3b - a^2b^2 - ab^3 = ab(12a^2 - ab - b^2)$
$= ab(4a+b)(3a-b)$

31. $2ax - 2a + 2bx - 2b = 2(ax - a + bx - b)$
$= 2[(ax-a) + (bx-b)]$
$= 2[a(x-1) + b(x-1)]$
$= 2(x-1)(a+b)$

33. $12a^3 - 12a^2 + 3a = 3a(4a^2 - 4a + 1)$
$= 3a(2a-1)^2$

35. $243 + 3a^2 = 3(81 + a^2)$

37. $12a^3 - 46a^2 + 40a = 2a(6a^2 - 23a + 20)$
$= 2a(2a-5)(3a-4)$

39. $x^3 - 2x^2 - x + 2 = (x^3 - 2x^2) - (x-2)$
$= x^2(x-2) - 1(x-2)$
$= (x-2)(x^2-1)$
$= (x-2)(x-1)(x+1)$

41. $4a^3 + 20a^2 + 25a = a(4a^2 + 20a + 25)$
$= a(2a+5)^2$

43. $27a^2b - 18ab + 3b = 3b(9a^2 - 6a + 1)$
$= 3b(3a-1)^2$

45. $48 - 12x - 6x^2 = 6(8 - 2x - x^2)$
$= 6(4+x)(2-x)$

47. $ax^2-4a+bx^2-4b=(ax^2-4a)+(bx^2-4b)$
$=a(x^2-4)+b(x^2-4)$
$=(x^2-4)(a+b)$
$=(x-2)(x+2)(a+b)$

49. $x^4-x^2y^2=x^2(x^2-y^2)$
$=x^2(x-y)(x+y)$

51. $18a^3+24a^2+8a=2a(9a^2+12a+4)$
$=2a(3a+2)^2$

53. $2b+ab-6a^2b=b(2+a-6a^2)$
$=b(1+2a)(2-3a)$

55. $4x-20-x^3+5x^2=(4x-20)-(x^3-5x^2)$
$=4(x-5)-x^2(x-5)$
$=(x-5)(4-x^2)$
$=(x-5)(2-x)(2+x)$

57. $72xy^2+48xy+8x=8x(9y^2+6y+1)$
$=8x(3y+1)^2$

59. $15y^2-2xy^2-x^2y^2=y^2(15-2x-x^2)$
$=y^2(5+x)(3-x)$

61. $y^3-9y=y(y^2-9)$
$=y(y-3)(y+3)$

63. $2x^4y^2-2x^2y^2=2x^2y^2(x^2-1)$
$=2x^2y^2(x-1)(x+1)$

65. $x^9-x^5=x^5(x^4-1)$
$=x^5(x^2-1)(x^2+1)$
$=x^5(x-1)(x+1)(x^2+1)$

67. $24x^3y+14x^2y-20xy=2xy(12x^2+7x-10)$
$=2xy(3x-2)(4x+5)$

69. $4x^4y^2-20x^3y^2+25x^2y^2$
$=x^2y^2(4x^2-20x+25)$
$=x^2y^2(2x-5)^2$

71. $m^4-256=(m^2-16)(m^2+16)$
$=(m-4)(m+4)(m^2+16)$

73. $y^8-81=(y^4-9)(y^4+9)$
$=(y^2-3)(y^2+3)(y^4+9)$

75. If the middle term has a negative coefficient, then b must be **less than** a.

77. $(2a+3)^2-25b^2=(2a+3-5b)(2a+3+5b)$

79. $(4x^2+12x+9)-4y^2=(2x+3)^2-4y^2$
$=(2x+3-2y)(2x+3+2y)$

81. $42\cdot 38=(40+2)(40-2)=1600-4=1596$
$84\cdot 76=(80+4)(80-4)=6400-16=6384$

83. $2ab+2bc+2cd+2da=2(ab+bc+cd+da)$
$=2[(ab+bc)+(cd+da)]$
$=2[b(a+c)+d(c+a)]$
$=2(a+c)(b+d)$

By trial and error, the largest product is obtained when the sum of a and c is 8 and when the sum of b and d is also 8. The greatest possible value is 128.

Section 8.6 Solving Equations

1. a. Yes, this equation is quadratic.

b. No. This equation is linear.

c. Yes, this equation is quadratic.

3. a. Yes. This equation can be solved without being rewritten.

b. Yes. This equation can be solved without being rewritten.

c. No. This equation cannot be solved without being rewritten.

d. Yes. This equation can be solved without being rewritten.

e. No. This equation cannot be solved without being rewritten.

f. Yes. This equation can be solved without being rewritten.

5. False. Since there is no equals sign, this is an expression.

7. The Principle of Zero Products states that if the product of two or more factors is zero, then at least one of the factors must be zero.

9. $0 = x^2 + x - 2$ is written in standard form.

11. $(y+3)(y+2)=0$

$y+3=0 \qquad y+2=0$

$y=-3 \qquad y=-2$

13. $(z-7)(z-3)=0$

$z-7=0 \qquad z-3=0$

$z=7 \qquad z=3$

15. $x(x-5)=0$

$x=0 \qquad x-5=0$

$x=5$

17. $a(a-9)=0$

$a=0 \qquad a-9=0$

$a=9$

19. $y(2y+3)=0$

$y=0 \qquad 2y+3=0$

$y=-\frac{3}{2}$

21. $2a(3a-2)=0$

$2a=0 \qquad 3a-2=0$

$a=0 \qquad a=\frac{2}{3}$

23. $9x^2-1=0$

$(3x-1)(3x+1)=0$

$3x-1=0 \qquad 3x+1=0$

$x=\frac{1}{3} \qquad x=-\frac{1}{3}$

25. $x^2+6x+8=0$

$(x+4)(x+2)=0$

$x+4=0 \qquad x+2=0$

$x=-4 \qquad x=-2$

27. $z^2+5z-14=0$

$(z+7)(z-2)=0$

$z+7=0 \qquad z-2=0$

$z=-7 \qquad z=2$

29. $x^2-5x+6=0$

$(x-3)(x-2)=0$

$x-3=0 \qquad x-2=0$

$x=3 \qquad x=2$

31. $2a^2-9a-5=0$

$(2a+1)(a-5)=0$

$2a+1=0 \qquad a-5=0$

$a=-\frac{1}{2} \qquad a=5$

33. $2x^2-6x-20=0$
$2(x^2-3x-10)$
$2(x-5)(x+2)=0$
$x-5=0 \qquad x+2=0$
$x=5 \qquad x=-2$

35. $x^2-7x=0$
$x(x-7)=0$
$x=0 \qquad x-7=0$
$x=7$

37. $a^2+5a=-4$
$a^2+5a+4=0$
$(a+4)(a+1)=0$
$a+4=0 \qquad a+1=0$
$a=-4 \qquad a=-1$

39. $y^2-5y=-6$
$y^2-5y+6=0$
$(y-3)(y-2)=0$
$y-3=0 \qquad y-2=0$
$y=3 \qquad y=2$

41. $2t^2+7t=4$
$2t^2+7t-4=0$
$(2t-1)(t+4)=0$
$2t-1=0 \qquad t+4=0$
$t=\frac{1}{2} \qquad t=-4$

43. $3t^2-13t=-4$
$3t^2-13t+4=0$
$(3t-1)(t-4)=0$
$3t-1=0 \qquad t-4=0$
$t=\frac{1}{3} \qquad t=4$

45. $x(x-12)=-27$
$x^2-12x+27=0$
$(x-9)(x-3)=0$
$x-9=0 \qquad x-3=0$
$x=9 \qquad x=3$

47. $y(y-7)=18$
$y^2-7y-18=0$
$(y-9)(y+2)=0$
$y-9=0 \qquad y+2=0$
$y=9 \qquad y=-2$

49. $p(p+3)=-2$
$p^2+3p+2=0$
$(p+1)(p+2)=0$
$p+1=0 \qquad p+2=0$
$p=-1 \qquad p=-2$

51. $y(y+4)=45$
$y^2+4y-45=0$
$(y+9)(y-5)=0$
$y+9=0 \qquad y-5=0$
$y=-9 \qquad y=5$

53. $(x+8)(x-3)=-30$
$x^2+5x-24=-30$
$x^2+5x+6=0$
$(x+3)(x+2)=0$
$x+3=0 \qquad x+2=0$
$x=-3 \qquad x=-2$

55. $(y+3)(y+10)=-10$
$y^2+13y+30=-10$
$y^2+13y+40=0$
$(y+5)(y+8)=0$
$y+5=0 \qquad y+8=0$
$y=-5 \qquad y=-8$

57. $(2x+5)(x+1)=-1$

$2x^2+7x+5=-1$

$2x^2+7x+6=0$

$(2x+3)(x+2)=0$

$2x+3=0 \qquad x+2=0$

$x=-\frac{3}{2} \qquad x=-2$

59. If the solutions include a positive and a negative solution, then c must be negative, so c is **less than** 0.

61. If the quadratic is a perfect-square trinomial, then both monomial factors are the same, so there is **one** solution.

63. a. Let x represent a positive integer. Then the next consecutive positive integer is $\underline{x+1}$, and the expression that represents the sum of the squares of the two integers is $x^2+(x+1)^2$.

b. An equation that can be used to find the integers is $x^2+(x+1)^2=113$.

65. Let x = a positive integer.

$x^2=6+5x$

$x^2-5x-6=0$

$(x-6)(x+1)=0$

$x-6=0 \qquad x+1=0$

$x=6 \qquad x=-1$

The integer is 6.

67. Let x = one of the numbers.

Then $6-x$ = the other number.

$x^2+(6-x)^2=20$

$x^2+36-12x+x^2=20$

$2x^2-12x+16=0$

$2(x^2-6x+8)=0$

$2(x-4)(x-2)=0$

$x-4=0 \qquad x-2=0$

$x=4 \qquad x=2$

The numbers are 2 and 4.

69. Let n and $n+1$ be the consecutive positive integers.

$n^2+(n+1)^2=41$

$n^2+n^2+2n+1-41=0$

$2n^2+2n-40=0$

$2(n^2+n-20)=0$

$2(n+5)(n-4)=0$

$n+5=0 \qquad n-4=0$

$n=-5 \qquad n=4$

$n+1=5$

The integers are 4 and 5.

71. Let n and $n+1$ be consecutive positive integers.

$n(n+1)=240$

$n^2+n-240=0$

$(n+16)(n-15)=0$

$n+16=0 \qquad n-15=0$

$n=-16 \qquad n=15$

$n+1=16$

The consecutive positive integers are 15 and 16.

73. Let h = the height. Then $3h$ = the base.

$$A = \frac{1}{2}(base)(height)$$
$$54 = \frac{1}{2}(3h)(h)$$
$$108 = 3h^2$$
$$0 = 3h^2 - 108$$
$$0 = 3(h^2 - 36)$$
$$0 = 3(h-6)(h+6)$$
$$h - 6 = 0 \qquad h + 6 = 0$$
$$h = 6 \qquad h = -6$$
$$3h = 18$$

The base is 18 feet and the height is 6 feet.

75. Let w = the width. Then $2w + 2$ = the length.

$$A = (length)(width)$$
$$144 = (2w+2)w$$
$$144 = 2w^2 + 2w$$
$$0 = 2w^2 + 2w - 144$$
$$0 = 2(w^2 + w - 72)$$
$$0 = 2(w+9)(w-8)$$
$$w + 9 = 0 \qquad w - 8 = 0$$
$$w = -9 \qquad w = 8$$
$$2w + 2 = 18$$

The width is 8 feet and the length is 18 feet.

77. Let x = the original side length.

Then $x + 4$ = the length of the extended side.

$$A = (side)^2$$
$$64 = (x+4)^2$$
$$64 = x^2 + 8x + 16$$
$$0 = x^2 + 8x - 48$$
$$0 = (x+12)(x-4)$$
$$x + 12 = 0 \qquad x - 4 = 0$$
$$x = -12 \qquad x = 4$$

The original side length was 4 m.

79. Let r = the original radius.

Then $r + 3$ = the increased radius.

$$A = \pi r^2$$
$$\pi r^2 + 100 = \pi(r+3)^2$$
$$\pi r^2 + 100 = \pi(r^2 + 6r + 9)$$
$$\pi r^2 + 100 = \pi r^2 + 6\pi r + 9\pi$$
$$0 = 6\pi r + 9\pi - 100$$
$$100 - 9\pi = 6\pi r$$
$$\frac{100 - 9\pi}{6\pi} = r$$
$$3.81 \approx r$$

The original radius was 3.81 inches.

81. Let x = the size of the border.

$$(6-x)(9-x) = 28$$
$$54 - 15x + x^2 = 28$$
$$x^2 - 15x + 26 = 0$$
$$(x-13)(x-2) = 0$$
$$x - 13 = 0 \qquad x - 2 = 0$$
$$x = 13 \qquad x = 2$$
$$6 - x = 4$$
$$9 - x = 7$$

The type area is 4 in by 7 in.

83. Let W = the width. Then $W + 3$ = the length.

$$A = (length)(width)$$
$$304 = (W+3)W$$
$$304 = W^2 + 3W$$
$$0 = W^2 + 3W - 304$$
$$0 = (W+19)(W-16)$$
$$W + 19 = 0 \qquad W - 16 = 0$$
$$W = -19 \qquad W = 16$$

The width is 16 feet.

85. $d = vt + 16t^2$

$320 = 16t + 16t^2$

$0 = 16t^2 + 16t - 320$

$0 = 16(t^2 + t - 20)$

$0 = 16(t-4)(t+5)$

$t - 4 = 0 \qquad t + 5 = 0$

$t = 4 \qquad t = -5$

It takes 4 seconds for the object to hit the ground.

87. $S = \dfrac{n^2 + n}{2}$

$120 = \dfrac{n^2 + n}{2}$

$240 = n^2 + n$

$0 = n^2 + n - 240$

$0 = (n+16)(n-15)$

$n + 16 = 0 \qquad n - 15 = 0$

$n = -16 \qquad n = 15$

15 natural numbers beginning with 1 will give a sum of 120.

89. $N = \dfrac{t^2 - t}{2}$

$45 = \dfrac{t^2 - t}{2}$

$90 = t^2 - t$

$0 = t^2 - t - 90$

$0 = (t-10)(t+9)$

$t - 10 = 0 \qquad t + 9 = 0$

$t = 10 \qquad t = -9$

There are 10 teams in the league.

91. $h = vt - 16t^2$

$0 = 48t - 16t^2$

$0 = 16t(3-t)$

$16t = 0 \qquad 3 - t = 0$

$t = 0 \qquad 3 = t$

The ball returns to the ground in 3 seconds.

93. $2y(y+4) = 3(y+4)$

$2y^2 + 8y = 3y + 12$

$2y^2 + 5y - 12 = 0$

$(2y-3)(y+4) = 0$

$2y - 3 = 0 \qquad y + 4 = 0$

$y = \dfrac{3}{2} \qquad y = -4$

95. $(b+5)^2 = 16$

$b^2 + 10b + 25 = 16$

$b^2 + 10b + 9 = 0$

$(b+9)(b+1) = 0$

$b + 9 = 0 \qquad b + 1 = 0$

$b = -9 \qquad b = -1$

97. $p^3 = 7p^2$

$p^3 - 7p^2 = 0$

$p^2(p-7) = 0$

$p^2 = 0 \qquad p - 7 = 0$

$p = 0 \qquad p = 7$

99. $(x+3)(2x-1) = (3-x)(5-3x)$

$2x^2 + 5x - 3 = 15 - 14x + 3x^2$

$0 = x^2 - 19x + 18$

$0 = (x-18)(x-1)$

$x - 18 = 0 \qquad x - 1 = 0$

$x = 18 \qquad x = 1$

101. $n(n+3) = 4$

$n^2 + 3n = 4$

$n^2 + 3n - 4 = 0$

$(n+4)(n-1) = 0$

$n + 4 = 0 \qquad n - 1 = 0$

$n = -4 \qquad n = 1$

$2n^3 = 2(-64) \qquad 2n^3 = 2(1)$

$2n^3 = -128 \qquad 2n^3 = 2$

103. One of the solutions of the original equation is $x = 0$. Therefore, the second line of the equation, where each side of the equation is divided by x, involves division by 0, which is undefined.

$$x^2 = x$$
$$x^2 - x = 0$$
$$x(x-1) = 0$$
$$x = 0 \qquad x - 1 = 0$$
$$x = 1$$

105. Let W = the width. Then $W + 10$ = the length. After cutting, the new width is $W - 4$ and the new length is $W + 10 - 4 = W + 6$. The depth of the box is 2 inches.

$$V = (length)(width)(height)$$
$$192 = (W+6)(W-4)(2)$$
$$192 = 2W^2 + 4W - 48$$
$$0 = 2W^2 + 4W - 240$$
$$0 = 2(W^2 + 2W - 120)$$
$$0 = 2(W+12)(W-10)$$
$$W + 12 = 0 \qquad W - 10 = 0$$
$$W = -12 \qquad W = 10$$
$$W + 10 = 20$$

The dimensions of the rectangle are 10 inches by 20 inches.

Chapter 8 Review Exercises

1. $14y^9 - 49y^6 + 7y^3 = 7y^3(2y^6 - 7y^3 + 1)$

2. $3a^2 - 12a + ab - 4b = (3a^2 - 12a) + (ab - 4b)$
$= 3a(a-4) + b(a-4)$
$= (a-4)(3a+b)$

3. $c^2 + 8c + 12 = (c+6)(c+2)$

4. $a^3 - 5a^2 + 6a = a(a^2 - 5a + 6)$
$= a(a-2)(a-3)$

5. $6x^2 - 29x + 28 = (2x-7)(3x-4)$

6. $3y^2 + 16y - 12 = (y+6)(3y-2)$

7. $18a^2 - 3a - 10 = (3a+2)(6a-5)$

8. $a^2b^2 - 1 = (ab-1)(ab+1)$

9. $4y^2 - 16y + 16 = 4(y^2 - 4y + 4)$
$= 4(y-2)^2$

10. $a(5a+1) = 0$
$a = 0 \qquad 5a + 1 = 0$
$a = -\frac{1}{5}$

11. $12a^2b + 3ab^2 = 3ab(4a+b)$

12. $b^2 - 13b + 30 = (b-10)(b-3)$

13. $10x^2 + 25x + 4xy + 10y$
$= (10x^2 + 25x) + (4xy + 10y)$
$= 5x(2x+5) + 2y(2x+5)$
$= (2x+5)(5x+2y)$

14. $3a^2 - 15a - 42 = 3(a^2 - 5a - 14)$
$= 3(a-7)(a+2)$

15. $n^4 - 2n^3 - 3n^2 = n^2(n^2 - 2n - 3)$
$= n^2(n-3)(n+1)$

16. $2x^2 - 5x + 6$ *nonfactorable*

17. $6x^2 - 7x + 2 = (2x-1)(3x-2)$

18. $16x^2+49$ *nonfactorable* **19.**

$(x-2)(2x-3)=0$

$x-2=0 \qquad 2x-3=0$

$x=2 \qquad x=\frac{3}{2}$

20. $7x^2-7=7(x^2-1)$

$=7(x-1)(x+1)$

21. $3x^5-9x^4-4x^3=x^3(3x^2-9x-4)$

22. $4x(x-3)-5(3-x)=4x(x-3)+5(x-3)$

$=(x-3)(4x+5)$

23. $a^2+5a-14=(a+7)(a-2)$

24. $y^2+5y-36=(y+9)(y-4)$

25. $5x^2-50x-120=5(x^2-10x-24)$

$=5(x-12)(x+2)$

26. $(x+1)(x-5)=16$

$x^2-4x-5=16$

$x^2-4x-21=0$

$(x-7)(x+3)=0$

$x-7=0 \qquad x+3=0$

$x=7 \qquad x=-3$

27. $7a^2+17a+6=(7a+3)(a+2)$

28. $4x^2+83x+60=(x+20)(4x+3)$

29. $9y^4-25z^2=(3y^2-5z)(3y^2+5z)$

30. $5x^2-5x-30=5(x^2-x-6)$

$=5(x+2)(x-3)$

31. $6-6y^2=5y$

$6-5y-6y^2=0$

$(3+2y)(2-3y)=0$

$3+2y=0 \qquad 2-3y=0$

$y=-\frac{3}{2} \qquad y=\frac{2}{3}$

32. $12b^3-58b^2+56b=2b(6b^2-29b+28)$

$=2b(2b-7)(3b-4)$

33. $5x^3+10x^2+35x=5x(x^2+2x+7)$

34. $x^2-23x+42=(x-21)(x-2)$

35. $a(3a+2)-7(3a+2)=(3a+2)(a-7)$

36. $8x^2-38x+45=(2x-5)(4x-9)$

37. $10a^2x-130ax+360x=10x(a^2-13a+36)$

$=10x(a-9)(a-4)$

38. $2a^2-19a-60=(2a+5)(a-12)$

39. $21ax-35bx-10by+6ay$

$=(21ax-35bx)-(10by-6ay)$

$=7x(3a-5b)-2y(5b-3a)$

$=7x(3a-5b)+2y(3a-5b)$

$=(3a-5b)(7x+2y)$

40. $a^6-100=(a^3-10)(a^3+10)$

41. $16a^2+8a+1=(4a+1)^2$

42. $4x^2+27x=7$
$4x^2+27x-7=0$
$(4x-1)(x+7)=0$
$4x-1=0 \quad x+7=0$
$x=\frac{1}{4} \quad x=-7$

43. $20a^2+10a-280=10(2a^2+a-28)$
$=10(2a-7)(a+4)$

44. $6x-18=6(x-3)$

45. $3x^4y+2x^3y+6x^2y=x^2y(3x^2+2x+6)$

46. $d^2+3d-40=(d+8)(d-5)$

47. $24x^2-12xy+10y-20x$
$=2(12x^2-6xy+5y-10x)$
$=2[(12x^2-6xy)+(5y-10x)]$
$=2[6x(2x-y)+5(y-2x)]$
$=2[6x(2x-y)-5(2x-y)]$
$=2(2x-y)(6x-5)$

48. $4x^3-20x^2-24x=4x(x^2-5x-6)$
$=4x(x-6)(x+1)$

49. $x^2-8x-20=0$
$(x-10)(x+2)=0$
$x-10=0 \quad x+2=0$
$x=10 \quad x=-2$

50. $3x^2-17x+10=(3x-2)(x-5)$

51. $16x^2-94x+33=(2x-11)(8x-3)$

52. $9x^2-30x+25=(3x-5)^2$

53. $12y^2+16y-3=(2y+3)(6y-1)$

54. $3x^2+36x+108=3(x^2+12x+36)$
$=3(x+6)^2$

55. Let W = the width. Then 2W = the length.
$A=(length)(width)$
$5000=(2W)W$
$5000=2W^2$
$0=2W^2-5000$
$0=2(W^2-2500)$
$0=2(W-50)(W+50)$
$W-50=0 \quad W+50=0$
$W=50 \quad W=-50$
$2W=100$
The length is 100 yards and the width is 50 yards.

56. Let W = the width.
Then 2W – 20 = the length.
$A=(length)(width)$
$6000=(2W-20)W$
$6000=2W^2-20W$
$0=2W^2-20W-6000$
$0=2(W^2-10W-3000)$
$0=2(W-60)(W+50)$
$W-60=0 \quad W+50=0$
$W=60 \quad W=-50$
$2W-20=100$
The length is 100 yards and the width is 60 yards.

57. Let n and n +1 be consecutive positive integers.

$n^2+(n+1)^2=41$
$n^2+n^2+2n+1=41$
$2n^2+2n-40=0$
$2(n^2+n-20)=0$
$2(n+5)(n-4)=0$
$n+5=0 \qquad n-4=0$
$n=-5 \qquad n=4$
$n+1=5$

The consecutive positive integers are 4 and 5.

58. $S=d^2$
$400=d^2$
$400-d^2=0$
$(20-d)(20+d)=0$
$20-d=0 \qquad 20+d=0$
$20=d \qquad d=-20$

The distance between the projector and the screen is 20 feet.

59. Let W = the width of the path.

$(15+2x)(12+2x)=270$
$180+54x+4x^2=270$
$4x^2+54x-90=0$
$2(2x^2+27x-45)=0$
$2(2x-3)(x+15)=0$
$2x-3=0 \qquad x+15=0$
$x=\frac{3}{2} \qquad x=-15$

The width of the larger rectangle is

$12+2\cdot\frac{3}{2}=12+3=15$ feet.

60. Let x = side length of original square.

$(x+4)^2=576$
$x^2+8x+16=576$
$x^2+8x-560=0$
$(x+28)(x-20)=0$
$x+28=0 \qquad x-20=0$
$x=-28 \qquad x=20$

The length of the side of the original square was 20 feet.

Chapter 8 Test

1. $6x^2y^2+9xy^2+12y^2=3y^2(2x^2+3x+4)$

2. $6x^3-8x^2+10x=2x(3x^2-4x+5)$

3. $p^2+5p+6=(p+2)(p+3)$ **4.**

$a(x-2)+b(2-x)=a(x-2)-b(x-2)$
$=(x-2)(a-b)$

5. $(2a-3)(a+7)=0$
$2a-3=0 \qquad x+7=0$
$a=\frac{3}{2} \qquad x=-7$

6. $a^2-19a+48=(a-3)(a-16)$

7. $x^3+2x^2-15x=x(x^2+2x-15)$
$=x(x+5)(x-3)$

8. $8x^2+20x-48=4(2x^2+5x-12)$
$=4(x+4)(2x-3)$

9. $ab+6a-3b-18=(ab+6a)-(3b+18)$
$=a(b+6)-3(b+6)$
$=(b+6)(a-3)$

10. $4x^2 - 1 = 0$

$(2x-1)(2x+1) = 0$

$2x - 1 = 0 \qquad 2x + 1 = 0$

$x = \frac{1}{2} \qquad x = -\frac{1}{2}$

11. $6x^2 + 19x + 8 = (2x+1)(3x+8)$

12. $x^2 - 9x - 36 = (x-12)(x+3)$

13. $2b^2 - 32 = 2(b^2 - 16)$
$= 2(b-4)(b+4)$

14. $4a^2 - 12ab + 9b^2 = (2a - 3b)^2$

15. $px + x - p - 1 = (px + x) - (p+1)$
$= x(p+1) - 1(p+1)$
$= (p+1)(x-1)$

16. $5x^2 - 45x - 15 = 5(x^2 - 9x - 3)$

17. $2x^2 + 4x - 5$ *nonfactorable*

18. $4x^2 - 49y^2 = (2x - 7y)(2x + 7y)$

19. $x(x-8) = -15$

$x^2 - 8x + 15 = 0$

$(x-5)(x-3) = 0$

$x - 5 = 0 \qquad x - 3 = 0$

$x = 5 \qquad x = 3$

20. $p^2 + 12p + 36 = (p+6)^2$

21. $18x^2 - 48xy + 32y^2 = 2(9x^2 - 24xy + 16y^2)$
$= 2(3x - 4y)^2$

22. $2y^4 - 14y^3 - 16y^2 = 2y^2(y^2 - 7y - 8)$
$= 2y^2(y-8)(y+1)$

23. Let W = the width. Then $2W + 3$ = the length.

$A = (length)(width)$

$90 = (2W+3)W$

$90 = 2W^2 + 3W$

$0 = 2W^2 + 3W - 90$

$0 = (2W+15)(W-6)$

$2W + 15 = 0 \qquad W - 6 = 0$

$W = -\frac{15}{2} \qquad W = 6$

$2W + 3 = 15$

The width is 6 cm and the length is 15 cm.

24. Let h = the height. Then $3h$ = the base.

$A = \frac{1}{2}(base)(height)$

$24 = \frac{1}{2}(3h)(h)$

$48 = 3h^2$

$0 = 3h^2 - 48$

$0 = 3(h^2 - 16)$

$0 = 3(h-4)(h+4)$

$h - 4 = 0 \qquad h + 4 = 0$

$h = 4 \qquad h = -4$

$3h = 12$

The length of the base is 12 inches.

25. Let n and $n + 1$ be two consecutive negative integers.

$n(n+1) = 156$

$n^2 + n - 156 = 0$

$(n+13)(n-12) = 0$

$n + 13 = 0 \qquad n - 12 = 0$

$n = -13 \qquad n = 12$

$n + 1 = -12$

The integers are -13 and -12.

Cumulative Review Exercises

1. $4-(-5)-6-11=4+5+(-6)+(-11)$
$=9+(-6)+(-11)$
$=3+(-11)=-8$

2. $\frac{0.372}{-0.046}=\frac{372}{-46}\approx -8.1$

$$\begin{array}{r} 8.08 \\ 46\overline{)372.00} \\ \underline{368} \\ 400 \\ \underline{368} \\ 32 \end{array}$$

3. $(3-7)^2\div(-2)-3\cdot(-4)=(-4)^2\div(-2)-3\cdot(-4)$
$=16\div(-2)-3\cdot(-4)$
$=-8-(-12)$
$=-8+12=4$

4. $-2a^2\div(2b)-c=-2(-4)^2\div(2\cdot 2)-(-1)$
$=-2(16)\div(2\cdot 2)-(-1)$
$=-32\div 4-(-1)$
$=-8+1$
$=-7$

5. Because the grouping has changed, this is the **Associative Property of Addition**.

6. $-\frac{3}{4}(-24x^2)=\frac{3}{\cancel{4}_1}\cdot\frac{\cancel{24}^6x^2}{1}=18x^2$

7. $-2[3x-4(3-2x)-8x]=-2[3x-12+8x-8x]$
$=-2[3x-12]$
$=-6x+24$

8. $-\frac{5}{7}x=-\frac{10}{21}$

$\left(-\frac{7}{5}\right)\left(-\frac{5}{7}x\right)=\left(-\frac{\cancel{7}^1}{\cancel{5}_1}\right)\left(-\frac{\cancel{10}^2}{\cancel{21}_3}\right)$

$x=\frac{2}{3}$

9. $4+3(x-2)=13$
$4+3x-6=13$
$3x-2=13$
$3x=15$
$x=5$

10. $3x-2=12-5x$
$3x-2+2=12-5x+2$
$3x=14-5x$
$3x+5x=14-5x+5x$
$8x=14$
$x=\frac{14}{8}=\frac{7}{4}$

11. $-2+4[3x-2(4-x)-3]=4x+2$
$-2+4[3x-8+2x-3]=4x+2$
$-2+4[5x-11]=4x+2$
$-2+20x-44=4x+2$
$20x-46=4x+2$
$16x=48$
$x=3$

12. $A=PB$
$42=1.20B$
$\frac{42}{1.20}=\frac{1.20B}{1.20}$
$35=B$

13. $-4x-2\ge 10$
$-4x\ge 12$
$\frac{-4x}{-4}\le\frac{12}{-4}$
$x\le -3$

14. $9-2(4x-5)<3(7-6x)$

$$9-8x+10<21-18x$$
$$-8x+19<21-18x$$
$$-8x<2-18x$$
$$10x<2$$
$$x<\frac{2}{10}$$
$$x<\frac{1}{5}$$

15. 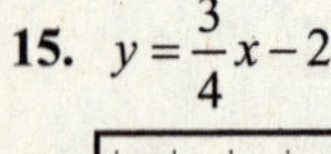 $y=\frac{3}{4}x-2$

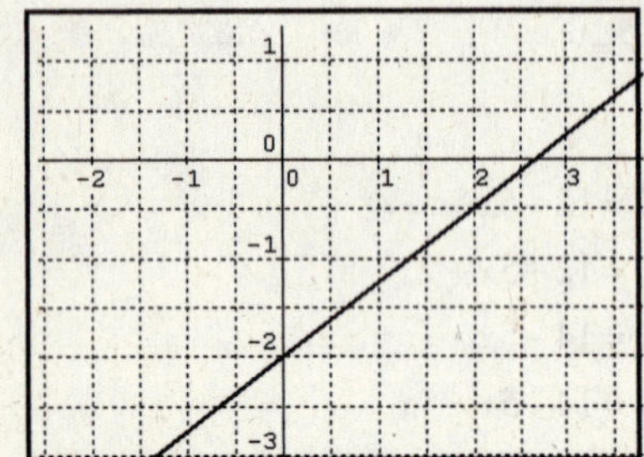

16. $f(x)=-3x-3$

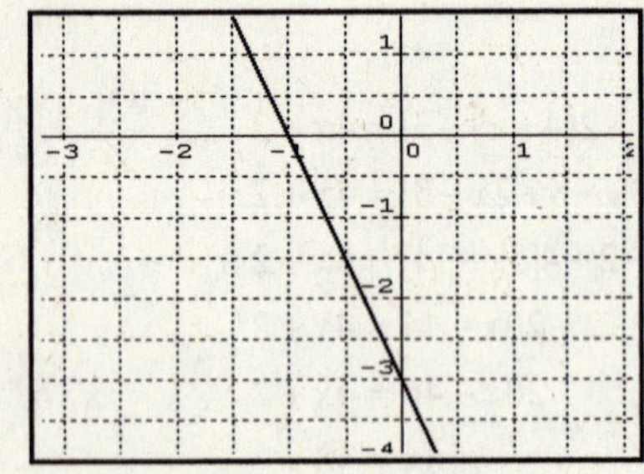

17. Domain: {-5, -3, -1, 1, 3}

Range: {-4, -2, 0, 2, 4}; Function

18. $f(x)=6x-5$

$$f(11)=6(11)-5$$
$$f(11)=66-5=61$$

19. $x+3y>2$

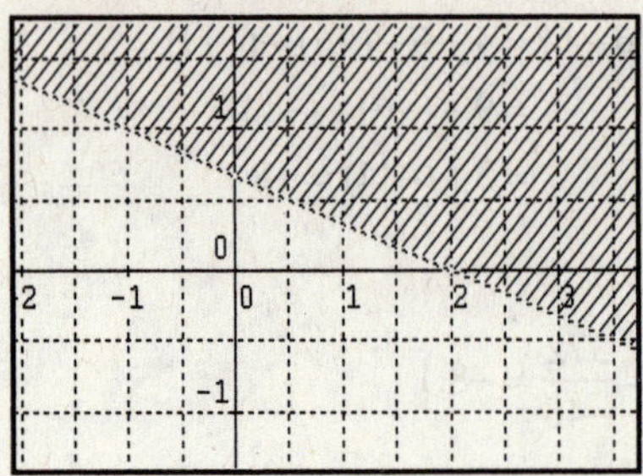

20. $x-3y=17 \Rightarrow x=3y+17$

$$6x+y=7 \Rightarrow 6(3y+17)+y=7$$
$$18y+102+y=7$$
$$19y=-95$$
$$y=-5$$
$$x=3y+17 \Rightarrow x=3(-5)+17$$
$$x=2$$

The solution is (2, -5).

$$2x-3y=-4$$
$$\underline{3(5x+\ y=\ 7)}$$

21. $2x-3y=-4$

$$\underline{15x+3y=21}$$
$$17x\qquad =17 \Rightarrow x=1$$
$$2x-3y=-4$$
$$2(1)-3y=-4$$
$$2-3y=-4$$
$$-3y=-6$$
$$y=2$$

The solution is (1, 2).

22. $(3y^3-5y^2-6)+(2y^2-8y+1)$

$$3y^3-5y^2+2y^2-8y-6+1$$
$$3y^3-3y^2-8y-5$$

23. $(-3a^4b^2)^3=-27a^{12}b^6$

24. $(x+2)(x^2-5x+4)$

$$=x^3-5x^2+4x+2x^2-10x+8$$
$$=x^3-3x^2-6x+8$$

25. $$\begin{array}{r} 4x+8+\dfrac{21}{2x-3} \\ 2x-3\overline{)\,8x^2+4x-3} \\ \underline{-8x^2+12x} \\ 16x-3 \\ \underline{-16x+24} \\ 21 \end{array}$$

26. $(x^{-4}y^2)^3 = \left(\dfrac{y^2}{x^4}\right)^3 = \dfrac{y^6}{x^{12}}$

27. $3a-3b-ax+bx = (3a-3b)-(ax-bx)$
$= 3(a-b)-x(a-b)$
$= (a-b)(3-x)$

28. $x^2+3xy-10y^2 = (x+5y)(x-2y)$

29. $6a^4+22a^3+12a^2 = 2a^2(3a^2+11a+6)$
$= 2a^2(3a+2)(a+3)$

30. $25a^2-36b^2 = (5a-6b)(5a+6b)$

31. $12x^2-36xy+27y^2 = 3(4x^2-12xy+9y^2)$
$= 3(2x-3y)^2$

32. $3x^2+11x-20=0$
$(3x-4)(x+5)=0$
$3x-4=0 \qquad x+5=0$
$x=\dfrac{4}{3} \qquad x=-5$

33. $f(x)=\dfrac{4}{5}x-3$
$f(-10)=-11 \qquad f(-5)=-7$
$f(0)=-3 \qquad f(5)=1$
$f(10)=5$

Range: {-11, -7, -3, 1, 5}

34. $Average = \dfrac{sum}{7}$
$= \dfrac{-4+(-7)+2+0+(-1)+(-6)+(-5)}{7}$
$= \dfrac{-21}{7} = -3°C$

35. Let L = the length. Then $0.40L$ = the width.
$P=2L+2W$
$42=2L+2(0.4L)$
$42=2L+0.8L$
$42=2.8L$
$15=L$
$0.4L=6$

The length is 15 cm and the width is 6 cm.

36. Let x = the length of the shorter piece.
Then $10-x$ = the length of the longer piece.
$4x=3(10-x)-2$
$4x=30-3x-2$
$4x=28-3x$
$7x=28$
$x=4$
$10-x=6$

The pieces are 4 feet and 6 feet long.

37. Company A: Cost = $40+0.35m$
Company B: Cost = $45+0.30m$
$6(40)+0.35m<6(45)+0.30m$
$240+0.35m<270+0.30m$
$0.05m<30$
$m<600$

Company A is cheaper if you drive 599 miles or less.

38. \$6500 was invested at 11%.

	P	·	R	=	I
Amount at 8%	4000	·	0.08	=	0.08(4,000)
Amount at 11%	x	·	0.11	=	$0.11(x)$
Total	x +4000	·		=	1035

$$\begin{aligned} 0.08(4000)+0.11x &= 1035 \\ 320+0.11x &= 1035 \\ 0.11x &= 715 \\ x &= 6500 \end{aligned}$$

39.
$$\begin{aligned} S &= R-rR \\ 99 &= 165-r(165) \\ -66 &= -165r \\ \frac{-66}{-165} &= r \\ 0.4 &= r \\ r &= 40\% \end{aligned}$$

The discount rate is 40%.

40. Let n, $n+2$, and $n+4$ be three consecutive even integers.

$$\begin{aligned} &5(n+2)=12+2(n+n+4) \\ &5n+10=12+2(2n+4) \\ &5n+10=12+4n+8 \\ &5n+10=4n+20 \\ &n=10 \\ &n+2=12 \\ &n+4=14 \end{aligned}$$

The consecutive even integers are 10, 12, and 14.

CHAPTER 9 Rational Expressions

Chapter 9 Prep Test

1. $12 = 2^2 \cdot 3$

$18 = 2 \cdot 3^2$

$LCM(12,18) = 2^2 \cdot 3^2 = 4 \cdot 9 = 36$

2. $\dfrac{9x^3y^4}{3x^2y^7} = \dfrac{9}{3} \cdot \dfrac{x^3}{x^2} \cdot \dfrac{y^4}{y^7} = \dfrac{3x}{y^3}$

3. $\dfrac{3}{4} - \dfrac{8}{9} = \dfrac{3}{4} \cdot \dfrac{9}{9} - \dfrac{8}{9} \cdot \dfrac{4}{4}$

$= \dfrac{27}{36} - \dfrac{32}{36} = -\dfrac{5}{36}$

4. $\left(-\dfrac{8}{11}\right) \div \left(\dfrac{4}{5}\right) = -\dfrac{\overset{2}{\cancel{8}}}{11} \cdot \dfrac{5}{\underset{1}{\cancel{4}}} = -\dfrac{10}{11}$

5. $\dfrac{2}{3}x - \dfrac{3}{4} = \dfrac{5}{6}$

$12 \cdot \dfrac{2}{3}x - 12 \cdot \dfrac{3}{4} = 12 \cdot \dfrac{5}{6}$

$8x - 9 = 10$

$8x = 19$

$x = \dfrac{19}{8}$

6. $a + 50° = 180°$

$a = 130°$

7. $x^2 - 4x - 12 = (x-6)(x+2)$

8. $2x^2 - x - 3 = (2x-3)(x+1)$

9. When Jean catches up their distances are equal.

Let t = Anthony's time.

Anthony: $d = 9t$

Jean: $d = 12(t - 10)$

$9t = 12(t-10)$

$9t = 12t - 120$

$-3t = -120$

$t = 40$

Jean catches up 40 minutes after Anthony started, at 9:40 A. M.

9.1 Multiplication and Division of Rational Expressions

1. True

3. False. If $x = 0$, then the expression is 0, which is a real number.

5. False. To divide, multiply the first fraction by the reciprocal of the second fraction.

7. A rational expression is in simplest form when all common factors of the numerator and denominator are divided out

9. The simplification is incorrect because terms can never be cancelled, only factors.

11. $\dfrac{x^2-4}{2x^2+4x} = \dfrac{(x+2)(x-2)}{2x(x+2)} = \dfrac{x-2}{2x}$

Common factor: $x + 2$

13. $\dfrac{16x^2y}{24xy^3} = \dfrac{2x}{3y^2}$

15. $\dfrac{(2x-1)^5}{(2x-1)^4} = 2x - 1$

17. $\dfrac{5-2x}{2x-5}=\dfrac{-1(2x-5)}{2x-5}=-1$

19. $\dfrac{12x^2(3-x)}{18x(3-x)}=\dfrac{2x}{3}$

21. $\dfrac{14x^3(7-3x)}{21x(3x-7)}=-\dfrac{2x^2}{3}$

23. $\dfrac{x^2-3x}{2x-6}=\dfrac{x(x-3)}{2(x-3)}=\dfrac{x}{2}$

25. $\dfrac{5xy-3y}{9-15x}=\dfrac{y(5x-3)}{-3(5x-3)}=-\dfrac{y}{3}$

27. $\dfrac{x^2+5x+6}{x^2+8x+15}=\dfrac{(x+2)(x+3)}{(x+5)(x+3)}=\dfrac{x+2}{x+5}$

29. $\dfrac{a^2+7a-8}{a^2+6a-7}=\dfrac{(a+8)(a-1)}{(a+7)(a-1)}=\dfrac{a+8}{a+7}$

31. $\dfrac{x^2+8x+16}{x^2-2x-24}=\dfrac{(x+4)(x+4)}{(x-6)(x+4)}=\dfrac{x+4}{x-6}$

33. $\dfrac{4-y^2}{y^2-3y-10}=\dfrac{(2-y)(2+y)}{(y-5)(y+2)}=\dfrac{2-y}{y-5}$

35. $\dfrac{3x^3-12x}{6x^3-24x^2+24x}=\dfrac{3x(x^2-4)}{6x(x^2-4x+4)}$

$=\dfrac{3x(x-2)(x+2)}{6x(x-2)(x-2)}$

$=\dfrac{x+2}{2(x-2)}$

37. $\dfrac{2n^2-9n+4}{2n^2-5n-12}=\dfrac{(2n-1)(n-4)}{(2n+3)(n-4)}=\dfrac{2n-1}{2n+3}$

39. $\dfrac{6a}{b}\cdot\dfrac{2}{d}=\dfrac{12a}{bd}$

41. $\dfrac{\overset{1}{\cancel{4}}\,a^2b^3}{\underset{3}{\cancel{15}}\,x^5y^2}\cdot\dfrac{\overset{5}{\cancel{25}}\,x^3y}{\underset{4}{\cancel{16}}\,ab}=\dfrac{5}{12}\cdot a^{2-1}b^{3-1}x^{3-5}y^{1-2}$

$=\dfrac{5}{12}ab^2x^{-2}y^{-1}$

$=\dfrac{5ab^2}{12x^2y}$

43. $\dfrac{\overset{2}{\cancel{18}}\,a^4b^2}{\underset{1}{\cancel{25}}\,x^2y^3}\cdot\dfrac{\overset{2}{\cancel{50}}\,x^5y^6}{\underset{3}{\cancel{27}}\,a^6b^2}=\dfrac{4}{3}\cdot\dfrac{a^4}{a^6}\cdot\dfrac{b^2}{b^2}\cdot\dfrac{x^5}{x^2}\cdot\dfrac{y^6}{y^3}=\dfrac{4x^3y^3}{3a^2}$

45. $\dfrac{8x-12}{14x+7}\cdot\dfrac{42x+21}{32x-48}=\dfrac{\overset{1}{\cancel{4}}\,\overset{1}{\cancel{(2x-3)}}}{\underset{1}{\cancel{7}}\,\underset{1}{\cancel{(2x+1)}}}\cdot\dfrac{\overset{3}{\cancel{21}}\,\overset{1}{\cancel{(2x+1)}}}{\underset{4}{\cancel{16}}\,\underset{1}{\cancel{(2x-3)}}}$

$=\dfrac{3}{4}$

47. $\dfrac{4a^2x-3a^2}{2by+5b}\cdot\dfrac{2b^3y+5b^3}{4ax-3a}$

$=\dfrac{\overset{a}{\cancel{a^2}}\,\overset{1}{\cancel{(4x-3)}}}{\underset{1}{\cancel{b}}\,\underset{1}{\cancel{(2y+5)}}}\cdot\dfrac{\overset{b^2}{\cancel{b^3}}\,\overset{1}{\cancel{(2y+5)}}}{\underset{1}{\cancel{a}}\,\underset{1}{\cancel{(4x-3)}}}$

$=ab^2$

49. $\dfrac{x^2+x-2}{xy^2}\cdot\dfrac{x^3y}{x^2+5x+6}$

$=\dfrac{\overset{1}{\cancel{(x+2)}}(x-1)}{\underset{1}{\cancel{x}}\,\underset{y}{\cancel{y^2}}}\cdot\dfrac{\overset{x^2}{\cancel{x^3}}\,\overset{1}{\cancel{y}}}{\underset{1}{\cancel{(x+2)}}(x+3)}$

$=\dfrac{x^2(x-1)}{y(x+3)}$

51. $\frac{x^5y^3}{x^2+13x+30}\cdot\frac{x^2+2x-3}{x^7y^2}$

$$=\frac{x^5y^3}{(x+10)(x+3)}\cdot\frac{(x+3)(x-1)}{x^7y^2}$$

$$=\frac{y(x-1)}{x^2(x+10)}$$

53. $\frac{8x^3+4x^2}{x^2-3x+2}\cdot\frac{x^2-4}{16x^2+8x}$

$$=\frac{4x^2(2x+1)}{(x-2)(x-1)}\cdot\frac{(x-2)(x+2)}{8x(2x+1)}$$

$$=\frac{x(x+2)}{2(x-1)}$$

55. $\frac{x^2-8x+7}{x^2+3x-4}\cdot\frac{x^2+3x-10}{x^2-9x+14}$

$$=\frac{(x-7)(x-1)}{(x+4)(x-1)}\cdot\frac{(x+5)(x-2)}{(x-7)(x-2)}$$

$$=\frac{x+5}{x+4}$$

57. $\frac{y^2+y-20}{y^2+2y-15}\cdot\frac{y^2+4y-21}{y^2+3y-28}$

$$=\frac{(y+5)(y-4)}{(y+5)(y-3)}\cdot\frac{(y+7)(y-3)}{(y+7)(y-4)}$$

$$=1$$

59. $\frac{25-n^2}{n^2-2n-35}\cdot\frac{n^2-8n-20}{n^2-3n-10}$

$$=\frac{(5-n)(5+n)}{(n-7)(n+5)}\cdot\frac{(n-10)(n+2)}{(n-5)(n+2)}$$

$$=-\frac{n-10}{n-7}$$

61. $\frac{x^2-11x+28}{x^2-13x+42}\cdot\frac{x^2+7x+10}{20-x-x^2}$

$$=\frac{(x-7)(x-4)}{(x-7)(x-6)}\cdot\frac{(x+5)(x+2)}{(4-x)(5+x)}$$

$$=-\frac{x+2}{x-6}$$

63. $\frac{x^2-4x-32}{x^2-8x-48}\cdot\frac{3x^2+17x+10}{3x^2-22x-16}$

$$=\frac{(x-8)(x+4)}{(x-12)(x+4)}\cdot\frac{(x+5)(3x+2)}{(x-8)(3x+2)}$$

$$=\frac{x+5}{x-12}$$

65. $\frac{3y^2+14y+8}{2y^2+7y-4}\cdot\frac{2y^2+9y-5}{3y^2+16y+5}$

$$=\frac{(3y+2)(y+4)}{(2y-1)(y+4)}\cdot\frac{(2y-1)(y+5)}{(3y+1)(y+5)}$$

$$=\frac{3y+2}{3y+1}$$

67. $\frac{9x^3y^4}{16a^4b^2}\div\frac{45x^4y^2}{14a^7b}=\frac{9x^3y^4}{16a^4b^2}\cdot\frac{14a^7b}{45x^4y^2}$

$$=\frac{7y^2a^3}{40xb}$$

69. $\frac{28x+14}{45x-30}\div\frac{14x+7}{30x-20}=\frac{28x+14}{45x-30}\cdot\frac{30x-20}{14x+7}$

$$=\frac{14(2x+1)}{15(3x-2)}\cdot\frac{10(3x-2)}{7(2x+1)}$$

$$=\frac{4}{3}$$

71. $$\frac{5a^2y+3a^2}{2x^3+5x^2}\div\frac{10ay+6a}{6x^3+15x^2}$$
$$=\frac{5a^2y+3a^2}{2x^3+5x^2}\cdot\frac{6x^3+15x^2}{10ay+6a}$$
$$=\frac{\cancel{a^2}\,\cancel{(5y+3)}}{\cancel{x^2}\,\cancel{(2x+5)}}\cdot\frac{3\cancel{x^2}\,\cancel{(2x+5)}}{2\cancel{a}\,\cancel{(5y+3)}}$$
$$=\frac{3a}{2}$$

73. $$\frac{x^3y^2}{x^2-3x-10}\div\frac{xy^4}{x^2-x-20}$$
$$=\frac{x^3y^2}{x^2-3x-10}\cdot\frac{x^2-x-20}{xy^4}$$
$$=\frac{\cancel{x^3}\,\cancel{y^2}}{\cancel{(x-5)}(x+2)}\cdot\frac{\cancel{(x-5)}(x+4)}{\cancel{x}\,\cancel{y^4}}$$
$$=\frac{x^2(x+4)}{y^2(x+2)}$$

75. $$\frac{x^2y^5}{x^2-11x+30}\div\frac{xy^6}{x^2-7x+10}$$
$$=\frac{x^2y^5}{x^2-11x+30}\cdot\frac{x^2-7x+10}{xy^6}$$
$$=\frac{\cancel{x^2}\,\cancel{y^5}}{\cancel{(x-5)}(x-6)}\cdot\frac{\cancel{(x-5)}(x-2)}{\cancel{x}\,\cancel{y^6}}$$
$$=\frac{x(x-2)}{y(x-6)}$$

77. $$\frac{x^2+3x-40}{x^2+2x-35}\div\frac{x^2+2x-48}{x^2+3x-18}$$
$$=\frac{x^2+3x-40}{x^2+2x-35}\cdot\frac{x^2+3x-18}{x^2+2x-48}$$
$$=\frac{\cancel{(x+8)}\,\cancel{(x-5)}}{(x+7)\,\cancel{(x-5)}}\cdot\frac{(x+6)(x-3)}{\cancel{(x+8)}(x-6)}$$
$$=\frac{(x+6)(x-3)}{(x+7)(x-6)}$$

79. $$\frac{y^2-y-56}{y^2+8y+7}\div\frac{y^2-13y+40}{y^2-4y-5}$$
$$=\frac{y^2-y-56}{y^2+8y+7}\cdot\frac{y^2-4y-5}{y^2-13y+40}$$
$$=\frac{\cancel{(y-8)}\,\cancel{(y+7)}}{\cancel{(y+1)}\,\cancel{(y+7)}}\cdot\frac{\cancel{(y-5)}\,\cancel{(y+1)}}{\cancel{(y-5)}\,\cancel{(y-8)}}$$
$$=1$$

81. $$\frac{x^2-x-2}{x^2-7x+10}\div\frac{x^2-3x-4}{40-3x-x^2}$$
$$=\frac{x^2-x-2}{x^2-7x+10}\cdot\frac{40-3x-x^2}{x^2-3x-4}$$
$$=\frac{\cancel{(x-2)}\,\cancel{(x+1)}}{\cancel{(x-2)}\,\cancel{(x-5)}}\cdot\frac{(8+x)\,\cancel{(5-x)}}{(x-4)\,\cancel{(x+1)}}$$
$$=-\frac{x+8}{x-4}$$

83. $$\frac{6n^2+13n+6}{4n^2-9}\div\frac{6n^2+n-2}{4n^2-1}$$
$$=\frac{6n^2+13n+6}{4n^2-9}\cdot\frac{4n^2-1}{6n^2+n-2}$$
$$=\frac{\cancel{(3n+2)}\,\cancel{(2n+3)}}{(2n-3)\,\cancel{(2n+3)}}\cdot\frac{\cancel{(2n-1)}(2n+1)}{\cancel{(3n+2)}\,\cancel{(2n-1)}}$$
$$=\frac{2n+1}{2n-3}$$

85. $$\frac{10+7x-12x^2}{8x^2-2x-15}\div\frac{6x^2-13x+5}{10x^2-13x+4}$$
$$=\frac{10+7x-12x^2}{8x^2-2x-15}\cdot\frac{10x^2-13x+4}{6x^2-13x+5}$$
$$=\frac{(5-4x)(2+3x)}{(4x+5)(2x-3)}\cdot\frac{(5x-4)\,\cancel{(2x-1)}}{(3x-5)\,\cancel{(2x-1)}}$$
$$=\frac{(5-4x)(2+3x)(5x-4)}{(4x+5)(2x-3)(3x-5)}$$

87. Yes.

89. **No.** $\frac{x-1}{x+1} \div \frac{x-4}{x+6} = \frac{x-1}{x+1} \cdot \frac{x+6}{x-4} = \frac{x^2+5x-6}{x^2-3x-4}$

91. $x-2=0 \qquad x+5=0$
$x=2 \qquad x=-5$

The expression is undefined when $x = 2$ and when $x = -5$.

93. $x^2-4x-5=0$
$(x-5)(x+1)=0$
$x-5=0 \qquad x+1=0$
$x=5 \qquad x=-1$

The expression is undefined when $x = -1$ and when $x = 5$.

95. $6x^2-5x-4=0$
$(3x-4)(2x+1)=0$
$3x-4=0 \qquad 2x+1=0$
$x=\frac{4}{3} \qquad x=-\frac{1}{2}$

The expression is undefined when $x = \frac{4}{3}$ and when $x = -\frac{1}{2}$.

97. $$\frac{Area\ triangle}{Area\ rect.} = \frac{\frac{1}{2}(2x)(x+4)}{(x+8)(x+4)} = \frac{x(x+4)}{(x+8)(x+4)} = \frac{x}{x+8}$$

99. The largest that the expression can be is 9. The smallest that the denominator can be is 1 so dividing 9 by values greater than 1 will always yield values less than or equal to 9. It is not possible for the expression to be greater than 10.

101. **a.** $\frac{660}{d+33} = \frac{660}{50+33} = \frac{660}{83} \approx 8\%$

b. As the depth increases, the recommended percent of oxygen decreases.

c. That is more than below the surface.

Section 9.2 Expressing Fractions in Terms of the LCD

1. **True.**

3. **False.** The LCM is x^8.

5. **True.**

7. $8x^3y = 2^3x^3y$
$12xy^2 = 2^2 \cdot 3xy^2$
$LCM = 2^3 \cdot 3x^3y^2 = 24x^3y^2$

9. $10x^4y^2 = 2 \cdot 5x^4y^2$
$15x^3y = 3 \cdot 5x^3y$
$LCM = 2 \cdot 3 \cdot 5x^4y^2 = 30x^4y^2$

11. $8x^2 = 2^3x^2$
$4x^2+8x = 2^2x(x+2)$
$LCM = 2^3x^2(x+2) = 8x^2(x+2)$

13. $2x^2y = 2x^2y$
$3x^2+12x = 3x(x+4)$
$LCM = 2 \cdot 3x^2y(x+4) = 6x^2y(x+4)$

15. $3x+3 = 3(x+1)$
$2x^2+4x+2 = 2(x+1)^2$
$LCM = 2 \cdot 3(x+1)^2 = 6(x+1)^2$

17. $(x-1)(x+2); (x-1)(x+3)$
$LCM = (x-1)(x+2)(x+3)$

19. $(2x+3)^2$; $(2x+3)(x-5)$
$LCM = (2x+3)^2(x-5)$

21. $x-1$; $x-2$; $(x-1)(x-2)$
$LCM = (x-1)(x-2)$

23. $x^2-x-6=(x-3)(x+2)$
$x^2+x-12=(x+4)(x-3)$
$LCM = (x-3)(x+2)(x+4)$

25. $x^2+5x+4=(x+4)(x+1)$
$x^2-3x-28=(x-7)(x+4)$
$LCM = (x+4)(x+1)(x-7)$

27. $x^2-2x-24=(x-6)(x+4)$
$x^2-36=(x-6)(x+6)$
$LCM = (x-6)(x+6)(x+4)$

29. $x^2-7x-30=(x-10)(x+3)$
$x^2-5x-24=(x-8)(x+3)$
$LCM = (x-10)(x+3)(x-8)$

31. $3x^2-11x+6=(3x-2)(x-3)$
$3x^2+4x-4=(3x-2)(x+2)$
$LCM = (3x-2)(x-3)(x+2)$

33. $15+2x-x^2=(5-x)(3+x)$
$(x-5)$
$(x+3)$
$LCM = (x+3)(x-5)$

35. **a. One.** $LCM = (x-3)(x+3)(x+4)$

b. None. $LCM = (x+3)^2(x-4)$

c. Two. $LCM = (x-3)^2(x+4)$

37. $LCM = y(y-3)^2$

39. To write the second fraction in terms of the LCM of the denominators, multiply its numerator and denominator by y.

41. $LCD = x^2$
$\frac{4}{x}=\frac{4x}{x^2}$
$\frac{3}{x^2}=\frac{3}{x^2}$

43. $LCD = 12y^2$
$\frac{x}{3y^2}=\frac{4x}{12y^2}$
$\frac{z}{4y}=\frac{3yz}{12y^2}$

45. $LCD = x^2(x-3)$
$\frac{y}{x(x-3)}=\frac{xy}{x^2(x-3)}$
$\frac{6}{x^2}=\frac{6(x-3)}{x^2(x-3)}=\frac{6x-18}{x^2(x-3)}$

47. $LCD = x(x-1)^2$
$\frac{9}{(x-1)^2}=\frac{9x}{x(x-1)^2}$
$\frac{6}{x(x-1)}=\frac{6(x-1)}{x(x-1)^2}=\frac{6x-6}{x(x-1)^2}$

49. $LCD = x(x-3)$
$\frac{3}{x-3}=\frac{3x}{x(x-3)}$
$-\frac{5}{x(3-x)}=\frac{5}{x(x-3)}$

51. $LCD = (x-5)^2$
$\frac{3}{(x-5)^2}=\frac{3}{(x-5)^2}$
$\frac{2}{5-x}=-\frac{2}{x-5}=-\frac{2(x-5)}{(x-5)^2}=-\frac{2x-10}{(x-5)^2}$

53. $LCD = x^2(x+2)$

$$\frac{3}{x^2+2x} = \frac{3}{x(x+2)} = \frac{3x}{x^2(x+2)}$$

$$\frac{4}{x^2} = \frac{4}{x^2} = \frac{4(x+2)}{x^2(x+2)} = \frac{4x+8}{x^2(x+2)}$$

55. $LCD = (x+3)(x-4)$

$$\frac{x-2}{x+3} = \frac{(x-2)(x-4)}{(x+3)(x-4)} = \frac{x^2-6x+8}{(x+3)(x-4)}$$

$$\frac{x}{x-4} = \frac{x(x+3)}{(x+3)(x-4)} = \frac{x^2+3x}{(x+3)(x-4)}$$

57. $LCD = (x+2)(x-1)$

$$\frac{3}{x^2+x-2} = \frac{3}{(x+2)(x-1)}$$

$$\frac{x}{x+2} = \frac{x}{x+2} = \frac{x(x-1)}{(x+2)(x-1)} = \frac{x^2-x}{(x+2)(x-1)}$$

59. $LCD = (2x-5)(x-2)$

$$\frac{5}{2x^2-9x+10} = \frac{5}{(2x-5)(x-2)}$$

$$\frac{x-1}{2x-5} = \frac{(x-1)(x-2)}{(2x-5)(x-2)} = \frac{x^2-3x+2}{(2x-5)(x-2)}$$

61. $LCD = (x+3)(x-3)(x-2)$

$$\frac{x}{x^2+x-6} = \frac{x(x-3)}{(x+3)(x-3)(x-2)}$$

$$= \frac{x^2-3x}{(x+3)(x-3)(x-2)}$$

$$\frac{2x}{x^2-9} = \frac{2x(x-2)}{(x+3)(x-3)(x-2)}$$

$$= \frac{2x^2-4x}{(x+3)(x-3)(x-2)}$$

63. $LCD = (x-3)^2(x+3)$

$$\frac{x}{9-x^2} = -\frac{x}{x^2-9} = -\frac{x(x-3)}{(x-3)^2(x+3)}$$

$$= -\frac{x^2-3x}{(x-3)^2(x+3)}$$

$$\frac{x-1}{x^2-6x+9} = \frac{(x-1)(x+3)}{(x-3)^2(x+3)}$$

$$= \frac{x^2+2x-3}{(x-3)^2(x+3)}$$

65. $LCD = 10^4$

$$\frac{3}{10^2} = \frac{3 \cdot 10^2}{10^4} = \frac{300}{10^4}$$

$$\frac{5}{10^4} = \frac{5}{10^4}$$

67. $LCD = b$

$$b = \frac{b^2}{b}$$

$$\frac{5}{b} = \frac{5}{b}$$

69. $LCD = y-1$

$$1 = \frac{y-1}{y-1}$$

$$\frac{y}{y-1} = \frac{y}{y-1}$$

71. $LCD = (x-1)^3$

$$\frac{x^2+1}{(x-1)^3} = \frac{x^2+1}{(x-1)^3}$$

$$\frac{x+1}{(x-1)^2} = \frac{(x+1)(x-1)}{(x-1)^3} = \frac{x^2-1}{(x-1)^3}$$

$$\frac{1}{x-1} = \frac{(x-1)^2}{(x-1)^3} = \frac{x^2-2x+1}{(x-1)^3}$$

73. $LCD=(x+y)(x+2)(x-2)$

$$\frac{1}{x^2+2x+xy+2y}=\frac{1}{(x+y)(x+2)}=\frac{x-2}{(x+y)(x+2)(x-2)}$$

$$\frac{1}{x^2+xy-2x-2y}=\frac{1}{(x+y)(x-2)}=\frac{x+2}{(x+y)(x+2)(x-2)}$$

75. The LCM of two expressions is equal to their product when the two expressions have no common factors.

Section 9.3 Addition and Subtraction of Rational Expressions

1. False. To add two fractions, add the numerators of like fractions and write the sum over the common denominator.

3. False. To add two rational expressions, rewrite both expressions with the LCM of the denominators.

5. $\frac{6a+1}{a-4}+\frac{2a-1}{a-4}=\frac{6a+1+2a-1}{a-4}=\frac{8a}{a-4}$

The denominators are the same; add the numerators.

7. $\frac{3}{y^2}+\frac{8}{y^2}=\frac{3+8}{y^2}=\frac{11}{y^2}$

9. $\frac{3}{x+4}-\frac{10}{x+4}=\frac{3-10}{x+4}=-\frac{7}{x+4}$

11. $\frac{3x}{2x+3}+\frac{5x}{2x+3}=\frac{3x+5x}{2x+3}=\frac{8x}{2x+3}$

13. $\frac{2x+1}{x-3}+\frac{3x+6}{x-3}=\frac{2x+1+3x+6}{x-3}=\frac{5x+7}{x-3}$

15. $\frac{5x-1}{x+9}-\frac{3x+4}{x+9}=\frac{5x-1-3x-4}{x+9}=\frac{2x-5}{x+9}$

17. $\frac{x-7}{2x+7}-\frac{4x-3}{2x+7}=\frac{x-7-4x+3}{2x+7}=\frac{-3x-4}{2x+7}$

19. $$\frac{x}{x^2+2x-15}-\frac{3}{x^2+2x-15}=\frac{x-3}{(x+5)(x-3)}=\frac{1}{x+5}$$

21. $$\frac{2x+3}{x^2-x-30}-\frac{x-2}{x^2-x-30}=\frac{2x+3-x+2}{(x-6)(x+5)}=\frac{x+5}{(x-6)(x+5)}=\frac{1}{x-6}$$

23. $$\frac{4y+7}{2y^2+7y-4}-\frac{y-5}{2y^2+7y-4}=\frac{4y+7-y+5}{(2y-1)(y+4)}=\frac{3y+12}{(2y-1)(y+4)}=\frac{3(y+4)}{(2y-1)(y+4)}=\frac{3}{2y-1}$$

25. $$\frac{2x^2+3x}{x^2-9x+20}+\frac{2x^2-3}{x^2-9x+20}-\frac{4x^2+2x+1}{x^2-9x+20}=\frac{2x^2+3x+2x^2-3-4x^2-2x-1}{x^2-9x+20}=\frac{x-4}{x^2-9x+20}=\frac{x-4}{(x-4)(x-5)}=\frac{1}{x-5}$$

27. Added. $\frac{8a}{15b}+\frac{2a}{15b}=\frac{10a}{15b}=\frac{2a}{3b}$

29. **Subtracted.** $\dfrac{x+4}{2x+1}-\dfrac{x-1}{2x+1}=\dfrac{x+4-x+1}{2x+1}$

$=\dfrac{5}{2x+1}$

31. To add or subtract rational expressions that have different denominators, you must first express each fraction in terms of a **common denominator**, which can be the **LCM** of the denominators of the rational expressions.

33. **False.** The common denominator is $24x^5$.

35. $\dfrac{4}{x}+\dfrac{5}{y}=\dfrac{4y}{xy}+\dfrac{5x}{xy}=\dfrac{4y+5x}{xy}$

37. $\dfrac{12}{x}-\dfrac{5}{2x}=\dfrac{24}{2x}-\dfrac{5}{2x}=\dfrac{19}{2x}$

39. $\dfrac{1}{2x}-\dfrac{5}{4x}+\dfrac{7}{6x}=\dfrac{6}{12x}-\dfrac{15}{12x}+\dfrac{14}{12x}=\dfrac{5}{12x}$

41. $\dfrac{5}{3x}-\dfrac{2}{x^2}+\dfrac{3}{2x}=\dfrac{10x}{6x^2}-\dfrac{12}{6x^2}+\dfrac{9x}{6x^2}=\dfrac{19x-12}{6x^2}$

43. $\dfrac{2}{x}-\dfrac{3}{2y}+\dfrac{3}{5x}-\dfrac{1}{4y}$

$=\dfrac{40y}{20xy}-\dfrac{30x}{20xy}+\dfrac{12y}{20xy}-\dfrac{5x}{20xy}$

$=\dfrac{52y-35x}{20xy}$

45. $\dfrac{2x+1}{3x}+\dfrac{x-1}{5x}=\dfrac{5(2x+1)}{15x}+\dfrac{3(x-1)}{15x}$

$=\dfrac{10x+5+3x-3}{15x}=\dfrac{13x+2}{15x}$

47. $\dfrac{x-3}{6x}+\dfrac{x+4}{8x}=\dfrac{4(x-3)}{24x}+\dfrac{3(x+4)}{24x}$

$=\dfrac{4x-12+3x+12}{24x}$

$=\dfrac{7x}{24x}=\dfrac{7}{24}$

49. $\dfrac{2x+9}{9x}-\dfrac{x-5}{5x}=\dfrac{5(2x+9)}{45x}-\dfrac{9(x-5)}{45x}$

$=\dfrac{10x+45-9x+45}{45x}$

$=\dfrac{x+90}{45x}$

51. $\dfrac{x+4}{2x}-\dfrac{x-1}{x^2}=\dfrac{x(x+4)}{2x^2}-\dfrac{2(x-1)}{2x^2}$

$=\dfrac{x^2+4x-2x+2}{2x^2}$

$=\dfrac{x^2+2x+2}{2x^2}$

53. $\dfrac{x-10}{4x^2}+\dfrac{x+1}{2x}=\dfrac{x-10}{4x^2}+\dfrac{2x(x+1)}{4x^2}$

$=\dfrac{x-10+2x^2+2x}{4x^2}$

$=\dfrac{2x^2+3x-10}{4x^2}$

55. $y+\dfrac{8}{3y}=\dfrac{3y^2}{3y}+\dfrac{8}{3y}=\dfrac{3y^2+8}{3y}$

57. $\dfrac{4}{x+4}+x=\dfrac{4}{x+4}+\dfrac{x(x+4)}{x+4}$

$=\dfrac{4+x^2+4x}{x+4}$

$=\dfrac{x^2+4x+4}{x+4}$

59. $5-\frac{x-2}{x+1}=\frac{5(x+1)}{x+1}-\frac{x-2}{x+1}$
$=\frac{5x+5-x+2}{x+1}$
$=\frac{4x+7}{x+1}$

61. $\frac{2x+1}{6x^2}-\frac{x-4}{4x}=\frac{2(2x+1)}{12x^2}-\frac{3x(x-4)}{12x^2}$
$=\frac{4x+2-3x^2+12x}{12x^2}$
$=\frac{-3x^2+16x+2}{12x^2}$

63. $\frac{x+2}{xy}-\frac{3x-2}{x^2y}=\frac{x(x+2)}{x^2y}-\frac{3x-2}{x^2y}$
$=\frac{x^2+2x-3x+2}{x^2y}$
$=\frac{x^2-x+2}{x^2y}$

65. $\frac{4x-3}{3x^2y}+\frac{2x+1}{4xy^2}=\frac{4y(4x-3)}{12x^2y^2}+\frac{3x(2x+1)}{12x^2y^2}$
$=\frac{16xy-12y+6x^2+3x}{12x^2y^2}$

67. $\frac{x-2}{8x^2}-\frac{x+7}{12xy}=\frac{3y(x-2)}{24x^2y}-\frac{2x(x+7)}{24x^2y}$
$=\frac{3xy-6y-2x^2-14x}{24x^2y}$

69. $\frac{4}{x-2}+\frac{5}{x+3}=\frac{4(x+3)}{(x-2)(x+3)}+\frac{5(x-2)}{(x-2)(x+3)}$
$=\frac{4x+12+5x-10}{(x-2)(x+3)}$
$=\frac{9x+2}{(x-2)(x+3)}$

71. $\frac{6}{x-7}-\frac{4}{x+3}=\frac{6(x+3)}{(x-7)(x+3)}-\frac{4(x-7)}{(x-7)(x+3)}$
$=\frac{6x+18-4x+28}{(x-7)(x+3)}$
$=\frac{2x+46}{(x-7)(x+3)}$
$=\frac{2(x+23)}{(x-7)(x+3)}$

73. $\frac{2x}{x+1}-\frac{1}{x-3}=\frac{2x(x-3)}{(x+1)(x-3)}-\frac{x+1}{(x+1)(x-3)}$
$=\frac{2x^2-6x-x-1}{(x+1)(x-3)}$
$=\frac{2x^2-7x-1}{(x+1)(x-3)}$

75. $\frac{4x}{2x-1}-\frac{5}{x-6}=\frac{4x(x-6)}{(2x-1)(x-6)}-\frac{5(2x-1)}{(2x-1)(x-6)}$
$=\frac{4x^2-24x-10x+5}{(2x-1)(x-6)}$
$=\frac{4x^2-34x+5}{(2x-1)(x-6)}$

77. $\frac{2a}{a-7}+\frac{5}{7-a}=\frac{2a}{a-7}-\frac{5}{a-7}=\frac{2a-5}{a-7}$

79. $\frac{x+1}{x-6}-\frac{x+1}{6-x}=\frac{x+1}{x-6}+\frac{x+1}{x-6}$
$=\frac{x+1+x+1}{x-6}$
$=\frac{2x+2}{x-6}=\frac{2(x+1)}{x-6}$

81. $$\frac{b+1}{b-1}+\frac{b-1}{b+1}=\frac{(b+1)(b+1)}{(b-1)(b+1)}+\frac{(b-1)(b-1)}{(b-1)(b+1)}$$
$$=\frac{b^2+2b+1+b^2-2b+1}{(b-1)(b+1)}$$
$$=\frac{2b^2+2}{(b-1)(b+1)}$$
$$=\frac{2(b^2+1)}{(b-1)(b+1)}$$

83. $$\frac{x}{x^2-9}+\frac{3}{x-3}=\frac{x}{(x-3)(x+3)}+\frac{3}{x-3}$$
$$=\frac{x}{(x-3)(x+3)}+\frac{3(x+3)}{(x-3)(x+3)}$$
$$=\frac{x+3x+9}{(x-3)(x+3)}$$
$$=\frac{4x+9}{(x-3)(x+3)}$$

85. $$\frac{2x}{x^2-x-6}-\frac{3}{x+2}=\frac{2x}{(x-3)(x+2)}-\frac{3}{x+2}$$
$$=\frac{2x}{(x-3)(x+2)}-\frac{3(x-3)}{(x-3)(x+2)}$$
$$=\frac{2x-3x+9}{(x-3)(x+2)}$$
$$=\frac{-x+9}{(x-3)(x+2)}$$

87. $$\frac{3x-1}{x^2-10x+25}-\frac{3}{x-5}=\frac{3x-1}{(x-5)^2}-\frac{3}{x-5}$$
$$=\frac{3x-1}{(x-5)^2}-\frac{3(x-5)}{(x-5)^2}$$
$$=\frac{3x-1-3x+15}{(x-5)^2}$$
$$=\frac{14}{(x-5)^2}$$

89. $$\frac{x+4}{x^2-x-42}+\frac{3}{7-x}=\frac{x+4}{(x-7)(x+6)}-\frac{3}{x-7}$$
$$=\frac{x+4}{(x-7)(x+6)}-\frac{3(x+6)}{(x-7)(x+6)}$$
$$=\frac{x+4-3x-18}{(x-7)(x+6)}$$
$$=\frac{-2x-14}{(x-7)(x+6)}$$
$$=\frac{-2(x+7)}{(x-7)(x+6)}$$

91. $$\frac{x}{2x+4}-\frac{2}{x^2+2x}=\frac{x}{2(x+2)}-\frac{2}{x(x+2)}$$
$$=\frac{x^2}{2x(x+2)}-\frac{4}{2x(x+2)}$$
$$=\frac{x^2-4}{2x(x+2)}$$
$$=\frac{(x+2)(x-2)}{2x(x+2)}$$
$$=\frac{x-2}{2x}$$

93. $$\frac{x-1}{x^2-x-2}+\frac{3}{x^2-3x+2}$$
$$=\frac{x-1}{(x-2)(x+1)}+\frac{3}{(x-2)(x-1)}$$
$$=\frac{(x-1)(x-1)}{(x-2)(x+1)(x-1)}+\frac{3(x+1)}{(x-2)(x+1)(x-1)}$$
$$=\frac{x^2-2x+1+3x+3}{(x-2)(x+1)(x-1)}$$
$$=\frac{x^2+x+4}{(x-2)(x+1)(x-1)}$$

95. $\frac{1}{x+1}+\frac{x}{x-6}-\frac{5x-2}{x^2-5x-6}$
$=\frac{1}{x+1}+\frac{x}{x-6}-\frac{5x-2}{(x-6)(x+1)}$
$=\frac{x-6}{(x+1)(x-6)}+\frac{x(x+1)}{(x+1)(x-6)}-\frac{5x-2}{(x-6)(x+1)}$
$=\frac{x-6+x^2+x-5x+2}{(x-6)(x+1)}$
$=\frac{x^2-3x-4}{(x-6)(x+1)}$
$=\frac{(x-4)(x+1)}{(x-6)(x+1)}$
$=\frac{x-4}{x-6}$

97. $\frac{a}{a-b}+\frac{b}{b-a}+1=\frac{a}{a-b}-\frac{b}{a-b}+\frac{a-b}{a-b}$
$=\frac{a-b+a-b}{a-b}$
$=\frac{2a-2b}{a-b}$
$=\frac{2(a-b)}{a-b}=2$

99. $b-3+\frac{5}{b+4}=\frac{b(b+4)}{b+4}-\frac{3(b+4)}{b+4}+\frac{5}{b+4}$
$=\frac{b^2+4b-3b-12+5}{b+4}$
$=\frac{b^2+b-7}{b+4}$

101. $\frac{(n+1)^2}{(n-1)^2}-1=\frac{(n+1)^2}{(n-1)^2}-\frac{(n-1)^2}{(n-1)^2}$
$=\frac{n^2+2n+1-n^2+2n-1}{(n-1)^2}$
$=\frac{4n}{(n-1)^2}$

103. $\frac{x^2+x-6}{x^2+2x-8}\cdot\frac{x^2+5x+4}{x^2+2x-3}-\frac{2}{x-1}$
$=\frac{(x+3)(x-2)}{(x+4)(x-2)}\cdot\frac{(x+4)(x+1)}{(x-1)(x+3)}-\frac{2}{x-1}$
$=\frac{x+1}{x-1}-\frac{2}{x-1}$
$=\frac{x+1-2}{x-1}=\frac{x-1}{x-1}=1$

105. $\frac{x^2-9}{x^2+6x+9}\div\frac{x^2+x-20}{x^2-x-12}+\frac{1}{x+1}$
$=\frac{(x-3)(x+3)}{(x+3)(x+3)}\cdot\frac{(x-4)(x+3)}{(x+5)(x-4)}+\frac{1}{x+1}$
$=\frac{x-3}{x+5}+\frac{1}{x+1}$
$=\frac{(x-3)(x+1)}{(x+5)(x+1)}+\frac{x+5}{(x+5)(x+1)}$
$=\frac{x^2-2x-3+x+5}{(x+5)(x+1)}$
$=\frac{x^2-x+2}{(x+5)(x+1)}$

107. $\frac{5b+4a}{ab}=\frac{5b}{ab}+\frac{4a}{ab}=\frac{5}{a}+\frac{4}{b}$

109. $\frac{3x^2+4xy}{x^2y^2}=\frac{3x^2}{x^2y^2}+\frac{4xy}{x^2y^2}=\frac{3}{y^2}+\frac{4}{xy}$

111. $\frac{1}{n}+\frac{1}{n+1}=\frac{n+1}{n(n+1)}+\frac{n}{n(n+1)}$
$=\frac{n+1+n}{n(n+1)}=\frac{2n+1}{n(n+1)}$

113. a. $\frac{x-1}{x+1}-\frac{1}{x}=\frac{x(x-1)}{x(x+1)}-\frac{x+1}{x(x+1)}$

$=\frac{x^2-x-x-1}{x(x+1)}$

$=\frac{x^2-2x-1}{x^2+x}$

b. $\frac{5}{2x}-\frac{2}{2x+1}=\frac{5(2x+1)}{2x(2x+1)}-\frac{2(2x)}{2x(2x+1)}$

$=\frac{10x+5-4x}{4x^2+2x}$

$=\frac{6x+5}{4x^2+2x}$

Section 9.4 Complex Fractions

1. Multiply by $\frac{2}{2}$.

3. Multiply by $\frac{(x-1)(x+4)}{(x-1)(x+4)}$.

5. **True**

7. To simplify the complex fraction, multiply the numerator and denominator of the complex fraction by the LCM of the denominators of the fractions $\frac{5}{x}, \frac{6}{x^2}$, and $\frac{4}{x^2}$. The LCD is x^2.

9. When you multiply the denominator of the complex fraction by x^2, the denominator of the complex fraction simplifies to x^2-4, which factors as $(x+2)(x-2)$.

11. $\frac{\frac{1}{3}+\frac{3}{x}}{\frac{1}{9}-\frac{9}{x^2}}=\left(\frac{\frac{1}{3}+\frac{3}{x}}{\frac{1}{9}-\frac{9}{x^2}}\right)\cdot\frac{9x^2}{9x^2}$

$=\frac{9x^2\cdot\frac{1}{3}+9x^2\cdot\frac{3}{x}}{9x^2\cdot\frac{1}{9}-9x^2\cdot\frac{9}{x^2}}$

$=\frac{3x^2+27x}{x^2-81}$

$=\frac{3x(x+9)}{(x-9)(x+9)}=\frac{3x}{x-9}$

13. $\frac{2-\frac{8}{x+4}}{3-\frac{12}{x+4}}=\left(\frac{2-\frac{8}{x+4}}{3-\frac{12}{x+4}}\right)\cdot\frac{x+4}{x+4}$

$=\frac{2(x+4)-\frac{8}{x+4}(x+4)}{3(x+4)-\frac{12}{(x+4)}(x+4)}$

$=\frac{2x+8-8}{3x+12-12}=\frac{2x}{3x}=\frac{2}{3}$

15. $\frac{1+\frac{5}{y-2}}{1-\frac{2}{y-2}}=\left(\frac{1+\frac{5}{y-2}}{1-\frac{2}{y-2}}\right)\cdot\frac{y-2}{y-2}$

$=\frac{1(y-2)+\frac{5}{y-2}(y-2)}{1(y-2)-\frac{2}{y-2}(y-2)}$

$=\frac{y-2+5}{y-2-2}=\frac{y+3}{y-4}$

17. $\frac{\frac{3}{x-2}+3}{\frac{4}{x-2}+4}=\left(\frac{\frac{3}{x-2}+3}{\frac{4}{x-2}+4}\right)\cdot\frac{x-2}{x-2}$

$=\frac{\frac{3}{x-2}(x-2)+3(x-2)}{\frac{4}{x-2}(x-2)+4(x-2)}$

$=\frac{3+3x-6}{4+4x-8}=\frac{3x-3}{4x-4}=\frac{3(x-1)}{4(x-1)}=\frac{3}{4}$

19. $$\frac{2-\frac{3}{x}-\frac{2}{x^2}}{2+\frac{5}{x}+\frac{2}{x^2}}=\left(\frac{2-\frac{3}{x}-\frac{2}{x^2}}{2+\frac{5}{x}+\frac{2}{x^2}}\right)\cdot\frac{x^2}{x^2}$$
$$=\frac{2(x^2)-\frac{3}{x}(x^2)-\frac{2}{x^2}(x^2)}{2(x^2)+\frac{5}{x}(x^2)+\frac{2}{x^2}(x^2)}$$
$$=\frac{2x^2-3x-2}{2x^2+5x+2}=\frac{(2x+1)(x-2)}{(2x+1)(x+2)}$$
$$=\frac{x-2}{x+2}$$

21. $$\frac{1-\frac{1}{x}-\frac{6}{x^2}}{1-\frac{9}{x^2}}=\left(\frac{1-\frac{1}{x}-\frac{6}{x^2}}{1-\frac{9}{x^2}}\right)\cdot\frac{x^2}{x^2}$$
$$=\frac{1(x^2)-\frac{1}{x}(x^2)-\frac{6}{x^2}(x^2)}{1(x^2)-\frac{9}{x^2}(x^2)}$$
$$=\frac{x^2-x-6}{x^2-9}=\frac{(x-3)(x+2)}{(x-3)(x+3)}=\frac{x+2}{x+3}$$

23. $$\frac{1-\frac{5}{x}-\frac{6}{x^2}}{1+\frac{6}{x}+\frac{5}{x^2}}=\left(\frac{1-\frac{5}{x}-\frac{6}{x^2}}{1+\frac{6}{x}+\frac{5}{x^2}}\right)\cdot\frac{x^2}{x^2}$$
$$=\frac{1(x^2)-\frac{5}{x}(x^2)-\frac{6}{x^2}(x^2)}{1(x^2)+\frac{6}{x}(x^2)+\frac{5}{x^2}(x^2)}$$
$$=\frac{x^2-5x-6}{x^2+6x+5}=\frac{(x-6)(x+1)}{(x+5)(x+1)}=\frac{x-6}{x+5}$$

25. $$\frac{1-\frac{6}{x}+\frac{8}{x^2}}{\frac{4}{x^2}+\frac{3}{x}-1}=\left(\frac{1-\frac{6}{x}+\frac{8}{x^2}}{\frac{4}{x^2}+\frac{3}{x}-1}\right)\cdot\frac{x^2}{x^2}$$
$$=\frac{1(x^2)-\frac{6}{x}(x^2)+\frac{8}{x^2}(x^2)}{\frac{4}{x^2}(x^2)+\frac{3}{x}(x^2)-1(x^2)}$$
$$=\frac{x^2-6x+8}{4+3x-x^2}=\frac{(x-4)(x-2)}{(1+x)(4-x)}=-\frac{x-2}{x+1}$$

27. $$\frac{x-\frac{4}{x+3}}{1+\frac{1}{x+3}}=\left(\frac{x-\frac{4}{x+3}}{1+\frac{1}{x+3}}\right)\cdot\frac{x+3}{x+3}$$
$$=\frac{x(x+3)-\frac{4}{x+3}(x+3)}{1(x+3)+\frac{1}{x+3}(x+3)}$$
$$=\frac{x^2+3x-4}{x+3+1}=\frac{(x+4)(x-1)}{x+4}=x-1$$

29. $$\frac{1-\frac{x}{2x+1}}{x-\frac{1}{2x+1}}=\left(\frac{1-\frac{x}{2x+1}}{x-\frac{1}{2x+1}}\right)\cdot\frac{2x+1}{2x+1}$$
$$=\frac{1(2x+1)-\frac{x}{2x+1}(2x+1)}{x(2x+1)-\frac{1}{2x+1}(2x+1)}$$
$$=\frac{2x+1-x}{2x^2+x-1}=\frac{x+1}{(x+1)(2x-1)}=\frac{1}{2x-1}$$

31. $$\frac{x-5+\frac{14}{x+4}}{x+3-\frac{2}{x+4}}=\left(\frac{x-5+\frac{14}{x+4}}{x+3-\frac{2}{x+4}}\right)\cdot\frac{x+4}{x+4}$$

$$=\frac{x(x+4)-5(x+4)+\frac{14}{x+4}(x+4)}{x(x+4)+3(x+4)-\frac{2}{x+4}(x+4)}$$

$$=\frac{x^2+4x-5x-20+14}{x^2+4x+3x+12-2}$$

$$=\frac{x^2-x-6}{x^2+7x+10}$$

$$=\frac{(x-3)(x+2)}{(x+5)(x+2)}=\frac{x-3}{x+5}$$

33. $$\frac{x+3-\frac{10}{x-6}}{x+2-\frac{20}{x-6}}=\left(\frac{x+3-\frac{10}{x-6}}{x+2-\frac{20}{x-6}}\right)\cdot\frac{x-6}{x-6}$$

$$=\frac{x(x-6)+3(x-6)-\frac{10}{x-6}(x-6)}{x(x-6)+2(x-6)-\frac{20}{x-6}(x-6)}$$

$$=\frac{x^2-6x+3x-18-10}{x^2-6x+2x-12-20}$$

$$=\frac{x^2-3x-28}{x^2-4x-32}=\frac{(x-7)(x+4)}{(x-8)(x+4)}=\frac{x-7}{x-8}$$

35. $$\frac{1-\frac{2}{x+1}}{1+\frac{1}{x-2}}=\left(\frac{1-\frac{2}{x+1}}{1+\frac{1}{x-2}}\right)\cdot\frac{(x+1)(x-2)}{(x+1)(x-2)}$$

$$=\frac{1(x+1)(x-2)-\frac{2}{x+1}(x+1)(x-2)}{1(x+1)(x-2)+\frac{1}{x-2}(x+1)(x-2)}$$

$$=\frac{x^2-x-2-2x+4}{x^2-x-2+x+1}=\frac{x^2-3x+2}{x^2-1}$$

$$=\frac{(x-2)(x-1)}{(x-1)(x+1)}=\frac{x-2}{x+1}$$

37. $$\frac{1-\frac{2}{x+4}}{1+\frac{3}{x-1}}=\left(\frac{1-\frac{2}{x+4}}{1+\frac{3}{x-1}}\right)\cdot\frac{(x+4)(x-1)}{(x+4)(x-1)}$$

$$=\frac{1(x+4)(x-1)-\frac{2}{x+4}(x+4)(x-1)}{1(x+4)(x-1)+\frac{3}{x-1}(x+4)(x-1)}$$

$$=\frac{x^2+3x-4-2x+2}{x^2+3x-4+3x+12}=\frac{x^2+x-2}{x^2+6x+8}$$

$$=\frac{(x+2)(x-1)}{(x+2)(x+4)}=\frac{x-1}{x+4}$$

39. $$\frac{\frac{1}{x}-\frac{2}{x-1}}{\frac{3}{x}+\frac{1}{x-1}}=\left(\frac{\frac{1}{x}-\frac{2}{x-1}}{\frac{3}{x}+\frac{1}{x-1}}\right)\cdot\frac{x(x-1)}{x(x-1)}$$

$$=\frac{\frac{1}{x}\cdot x(x-1)-\frac{2}{x-1}\cdot x(x-1)}{\frac{3}{x}\cdot x(x-1)+\frac{1}{x-1}\cdot x(x-1)}$$

$$=\frac{x-1-2x}{3x-3+x}=\frac{-x-1}{4x-3}$$

41. $$\frac{\frac{3}{b-4}-\frac{2}{b+1}}{\frac{5}{b+1}-\frac{1}{b-4}}=\left(\frac{\frac{3}{b-4}-\frac{2}{b+1}}{\frac{5}{b+1}-\frac{1}{b-4}}\right)\cdot\frac{(b+1)(b-4)}{(b+1)(b-4)}$$

$$=\frac{\frac{3}{b-4}(b+1)(b-4)-\frac{2}{b+1}(b+1)(b-4)}{\frac{5}{b+1}(b+1)(b-4)-\frac{1}{b-4}(b+1)(b-4)}$$

$$=\frac{3b+3-2b+8}{5b-20-b-1}=\frac{b+11}{4b-21}$$

43. **True.** Reciprocal of $\frac{1}{1-\frac{1}{x}}$:

$$\frac{1-\frac{1}{x}}{1}=1-\frac{1}{x}=\frac{x}{x}-\frac{1}{x}=\frac{x-1}{x}.$$

45. $1+\dfrac{1}{1+\dfrac{1}{2}}=1+\dfrac{1}{\dfrac{3}{2}}=1+\dfrac{2}{3}=\dfrac{5}{3}$

47. $1-\dfrac{1}{1-\dfrac{1}{x}}=1-\dfrac{1}{\dfrac{x-1}{x}}$

$=1-\dfrac{x}{x-1}=\dfrac{x-1}{x-1}-\dfrac{x}{x-1}$

$=\dfrac{-1}{x-1}$

49. $\dfrac{a^{-1}-b^{-1}}{a^{-2}-b^{-2}}=\left(\dfrac{\dfrac{1}{a}-\dfrac{1}{b}}{\dfrac{1}{a^2}-\dfrac{1}{b^2}}\right)\cdot\dfrac{a^2b^2}{a^2b^2}$

$=\dfrac{\dfrac{1}{a}a^2b^2-\dfrac{1}{b}a^2b^2}{\dfrac{1}{a^2}a^2b^2-\dfrac{1}{b^2}a^2b^2}=\dfrac{ab^2-a^2b}{b^2-a^2}$

$=\dfrac{ab(b-a)}{(b-a)(b+a)}=\dfrac{ab}{b+a}$

51. Multiplying the numerator and denominator of a complex fraction by the LCD of the fractions in the numerator and denominator eliminates fractions from the numerator and denominator. The result is a rational expression that is not a complex fraction.

53. $\dfrac{1}{\dfrac{1}{r_1}+\dfrac{1}{r_2}}=\dfrac{1}{\dfrac{1}{2}+\dfrac{1}{3}}\cdot\dfrac{6}{6}$

$=\dfrac{1\cdot 6}{\dfrac{1}{2}\cdot 6+\dfrac{1}{3}\cdot 6}=\dfrac{6}{3+2}=\dfrac{6}{5}\,ohms$

$\dfrac{r_1r_2}{r_1+r_2}=\dfrac{2(3)}{2+3}=\dfrac{6}{5}\,ohms$

55. The form $\dfrac{r_1r_2}{r_1+r_2}$ is easier to work with because when you substitute for r_1 and r_2 the result is a simple fraction. In the original form of the equation, after substituting the values, you still have a complex fraction that must be simplified.

Section 9.5 Equations Containing Fractions

1. The process of clearing denominators in an equation containing fractions is an application of the **Multiplication Property of Equations**.

3. **a.** These denominators can be cleared because this is an equation.

b. These denominators cannot be cleared because this is an expression.

5. A ratio is a quotient (fraction). A proportion is an equation that states the equality of two ratios (or rates).

7. **a.** The corresponding part for side *RP* is ***ZX***.

b. The corresponding part for side *YX* is ***QP***.

c. The corresponding part for $\angle X$ is $\angle \boldsymbol{P}$.

9. Sometimes, when we are clearing denominators, we multiply both sides of an equation by a factor containing a variable. This can result in a value of the variable that does not satisfy the original equation. Therefore, the solutions must be checked.

11. **a.** When you multiply the left side of the equation by the LCM $x(x-4)$, the left side simplifies to $7(x-4)$.

b. When you multiply the right side of the equation by the LCM $x(x-4)$, the right side simplifies to $3x$.

13. To check the solution for Exercise 12, substitute 7 for x in the original equation.

$\frac{7}{7}=\frac{3}{7-4}$ The solution checks.

$\frac{7}{7}=\frac{3}{3}$

$1=1$

15. $x-9=0$ *yields* $x=9$

$x^2-9x=x(x-9)=0$ *yields* $x=0,9$

17. $2+\frac{5}{x}=7$

$x\left(2+\frac{5}{x}\right)=x(7)$

$2x+5=7x$

$5=5x$

$1=x$

The solution checks.

19. $1-\frac{9}{x}=4$

$x\left(1-\frac{9}{x}\right)=x(4)$

$x-9=4x$

$-9=3x$

$-3=x$

The solution checks.

21. $\frac{2}{y}+5=9$

$y\left(\frac{2}{y}+5\right)=y(9)$

$2+5y=9y$

$2=4y$

$\frac{2}{4}=y$

$y=\frac{1}{2}$

The solution checks.

23. $\frac{4x}{x-4}+5=\frac{5x}{x-4}$

$(x-4)\left(\frac{4x}{x-4}+5\right)=(x-4)\left(\frac{5x}{x-4}\right)$

$4x+5(x-4)=5x$

$4x+5x-20=5x$

$9x-20=5x$

$4x=20$

$x=5$

The solution checks.

25. $2+\frac{3}{a-3}=\frac{a}{a-3}$

$(a-3)\left(2+\frac{3}{a-3}\right)=(a-3)\left(\frac{a}{a-3}\right)$

$2a-6+3=a$

$2a-3=a$

$-3=-a$

$a=3$

The solution does not check. The equation has no solution.

27. $\dfrac{x}{x-1}=\dfrac{8}{x+2}$

$(x-1)(x+2)\left(\dfrac{x}{x-1}\right)=(x-1)(x+2)\left(\dfrac{8}{x+2}\right)$

$x^2+2x=8x-8$

$x^2-6x+8=0$

$(x-4)(x-2)=0$

$x-4=0 \qquad x-2=0$

$x=4 \qquad x=2$

Both solutions check.

29. $\dfrac{2x}{x+4}=\dfrac{3}{x-1}$

$(x+4)(x-1)\cdot\dfrac{2x}{x+4}=(x+4)(x-1)\cdot\dfrac{3}{x-1}$

$2x^2-2x=3x+12$

$2x^2-5x-12=0$

$(2x+3)(x-4)=0$

$2x+3=0 \qquad x-4=0$

$x=-\dfrac{3}{2} \qquad x=4$

Both solutions check.

31. $x+\dfrac{6}{x-2}=\dfrac{3x}{x-2}$

$(x-2)\left(x+\dfrac{6}{x-2}\right)=(x-2)\left(\dfrac{3x}{x-2}\right)$

$x^2-2x+6=3x$

$0=x^2-5x+6=(x-3)(x-2)$

$x-3=0 \qquad x-2=0$

$x=3 \qquad x=2$

The solution $x = 2$ does not work. The only solution is $x = 3$.

33. $\dfrac{x}{x+2}+\dfrac{2}{x-2}=\dfrac{x+6}{x^2-4}$

$(x+2)(x-2)\left(\dfrac{x}{x+2}+\dfrac{2}{x-2}\right)$

$=(x+2)(x-2)\left(\dfrac{x+6}{(x+2)(x-2)}\right)$

$x^2-2x+2x+4=x+6$

$x^2-x-2=0$

$(x+1)(x-2)=0$

$x+1=0 \qquad x-2=0$

$x=-1 \qquad x=2$

$x = 2$ does not check. The only solution is $x = -1$.

35. $\dfrac{8}{y}=\dfrac{2}{y-2}+1$

$y(y-2)\left(\dfrac{8}{y}\right)=y(y-2)\left(\dfrac{2}{y-2}+1\right)$

$8y-16=2y+y^2-2y$

$0=y^2-8y+16=(y-4)^2$

$y-4=0$

$y=4$

The solution checks.

37. The quotient of two quantities that have the same unit is called a **ratio**. The quotient of two quantities that have different units is called a **rate**.

39. $\dfrac{x}{12}=\dfrac{3}{4}$

$12\cdot\dfrac{x}{12}=12\cdot\dfrac{3}{4}$

$x=9$

41. $\dfrac{4}{9}=\dfrac{x}{27}$

$27\cdot\dfrac{4}{9}=27\cdot\dfrac{x}{27}$

$12=x$

43. $\frac{x+3}{12}=\frac{5}{6}$

$12\cdot\frac{x+3}{12}=12\cdot\frac{5}{6}$

$x+3=10$

$x=7$

45. $\frac{18}{x+4}=\frac{9}{5}$

$5(x+4)\cdot\frac{18}{x+4}=5(x+4)\cdot\frac{9}{5}$

$90=9x+36$

$54=9x$

$6=x$

47. $\frac{2}{x}=\frac{4}{x+1}$

$x(x+1)\cdot\frac{2}{x}=x(x+1)\cdot\frac{4}{x+1}$

$2x+2=4x$

$2=2x$

$x=1$

49. $\frac{x+3}{4}=\frac{x}{8}$

$8\cdot\frac{x+3}{4}=8\cdot\frac{x}{8}$

$2x+6=x$

$x=-6$

51. $\frac{2}{x-1}=\frac{6}{2x+1}$

$(x-1)(2x+1)\cdot\frac{2}{x-1}=(x-1)(2x+1)\cdot\frac{6}{2x+1}$

$4x+2=6x-6$

$-2x=-8$

$x=4$

53. $\frac{2x}{7}=\frac{x-2}{14}$

$14\cdot\frac{2x}{7}=14\cdot\frac{x-2}{14}$

$4x=x-2$

$3x=-2$

$x=-\frac{2}{3}$

55. **True.**

57. 8 cm would be less than 30 mi.

59. $\frac{3}{500}=\frac{n}{2000}$

$2000\cdot\frac{3}{500}=2000\cdot\frac{n}{2000}$

$12=n$

12 transistors were defective.

61. $\frac{1}{6}=\frac{n}{75}$

$150\cdot\frac{1}{6}=150\cdot\frac{n}{75}$

$25=2n$

$12.5=n$

12.5 million children live in poverty.

63. $\frac{210}{300}=\frac{n}{25{,}000}$

$75{,}000\cdot\frac{210}{300}=75{,}000\cdot\frac{n}{25{,}000}$

$52{,}500=3n$

$17{,}500=n$

17,500 homes have Wi-Fi access.

65. $8.2\,ft = 98.4in$

$\frac{18}{98.4} = \frac{n}{7}$

$98.4n = 18(7)$

$98.4n = 126$

$n = 1.28$

The claw is approximately 1.28 inches long.

67. $\frac{3}{10} = \frac{n}{600}$

$10n = 3(600)$

$10n = 1800$

$n = 180$

180 panels are needed.

69. $\frac{2}{100} = \frac{n}{10{,}000}$

$100n = 20{,}000$

$n = 200$

No, the shipment will not be accepted because there are more than the maximum defects allowed.

71. $\frac{3}{4} = \frac{48}{n}$

$3n = 192$

$n = 64$

The standing height is 64 inches.

73. $\frac{0.375}{25} = \frac{2.625}{n}$

$0.375n = 65.625$

$n = 175$

The cities are 175 miles apart.

75. $\frac{170{,}000}{60} = \frac{n}{45}$

$60n = 45(170{,}000)$

$60n = 7{,}650{,}000$

$n = 127{,}500$

127,500 pounds of fuel are needed.

77. $\frac{3}{8} = \frac{n}{60}$

$8n = 3(60)$

$8n = 180$

$n = 22.5$

The mixture needs 22.5 gallons of yellow paint.

79. $\frac{5}{1200} = \frac{5+n}{1680}$

$1200(5+n) = 5(1680)$

$6000 + 1200n = 8400$

$1200n = 2400$

$n = 2$

2 additional gallons are needed.

81. $\frac{1170}{18{,}000} = \frac{1170+n}{19{,}500}$

$1170(19{,}500) = 18{,}000(1170+n)$

$22{,}815{,}000 = 21{,}060{,}000 + 18{,}000n$

$1{,}755{,}000 = 18{,}000n$

$97.5 = n$

Sales tax is \$97.50 higher.

83. $\frac{AC}{DF} = \frac{AB}{DE}$

$\frac{AC}{15} = \frac{4}{9}$

$9AC = 15(4)$

$9AC = 60$

$AC = 6.7cm$

85. $\frac{h}{H} = \frac{CB}{FE}$

$\frac{h}{7} = \frac{5}{12}$

$12h = 5(7)$

$12h = 35$

$h = 2.9m$

87. $\dfrac{DF}{AC}=\dfrac{EF}{BC}$

$\dfrac{DF}{5}=\dfrac{9}{6}$

$6DF=9(5)$

$6DF=45$

$DF=7.5ft$

$Perimeter=6+9+7.5=22.5ft^2$

89. $\dfrac{h}{H}=\dfrac{AB}{DE}$

$\dfrac{h}{12}=\dfrac{12}{18}$

$18h=12(12)$

$18h=144$

$h=8m$

$A=\dfrac{1}{2}bh$

$A=\dfrac{1}{2}(12)(8)=48m^2$

91. True.

93. a. Triangle *CBD* is similar to triangle *CAE*.

b. $\dfrac{BD}{AE}=\dfrac{CD}{CE}$

95. $\dfrac{BD}{AE}=\dfrac{BC}{AC}$

$\dfrac{5}{8}=\dfrac{BC}{10}$

$8BC=5(10)$

$8BC=50$

$BC=6.25cm$

97. $\dfrac{DE}{AC}=\dfrac{DB}{AB}$

$\dfrac{6}{10}=\dfrac{DB}{15}$

$10DB=6(15)$

$10DB=90$

$DB=9$

$AD+DB=AB$

$AD+9=15$

$AD=6in$

99. $(MO)^2=(MN)^2+(NO)^2$

$(MO)^2=10^2+24^2=100+576$

$(MO)^2=676$

$MO=26$

$MO+OP=MP$

$26+OP=39$

$OP=13cm$

101. $\dfrac{AB}{DC}=\dfrac{BO}{CO}$

$\dfrac{14}{DC}=\dfrac{8}{20}$

$8DC=14(20)$

$8DC=280$

$DC=35m$

103. $n+\dfrac{1}{n}=\dfrac{26}{5}$

$5n\left(n+\dfrac{1}{n}\right)=5n\cdot\dfrac{26}{5}$

$5n^2+5=26n$

$5n^2-26n+5=0$

$(n-5)(5n-1)=0$

$n-5=0 \qquad 5n-1=0$

$n=5 \qquad n=\dfrac{1}{5}$

105. $\frac{1}{6}=\frac{42}{x}$

$x=6(42)$

$x=252$ *attempts*

$252-42=210$ *shots made*

107. **a.** $\frac{7.5}{360}=\frac{520}{C}$

$7.5C=360(520)$

$7.5C=187{,}200$

$C=24{,}960mi$

b. $difference=24{,}960-24{,}800=160mi$

Section 9.6 Variation

1. **a.** Two quantities are directly proportional when an increase in one quantity leads to a proportional increase in the other quantity.

b. Two quantities are inversely proportional when an increase in one quantity leads to a proportional decrease in the other quantity.

3. **a.** True.

b. False. Since this is inverse variation, if x increases, then y decreases.

5. $W=\frac{k}{L}$

7. $t=ks$

$24=k(120)$

$k=\frac{24}{120}=0.2$

$t=0.2s$

9. $T=\frac{k}{S}$

$0.2=\frac{k}{8}$

$1.6=k$

$T=\frac{1.6}{S}$

11. $L=kP$

$24=k(21)$

$k=\frac{24}{21}=\frac{8}{7}$

$L=kP$

$L=\frac{8}{7}P$

$80=\frac{8}{7}P$

$P=\frac{7}{8}\cdot 80=70$

13. $T=kS$

$36=k(9)$

$k=4$

$T=kS$

$T=4S$

$T=4(2)$

$T=8$

15. $A=kB$

$6=k(18)$

$k=\frac{6}{18}=\frac{1}{3}$

$A=kB$

$A=\frac{1}{3}B$

$A=\frac{1}{3}(21)=7$

17. $W=\frac{k}{L}$

$20=\frac{k}{12}$

$k=20(12)=240$

$W=\frac{k}{L}$

$90=\frac{240}{L}$

$L(90)=240$

$L=2.67$

19. $C=\frac{k}{D}$

$4.5=\frac{k}{2}$

$9=k$

$C=\frac{k}{D}$

$C=\frac{9}{D}$

$C=\frac{9}{9}=1$

21. The variation equation is $d=4t$. The constant of variation is 4.

23. Inverse variation; as *x* increases, *y* decreases.

25. Since $\frac{101}{26}\approx 4$, the closest answer is ***i***.

27. $d=kt$

$8920=k(8)$

$k=\frac{8920}{8}=1115$

$d=kt$

$d=1115(3)$

$d=3345\,ft$

29. $P=kd$

$2.25=k(5)$

$k=\frac{2.25}{5}=0.45$

$P=kd$

$P=0.45d$

$P=0.45(12)$

$P=5.4\,psi$

31. $I=kV$

$4=k(100)$

$k=\frac{4}{100}=0.04$

$I=kV$

$I=0.04V$

$I=0.04(75)$

$I=3amperes$

33. $N=\frac{k}{C}$

$390=\frac{k}{0.50}$

$195=k$

$N=\frac{k}{C}=\frac{195}{C}$

$N=\frac{195}{0.20}=975\ items$

35. $t=\frac{k}{v}$

$4=\frac{k}{65}$

$260=k$

$t=\frac{k}{v}$

$5=\frac{260}{v}$

$v=\frac{260}{5}=52mph$

37. $P=\dfrac{k}{V}$

$25=\dfrac{k}{400}$

$10{,}000=k$

$P=\dfrac{k}{V}=\dfrac{10{,}000}{V}$

$P=\dfrac{10{,}000}{150}=66.\overline{6}psi$

39. $S=\dfrac{k}{P}$

$1800=\dfrac{k}{1800}$

$3{,}240{,}000=k$

$S=\dfrac{k}{P}=\dfrac{3{,}240{,}000}{P}$

$S=\dfrac{3{,}240{,}000}{1500}=2160 computers$

41. The formula for the area of a circle is $A=\pi r^2$. In this formula, an increase in the radius r of the circle leads to a proportional increase in the area A. Therefore, the formula represents a direct variation.

43. **a.** $i=kb$

$6.65=k(475)$

$\dfrac{6.65}{475}=k$

$k=0.014$

b. k represents the monthly interest rate, 1.4%.

c. $i=kb$

$i=0.014b$

$4.06=0.014b$

$\dfrac{4.06}{0.014}=b$

$b=\$290$

45. $R=\dfrac{k}{T}$

$20=\dfrac{k}{48}$

$k=960$

$R=\dfrac{k}{T}$

$R=\dfrac{960}{30}=32rpm$

47. $R=\dfrac{k}{T}$

$150=\dfrac{k}{30}$

$k=4500$

$R=\dfrac{k}{T}$

$R=\dfrac{4500}{12}=375rpm$

Section 9.7 Literal Equations

1. True

3. True

5. In solving the equation for R, the goal is to get R alone on one side of the equation.

7. $A=\dfrac{1}{2}bh$

$2A=2\cdot\dfrac{1}{2}bh$

$2A=bh$

$\dfrac{2A}{b}=\dfrac{bh}{b}$

$h=\dfrac{2A}{b}$

9. $d = rt$

$\frac{d}{r} = \frac{rt}{r}$

$t = \frac{d}{r}$

11. $PV = nRT$

$\frac{PV}{nR} = \frac{nRT}{nR}$

$T = \frac{PV}{nR}$

13. $P = 2L + 2W$

$P - 2W = 2L + 2W - 2W$

$P - 2W = 2L$

$\frac{P-2W}{2} = \frac{2L}{2}$

$L = \frac{P-2W}{2}$

15. $A = \frac{1}{2}h(b_1 + b_2)$

$2A = 2 \cdot \frac{1}{2}h(b_1 + b_2)$

$2A = hb_1 + hb_2$

$2A - hb_2 = hb_1$

$\frac{2A - hb_2}{h} = \frac{hb_1}{h}$

$b_1 = \frac{2A - hb_2}{h}$

17. $V = \frac{1}{3}Ah$

$3V = 3 \cdot \frac{1}{3}Ah$

$3V = Ah$

$\frac{3V}{A} = \frac{Ah}{A}$

$h = \frac{3V}{A}$

19. $R = \frac{C-S}{t}$

$tR = t \cdot \frac{C-S}{t}$

$tR = C - S$

$tR - C = C - S - C$

$tR - C = -S$

$\frac{tR - C}{-1} = \frac{-S}{-1}$

$S = C - tR$

21. $A = P + \mathrm{Pr}t$

$A = P(1 + rt)$

$\frac{A}{1+rt} = \frac{P(1+rt)}{1+rt}$

$P = \frac{A}{1+rt}$

23. $A = Sw + w$

$A = w(S+1)$

$\frac{A}{S+1} = \frac{w(S+1)}{S+1}$

$w = \frac{A}{S+1}$

25. Yes. $A = P(1+i)$ or $A = P(1+i)$

$A = P + Pi$ | $\frac{A}{P} = 1 + i$

$A - P = Pi$ | $\frac{A}{P} - 1 = i$

$\frac{A-P}{P} = i$

27. $cx - y = bx + 5$

$cx = bx + 5 + y$

$cx - bx = 5 + y$

$x(c - b) = y + 5$

$x = \frac{y+5}{c-b}$

29. a. $S = C + rC$

$S - C = rC$

$\frac{S-C}{C} = r$

b. $r = \frac{S-C}{C}$

$r = \frac{180-108}{108}$

$r = \frac{72}{108} = 0.\overline{6}6\overline{6} = 66.\overline{6}\%$

c. $r = \frac{S-C}{C}$

$r = \frac{150-110}{110}$

$r = \frac{40}{110} = 0.3\overline{636} = 36.\overline{36}\%$

31. a. $R = \frac{1}{\frac{1}{R_1} + \frac{1}{R_2}}$

$R = \frac{R_1R_2}{R_1R_2} \cdot \frac{1}{\frac{1}{R_1} + \frac{1}{R_2}}$

$R = \frac{R_1R_2}{R_1R_2 \cdot \frac{1}{R_1} + R_1R_2 \cdot \frac{1}{R_2}}$

$R = \frac{R_1R_2}{R_1 + R_2}$

$(R_1 + R_2)R = \frac{R_1R_2}{R_1 + R_2}(R_1 + R_2)$

$R_1R + R_2R = R_1R_2$

$R_2R = R_1R_2 - R_1R$

$R_2R = R_1(R_2 - R)$

$\frac{R_2R}{R_2 - R} = R_1$

b. $R_1 = \frac{R_2R}{R_2 - R}$

$R_1 = \frac{30(12)}{30-12} = \frac{360}{18}$

$R_1 = 20 ohms$

c. $R_1 = \frac{R_2R}{R_2 - R}$

$R_1 = \frac{15(6)}{15-6} = \frac{90}{9}$

$R_1 = 10 ohms$

Section 9.8 Application Problems

1. If it takes a janitorial crew 5 h to clean a company's offices, then in x hours the crew has completed $\frac{x}{5}$ of the job.

3. If only two people worked on a job and together they completed it, and one person completed $\frac{t}{30}$ of the job and the other person completed $\frac{t}{20}$ of the job, then $\frac{t}{30} + \frac{t}{20} = 1$.

5. Jen has the greater rate of work because she finishes the job faster.

7. a. $8 + 4 = 12 mph$

b. $8 - 4 = 4 mph$

9.

	Rate of work	·	Time worked	=	Part of task completed
Electrician	$\frac{1}{10}$	·	t	=	$\frac{t}{10}$
Assistant	$\frac{1}{12}$	·	t	=	$\frac{t}{12}$

11. It will take 2 hours to fill the pool with both sprinklers working.

	Rate	·	Time	=	Part of task completed
Large sprinkler	$\frac{1}{3}$	·	t	=	$\frac{t}{3}$
Small sprinkler	$\frac{1}{6}$	·	t	=	$\frac{t}{6}$

$$\frac{t}{3}+\frac{t}{6}=1$$
$$6\left(\frac{t}{3}+\frac{t}{6}\right)=6(1)$$
$$2t+t=6$$
$$3t=6$$
$$t=2$$

13. Both skiploaders can complete the job in 3 hours.

	Rate	·	Time	=	Part of task completed
Skiploader	$\frac{1}{12}$	·	t	=	$\frac{t}{12}$
Large skiploader	$\frac{1}{4}$	·	t	=	$\frac{t}{4}$

$$\frac{t}{12}+\frac{t}{4}=1$$
$$12\left(\frac{t}{12}+\frac{t}{4}\right)=12(1)$$
$$t+3t=12$$
$$4t=12$$
$$t=3$$

15. It will take 30 hours if the computers work together.

	Rate	·	Time	=	Part of task completed
Computer #1	$\frac{1}{75}$	·	t	=	$\frac{t}{75}$
Computer #2	$\frac{1}{50}$	·	t	=	$\frac{t}{50}$

$$\frac{t}{75}+\frac{t}{50}=1$$
$$150\left(\frac{t}{75}+\frac{t}{50}\right)=150(1)$$
$$2t+3t=150$$
$$5t=150$$
$$t=30$$

17. It takes 24 minutes with both air conditioners running.

	Rate	·	Time	=	Part of task completed
Small AC	$\frac{1}{60}$	·	t	=	$\frac{t}{60}$
Large AC	$\frac{1}{40}$	·	t	=	$\frac{t}{40}$

$$\frac{t}{60}+\frac{t}{40}=1$$
$$120\left(\frac{t}{60}+\frac{t}{40}\right)=120(1)$$
$$2t+3t=120$$
$$5t=120$$
$$t=24$$

19. The job takes the second welder 15 hours working alone.

	Rate	·	Time	=	Part of task completed
Welder #1	$\frac{1}{10}$	·	6	=	$6\cdot\frac{1}{10}=\frac{3}{5}$
Welder #2	$\frac{1}{x}$	·	6	=	$\frac{6}{x}$

$$\frac{3}{5}+\frac{6}{x}=1$$
$$5x\left(\frac{3}{5}+\frac{6}{x}\right)=5x(1)$$
$$3x+30=5x$$
$$30=2x$$
$$x=15$$

21. It will take 90 minutes for the second pipeline to fill the tank working alone.

	Rate	·	Time	=	Part of task completed
Pipeline #1	$\frac{1}{45}$	·	30	=	$30\cdot\frac{1}{45}=\frac{2}{3}$
Pipeline #2	$\frac{1}{x}$	·	30	=	$\frac{30}{x}$

$$\frac{2}{3}+\frac{30}{x}=1$$
$$3x\left(\frac{2}{3}+\frac{30}{x}\right)=3x(1)$$
$$2x+90=3x$$
$$90=x$$

23. It will take the older reaper 3 hours working alone.

	Rate	·	Time	=	Part of task completed
Old reaper	$\frac{1}{x}$	·	1	=	$\frac{1}{x}$
New reaper	$\frac{1}{1.5}$	·	1	=	$\frac{1}{1.5}=\frac{2}{3}$

$$\frac{1}{x}+\frac{2}{3}=1$$
$$3x\left(\frac{1}{x}+\frac{2}{3}\right)=3x(1)$$
$$3+2x=3x$$
$$3=x$$

25. The second technician works 3 hours.

	Rate	·	Time	=	Part of task completed
Tech #1	$\frac{1}{4}$	·	2	=	$2\cdot\frac{1}{4}=\frac{1}{2}$
Tech #2	$\frac{1}{6}$	·	t	=	$\frac{t}{6}$

$$\frac{1}{2}+\frac{t}{6}=1$$
$$6\left(\frac{1}{2}+\frac{t}{6}\right)=6(1)$$
$$3+t=6$$
$$t=3$$

27. It takes the small unit $14\frac{2}{3}$ hours to heat the pool.

	Rate	·	Time	=	Part of task completed
Large unit	$\frac{1}{8}$	·	2	=	$2\cdot\frac{1}{8}=\frac{1}{4}$
Small unit	$\frac{1}{x}$	·	11	=	$\frac{11}{x}$

$$\frac{1}{4}+\frac{11}{x}=1$$
$$4x\left(\frac{1}{4}+\frac{11}{x}\right)=4x(1)$$
$$x+44=4x$$
$$44=3x$$
$$x=\frac{44}{3}=14\frac{2}{3}$$

29. It takes the apprentice 3 hours to make the repair.

	Rate	·	Time	=	Part of task completed
Mechanic	$\frac{1}{2}$	·	1	=	$\frac{1}{2}$
Apprentice	$\frac{1}{6}$	·	t	=	$\frac{t}{6}$

$$\frac{1}{2}+\frac{t}{6}=1$$
$$6\left(\frac{1}{2}+\frac{t}{6}\right)=6(1)$$
$$3+t=6$$
$$t=3$$

31. When they work together t will be less time than for either of them working alone, so t will be less than k.

33. a.

	Distance	÷	Rate	=	Time
Against the wind	1440	÷	$380 - r$	=	$\frac{1440}{380-r}$
With the wind	1600	÷	$380 + r$	=	$\frac{1600}{380+r}$

b. $\frac{1440}{380-r}=\frac{1600}{380+r}$

35. The camper hiked at a rate of 4 mph.

	Distance	÷	Rate	=	Time
Driving	80	÷	$10r$	=	$\frac{80}{10r}=\frac{8}{r}$
Hiking	4	÷	r	=	$\frac{4}{r}$

$$\frac{8}{r}+\frac{4}{r}=3$$
$$r\left(\frac{8}{r}+\frac{4}{r}\right)=r(3)$$
$$8+4=3r$$
$$12=3r$$
$$r=4$$

37. The rate of the jet was 360 mph.

	Distance	÷	Rate	=	Time
Jet	1080	÷	$4r$	=	$\frac{1080}{4r}=\frac{270}{r}$
Helicopter	180	÷	r	=	$\frac{180}{r}$

$$\frac{270}{r}+\frac{180}{r}=5$$
$$r\left(\frac{270}{r}+\frac{180}{r}\right)=r(5)$$
$$270+180=5r$$
$$450=5r$$
$$r=90$$
$$4r=360$$

39. The rate of the boat on the first leg was 7.5 mph.

	Distance	÷	Rate	=	Time
Leg #1	15	÷	r	=	$\frac{15}{r}$
Leg #2	19	÷	$r+2$	=	$\frac{19}{r+2}$

$$\frac{15}{r}+\frac{19}{r+2}=4$$
$$r(r+2)\left(\frac{15}{r}+\frac{19}{r+2}\right)=r(r+2)(4)$$
$$15r+30+19r=4r^2+8r$$
$$34r+30=4r^2+8r$$
$$0=4r^2-26r-30$$
$$0=2r^2-13r-15$$
$$0=(2r-15)(r+1)$$
$$2r-15=0 \qquad r+1=0$$
$$r=7.5 \qquad r=-1$$

41. The family can go 21.6 miles down the river.

	Distance	÷	Rate	=	Time
With the current	d	÷	$15+3=18$	=	$\frac{d}{18}$
Against the current	d	÷	$15-3=12$	=	$\frac{d}{12}$

$$\frac{d}{18}+\frac{d}{12}=3$$
$$36\left(\frac{d}{18}+\frac{d}{12}\right)=36(3)$$
$$2d+3d=108$$
$$5d=108$$
$$d=21.6$$

43. The technician's speed in congested traffic was 20 mph.

	Distance	÷	Rate	=	Time
Congested traffic	10	÷	r	=	$\frac{10}{r}$
Expressway	20	÷	$r+20$	=	$\frac{20}{r+20}$

$$\frac{10}{r}+\frac{20}{r+20}=1$$
$$r(r+20)\left(\frac{10}{r}+\frac{20}{r+20}\right)=r(r+20)(1)$$
$$10(r+20)+20r=r^2+20r$$
$$10r+200+20r=r^2+20r$$
$$30r+200=r^2+20r$$
$$0=r^2-10r-200$$
$$0=(r-20)(r+10)$$
$$r-20=0 \qquad r+10=0$$
$$r=20 \qquad r=-10$$

45. The rate of the current is 5mph.

	Distance	÷	Rate	=	Time
With the current	75	÷	$20+c$	=	$\frac{75}{20+c}$
Against the current	45	÷	$20-c$	=	$\frac{45}{20-c}$

$$\frac{75}{20+c}=\frac{45}{20-c}$$
$$(20+c)(20-c)\frac{75}{20+c}=(20+c)(20-c)\frac{45}{(20-c)}$$
$$75(20-c)=45(20+c)$$
$$1500-75c=900+45c$$
$$600=120c$$
$$c=5$$

47. The rate of the freight train was 30 mph and the rate of the express train was 50 mph.

	Distance	÷	Rate	=	Time
Express train	600	÷	$r+20$	=	$\frac{600}{r+20}$
Freight train	360	÷	r	=	$\frac{360}{r}$

$$\frac{600}{r+20}=\frac{360}{r}$$
$$r(r+20)\left(\frac{600}{r+20}\right)=r(r+20)\left(\frac{360}{r}\right)$$
$$600r=360r+7200$$
$$240r=7200$$
$$r=30$$
$$r+20=50$$

49. The rate of the current is 2 mph.

	Distance	÷	Rate	=	Time
With the current	24	÷	$6+c$	=	$\frac{24}{6+c}$
Against the current	12	÷	$6-c$	=	$\frac{12}{6-c}$

$$\frac{24}{6+c}=\frac{12}{6-c}$$
$$(6+c)(6-c)\left(\frac{24}{6+c}\right)=(6+c)(6-c)\left(\frac{12}{6-c}\right)$$
$$144-24c=72+12c$$
$$72=36c$$
$$c=2$$

51. The rate of the jet stream is 50 mph.

	Distance	÷	Rate	=	Time
With the jet stream	2400	÷	$550+j$	=	$\frac{2400}{550+j}$
Against the jet stream	2000	÷	$550-j$	=	$\frac{2000}{550-j}$

$$\frac{2400}{550+j}=\frac{2000}{550-j}$$
$$(550+j)(550-j)\left(\frac{2400}{550+j}\right)$$
$$=(550+j)(550-j)\left(\frac{2000}{550-j}\right)$$
$$1{,}320{,}000-2400j=1{,}100{,}000+2000j$$
$$220{,}000=4400j$$
$$j=50$$

53. The rate of the current is 5 mph.

	Distance	÷	Rate	=	Time
With the current	25	÷	$20+c$	=	$\frac{25}{20+c}$
Against the current	15	÷	$20-c$	=	$\frac{15}{20-c}$

$$\frac{25}{20+c}=\frac{15}{20-c}$$
$$(20+c)(20-c)\left(\frac{25}{20+c}\right)$$
$$=(20+c)(20-c)\left(\frac{15}{20-c}\right)$$
$$500-25c=300+15c$$
$$200=40c$$
$$c=5$$

55. The expression $380 + r$ represents the plane's speed when it is flying in the same direction as the wind. The plane's speed is increased by the wind. The expression $380 - r$ represents the plane's speed when it is flying in the opposite direction of (against) the wind. The plane's speed is decreased by the wind.

57. Together the pipes can fill the tank in $1\frac{1}{19}$ hr.

	Rate	·	Time	=	Part of task completed
Pipe #1	$\frac{1}{2}$	·	t	=	$\frac{t}{2}$
Pipe #2	$\frac{1}{4}$	·	t	=	$\frac{t}{4}$
Pipe #3	$\frac{1}{5}$	·	t	=	$\frac{t}{5}$

$$\frac{t}{2}+\frac{t}{4}+\frac{t}{5}=1$$
$$20\left(\frac{t}{2}+\frac{t}{4}+\frac{t}{5}\right)=20(1)$$
$$10t+5t+4t=20$$
$$19t=20$$
$$t=\frac{20}{19}=1\frac{1}{19}$$

59. The time spent in the canoe was 2 hours.

	Distance	÷	Rate	=	Time
Canoe	18	÷	$3r$	=	$\frac{18}{3r}=\frac{6}{r}$
Hike	3	÷	r	=	$\frac{3}{r}$

$$\frac{6}{r}=\frac{3}{r}+1$$
$$r\left(\frac{6}{r}\right)=r\left(\frac{3}{r}+1\right)$$
$$6=3+r$$
$$r=3$$
$$\frac{3}{r}+1=2$$

61. The bus usually travels at 60 mph.

	Distance	÷	Rate	=	Time
Good weather	150	÷	r	=	$\frac{150}{r}$
Bad weather	150	÷	r - 10	=	$\frac{150}{r-10}$

$$\frac{150}{r}+\frac{1}{2}=\frac{150}{r-10}$$
$$2r(r-10)\left(\frac{150}{r}+\frac{1}{2}\right)=2r(r-10)\left(\frac{150}{r-10}\right)$$
$$300r-3000+r^2-10r=300r$$
$$r^2-10r-3000=0$$
$$(r-60)(r+50)=0$$
$$r-60=0 \quad r+50=0$$
$$r=60 \quad r=-50$$

Chapter 9 Review Exercises

1. $\frac{8ab^2}{15x^3y}\cdot\frac{5xy^4}{16a^2b}=\frac{8ab\cdot b\cdot 5xy\cdot y^3}{3x^2\cdot 5xy\cdot 2\cdot 8ab\cdot a}=\frac{by^3}{6ax^2}$

2. $\frac{5}{3x-4}+\frac{4}{2x+3}=\frac{5(2x+3)+4(3x-4)}{(2x+3)(3x-4)}$

$=\frac{10x+15+12x-16}{(2x+3)(3x-4)}$

$=\frac{22x-1}{(2x+3)(3x-4)}$

3. $4x+3y=12$

$4x=-3y+12$

$x=\frac{-3y+12}{4}$

$x=-\frac{3}{4}y+3$

4. $\frac{16x^5y^3}{24xy^{10}}=\frac{8xy^3\cdot 2x^4}{8xy^3\cdot 3y^7}=\frac{2x^4}{3y^7}$

5. $\frac{20x^2-45x}{6x^3+4x^2}\div\frac{40x^3-90x^2}{12x^2+8x}$

$=\frac{20x^2-45x}{6x^3+4x^2}\cdot\frac{12x^2+8x}{40x^3-90x^2}$

$=\frac{5x(4x-9)}{2x^2(3x+2)}\cdot\frac{4x(3x+2)}{10x^2(4x-9)}=\frac{1}{x^2}$

6. $\frac{x-\frac{16}{5x-2}}{3x-4-\frac{88}{5x-2}}=\frac{5x-2}{5x-2}\cdot\left(\frac{x-\frac{16}{5x-2}}{3x-4-\frac{88}{5x-2}}\right)$

$=\frac{(5x-2)x-16}{3x(5x-2)-4(5x-2)-88}$

$=\frac{5x^2-2x-16}{15x^2-6x-20x+8-88}$

$=\frac{(x-2)(5x+8)}{15x^2-26x-80}=\frac{(x-2)(5x+8)}{(5x+8)(3x-10)}$

$=\frac{x-2}{3x-10}$

7. $24a^2b^5=2^3\cdot 3a^2b^5$

$36a^3b=2^2\cdot 3^2a^3b$

$LCM=2^3\cdot 3^2a^3b^5=72a^3b^5$

8. $\frac{5x}{3x+7}-\frac{x}{3x+7}=\frac{5x-x}{3x+7}=\frac{4x}{3x+7}$

9. $16x=2^4x$

$8x^2=2^3x^2$

$LCD=2^4x^2=16x^2$

$\frac{3}{16x}=\frac{3x}{16x^2}$

$\frac{5}{8x^2}=\frac{10}{16x^2}$

10. $\frac{2x^2-13x-45}{2x^2-x-15}=\frac{(x-9)(2x+5)}{(2x+5)(x-3)}=\frac{x-9}{x-3}$

11. $\frac{x^2-5x-14}{x^2-3x-10}\div\frac{x^2-4x-21}{x^2-9x+20}$

$=\frac{x^2-5x-14}{x^2-3x-10}\cdot\frac{x^2-9x+20}{x^2-4x-21}$

$=\frac{(x-7)(x+2)}{(x-5)(x+2)}\cdot\frac{(x-4)(x-5)}{(x-7)(x+3)}$

$=\frac{x-4}{x+3}$

12. $\frac{2y}{5y-7}+\frac{3}{7-5y}=\frac{2y}{5y-7}-\frac{3}{5y-7}=\frac{2y-3}{5y-7}$

13. $\frac{3x^3+10x^2}{10x-2}\cdot\frac{20x-4}{6x^4+20x^3}$

$=\frac{x^2(3x+10)}{2(5x-1)}\cdot\frac{4(5x-1)}{2x^3(3x+10)}=\frac{1}{x}$

14. $\frac{5x+3}{2x^2+5x-3}-\frac{3x+4}{2x^2+5x-3}=\frac{5x+3-3x-4}{2x^2+5x-3}$

$=\frac{2x-1}{(2x-1)(x+3)}$

$=\frac{1}{x+3}$

15. $5x^4(x-7)^2 = 5x^4(x-7)^2$

$15x(x-7) = 3\cdot 5x(x-7)$

$LCM = 3\cdot 5x^4(x-7)^2 = 15x^4(x-7)^2$

16. $\dfrac{6}{x-7} = \dfrac{8}{x-6}$

$(x-7)(x-6)\cdot\dfrac{6}{x-7} = (x-7)(x-6)\cdot\dfrac{8}{x-6}$

$6x-36 = 8x-56$

$20 = 2x$

$x = 10$

The solution checks.

17. $\dfrac{x+8}{x+4} = 1+\dfrac{5}{x+4}$

$(x+4)\left(\dfrac{x+8}{x+4}\right) = (x+4)\left(1+\dfrac{5}{x+4}\right)$

$x+8 = x+4+5$

$x+8 = x+9$

$8 \neq 9$

The equation has no solution.

18. $\dfrac{12a^2b(4x-7)}{15ab^2(7-4x)} = \dfrac{3\cdot 4a^2b(4x-7)}{-3\cdot 5ab^2(4x-7)} = -\dfrac{4a}{5b}$

19. $\dfrac{5x-1}{x^2-9}+\dfrac{4x-3}{x^2-9}-\dfrac{8x-1}{x^2-9} = \dfrac{5x-1+4x-3-8x+1}{x^2-9}$

$= \dfrac{x-3}{(x-3)(x+3)}$

$= \dfrac{1}{x+3}$

20. $\dfrac{AC}{DF} = \dfrac{BC}{EF}$

$\dfrac{AC}{12} = \dfrac{6}{9}$

$9AC = 72$

$AC = 8$

$P = 8+6+10 = 24in$

21. $\dfrac{20}{2x+3} = \dfrac{17x}{2x+3}-5$

$(2x+3)\cdot\left(\dfrac{20}{2x+3}\right) = (2x+3)\cdot\left(\dfrac{17x}{2x+3}-5\right)$

$20 = 17x-10x-15$

$20 = 7x-15$

$35 = 7x$

$x = 5$

The solution checks.

22. $\dfrac{\dfrac{5}{x-1}-\dfrac{3}{x+3}}{\dfrac{6}{x+3}+\dfrac{2}{x-1}} = \dfrac{(x-1)(x+3)}{(x-1)(x+3)}\cdot\left(\dfrac{\dfrac{5}{x-1}-\dfrac{3}{x+3}}{\dfrac{6}{x+3}+\dfrac{2}{x-1}}\right)$

$= \dfrac{5(x+3)-3(x-1)}{6(x-1)+2(x+3)}$

$= \dfrac{5x+15-3x+3}{6x-6+2x+6} = \dfrac{2x+18}{8x}$

$= \dfrac{2(x+9)}{8x} = \dfrac{x+9}{4x}$

23. $\dfrac{x-1}{x+2}+\dfrac{3x-2}{5-x}+\dfrac{5x^2+15x-11}{x^2-3x-10}$

$= \dfrac{x-1}{x+2}-\dfrac{3x-2}{x-5}+\dfrac{5x^2+15x-11}{(x-5)(x+2)}$

$= \dfrac{(x-1)(x-5)-(x+2)(3x-2)+(5x^2+15x-11)}{(x-5)(x+2)}$

$= \dfrac{x^2-6x+5-3x^2-4x+4+5x^2+15x-11}{(x-5)(x+2)}$

$= \dfrac{3x^2+5x-2}{(x-5)(x+2)} = \dfrac{(3x-1)(x+2)}{(x-5)(x+2)} = \dfrac{3x-1}{x-5}$

24. $\dfrac{18x^2+25x-3}{9x^2-28x+3} \div \dfrac{2x^2+9x+9}{x^2-6x+9}$

$= \dfrac{18x^2+25x-3}{9x^2-28x+3}\cdot\dfrac{x^2-6x+9}{2x^2+9x+9}$

$= \dfrac{(9x-1)(2x+3)}{(9x-1)(x-3)}\cdot\dfrac{(x-3)(x-3)}{(2x+3)(x+3)}$

$= \dfrac{x-3}{x+3}$

25. $\dfrac{x^2+x-30}{15+2x-x^2}=\dfrac{(x+6)(x-5)}{(5-x)(3+x)}=-\dfrac{x+6}{x+3}$

26. $\dfrac{5}{7}+\dfrac{x}{2}=2-\dfrac{x}{7}$

$14\cdot\dfrac{5}{7}+14\cdot\dfrac{x}{2}=14\cdot 2-14\cdot\dfrac{x}{7}$

$10+7x=28-2x$

$9x=18$

$x=2$

27. $\dfrac{x+\dfrac{6}{x-5}}{1+\dfrac{2}{x-5}}=\dfrac{x-5}{x-5}\cdot\left(\dfrac{x+\dfrac{6}{x-5}}{1+\dfrac{2}{x-5}}\right)$

$=\dfrac{x(x-5)+6}{x-5+2}=\dfrac{x^2-5x+6}{x-3}$

$=\dfrac{(x-2)(x-3)}{x-3}=x-2$

28. $\dfrac{3x^2+4x-15}{x^2-11x+28}\cdot\dfrac{x^2-5x-14}{3x^2+x-10}$

$=\dfrac{(3x-5)(x+3)}{(x-7)(x-4)}\cdot\dfrac{(x-7)(x+2)}{(3x-5)(x+2)}$

$=\dfrac{x+3}{x-4}$

29. $\dfrac{x}{5}=\dfrac{x+12}{9}$

$45\left(\dfrac{x}{5}\right)=45\left(\dfrac{x+12}{9}\right)$

$9x=5(x+12)$

$9x=5x+60$

$4x=60$

$x=15$

30. $\dfrac{3}{20}=\dfrac{x}{80}$

$20x=3\cdot 80$

$20x=240$

$x=12$

31. $\dfrac{1-\dfrac{1}{x}}{1-\dfrac{8x-7}{x^2}}=\dfrac{x^2}{x^2}\cdot\left(\dfrac{1-\dfrac{1}{x}}{1-\dfrac{8x-7}{x^2}}\right)$

$=\dfrac{x^2-x}{x^2-8x+7}$

$=\dfrac{x(x-1)}{(x-7)(x-1)}=\dfrac{x}{x-7}$

32. $x-2y=15$

$x=2y+15$

33. $\dfrac{6}{a}+\dfrac{9}{b}=\dfrac{6b+9a}{ab}=\dfrac{3(2b+3a)}{ab}=\dfrac{3(3a+2b)}{ab}$

34. $10x^2-11x+3=(2x-1)(5x-3)$

$20x^2-17x+3=(4x-1)(5x-3)$

$LCM=(2x-1)(5x-3)(4x-1)$

35. $y=kx$

$5=k(20)$

$k=\dfrac{5}{20}=\dfrac{1}{4}$

$y=kx=\dfrac{1}{4}x$

$y=\dfrac{1}{4}\cdot 12=3$

36. $y=\dfrac{k}{x}$

$400=\dfrac{k}{5}$

$2000=k$

$y=\dfrac{k}{x}=\dfrac{2000}{x}$

$y=\dfrac{2000}{20}=100$

37. $i=\dfrac{100m}{c}$

$ci=100m$

$\dfrac{ci}{i}=\dfrac{100m}{i}$

$c=\dfrac{100m}{i}$

38. $\dfrac{15}{x}=\dfrac{3}{8}$

$3x=15\cdot 8$

$3x=120$

$x=40$

The solution checks.

39. $\dfrac{22}{2x+5}=2$

$22=2(2x+5)$

$22=4x+10$

$12=4x$

$x=3$

The solution checks.

40. $\dfrac{x+7}{15x}+\dfrac{x-2}{20x}=\dfrac{4(x+7)+3(x-2)}{60x}$

$=\dfrac{4x+28+3x-6}{60x}=\dfrac{7x+22}{60x}$

41. $\dfrac{16a^2-9}{16a^2-24a+9}\cdot\dfrac{8a^2-13a-6}{4a^2-5a-6}$

$=\dfrac{(4a-3)(4a+3)}{(4a-3)(4a-3)}\cdot\dfrac{(8a+3)(a-2)}{(4a+3)(a-2)}$

$=\dfrac{8a+3}{4a-3}$

42. $12x^2+16x-3=(6x-1)(2x+3)$

$6x^2+7x-3=(3x-1)(2x+3)$

$LCM=(6x-1)(2x+3)(3x-1)$

$\dfrac{x}{12x^2+16x-3}=\dfrac{x}{(6x-1)(2x+3)}$

$=\dfrac{x(3x-1)}{(6x-1)(2x+3)(3x-1)}$

$=\dfrac{3x^2-x}{(6x-1)(2x+3)(3x-1)}$

$\dfrac{4x^2}{6x^2+7x-3}=\dfrac{4x^2}{(3x-1)(2x+3)}$

$=\dfrac{4x^2(6x-1)}{(6x-1)(2x+3)(3x-1)}$

$=\dfrac{24x^3-4x^2}{(6x-1)(2x+3)(3x-1)}$

43. $\dfrac{3}{4ab}+\dfrac{5}{4ab}=\dfrac{8}{4ab}=\dfrac{2}{ab}$

44. $\dfrac{20}{x+2}=\dfrac{5}{16}$

$5(x+2)=20(16)$

$5x+10=320$

$5x=310$

$x=62$

The solution checks.

45. $\dfrac{5x}{3}-\dfrac{2}{5}=\dfrac{8x}{5}$

$15\left(\dfrac{5x}{3}-\dfrac{2}{5}\right)=15\left(\dfrac{8x}{5}\right)$

$25x-6=24x$

$-6=-x$

$x=6$

46. $\dfrac{6a^2b^7}{25x^3y}\div\dfrac{12a^3b^4}{5x^2y^2}=\dfrac{\overset{1}{\cancel{6}}\,a^2b^7}{\underset{5}{\cancel{25}}\,x^3y}\cdot\dfrac{\overset{1}{\cancel{5}}\,x^2y^2}{\underset{2}{\cancel{12}}\,a^3b^4}=\dfrac{b^3y}{10ax}$

47. The apprentice can construct the patio in 6 hours.

	Rate	·	Time	=	Part completed
Mason	$\frac{1}{3}$	·	2	=	$\frac{2}{3}$
Apprentice	$\frac{1}{a}$	·	2	=	$\frac{2}{a}$

$$\frac{2}{3}+\frac{2}{a}=1$$
$$3a\left(\frac{2}{3}+\frac{2}{a}\right)=3a(1)$$
$$2a+6=3a$$
$$a=6$$

48. $\frac{21}{14}=\frac{12}{x}$
$$21x=12(14)$$
$$21x=168$$
$$x=8$$

A 12 pound force stretches the spring 8 inches.

49. The rate of the wind is 20 mph.

	Distance	÷	Rate	=	Time
Against the wind	1900	÷	400 - r	=	$\frac{1900}{400-r}$
With the wind	2100	÷	400 + r	=	$\frac{2100}{400+r}$

$$\frac{1900}{400-r}=\frac{2100}{400+r}$$
$$(400-r)(400+r)\left(\frac{1900}{400-r}\right)$$
$$=(400-r)(400+r)\left(\frac{2100}{400+r}\right)$$
$$760{,}000+1900r=840{,}000-2100r$$
$$4000r=80{,}000$$
$$r=20$$

50. $\frac{4}{2}=\frac{n+4}{10}$
$$2(n+4)=40$$
$$2n+8=40$$
$$2n=32$$
$$n=16$$

The gardener needs 16 additional ounces of insecticide.

51. Both hoses can fill the pool in 6 hours working together.

	Rate	·	Time	=	Part of task completed
Hose #1	$\frac{1}{15}$	·	t	=	$\frac{t}{15}$
Hose #2	$\frac{1}{10}$	·	t	=	$\frac{t}{10}$

$$\frac{t}{15}+\frac{t}{10}=1$$
$$30(\frac{t}{15}+\frac{t}{10})=30(1)$$
$$2t+3t=30$$
$$5t=30$$
$$t=6$$

52. The rate of the car is 45 mph.

	Distance	÷	Rate	=	Time
Car	315	÷	r	=	$\frac{315}{r}$
Bus	245	÷	r - 10	=	$\frac{245}{r-10}$

$$\frac{315}{r}=\frac{245}{r-10}$$
$$r(r-10)\left(\frac{315}{r}\right)=r(r-10)\left(\frac{245}{r-10}\right)$$
$$315r-3150=245r$$
$$70r=3150$$
$$r=45$$

53. $w = kh$

$82 = k(4)$

$k = 20.5$

$w = kh = 20.5h$

$w = 20.5(20)$

$w = \$410$

54. $t = \dfrac{k}{r}$

$4 = \dfrac{k}{55}$

$k = 220$

$t = \dfrac{k}{r} = \dfrac{220}{r}$

$5 = \dfrac{220}{r}$

$5r = 220$

$r = 44$

Chapter 9 Test

1. $\dfrac{x^2+3x+2}{x^2+5x+4} \div \dfrac{x^2-x-6}{x^2+2x-15} = \dfrac{x^2+3x+2}{x^2+5x+4} \cdot \dfrac{x^2+2x-15}{x^2-x-6}$

$= \dfrac{(x+2)(x+1)}{(x+4)(x+1)} \cdot \dfrac{(x+5)(x-3)}{(x+2)(x-3)}$

$= \dfrac{x+5}{x+4}$

2. $\dfrac{2x}{x^2+3x-10} - \dfrac{4}{x^2+3x-10} = \dfrac{2x-4}{x^2+3x-10}$

$= \dfrac{2(x-2)}{(x+5)(x-2)} = \dfrac{2}{x+5}$

3. $6x-3 = 3(2x-1)$

$2x^2+x-1 = (2x-1)(x+1)$

$LCM = 3(2x-1)(x+1)$

4. $\dfrac{3}{x+4} = \dfrac{5}{x+6}$

$3(x+6) = 5(x+4)$

$3x+18 = 5x+20$

$-2x = 2$

$x = -1$

The solution checks.

5. $\dfrac{x^3y^4}{x^2-4x+4} \cdot \dfrac{x^2-x-2}{x^6y^4} = \dfrac{x^3y^4(x-2)(x+1)}{(x-2)(x-2)x^3y^4 \cdot x^3}$

$= \dfrac{x+1}{x^3(x-2)}$

6. $\dfrac{1+\frac{1}{x}-\frac{12}{x^2}}{1+\frac{2}{x}-\frac{8}{x^2}} = \dfrac{x^2}{x^2} \cdot \dfrac{1+\frac{1}{x}-\frac{12}{x^2}}{1+\frac{2}{x}-\frac{8}{x^2}} = \dfrac{x^2+x-12}{x^2+2x-8}$

$= \dfrac{(x+4)(x-3)}{(x+4)(x-2)} = \dfrac{x-3}{x-2}$

7. $x^2-2x = x(x-2)$

$x^2-4 = (x-2)(x+2)$

$LCM = x(x-2)(x+2)$

$\dfrac{3}{x^2-2x} = \dfrac{3}{x(x-2)} = \dfrac{2(x+2)}{x(x-2)(x+2)}$

$= \dfrac{2x+4}{x(x-2)(x+2)}$

$\dfrac{x}{x^2-4} = \dfrac{x}{(x-2)(x+2)}$

$= \dfrac{x^2}{x(x-2)x+2)}$

8. $3x+5y+15 = 0$

$3x = -5y-15$

$x = -\dfrac{5}{3}y - 5$

9. $\dfrac{6}{x}-2=1$

$\dfrac{6}{x}=3$

$3x=6$

$x=2$

The solution checks.

10. $\dfrac{2}{2x-1}-\dfrac{3}{3x+1}=\dfrac{2(3x+1)-3(2x-1)}{(2x-1)(3x+1)}$

$=\dfrac{6x+2-6x+3}{(2x-1)(3x+1)}$

$=\dfrac{5}{(2x-1)(3x+1)}$

11. $\dfrac{x^2-x-56}{x^2+8x+7}\div\dfrac{x^2-13x+40}{x^2-4x-5}$

$=\dfrac{x^2-x-56}{x^2+8x+7}\cdot\dfrac{x^2-4x-5}{x^2-13x+40}$

$=\dfrac{(x-8)(x+7)}{(x+1)(x+7)}\cdot\dfrac{(x-5)(x+1)}{(x-5)(x-8)}$

$=1$

12. $\dfrac{3x}{x^2+5x-24}-\dfrac{9}{x^2+5x-24}=\dfrac{3x-9}{x^2+5x-24}$

$=\dfrac{3(x-3)}{(x+8)(x-3)}=\dfrac{3}{x+8}$

13. $3x^2+6x=3x(x+2)$

$2x^2+8x+8=2(x+2)^2$

$LCM=6x(x+2)^2$

14. $\dfrac{x^2-7x+10}{25-x^2}=\dfrac{(x-5)(x-2)}{(5-x)(5+x)}=-\dfrac{x-2}{x+5}$

15. $\dfrac{3x}{x-3}-2=\dfrac{10}{x-3}$

$(x-3)\left(\dfrac{3x}{x-3}-2\right)=(x-3)\left(\dfrac{10}{x-3}\right)$

$3x-2x+6=10$

$x+6=10$

$x=4$

The solution checks.

16. $f=v+at$

$f-v=at$

$\dfrac{f-v}{a}=\dfrac{at}{a}$

$\dfrac{f-v}{a}=t$

17. $\dfrac{12x^4y^2}{18xy^7}=\dfrac{6xy^2\cdot 2x^3}{6xy^2\cdot 3y^5}=\dfrac{2x^3}{3y^5}$

18. $\dfrac{2}{2x-1}-\dfrac{1}{x+1}=\dfrac{2(x+1)-1(2x-1)}{(2x-1)(x+1)}$

$=\dfrac{2x+2-2x+1}{(2x-1)(x+1)}$

$=\dfrac{3}{(2x-1)(x+1)}$

19. $\dfrac{2}{x-2}=\dfrac{12}{x+3}$

$2(x+3)=12(x-2)$

$2x+6=12x-24$

$-10x=-30$

$x=3$

The solution checks.

20. $\dfrac{x^5y^3}{x^2-x-6}\cdot\dfrac{x^2-9}{x^2y^4}=\dfrac{x^2y^3\cdot x^3(x-3)(x+3)}{(x-3)(x+2)x^2y^3\cdot y}$

$=\dfrac{x^3(x+3)}{y(x+2)}$

21. $x(1-x)=-x(x-1)$

$LCM=x(x-1)(x+1)$

$$\frac{3y}{x(1-x)}=-\frac{3y(x+1)}{x(x-1)(x+1)}=-\frac{3xy+3y}{x(x-1)(x+1)}$$

$$\frac{x}{(x+1)(x-1)}=\frac{x^2}{x(x+1)(x-1)}$$

22. $$\frac{1-\frac{2}{x}-\frac{15}{x^2}}{1-\frac{25}{x^2}}=\frac{x^2}{x^2}\cdot\left(\frac{1-\frac{2}{x}-\frac{15}{x^2}}{1-\frac{25}{x^2}}\right)$$

$$=\frac{x^2-2x-15}{x^2-25}$$

$$=\frac{(x-5)(x+3)}{(x-5)(x+5)}=\frac{x+3}{x+5}$$

23. $d=kw$

$2=k(5)$

$k=0.4$

$d=kw=0.4w$

$d=0.4(28)$

$d=11.2in$

24. $\frac{4}{10}=\frac{4+n}{15}$

$10(4+n)=4(15)$

$40+10n=60$

$10n=20$

$n=2$

The mixture needs an additional 2 pounds of salt.

25. The rate of the wind is 20 mph

	Distance	÷	Rate	=	Time
Against the wind	180	÷	110 - r	=	$\frac{180}{110-r}$
With the wind	260	÷	110 + r	=	$\frac{260}{110+r}$

$$\frac{180}{110-r}=\frac{260}{110+r}$$

$180(110+r)=260(110-r)$

$19{,}800+180r=28{,}600-260r$

$440r=8800$

$r=20$

26. It takes 6 hours to fill the tank when both pipes work together.

	Rate	·	Time	=	Part of task completed
Pipe #1	$\frac{1}{9}$	·	t	=	$\frac{t}{9}$
Pipe #2	$\frac{1}{18}$	·	t	=	$\frac{t}{18}$

$$\frac{t}{9}+\frac{t}{18}=1$$

$$18\left(\frac{t}{9}+\frac{t}{18}\right)=18(1)$$

$2t+t=18$

$3t=18$

$t=6$

27. $V=\frac{k}{P}$

$2.5=\frac{k}{6}$

$k=15$

$V=\frac{k}{P}$

$V=\frac{15}{P}=\frac{15}{12}$

$V=1.25\,psi$

Cumulative Review Exercises

1. $-|-17|=-17$

2. $-\frac{3}{4}\cdot(2)^3=-\frac{3}{4}\cdot\frac{8}{1}=-6$

3. $\left(\frac{2}{3}\right)^2\div\left(\frac{3}{2}-\frac{2}{3}\right)+\frac{1}{2}=\left(\frac{2}{3}\right)^2\div\left(\frac{9}{6}-\frac{4}{6}\right)+\frac{1}{2}$

$=\left(\frac{2}{3}\right)^2\div\left(\frac{5}{6}\right)+\frac{1}{2}$

$=\left(\frac{2}{3}\right)^2\cdot\frac{6}{5}+\frac{1}{2}=\frac{4}{9}\cdot\frac{6}{5}+\frac{1}{2}$

$=\frac{8}{15}+\frac{1}{2}=\frac{16}{30}+\frac{15}{30}=\frac{31}{30}$

4. $-a^2+(a-b)^2=-(-2)^2+(-2-3)^2$

$=-(-2)^2+(-5)^2$

$=-4+25=21$

5. $-2x-(-3y)+7x-5y=-2x+7x+3y-5y$

$=5x-2y$

6. $2[3x-7(x-3)-8]=2[3x-7x+21-8]$

$=2[-4x+13]$

$=-8x+26$

7. $3-\frac{1}{4}x=8$

$-\frac{1}{4}x=5$

$\left(-\frac{4}{1}\right)\left(-\frac{1}{4}x\right)=\left(-\frac{4}{1}\right)5$

$x=-20$

8. $3[x-2(x-3)]=2(3-2x)$

$3[x-2x+6]=6-4x$

$3[-x+6]=6-4x$

$-3x+18=6-4x$

$x=-12$

9. $A=PB$

$A=\frac{1}{6}\cdot 60$

$A=10$

10. $\frac{5}{9}x<1$

$\frac{9}{5}\cdot\frac{5}{9}x<\frac{9}{5}\cdot 1$

$x<\frac{9}{5}$

11. $x-2\geq 4x-47$

$-3x\geq -45$

$\frac{-3x}{-3}\leq\frac{-45}{-3}$

$x\leq 15$

12. $y=2x-1$

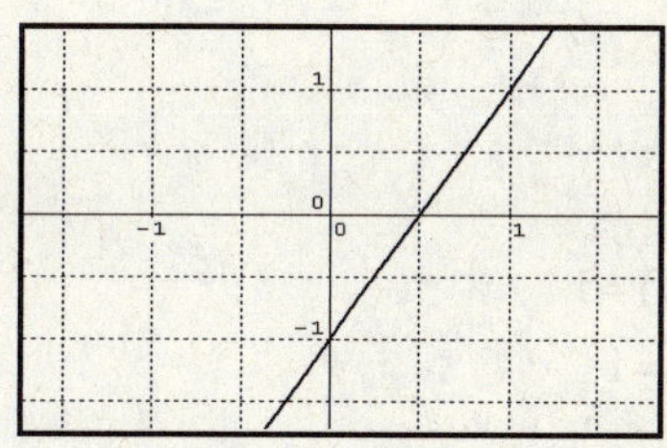

13. $f(x)=3x+2$

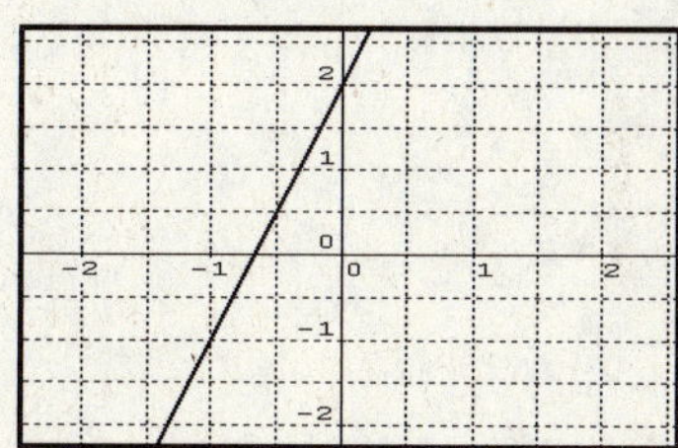

14. $x=3$

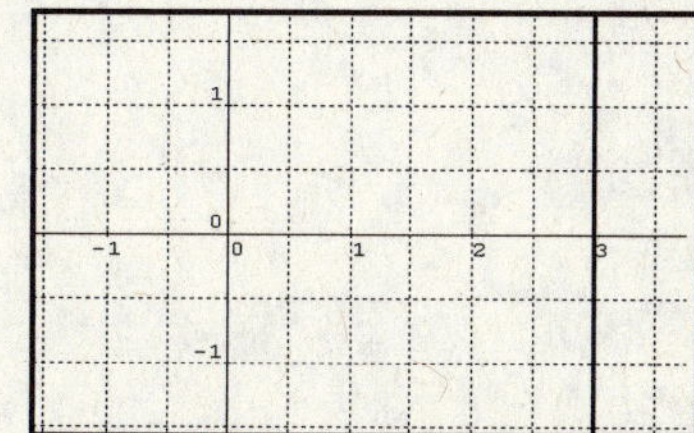

15. $5x+2y<6$

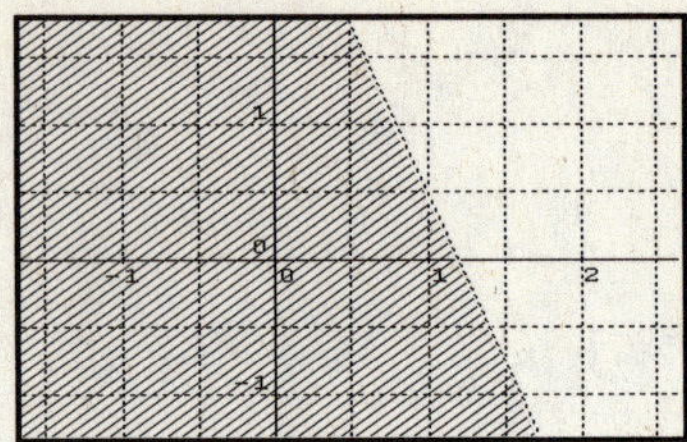

16. $f(x)=-2x+11$

$f(-8)=-2(-8)+11=27$

$f(-4)=-2(-4)+11=19$

$f(0)=-2(0)+11=11$

$f(3)=-2(3)+11=5$

$f(7)=-2(7)+11=-3$

Range: {-3, 5, 11, 19, 27}

17. $f(x)=4x-3$

$f(10)=4(10)-3=37$

18. $6x-y=1$

$y=3x+1$

$6x-(3x+1)=1$

$6x-3x-1=1$

$3x-1=1$

$3x=2$

$x=\frac{2}{3}$

$y=3x+1$

$y=3\cdot\frac{2}{3}+1$

$y=3$

The solution is $\left(\frac{2}{3},3\right)$.

19.

$$\begin{aligned} 2x-3y&=4 \\ 3(4x+y)&=1(3) \end{aligned}$$

$$\begin{aligned} 2x-3y&=4 \\ 12x+3y&=3 \\ \hline 14x\qquad&=7 \end{aligned}$$

$x=\frac{1}{2}$

$4x+y=1$

$4\cdot\frac{1}{2}+y=1$

$2+y=1$

$y=-1$

The solution is $\left(\frac{1}{2},-1\right)$.

20. $(3xy^4)(-2x^3y)=-6x^4y^5$

21. $(a^4b^3)^5=a^{20}b^{15}$

22. $\frac{a^2b^{-5}}{a^{-1}b^{-3}}=a^{2-(-1)}b^{-5-(-3)}=a^{2+1}b^{-5+3}=a^3b^{-2}=\frac{a^3}{b^2}$

23. $(a-3b)(a+4b)=a^2+ab-12b^2$

24. $\frac{15b^4-5b^2+10b}{5b}=\frac{15b^4}{5b}-\frac{5b^2}{5b}+\frac{10b}{5b}=3b^3-b+2$

25.

$$\begin{array}{r} x^2+2x+4 \\ x-2\overline{)x^3+0x^2+0x-8} \\ \underline{x^3-2x^2} \qquad\qquad \\ 2x^2+0x \qquad \\ \underline{2x^2-4x} \qquad \\ 4x-8 \\ \underline{4x-8} \\ 0 \end{array}$$

$(x^3-8)\div(x-2)=x^2+2x+4$

26. $12x^2-x-1=(4x+1)(3x-1)$

27. $y^2-7y+6=(y-6)(y-1)$

28. $2a^3+7a^2-15a=a(2a^2+7a-15)$
$=a(2a-3)(a+5)$

29. $4b^2-100=4(b^2-25)=4(b-5)(b+5)$

30. $(x+3)(2x-5)=0$
$x+3=0 \quad 2x-5=0$
$x=-3 \quad x=\frac{5}{2}$

31. $\frac{x^2+3x-28}{16-x^2}=\frac{(x+7)(x-4)}{(4-x)(4+x)}=-\frac{x+7}{x+4}$

32. $\frac{x^2-3x-10}{x^2-4x-12}\div\frac{x^2-x-20}{x^2-2x-24}$
$=\frac{x^2-3x-10}{x^2-4x-12}\cdot\frac{x^2-2x-24}{x^2-x-20}$
$=\frac{(x-5)(x+2)}{(x-6)(x+2)}\cdot\frac{(x-6)(x+4)}{(x-5)(x+4)}=1$

33. $\frac{6}{3x-1}-\frac{2}{x+1}=\frac{6(x+1)-2(3x-1)}{(3x-1)(x+1)}$
$=\frac{6x+6-6x+2}{(3x-1)(x+1)}$
$=\frac{8}{(3x-1)(x+1)}$

34. $\frac{4x}{x-3}-2=\frac{8}{x-3}$
$(x-3)\left(\frac{4x}{x-3}-2\right)=(x-3)\left(\frac{8}{x-3}\right)$
$4x-2x+6=8$
$2x+6=8$
$2x=2$
$x=1$

35. $f=v+at$
$f-v=at$
$\frac{f-v}{t}=\frac{at}{t}$
$\frac{f-v}{t}=a$

36. $5x-18=-3$
$5x-18+18=-3+18$
$5x=15$
$x=3$

37. \$5000 is invested at 11%.

	P	·	R	=	I
Amount at 7%	5000	·	0.07	=	0.07(5000)
Amount at 11%	x	·	0.11	=	$0.11x$
Total	x + 5000	·	0.09	=	0.09(x+5000)

$0.07(5000)+0.11x=0.09(x+5000)$
$350+0.11x=0.09x+450$
$0.02x=100$
$x=5000$

38. The second alloy is 70% silver.

	Base	·	Percentage	=	Amount
40% silver	60	·	0.40	=	0.4(60)
Other silver	120		x		$120x$
Total	180	·	0.60	=	0.6(180)

$0.4(60)+120x=0.6(180)$
$24+120x=108$
$120x=84$
$x=0.7=70\%$

39. Let h = the height of the triangle.

Then the base = $2h - 2$.

$$A = \frac{1}{2}bh$$
$$30 = \frac{1}{2}(2h-2)h$$
$$30 = (h-1)h$$
$$30 = h^2 - h$$
$$0 = h^2 - h - 30$$
$$0 = (h-6)(h+5)$$
$$h-6=0 \qquad h+5=0$$
$$h=6 \qquad h=-5$$
$$2h-2=10$$

The base is 10 inches and the height is 6 inches.

40. It will take both pipes 8 minutes to fill the tank.

	Rate	·	Time	=	Part of task completed
Pipe #1	$\frac{1}{12}$	·	t	=	$\frac{t}{12}$
Pipe #2	$\frac{1}{24}$	·	t	=	$\frac{t}{24}$

$$\frac{t}{12} + \frac{t}{24} = 1$$
$$24\left(\frac{t}{12} + \frac{t}{24}\right) = 1(24)$$
$$2t + t = 24$$
$$3t = 24$$
$$t = 8$$

CHAPTER 10 Radical Expressions

Chapter 10 Prep Test

1. $-|-14| = -14$

2. $3x^2y - 4xy^2 - 5x^2y = 3x^2y - 5x^2y - 4xy^2$
$= -2x^2y - 4xy^2$

3. $1.5h = 21$
$\frac{1.5h}{1.5} = \frac{21}{1.5}$
$h = 14$

4. $3x - 2 = 5 - 2x$
$3x = 7 - 2x$
$5x = 7$
$x = \frac{7}{5}$

5. $x^3 \cdot x^3 = x^{3+3} = x^6$

6. $(x+y)^2 = x^2 + 2xy + y^2$

7. $(2x-3)^2 = 4x^2 - 12x + 9$

8. $(a-5)(a+5) = a^2 - 25$

9. $(2-3v)(2+3v) = 4 - 9v^2$

10. $\frac{2x^4y^3}{18x^2y} = \frac{2}{18} \cdot \frac{x^4}{x^2} \cdot \frac{y^3}{y} = \frac{x^2y^2}{9}$

Section 10.1 Introduction to Radical Expressions

1. The symbol $\sqrt{\ }$ is called a **radical sign**.

3. The square of an integer is a **perfect square**.

5. The property that states that if a and b are positive real numbers, then $\sqrt{ab} = \sqrt{a} \cdot \sqrt{b}$ is called the **Product** Property of Square Roots.

7. A square root of 36 is 6 because $6^2 = 36$. A square root of 36 is -6 because $(-6)^2 = 36$.

9. The numbers 2, 20, and 50 are not perfect squares because there is no integer whose square is equal to them.

11. The square of an integer is a perfect square. A perfect-square factor of a number is a perfect square that divides evenly into the number.

13. In the expression $\sqrt{a}$, the symbol $\sqrt{\ }$ is called the **radical sign** and a is called the **radicand**,

15. $\sqrt{16} = 4$

17. $\sqrt{49} = 7$

19. $\sqrt{32} = \sqrt{16 \cdot 2} = \sqrt{16}\sqrt{2} = 4\sqrt{2}$

21. $\sqrt{8} = \sqrt{4 \cdot 2} = \sqrt{4}\sqrt{2} = 2\sqrt{2}$

23. $-6\sqrt{18} = -6\sqrt{9 \cdot 2} = -6\sqrt{9}\sqrt{2}$
$= -6(3)\sqrt{2} = -18\sqrt{2}$

25. $5\sqrt{40} = 5\sqrt{4 \cdot 10} = 5\sqrt{4}\sqrt{10} = 5(2)\sqrt{10} = 10\sqrt{10}$

27. $\sqrt{15}$ does not simplify

29. $\sqrt{29}$ does not simplify

31. $-9\sqrt{72} = -9\sqrt{36 \cdot 2} = -9\sqrt{36}\sqrt{2}$
$= -9(6)\sqrt{2} = -54\sqrt{2}$

33. $\sqrt{45} = \sqrt{9 \cdot 5} = \sqrt{9}\sqrt{5} = 3\sqrt{5}$

35. $6\sqrt{128} = 6\sqrt{64 \cdot 2} = 6\sqrt{64}\sqrt{2} = 6(8)\sqrt{2} = 48\sqrt{2}$

37. $\sqrt{300} = \sqrt{100 \cdot 3} = \sqrt{100}\sqrt{3} = 10\sqrt{3}$

39. $7\sqrt{98} = 7\sqrt{49 \cdot 2} = 7\sqrt{49}\sqrt{2} = 7(7)\sqrt{2} = 49\sqrt{2}$

41. $\sqrt{120} = \sqrt{4 \cdot 30} = \sqrt{4}\sqrt{30} = 2\sqrt{30}$

43. $\sqrt{160} = \sqrt{16 \cdot 10} = \sqrt{16}\sqrt{10} = 4\sqrt{10}$

45. **a.** Rational
b. Irrational
c. Irrational
d. Not real

47. $\sqrt{240} = 15.492$

49. $\sqrt{288} = 16.971$

51. $\sqrt{245} = 15.652$

53. $\sqrt{352} = 18.762$

55. A variable exponential expression is a perfect square when each exponent is an even number.

57. $16x^6y^8$ is a perfect square because $16 = (4)^2$, $x^6 = (x^3)^2$, and $y^8 = (y^4)^2$.

59. $\sqrt{x^6} = x^3$

61. $\sqrt{y^{15}} = \sqrt{y^{14}y} = \sqrt{y^{14}}\sqrt{y} = y^7\sqrt{y}$

63. $\sqrt{a^{20}} = a^{10}$

65. $\sqrt{x^4y^4} = x^2y^2$

67. $\sqrt{4x^4} = 2x^2$

69. $\sqrt{24x^2} = \sqrt{4x^2 \cdot 6} = \sqrt{4x^2}\sqrt{6} = 2x\sqrt{6}$

71. $\sqrt{x^3y^7} = \sqrt{x^2y^6xy} = \sqrt{x^2y^6}\sqrt{xy} = xy^3\sqrt{xy}$

73. $\sqrt{a^3b^{11}} = \sqrt{a^2b^{10}ab} = \sqrt{a^2b^{10}}\sqrt{ab} = ab^5\sqrt{ab}$

75. $\sqrt{60x^5} = \sqrt{4x^4 \cdot 15x} = \sqrt{4x^4}\sqrt{15x} = 2x^2\sqrt{15x}$

77. $\sqrt{49a^4b^8} = 7a^2b^4$

79. $\sqrt{18x^5y^7} = \sqrt{9x^4y^6 \cdot 2xy} = \sqrt{9x^4y^6}\sqrt{2xy}$
$= 3x^2y^3\sqrt{2xy}$

81. $\sqrt{40x^{11}y^7} = \sqrt{4x^{10}y^6 \cdot 10xy}$
$= \sqrt{4x^{10}y^6}\sqrt{10xy} = 2x^5y^3\sqrt{10xy}$

83. $\sqrt{80a^9b^{10}} = \sqrt{16a^8b^{10} \cdot 5a}$
$= \sqrt{16a^8b^{10}}\sqrt{5a} = 4a^4b^5\sqrt{5a}$

85. $-2\sqrt{16a^2b^3} = -2\sqrt{16a^2b^2b} = -2\sqrt{16a^2b^2}\sqrt{b}$
$= -2(4ab)\sqrt{b} = -8ab\sqrt{b}$

87. $x\sqrt{x^4y^2} = x \cdot x^2y = x^3y$

89. $-4\sqrt{20a^4b^7} = -4\sqrt{4a^4b^6 \cdot 5b}$
$= -4\sqrt{4a^4b^6}\sqrt{5b} = -4(2a^2b^3)\sqrt{5b}$
$= -8a^2b^3\sqrt{5b}$

91. $3x\sqrt{12x^2y^7} = 3x\sqrt{4x^2y^6 \cdot 3y}$
$= 3x\sqrt{4x^2y^6}\sqrt{3y} = 3x(2xy^3)\sqrt{3y}$
$= 6x^2y^3\sqrt{3y}$

93. $2x^2\sqrt{8x^2y^3} = 2x^2\sqrt{4x^2y^2 \cdot 2y}$
$= 2x^2\sqrt{4x^2y^2}\sqrt{2y} = 2x^2(2xy)\sqrt{2y}$
$= 4x^3y\sqrt{2y}$

95. $\sqrt{25(a+4)^2} = 5(a+4) = 5a+20$

97. $\sqrt{4(x+2)^4} = 2(x+2)^2 = 2x^2+8x+8$

99. $\sqrt{x^2+4x+4} = \sqrt{(x+2)^2} = x+2$

101. $\sqrt{y^2+2y+1} = \sqrt{(y+1)^2} = y+1$

103. Rational

105. Irrational

107. a. $N = 2.3\sqrt{S} = 2.3\sqrt{1} = 2.3$ *credit cards*

b. $N = 2.3\sqrt{S} = 2.3\sqrt{2} = 3.3$ *credit cards*

c. $N = 2.3\sqrt{S} = 2.3\sqrt{3} = 4.0$ *credit cards*

d. $N = 2.3\sqrt{S} = 2.3\sqrt{4} = 4.6$ *credit cards*

109. $d = 1.2\sqrt{h} = 1.2\sqrt{5000} = 84.9mi$

111. $A = s^2$
$76 = s^2$
$\sqrt{76} = s$
$8.7cm \approx s$

113. $\sqrt{4x}$ is real for $x \geq 0$.

115. $\sqrt{x+5}$ is real for $x \geq$ -5.
$x+5 \geq 0$
$x \geq -5$

117. $\sqrt{5-2x}$ is real for $x \leq \frac{5}{2}$.
$5-2x \geq 0$
$-2x \geq -5$
$x \leq \frac{5}{2}$

119. $\sqrt{x^2+1}$ is real for all real numbers.

121. $2\sqrt{2}$ is in simplest form because the radicand does not have a perfect square factor greater than 1. $\sqrt{8}$ is not in simplest form because the radicand does have a perfect-square factor: 4, which is a perfect square, is a factor of 8.

123. a. The Xmin and Xmax values represent the minimum and maximum radii of an unbanked curve. The Ymin and Ymax values represent the minimum and maximum speeds of a car.

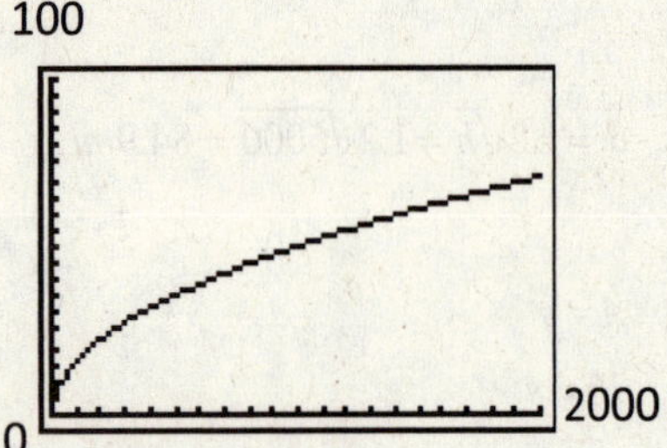

b. As the radius increases, the maximum safe speed increases. On the graph, as r increases, v increases.

c. From the graph, when r is 100 ft, v is about 16 mph.

d. From the graph, when v is 40 mph, r is about 640 ft.

e. At a speed of 161.623 mph, r is about 10,500 ft, which is more than 1 mile.

Section 10.2 Addition and Subtraction of Algebraic Expressions

1. Yes. The radical parts are exactly the same.

3. Yes. The radical parts are exactly the same.

5. No. The radical parts are not the same.

7. $5\sqrt{5}-8\sqrt{5}=(5-8)\sqrt{5}=-3\sqrt{5}$

9. $3\sqrt{5}+8\sqrt{5}=11\sqrt{5}$

11. $4\sqrt{5}-10\sqrt{5}=-6\sqrt{5}$

13. $-3\sqrt{3}-5\sqrt{3}=-8\sqrt{3}$

15. $3\sqrt{y}+2\sqrt{y}=5\sqrt{y}$

17. $-5\sqrt{2a}+2\sqrt{2a}=-3\sqrt{2a}$

19. $-7\sqrt{5a}-5\sqrt{5a}=-12\sqrt{5a}$

21. $2y\sqrt{3}-9y\sqrt{3}=-7y\sqrt{3}$

23. $-5b\sqrt{3x}-2b\sqrt{3x}=-7b\sqrt{3x}$

25. $-4\sqrt{xy}+6\sqrt{xy}=2\sqrt{xy}$

27. $$\begin{aligned}\sqrt{32}-\sqrt{98}&=\sqrt{16\cdot 2}-\sqrt{49\cdot 2}\\&=4\sqrt{2}-7\sqrt{2}\\&=-3\sqrt{2}\end{aligned}$$

29. $$\begin{aligned}4\sqrt{128}-3\sqrt{32}&=4\sqrt{64\cdot 2}-3\sqrt{16\cdot 2}\\&=4(8)\sqrt{2}-3(4)\sqrt{2}\\&=32\sqrt{2}-12\sqrt{2}\\&=20\sqrt{2}\end{aligned}$$

31. $$\begin{aligned}5\sqrt{75}-2\sqrt{18}&=5\sqrt{25\cdot 3}-2\sqrt{9\cdot 2}\\&=5(5)\sqrt{3}-2(3)\sqrt{2}\\&=25\sqrt{3}-6\sqrt{2}\end{aligned}$$

33. $$\begin{aligned}-3\sqrt{25y}+8\sqrt{49y}&=-3(5)\sqrt{y}+8(7)\sqrt{y}\\&=-15\sqrt{y}+56\sqrt{y}\\&=41\sqrt{y}\end{aligned}$$

35. $$\begin{aligned}3\sqrt{3x^2}-5\sqrt{27x^2}&=3x\sqrt{3}-5\sqrt{9x^2\cdot 3}\\&=3x\sqrt{3}-5(3x)\sqrt{3}\\&=3x\sqrt{3}-15x\sqrt{3}\\&=-12x\sqrt{3}\end{aligned}$$

37. $2x\sqrt{xy^2}-3y\sqrt{x^2y}=2xy\sqrt{x}-3xy\sqrt{y}$

39. $3x\sqrt{12x}-5\sqrt{27x^3}=3x\sqrt{4\cdot 3x}-5\sqrt{9x^2\cdot 3x}$
$=3x(2)\sqrt{3x}-5(3x)\sqrt{3x}$
$=6x\sqrt{3x}-15x\sqrt{3x}$
$=-9x\sqrt{3x}$

41. $4y\sqrt{8y^3}-7\sqrt{18y^5}=4y\sqrt{4y^2\cdot 2y}-7\sqrt{9y^4\cdot 2y}$
$=4y(2y)\sqrt{2y}-7(3y^2)\sqrt{2y}$
$=8y^2\sqrt{2y}-21y^2\sqrt{2y}$
$=-13y^2\sqrt{2y}$

43. $b^2\sqrt{a^5b}+3a^2\sqrt{ab^5}=b^2\sqrt{a^4\cdot ab}+3a^2\sqrt{b^4\cdot ab}$
$=a^2b^2\sqrt{ab}+3a^2b^2\sqrt{ab}$
$=4a^2b^2\sqrt{ab}$

45. $4\sqrt{2}-5\sqrt{2}+8\sqrt{2}=7\sqrt{2}$

47. $5\sqrt{x}-8\sqrt{x}+9\sqrt{x}=6\sqrt{x}$

49. $8\sqrt{2}-3\sqrt{y}-8\sqrt{2}=-3\sqrt{y}$

51. $8\sqrt{8}-4\sqrt{32}-9\sqrt{50}=8\sqrt{4\cdot 2}-4\sqrt{16\cdot 2}-9\sqrt{25\cdot 2}$
$=8(2)\sqrt{2}-4(4)\sqrt{2}-9(5)\sqrt{2}$
$=16\sqrt{2}-16\sqrt{2}-45\sqrt{2}$
$=-45\sqrt{2}$

53. $-2\sqrt{3}+5\sqrt{27}-4\sqrt{45}=-2\sqrt{3}+5\sqrt{9\cdot 3}-4\sqrt{9\cdot 5}$
$=-2\sqrt{3}+5(3)\sqrt{3}-4(3)\sqrt{5}$
$=-2\sqrt{3}+15\sqrt{3}-12\sqrt{5}$
$=13\sqrt{3}-12\sqrt{5}$

55. $4\sqrt{75}+3\sqrt{48}-\sqrt{99}$
$=4\sqrt{25\cdot 3}+3\sqrt{16\cdot 3}-\sqrt{9\cdot 11}$
$=4(5)\sqrt{3}+3(4)\sqrt{3}-3\sqrt{11}$
$=20\sqrt{3}+12\sqrt{3}-3\sqrt{11}$
$=32\sqrt{3}-3\sqrt{11}$

57. $\sqrt{25x}-\sqrt{9x}+\sqrt{16x}=5\sqrt{x}-3\sqrt{x}+4\sqrt{x}$
$=6\sqrt{x}$

59. $3\sqrt{3x}+\sqrt{27x}-8\sqrt{75x}$
$=3\sqrt{3x}+\sqrt{9\cdot 3x}-8\sqrt{25\cdot 3x}$
$=3\sqrt{3x}+3\sqrt{3x}-8(5)\sqrt{3x}$
$=3\sqrt{3x}+3\sqrt{3x}-40\sqrt{3x}$
$=-34\sqrt{3x}$

61. $2a\sqrt{75b}-a\sqrt{20b}+4a\sqrt{45b}$
$=2a\sqrt{25\cdot 3b}-a\sqrt{4\cdot 5b}+4a\sqrt{9\cdot 5b}$
$=2a(5)\sqrt{3b}-a(2)\sqrt{5b}+4a(3)\sqrt{5b}$
$=10a\sqrt{3b}-2a\sqrt{5b}+12a\sqrt{5b}$
$=10a\sqrt{3b}+10a\sqrt{5b}$

63. $x\sqrt{3y^2}-2y\sqrt{12x^2}+xy\sqrt{3}$
$=x\sqrt{y^2\cdot 3}-2y\sqrt{4x^2\cdot 3}+xy\sqrt{3}$
$=xy\sqrt{3}-2y(2x)\sqrt{3}+xy\sqrt{3}$
$=xy\sqrt{3}-4xy\sqrt{3}+xy\sqrt{3}$
$=-2xy\sqrt{3}$

65. $3\sqrt{ab^3}+4a\sqrt{a^2b}-5b\sqrt{4ab}$
$=3\sqrt{b^2\cdot ab}+4a\sqrt{a^2b}-5b\sqrt{4ab}$
$=3b\sqrt{ab}+4a^2\sqrt{b}-5b(2)\sqrt{ab}$
$=3b\sqrt{ab}+4a^2\sqrt{b}-10b\sqrt{ab}$
$=-7b\sqrt{ab}+4a^2\sqrt{b}$

67. $3a\sqrt{2ab^2}-\sqrt{a^2b^2}+4b\sqrt{3a^2b}$
$=3ab\sqrt{2a}-ab+4ab\sqrt{3b}$

69. **a.** False

b. True

c. False

d. True

71. $5\sqrt{x+2}+3\sqrt{x+2}=8\sqrt{x+2}$

73. $$\frac{1}{2}\sqrt{8x^2y}+\frac{1}{3}\sqrt{18x^2y}=\frac{1}{2}\sqrt{4x^2\cdot 2y}+\frac{1}{3}\sqrt{9x^2\cdot 2y}$$
$$=\frac{1}{2}(2x)\sqrt{2y}+\frac{1}{3}(3x)\sqrt{2y}$$
$$=x\sqrt{2y}+x\sqrt{2y}$$
$$=2x\sqrt{2y}$$

75. $$\frac{a}{3}\sqrt{54ab^3}+\frac{b}{4}\sqrt{96a^3b}$$
$$=\frac{a}{3}\sqrt{9b^2\cdot 6ab}+\frac{b}{4}\sqrt{16a^2\cdot 6ab}$$
$$=\frac{a}{3}(3b)\sqrt{6ab}+\frac{b}{4}(4a)\sqrt{6ab}$$
$$=ab\sqrt{6ab}+ab\sqrt{6ab}$$
$$=2ab\sqrt{6ab}$$

77. $P=4\sqrt{3}+2\sqrt{3}+2\sqrt{15}$

$P=6\sqrt{3}+2\sqrt{15}$

The perimeter is ($6\sqrt{3}+2\sqrt{15}$) cm.

79. $P=2(4\sqrt{5})+2(\sqrt{5})$

$P=8\sqrt{5}+2\sqrt{5}$

$P=10\sqrt{5}\approx 22.4$

The perimeter is 22.4 cm.

81. Write each radicand as the product of a perfect square and factors that do not contain a perfect square.

$$a\sqrt{32ab^2}+b\sqrt{50a^3}=a\sqrt{16b^2\cdot 2a}+b\sqrt{25a^2\cdot 2a}$$

Use the Product Property of Square Roots. Write the perfect square under the first radical sign and all the remaining factors under the second radical sign.

$$=a\sqrt{16b^2}\sqrt{2a}+b\sqrt{25a^2}\sqrt{2a}$$

Take the square roots of the perfect squares. Simplify.

$$=a\cdot 4b\sqrt{2a}+b\cdot 5a\sqrt{2a}$$
$$=4ab\sqrt{2a}+5ab\sqrt{2a}$$

The radical expressions are like variable expressions and therefore can be added together.

$$=9ab\sqrt{2a}$$

Section 10.3 Multiplication and Division of Radical Expressions

1. Conjugate: $3-\sqrt{5}$

3. Conjugate: $\sqrt{2a}+8$

5. $(5-\sqrt{y})(5+\sqrt{y})=5^2-\left(\sqrt{y}\right)^2=25-y$

7. Multiply by $\frac{\sqrt{x}}{\sqrt{x}}$.

9. $\sqrt{3}\cdot\sqrt{5}=\sqrt{15}$

11. $\sqrt{5}\sqrt{5}=\sqrt{25}=5$

13. $\sqrt{3}\sqrt{12}=\sqrt{36}=6$

15. $\left(\sqrt{7y}\right)^2 = 7y$

17. $\sqrt{xy^3}\sqrt{x^5y} = \sqrt{x^6y^4} = x^3y^2$

19. $\sqrt{3a^2b^5}\sqrt{6ab^7} = \sqrt{18a^3b^{12}}$
$= \sqrt{9a^2b^{12}\cdot 2a} = 3ab^6\sqrt{2a}$

21. $\sqrt{6a^3b^2}\sqrt{24a^5b} = \sqrt{144a^8b^2b} = 12a^4b\sqrt{b}$

23. $\sqrt{2ac}\sqrt{5ab}\sqrt{10cb} = \sqrt{100a^2b^2c^2} = 10abc$

25. $\sqrt{2}(\sqrt{2}-\sqrt{3}) = \sqrt{4}-\sqrt{6} = 2-\sqrt{6}$

27. $\sqrt{8}(\sqrt{2}-\sqrt{5}) = \sqrt{16}-\sqrt{40}$
$= 4-\sqrt{4\cdot 10} = 4-2\sqrt{10}$

29. $\sqrt{5}(\sqrt{10}-\sqrt{x}) = \sqrt{50}-\sqrt{5x}$
$= \sqrt{25\cdot 2}-\sqrt{5x}$
$= 5\sqrt{2}-\sqrt{5x}$

31. $\sqrt{x}(\sqrt{x}-\sqrt{y}) = x-\sqrt{xy}$

33. $\sqrt{3a}(\sqrt{3a}-\sqrt{3b}) = 3a-\sqrt{9ab} = 3a-3\sqrt{ab}$

35. $(\sqrt{x}-3)^2 = (\sqrt{x}-3)(\sqrt{x}-3) = x-6\sqrt{x}+9$

37. $(\sqrt{5}+3)(\sqrt{5}+7) = 5+7\sqrt{5}+3\sqrt{5}+21$
$= 26+10\sqrt{5}$

39. $(2\sqrt{x}-5)(\sqrt{x}-2) = 2x-4\sqrt{x}-5\sqrt{x}+10$
$= 2x-9\sqrt{x}+10$

41. $(3\sqrt{x}-2y)(5\sqrt{x}-4y)$
$= 15x-12y\sqrt{x}-10y\sqrt{x}+8y^2$
$= 15x-22y\sqrt{x}+8y^2$

43. $(\sqrt{2}-\sqrt{y})(\sqrt{2}+\sqrt{y}) = 2-y$

45. $(5+\sqrt{6})(5-\sqrt{6}) = 25-6 = 19$

47. $(2\sqrt{x}+\sqrt{y})(5\sqrt{x}+4\sqrt{y})$
$= 10x+8\sqrt{xy}+5\sqrt{xy}+4y$
$= 10x+13\sqrt{xy}+4y$

49. For $a > 0$, $(\sqrt{a}-1)(\sqrt{a}+1) = a-1$, which is **less than** a.

51. The expression $\frac{\sqrt{5}}{5}$ has no radical in the denominator and no fraction under the radical sign. The radical in the numerator has no perfect-square factors other than 1. This expression is in simplest form. The expression $\frac{1}{\sqrt{5}}$ has a radical in the denominator, so it is not in simplest from.

53. **a.** No. This fraction has a radical in the denominator.

b. Yes

c. No. This fraction has a fraction under the radical sign.

d. No. This fraction has a radical in the denominator.

55. $\frac{\sqrt{32}}{\sqrt{2}} = \sqrt{\frac{32}{2}} = \sqrt{16} = 4$

57. $\dfrac{\sqrt{98}}{\sqrt{2}}=\sqrt{\dfrac{98}{2}}=\sqrt{49}=7$

59. $\dfrac{\sqrt{27a}}{\sqrt{3a}}=\sqrt{\dfrac{27a}{3a}}=\sqrt{9}=3$

61. $\dfrac{\sqrt{15x^3y}}{\sqrt{3xy}}=\sqrt{\dfrac{15x^3y}{3xy}}=\sqrt{5x^2}=x\sqrt{5}$

63. $\dfrac{\sqrt{2a^5b^4}}{\sqrt{98ab^4}}=\sqrt{\dfrac{2a^5b^4}{98ab^4}}=\sqrt{\dfrac{a^4}{49}}=\dfrac{a^2}{7}$

65. $\dfrac{1}{\sqrt{3}}=\dfrac{1}{\sqrt{3}}\cdot\dfrac{\sqrt{3}}{\sqrt{3}}=\dfrac{\sqrt{3}}{3}$

67. $\dfrac{15}{\sqrt{75}}=\dfrac{15}{\sqrt{25\cdot 3}}=\dfrac{15}{5\sqrt{3}}=\dfrac{3}{\sqrt{3}}\cdot\dfrac{\sqrt{3}}{\sqrt{3}}=\dfrac{3\sqrt{3}}{3}=\sqrt{3}$

69. $\dfrac{3}{\sqrt{x}}=\dfrac{3}{\sqrt{x}}\cdot\dfrac{\sqrt{x}}{\sqrt{x}}=\dfrac{3\sqrt{x}}{x}$

71. $\dfrac{6}{\sqrt{12x}}=\dfrac{6}{\sqrt{4\cdot 3x}}=\dfrac{6}{2\sqrt{3x}}$

$=\dfrac{3}{\sqrt{3x}}\cdot\dfrac{\sqrt{3x}}{\sqrt{3x}}=\dfrac{3\sqrt{3x}}{3x}=\dfrac{\sqrt{3x}}{x}$

73. $\dfrac{8}{\sqrt{32x}}=\dfrac{8}{\sqrt{16\cdot 2x}}=\dfrac{8}{4\sqrt{2x}}$

$=\dfrac{2}{\sqrt{2x}}\cdot\dfrac{\sqrt{2x}}{\sqrt{2x}}=\dfrac{2\sqrt{2x}}{2x}=\dfrac{\sqrt{2x}}{x}$

75. $\dfrac{\sqrt{4x^2}}{\sqrt{9y}}=\dfrac{2x}{3\sqrt{y}}\cdot\dfrac{\sqrt{y}}{\sqrt{y}}=\dfrac{2x\sqrt{y}}{3y}$

77. $\dfrac{5\sqrt{8}}{4\sqrt{50}}=\dfrac{5}{4}\cdot\sqrt{\dfrac{8}{50}}=\dfrac{5}{4}\cdot\sqrt{\dfrac{4}{25}}=\dfrac{5}{4}\cdot\dfrac{2}{5}=\dfrac{1}{2}$

79. $\dfrac{\sqrt{12a^3b}}{\sqrt{24a^2b^2}}=\sqrt{\dfrac{12a^3b}{24a^2b^2}}=\sqrt{\dfrac{a}{2b}}$

$=\dfrac{\sqrt{a}}{\sqrt{2b}}\cdot\dfrac{\sqrt{2b}}{\sqrt{2b}}=\dfrac{\sqrt{2ab}}{2b}$

81. $\dfrac{\sqrt{9xy^2}}{\sqrt{27x}}=\sqrt{\dfrac{9xy^2}{27x}}=\sqrt{\dfrac{y^2}{3}}=\dfrac{y}{\sqrt{3}}\cdot\dfrac{\sqrt{3}}{\sqrt{3}}=\dfrac{y\sqrt{3}}{3}$

83. $\dfrac{1}{\sqrt{2}-3}\cdot\dfrac{\sqrt{2}+3}{\sqrt{2}+3}=\dfrac{\sqrt{2}+3}{2-9}=-\dfrac{\sqrt{2}+3}{7}$

85. $\dfrac{3}{5+\sqrt{5}}\cdot\dfrac{5-\sqrt{5}}{5-\sqrt{5}}=\dfrac{15-3\sqrt{5}}{25-5}=\dfrac{15-3\sqrt{5}}{20}$

87. $\dfrac{\sqrt{xy}}{\sqrt{x}-\sqrt{y}}\cdot\dfrac{\sqrt{x}+\sqrt{y}}{\sqrt{x}+\sqrt{y}}=\dfrac{x\sqrt{y}+y\sqrt{x}}{x-y}$

89. $\dfrac{\sqrt{16x^3y^2}}{\sqrt{8x^3y}}=\sqrt{\dfrac{16x^3y^2}{8x^3y}}=\sqrt{2y}$

91. $\dfrac{\sqrt{5}}{\sqrt{2}-\sqrt{5}}\cdot\dfrac{\sqrt{2}+\sqrt{5}}{\sqrt{2}+\sqrt{5}}=\dfrac{\sqrt{10}+5}{2-5}=-\dfrac{\sqrt{10}+5}{3}$

93. $\dfrac{\sqrt{x}}{\sqrt{x}+3}\cdot\dfrac{\sqrt{x}-3}{\sqrt{x}-3}=\dfrac{x-3\sqrt{x}}{x-9}$

95. $\dfrac{5\sqrt{3}-7\sqrt{3}}{4\sqrt{3}}=\dfrac{-2\sqrt{3}}{4\sqrt{3}}=-\dfrac{1}{2}$

97. $\dfrac{5\sqrt{8}-3\sqrt{2}}{\sqrt{2}}=\dfrac{5\sqrt{4\cdot 2}-3\sqrt{2}}{\sqrt{2}}$

$=\dfrac{5(2)\sqrt{2}-3\sqrt{2}}{\sqrt{2}}=\dfrac{7\sqrt{2}}{\sqrt{2}}=7$

99. $\dfrac{3\sqrt{2}-8\sqrt{2}}{\sqrt{2}}=\dfrac{-5\sqrt{2}}{\sqrt{2}}=-5$

101. $\frac{2+\sqrt{3}}{2-\sqrt{3}}\cdot\frac{2+\sqrt{3}}{2+\sqrt{3}}=\frac{4+4\sqrt{3}+3}{4-3}=7+4\sqrt{3}$

103. $\frac{3-\sqrt{6}}{5-2\sqrt{6}}\cdot\frac{5+2\sqrt{6}}{5+2\sqrt{6}}=\frac{15+\sqrt{6}-12}{25-24}=3+\sqrt{6}$

105. $\frac{\sqrt{2}+2\sqrt{6}}{2\sqrt{2}-3\sqrt{6}}\cdot\frac{2\sqrt{2}+3\sqrt{6}}{2\sqrt{2}+3\sqrt{6}}$

$=\frac{4+3\sqrt{12}+4\sqrt{12}+36}{8-54}$

$=\frac{40+7\sqrt{4\cdot 3}}{-46}=-\frac{40+14\sqrt{3}}{46}$

$=-\frac{2(20+7\sqrt{3})}{2\cdot 23}=-\frac{20+7\sqrt{3}}{23}$

107. $\frac{3+\sqrt{x}}{2-\sqrt{x}}\cdot\frac{2+\sqrt{x}}{2+\sqrt{x}}=\frac{6+5\sqrt{x}+x}{4-x}$

109. $\frac{3+2\sqrt{y}}{2-\sqrt{y}}\cdot\frac{2+\sqrt{y}}{2+\sqrt{y}}=\frac{6+7\sqrt{y}+2y}{4-y}$

111. $\frac{\sqrt{x}+\sqrt{y}}{\sqrt{x}-\sqrt{y}}\cdot\frac{\sqrt{x}+\sqrt{y}}{\sqrt{x}+\sqrt{y}}=\frac{x+2\sqrt{xy}+y}{x-y}$

113. $\sqrt{\frac{5}{8}}\sqrt{\frac{5}{8}}=\frac{5}{8}$

115. $\sqrt{1\frac{9}{16}}=\sqrt{\frac{25}{16}}=\frac{5}{4}$

117. $-\sqrt{6\frac{1}{4}}=-\sqrt{\frac{25}{4}}=-\frac{5}{2}$

119. $A=\frac{1}{2}(10+\sqrt{8})(12+\sqrt{2})$

$A=\frac{1}{2}(10+2\sqrt{2})(12+\sqrt{2})$

$A=\frac{1}{2}(120+10\sqrt{2}+24\sqrt{2}+4)$

$A=\frac{1}{2}(124+34\sqrt{2})=(62+17\sqrt{2})m^2$

121. $\sqrt{x+2}+\sqrt{x-1}=3$

$\sqrt{2+2}+\sqrt{2-1}\stackrel{?}{=}3$

$\sqrt{4}+\sqrt{1}\stackrel{?}{=}3$

$2+1\stackrel{?}{=}3$

$3=3$

123. To rationalize a single term denominator, multiply the numerator and denominator of the fraction by the radical in the denominator. To rationalize a binomial denominator, multiply the numerator and denominator by the conjugate of the denominator.

Section 10.4 Solving Equations Containing Radical Expressions

1. a. This is **not a radical equation** because the radical does not contain a variable.

b. This is a **radical equation**.

c. This is a **radical equation**.

d. This is **not a radical equation** because the radical does not contain a variable.

3. It is **sometimes true** that we can square both sides of an equation without changing the solutions of the equation.

5. It is **sometimes true** that the first step in solving an equation is to square both sides of the equation.

7. $\sqrt{x-2}=5$

$\left(\sqrt{x-2}\right)^2=5^2$

$x-2=25$

$x=27$

9. $\sqrt{x}=5$

$\left(\sqrt{x}\right)^2=5^2$

$x=25$

Check : $\sqrt{x}=5$

$\sqrt{25}\overset{?}{=}5$

$5=5$

11. $\sqrt{a}=12$

$\left(\sqrt{a}\right)^2=12^2$

$a=144$

Check : $\sqrt{a}=12$

$\sqrt{144}\overset{?}{=}12$

$12=12$

13. $\sqrt{5x}=5$

$\left(\sqrt{5x}\right)^2=5^2$

$5x=25$

$x=5$

Check : $\sqrt{5x}=5$

$\sqrt{5\cdot 5}\overset{?}{=}5$

$\sqrt{25}\overset{?}{=}5$

$5=5$

15. $\sqrt{4x}=8$

$\left(\sqrt{4x}\right)^2=8^2$

$4x=64$

$x=16$

Check : $\sqrt{4x}=8$

$\sqrt{4\cdot 16}\overset{?}{=}8$

$\sqrt{64}\overset{?}{=}8$

$8=8$

17. $\sqrt{2x}-4=0$

$\sqrt{2x}=4$

$\left(\sqrt{2x}\right)^2=4^2$

$2x=16$

$x=8$

Check : $\sqrt{2x}-4=0$

$\sqrt{2\cdot 8}-4\overset{?}{=}0$

$\sqrt{16}-4\overset{?}{=}0$

$4-4\overset{?}{=}0$

$0=0$

19. $\sqrt{4x}+5=2$

$\sqrt{4x}=-3$

$\left(\sqrt{4x}\right)^2=(-3)^2$

$4x=9$

$x=\dfrac{9}{4}$

Check : $\sqrt{4x}+5=2$

$\sqrt{4\cdot\dfrac{9}{4}}+5\overset{?}{=}2$

$\sqrt{9}+5\overset{?}{=}2$

$3+5\overset{?}{=}2$

$8\neq 2$

The solution does not check so the equation has no solution.

21. $\sqrt{3x-2}=4$

$\left(\sqrt{3x-2}\right)^2=4^2$

$3x-2=16$

$3x=18$

$x=6$

$Check: \sqrt{3x-2}=4$

$\sqrt{3\cdot 6-2}\overset{?}{=}4$

$\sqrt{18-2}\overset{?}{=}4$

$\sqrt{16}\overset{?}{=}4$

$4=4$

23. $\sqrt{2x+1}=7$

$\left(\sqrt{2x+1}\right)^2=7^2$

$2x+1=49$

$2x=48$

$x=24$

$Check: \sqrt{2x+1}=7$

$\sqrt{2\cdot 24+1}\overset{?}{=}7$

$\sqrt{49}\overset{?}{=}7$

$7=7$

25. $\sqrt{5x+2}=0$

$\left(\sqrt{5x+2}\right)^2=0^2$

$5x+2=0$

$5x=-2$

$x=-\frac{2}{5}$

$Check: \sqrt{5x+2}=0$

$\sqrt{5\left(-\frac{2}{5}\right)+2}\overset{?}{=}0$

$\sqrt{-2+2}\overset{?}{=}0$

$\sqrt{0}\overset{?}{=}0$

$0=0$

27. $\sqrt{3x}-6=-4$

$\sqrt{3x}=2$

$\left(\sqrt{3x}\right)^2=2^2$

$3x=4$

$x=\frac{4}{3}$

$Check: \sqrt{3x}-6=-4$

$\sqrt{3\cdot\frac{4}{3}}-6\overset{?}{=}-4$

$\sqrt{4}-6\overset{?}{=}-4$

$2-6\overset{?}{=}-4$

$-4=-4$

29. $0=2-\sqrt{3-x}$

$\sqrt{3-x}=2$

$\left(\sqrt{3-x}\right)^2=2^2$

$3-x=4$

$-x=1$

$x=-1$

$Check: 0=2-\sqrt{3-x}$

$0\overset{?}{=}2-\sqrt{3-(-1)}$

$0\overset{?}{=}2-\sqrt{4}$

$0\overset{?}{=}2-2$

$0=0$

31.
$$0=\sqrt{3x-9}-6$$
$$6=\sqrt{3x-9}$$
$$6^2=(\sqrt{3x-9})^2$$
$$36=3x-9$$
$$45=3x$$
$$15=x$$
$$\textit{Check}: 0=\sqrt{3x-9}-6$$
$$0\overset{?}{=}\sqrt{3\cdot 15-9}-6$$
$$0\overset{?}{=}\sqrt{45-9}-6$$
$$0\overset{?}{=}\sqrt{36}-6$$
$$0\overset{?}{=}6-6$$
$$0=0$$

33.
$$\sqrt{5x-1}=\sqrt{3x+9}$$
$$\left(\sqrt{5x-1}\right)^2=\left(\sqrt{3x+9}\right)^2$$
$$5x-1=3x+9$$
$$2x=10$$
$$x=5$$
$$\textit{Check}: \sqrt{5x-1}=\sqrt{3x+9}$$
$$\sqrt{5\cdot 5-1}\overset{?}{=}\sqrt{3\cdot 5+9}$$
$$\sqrt{25-1}\overset{?}{=}\sqrt{15+9}$$
$$\sqrt{24}=\sqrt{24}$$

35.
$$\sqrt{5x-3}=\sqrt{4x-2}$$
$$\left(\sqrt{5x-3}\right)^2=\left(\sqrt{4x-2}\right)^2$$
$$5x-3=4x-2$$
$$x=1$$
$$\textit{Check}: \sqrt{5x-3}=\sqrt{4x-2}$$
$$\sqrt{5\cdot 1-3}\overset{?}{=}\sqrt{4\cdot 1-2}$$
$$\sqrt{2}=\sqrt{2}$$

37.
$$\sqrt{x^2-5x+6}=\sqrt{x^2-8x+9}$$
$$\left(\sqrt{x^2-5x+6}\right)^2=\left(\sqrt{x^2-8x+9}\right)^2$$
$$x^2-5x+6=x^2-8x+9$$
$$-5x+6=-8x+9$$
$$3x=3$$
$$x=1$$
$$\textit{Check}: \sqrt{x^2-5x+6}=\sqrt{x^2-8x+9}$$
$$\sqrt{1^2-5\cdot 1+6}\overset{?}{=}\sqrt{1^2-8\cdot 1+9}$$
$$\sqrt{1-5+6}\overset{?}{=}\sqrt{1-8+9}$$
$$\sqrt{2}=\sqrt{2}$$

39.
$$\sqrt{x}=\sqrt{x+3}-1$$
$$\left(\sqrt{x}\right)^2=\left(\sqrt{x+3}-1\right)^2$$
$$x=x+3-2\sqrt{x+3}+1$$
$$x=x+4-2\sqrt{x+3}$$
$$2\sqrt{x+3}=4$$
$$\sqrt{x+3}=2$$
$$\left(\sqrt{x+3}\right)^2=2^2$$
$$x+3=4$$
$$x=1$$
$$\textit{Check}: \sqrt{x}=\sqrt{x+3}-1$$
$$\sqrt{1}\overset{?}{=}\sqrt{1+3}-1$$
$$1\overset{?}{=}\sqrt{4}-1$$
$$1\overset{?}{=}2-1$$
$$1=1$$

41. $\sqrt{2x+5}=5-\sqrt{2x}$

$\left(\sqrt{2x+5}\right)^2=\left(5-\sqrt{2x}\right)^2$

$2x+5=25-10\sqrt{2x}+2x$

$10\sqrt{2x}=20$

$\sqrt{2x}=2$

$\left(\sqrt{2x}\right)^2=2^2$

$2x=4$

$x=2$

Check : $\sqrt{2x+5}=5-\sqrt{2x}$

$\sqrt{2\cdot 2+5}\stackrel{?}{=}5-\sqrt{2\cdot 2}$

$\sqrt{4+5}\stackrel{?}{=}5-\sqrt{4}$

$\sqrt{9}\stackrel{?}{=}5-2$

$3=3$

43. $\sqrt{3x}-\sqrt{3x+7}=1$

$\left(\sqrt{3x}\right)^2=\left(1+\sqrt{3x+7}\right)^2$

$3x=1+2\sqrt{3x+7}+3x+7$

$3x=8+2\sqrt{3x+7}+3x$

$-8=2\sqrt{3x+7}$

$-4=\sqrt{3x+7}$

$(-4)^2=\left(\sqrt{3x+7}\right)^2$

$16=3x+7$

$9=3x$

$3=x$

Check : $\sqrt{3x}-\sqrt{3x+7}=1$

$\sqrt{3\cdot 3}-\sqrt{3\cdot 3+7}\stackrel{?}{=}1$

$\sqrt{9}-\sqrt{9+7}\stackrel{?}{=}1$

$3-4\stackrel{?}{=}1$

$-1\neq 1$

The solution does not check so the equation has no solution.

45. Equation ***iii*** has no solution because a square root is never negative.

47. $5+\sqrt{4n}=7$

$\sqrt{4n}=2$

$\left(\sqrt{4n}\right)^2=2^2$

$4n=4$

$n=1$

The number is 1.

49. In a right triangle, the hypotenuse is the side opposite the **right** angle. The other two sides are called **legs**.

51. The Pythagorean Theorem states that, in a right triangle, the square of the hypotenuse is equal to the sum of the squares of the legs.

53. $c^2=a^2+b^2$

$c^2=5^2+9^2$

$c^2=25+81$

$c^2=106$

$c=\sqrt{106}\approx 10.30\ cm$

55. $c^2=a^2+b^2$

$12^2=7^2+b^2$

$144=49+b^2$

$95=b^2$

$b=\sqrt{95}\approx 9.75\ ft$

57. $c^2=a^2+b^2$

$c^2=5^2+11^2$

$c^2=25+121$

$c^2=146$

$c=\sqrt{146}\approx 12.08\ mi$

59. **a.** $\frac{C^2}{H}=32 \Rightarrow C^2=32H \Rightarrow C=\sqrt{32H}$

b. $C=4\sqrt{2H} \Rightarrow C^2=16(2H)$

$\Rightarrow C^2=32H \Rightarrow C=\sqrt{32H}$

c. $\frac{C^2}{32}=H \Rightarrow C^2=32H \Rightarrow C=\sqrt{32H}$

d. $\left(\frac{C}{4}\right)^2=2H \Rightarrow \frac{C^2}{16}=2H$

$\Rightarrow C^2=32H \Rightarrow C=\sqrt{32H}$

All are equivalent.

61. $C=\sqrt{32H}$

$20=\sqrt{32H}$

$20^2=\left(\sqrt{32H}\right)^2$

$400=32H$

$12.5=H$

The depth is 12.5 ft.

63. $c^2=a^2+b^2$

$c^2=90^2+90^2$

$c^2=8100+8100$

$c^2=16,200$

$c=\sqrt{16,200}\approx 127.3ft$

The pitcher is less than halfway between home plate and second base.

65. $c^2=a^2+b^2$

$c^2=2.5^2+4^2$

$c^2=6.25+16$

$c^2=22.25$

$c=\sqrt{22.25}\approx 4.7mi$

Yes, she will be able to call her friend.

67. $c^2=a^2+b^2$

$c^2=6^2+24^2$

$c^2=36+576$

$c^2=612$

$c=\sqrt{612}\approx 24.7ft$

The ladder is long enough.

69. $T=2\pi\sqrt{\frac{L}{32}}$

$3=2\pi\sqrt{\frac{L}{32}}$

$\frac{3}{2\pi}=\sqrt{\frac{L}{32}}$

$\left(\frac{3}{2\pi}\right)^2=\left(\sqrt{\frac{L}{32}}\right)^2$

$\frac{9}{4\pi^2}=\frac{L}{32}$

$L=\frac{9(32)}{4\pi^2}\approx 7.30ft$

The length is 7.30 feet.

71. $T=\sqrt{\frac{d}{16}}$

$2=\sqrt{\frac{d}{16}}$

$2^2=\left(\sqrt{\frac{d}{16}}\right)^2$

$4=\frac{d}{16}$

$64=d$

The bridge is 64 feet high.

73. $c^2=a^2+b^2$

$41^2=20.5^2+b^2$

$1681=420.25+b^2$

$1260.75=b^2$

$b=\sqrt{1260.75}\approx 35.5in$

The screen is 35.5 inches high.

75. $\sqrt{\frac{3y}{5}}-1=2$

$\sqrt{\frac{3y}{5}}=3$

$\left(\sqrt{\frac{3y}{5}}\right)^2=3^2$

$\frac{3y}{5}=9$

$3y=45$

$y=15$

77. The leg from (0, 0) to (5, 0) is 5 units long. The leg from (5, 0) to (5, 12) is 12 units long.

$c^2=a^2+b^2$

$c^2=5^2+12^2$

$c^2=25+144$

$c^2=169$

$c=\sqrt{169}=13$

$P=a+b+c$

$P=5+12+13$

$P=30$

The perimeter is 30 units.

79. Circle: $A=\pi r^2$

Square: $A=4\cdot\frac{1}{2}(r\cdot r)$

$A=2r^2$

Shaded portion: $A=\pi r^2-2r^2$

$A=r^2(\pi-2)$

81. No, because the Pythagorean Theorem can only be applied to right triangles.

83. **a.** Each small triangle has legs of x. The hypotenuse of the triangle is the length of the side of the inner square.

$c^2=x^2+x^2$

$c^2=2x^2$

$c=\sqrt{2x^2}$

$c=x\sqrt{2}$

b. $A=(x\sqrt{2})^2$

$A=2x^2 sq.in.$

85. $a^2+b^2+a^2+b^2=100$

$2a^2+2b^2=100$

$a^2+b^2=50$

$\sqrt{a^2+b^2}=\sqrt{50}$

$\sqrt{a^2+b^2}=\sqrt{25\cdot 2}$

$\sqrt{a^2+b^2}=5\sqrt{2}units$

Chapter 10 Review Exercises

1. $5\sqrt{3}-16\sqrt{3}=-11\sqrt{3}$

2. $\frac{2x}{\sqrt{3}-\sqrt{5}}\cdot\frac{\sqrt{3}+\sqrt{5}}{\sqrt{3}+\sqrt{5}}=\frac{2x(\sqrt{3}+\sqrt{5})}{3-5}$

$=\frac{2x(\sqrt{3}+\sqrt{5})}{-2}$

$=-x\sqrt{3}-x\sqrt{5}$

3. $\sqrt{x^2+16x+64}=\sqrt{(x+8)^2}=x+8$

4. $2-\sqrt{7x}=5$

$-\sqrt{7x}=3$

$\sqrt{7x}=-3$

The equation has no solution because a square root cannot be equal to a negative number.

5. $\dfrac{5\sqrt{y}-2\sqrt{y}}{3\sqrt{y}}=\dfrac{3\sqrt{y}}{3\sqrt{y}}=1$

6. $6\sqrt{7}+\sqrt{7}=7\sqrt{7}$

7. $(6\sqrt{a}+5\sqrt{b})(2\sqrt{a}+3\sqrt{b})=12a+28\sqrt{ab}+15b$

8. $\sqrt{49(x+3)^4}=7(x+3)^2=7(x^2+6x+9)$

$=7x^2+42x+63$

9. $2\sqrt{36}=2(6)=12$

10. $\sqrt{b}=4$

$\left(\sqrt{b}\right)^2=4^2$

$b=16$

11. $9x\sqrt{5}-5x\sqrt{5}=4x\sqrt{5}$

12. $(\sqrt{5ab}-\sqrt{7})(\sqrt{5ab}+\sqrt{7})=5ab-7$

13. $\sqrt{2x+9}=\sqrt{8x-9}$

$\left(\sqrt{2x+9}\right)^2=\left(\sqrt{8x-9}\right)^2$

$2x+9=8x-9$

$-6x=-18$

$x=3$

$Check:\sqrt{2x+9}=\sqrt{8x-9}$

$\sqrt{6+9}\stackrel{?}{=}\sqrt{24-9}$

$\sqrt{15}=\sqrt{15}$

14. $\sqrt{35}$ does not simplify

15. $2x\sqrt{60x^3y^3}+3x^2y\sqrt{15xy}$

$=2x\sqrt{4x^2y^2\cdot 15xy}+3x^2y\sqrt{15xy}$

$=2x(2xy)\sqrt{15xy}+3x^2y\sqrt{15xy}$

$=4x^2y\sqrt{15xy}+3x^2y\sqrt{15xy}$

$=7x^2y\sqrt{15xy}$

16. $\dfrac{\sqrt{3x^3y}}{\sqrt{27xy^5}}=\sqrt{\dfrac{3x^3y}{27xy^5}}=\sqrt{\dfrac{x^2}{9y^4}}=\dfrac{x}{3y^2}$

17. $(3\sqrt{x}-\sqrt{y})^2=9x-6\sqrt{xy}+y$

18. $\sqrt{(a+4)^2}=a+4$

19. $5\sqrt{48}=5\sqrt{16\cdot 3}=5(4)\sqrt{3}=20\sqrt{3}$

20. $3\sqrt{12x}+5\sqrt{48x}=3\sqrt{4\cdot 3x}+5\sqrt{16\cdot 3x}$

$=3\cdot 2\sqrt{3x}+5\cdot 4\sqrt{3x}$

$=6\sqrt{3x}+20\sqrt{3x}$

$=26\sqrt{3x}$

21. $\dfrac{8}{\sqrt{x}-3}\cdot\dfrac{\sqrt{x}+3}{\sqrt{x}+3}=\dfrac{8\sqrt{x}+24}{x-9}$

22. $\sqrt{6a}(\sqrt{3a}+\sqrt{2a})=\sqrt{18a^2}+\sqrt{12a^2}$

$=\sqrt{9a^2\cdot 2}+\sqrt{4a^2\cdot 3}$

$=3a\sqrt{2}+2a\sqrt{3}$

23. $3\sqrt{18a^5b}=3\sqrt{9a^4 2ab}=3(3a^2)\sqrt{2ab}=9a^2\sqrt{2ab}$

24. $-3\sqrt{120}=-3\sqrt{4\cdot 30}=-3(2)\sqrt{30}=-6\sqrt{30}$

25. $\sqrt{20a^5b^9} - 2ab^2\sqrt{45a^3b^5}$
$= \sqrt{4a^4b^8 \cdot 5ab} - 2ab^2\sqrt{9a^2b^4 \cdot 5ab}$
$= 2a^2b^4\sqrt{5ab} - 2ab^2 \cdot 3ab^2\sqrt{5ab}$
$= 2a^2b^4\sqrt{5ab} - 6a^2b^4\sqrt{5ab}$
$= -4a^2b^4\sqrt{5ab}$

26. $\dfrac{\sqrt{98x^7y^9}}{\sqrt{2x^3y}} = \sqrt{\dfrac{98x^7y^9}{2x^3y}} = \sqrt{49x^4y^8} = 7x^2y^4$

27. $\sqrt{5x+1} = \sqrt{20x-8}$
$\left(\sqrt{5x+1}\right)^2 = \left(\sqrt{20x-8}\right)^2$
$5x+1 = 20x-8$
$-15x = -9$
$x = \dfrac{9}{15} = \dfrac{3}{5}$
Check: $\sqrt{5x+1} = \sqrt{20x-8}$
$\sqrt{5\cdot\dfrac{3}{5}+1} \stackrel{?}{=} \sqrt{20\cdot\dfrac{3}{5}-8}$
$\sqrt{3+1} \stackrel{?}{=} \sqrt{12-8}$
$\sqrt{4} = \sqrt{4}$

28. $\sqrt{c^{18}} = c^9$

29. $\sqrt{450} = \sqrt{225\cdot 2} = 15\sqrt{2}$

30. $6a\sqrt{80b} - \sqrt{180a^2b} + 5a\sqrt{b}$
$= 6a\sqrt{16\cdot 5b} - \sqrt{36a^2\cdot 5b} + 5a\sqrt{b}$
$= 6a\cdot 4\sqrt{5b} - 6a\sqrt{5b} + 5a\sqrt{b}$
$= 24a\sqrt{5b} - 6a\sqrt{5b} + 5a\sqrt{b}$
$= 18a\sqrt{5b} + 5a\sqrt{b}$

31. $\dfrac{16}{\sqrt{a}}\cdot\dfrac{\sqrt{a}}{\sqrt{a}} = \dfrac{16\sqrt{a}}{a}$

32. $6 - \sqrt{2y} = 2$
$4 = \sqrt{2y}$
$4^2 = \left(\sqrt{2y}\right)^2$
$16 = 2y$
$8 = y$
Check: $6 - \sqrt{2y} = 2$
$6 - \sqrt{2(8)} \stackrel{?}{=} 2$
$6 - \sqrt{16} \stackrel{?}{=} 2$
$6 - 4 \stackrel{?}{=} 2$
$2 = 2$

33. $\sqrt{a^3b^4c}\sqrt{a^7b^2c^3} = \sqrt{a^{10}b^6c^4} = a^5b^3c^2$

34. $7\sqrt{630} = 7\sqrt{9\cdot 70} = 7\cdot 3\sqrt{70} = 21\sqrt{70}$

35. $y\sqrt{24y^6} = y\sqrt{4y^6\cdot 6} = y\cdot 2y^3\sqrt{6} = 2y^4\sqrt{6}$

36. $\dfrac{\sqrt{250}}{\sqrt{10}} = \sqrt{\dfrac{250}{10}} = \sqrt{25} = 5$

37. $\sqrt{x^2+5x+4} = \sqrt{x^2+7x-6}$
$\sqrt{\left(x^2+5x+4\right)}^2 = \left(\sqrt{x^2+7x-6}\right)^2$
$x^2+5x+4 = x^2+7x-6$
$5x+4 = 7x-6$
$-2x = -10$
$x = 5$
Check: $\sqrt{x^2+5x+4} = \sqrt{x^2+7x-6}$
$\sqrt{5^2+5\cdot 5-4} \stackrel{?}{=} \sqrt{5^2+7\cdot 5-6}$
$\sqrt{25+25-4} \stackrel{?}{=} \sqrt{25+35-6}$
$\sqrt{4} = \sqrt{4}$

38. $(4\sqrt{y} - \sqrt{5})(2\sqrt{y} + 3\sqrt{5}) = 8y + 10\sqrt{5y} - 15$

39. $\sqrt{9900} \approx 99.499$

40. $\sqrt{7}\sqrt{7} = 7$

41. $5x\sqrt{150x^7} = 5x\sqrt{25x^6 \cdot 6x}$
$= 5x \cdot 5x^3\sqrt{6x} = 25x^4\sqrt{6x}$

42. $2x^2\sqrt{18x^2y^5} + 6y\sqrt{2x^6y^3} - 9xy^2\sqrt{8x^4y}$
$= 2x^2\sqrt{9x^2y^4 \cdot 2y} + 6y\sqrt{x^6y^2 \cdot 2y} - 9xy^2\sqrt{4x^4 \cdot 2y}$
$= 2x^2(3xy^2)\sqrt{2y} + 6y(x^3y)\sqrt{2y} - 9xy^2(2x^2)\sqrt{2y}$
$= 6x^3y^2\sqrt{2y} + 6x^3y^2\sqrt{2y} - 18x^3y^2\sqrt{2y}$
$= -6x^3y^2\sqrt{2y}$

43. $\dfrac{\sqrt{54a^3}}{\sqrt{6a}} = \sqrt{\dfrac{54a^3}{6a}} = \sqrt{9a^2} = 3a$

44. $4\sqrt{250} = 4\sqrt{25 \cdot 10} = 4 \cdot 5\sqrt{10} = 20\sqrt{10}$

45. $\sqrt{5x} = 10$
$\left(\sqrt{5x}\right)^2 = 10^2$
$5x = 100$
$x = 20$
Check : $\sqrt{5x} = 10$
$\sqrt{5 \cdot 20} \stackrel{?}{=} 10$
$\sqrt{100} \stackrel{?}{=} 10$
$10 = 10$

46. $\dfrac{3a\sqrt{3} + 2\sqrt{12a^2}}{\sqrt{27}} = \dfrac{3a\sqrt{3} + 2\sqrt{4a^2 \cdot 3}}{\sqrt{9 \cdot 3}}$
$= \dfrac{3a\sqrt{3} + 4a\sqrt{3}}{3\sqrt{3}}$
$= \dfrac{7a\sqrt{3}}{3\sqrt{3}} = \dfrac{7a}{3}$

47. $\sqrt{2}\sqrt{50} = \sqrt{100} = 10$

48. $\sqrt{36x^4y^5} = \sqrt{36x^4y^4y} = 6x^2y^2\sqrt{y}$

49. $4y\sqrt{243x^{17}y^9} = 4y\sqrt{81x^{16}y^8 \cdot 3xy}$
$= 4y(9x^8y^4)\sqrt{3xy}$
$= 36x^8y^5\sqrt{3xy}$

50. $\sqrt{x+1} - \sqrt{x-2} = 1$
$\sqrt{x+1} = \sqrt{x-2} + 1$
$\left(\sqrt{x+1}\right)^2 = \left(\sqrt{x-2} + 1\right)^2$
$x + 1 = x - 2 + 2\sqrt{x-2} + 1$
$x + 1 = x - 1 + 2\sqrt{x-2}$
$2 = 2\sqrt{x-2}$
$1 = \sqrt{x-2}$
$1^2 = \left(\sqrt{x-2}\right)^2$
$1 = x - 2$
$3 = x$
Check : $\sqrt{x+1} - \sqrt{x-2} = 1$
$\sqrt{3+1} - \sqrt{3-2} \stackrel{?}{=} 1$
$\sqrt{4} - \sqrt{1} \stackrel{?}{=} 1$
$2 - 1 \stackrel{?}{=} 1$
$1 = 1$

51. $\sqrt{400} = 20$

52. $-4\sqrt{8x} + 7\sqrt{18x} - 3\sqrt{50x}$
$= -4\sqrt{4 \cdot 2x} + 7\sqrt{9 \cdot 2x} - 3\sqrt{25 \cdot 2x}$
$= -4(2)\sqrt{2x} + 7(3)\sqrt{2x} - 3(5)\sqrt{2x}$
$= -8\sqrt{2x} + 21\sqrt{2x} - 15\sqrt{2x}$
$= -2\sqrt{2x}$

53. $0 = \sqrt{10x+4} - 8$

$8 = \sqrt{10x+4}$

$8^2 = \left(\sqrt{10x+4}\right)^2$

$64 = 10x+4$

$60 = 10x$

$6 = x$

$Check: 0 = \sqrt{10x+4} - 8$

$0 \stackrel{?}{=} \sqrt{10 \cdot 6+4} - 8$

$0 \stackrel{?}{=} \sqrt{60+4} - 8$

$0 \stackrel{?}{=} \sqrt{64} - 8$

$0 \stackrel{?}{=} 8 - 8$

$0 = 0$

54. $\sqrt{3}(\sqrt{12} - \sqrt{3}) = \sqrt{36} - \sqrt{9} = 6 - 3 = 3$

55. $\sqrt{n+n+2} = 10$

$\sqrt{2n+2} = 10$

$(\sqrt{2n+2})^2 = 10^2$

$2n+2 = 100$

$2n = 98$

$n = 49$

$n+2 = 51$

The larger integer is 51.

56. $d = 4000\sqrt{\dfrac{W_0}{W_a}}$

$8000 = 4000\sqrt{\dfrac{W_0}{36}}$

$\dfrac{8000}{4000} = \dfrac{\sqrt{W_0}}{\sqrt{36}}$

$2 = \dfrac{\sqrt{W_0}}{6}$

$2(6) = \sqrt{W_0}$

$12^2 = \left(\sqrt{W_0}\right)^2$

$144 = W_0$

He weighs 144 pounds on Earth.

57. $c^2 = a^2 + b^2$

$18^2 = 11^2 + b^2$

$324 = 121 + b^2$

$203 = b^2$

$b = \sqrt{203} \approx 14.25cm$

58. $v = 4\sqrt{r}$

$20 = 4\sqrt{r}$

$5 = \sqrt{r}$

$5^2 = (\sqrt{r})^2$

$25 ft = r$

59. $v = 3\sqrt{d}$

$30 = 3\sqrt{d}$

$10 = \sqrt{d}$

$10^2 = (\sqrt{d})^2$

$d = 100 ft$

60. $c^2 = a^2 + b^2$

$c^2 = 8^2 + 25^2$

$c^2 = 64 + 625$

$c^2 = 689$

$c = \sqrt{689} \approx 26.25 ft$

Chapter 10 Test

1. $\sqrt{121x^8y^2} = 11x^4y$

2. $5\sqrt{8} - 3\sqrt{50} = 5\sqrt{4 \cdot 2} - 3\sqrt{25 \cdot 2}$

$= 5(2)\sqrt{2} - 3(5)\sqrt{2}$

$= 10\sqrt{2} - 15\sqrt{2} = -5\sqrt{2}$

3. $\sqrt{3x^2y}\sqrt{6x^2}\sqrt{2x} = \sqrt{36x^5y} = \sqrt{36x^4xy} = 6x^2\sqrt{xy}$

4. $\sqrt{45} = \sqrt{9 \cdot 5} = 3\sqrt{5}$

5. $\sqrt{72x^7y^2} = \sqrt{36x^6y^2 \cdot 2x} = 6x^3y\sqrt{2x}$

6. $3\sqrt{8y} - 2\sqrt{72x} + 5\sqrt{18x}$
$= 3\sqrt{4 \cdot 2y} - 2\sqrt{36 \cdot 2x} + 5\sqrt{9 \cdot 2x}$
$= 3(2)\sqrt{2y} - 2(6)\sqrt{2x} + 5(3)\sqrt{2x}$
$= 6\sqrt{2y} - 12\sqrt{2x} + 15\sqrt{2x}$
$= 6\sqrt{2y} + 3\sqrt{2x}$

7. $\left(\sqrt{y}+3\right)\left(\sqrt{y}+5\right) = y + 8\sqrt{y} + 15$

8. $\frac{4}{\sqrt{8}} = \frac{4}{\sqrt{4 \cdot 2}} = \frac{4}{2\sqrt{2}} = \frac{2}{\sqrt{2}} \cdot \frac{\sqrt{2}}{\sqrt{2}} = \frac{2\sqrt{2}}{2} = \sqrt{2}$

9. $\frac{\sqrt{162}}{\sqrt{2}} = \sqrt{\frac{162}{2}} = \sqrt{81} = 9$

10. $\sqrt{5x-6} = 7$
$\left(\sqrt{5x-6}\right)^2 = 7^2$
$5x - 6 = 49$
$5x = 55$
$x = 11$
Check: $\sqrt{5x-6} = 7$
$\sqrt{5(11)-6} \stackrel{?}{=} 7$
$\sqrt{55-6} \stackrel{?}{=} 7$
$\sqrt{49} \stackrel{?}{=} 7$
$7 = 7$

11. $\sqrt{500} \approx 22.361$

12. $\sqrt{32a^5b^{11}} = \sqrt{16a^4b^{10} \cdot 2ab} = 4a^2b^5\sqrt{2ab}$

13. $\sqrt{a}\left(\sqrt{a} - \sqrt{b}\right) = a - \sqrt{ab}$

14. $\sqrt{8x^3y}\sqrt{10xy^4} = \sqrt{80x^4y^5}$
$= \sqrt{16x^4y^4 \cdot 5y} = 4x^2y^2\sqrt{5y}$

15. $\frac{\sqrt{98a^6b^4}}{\sqrt{2a^3b^2}} = \sqrt{\frac{98a^6b^4}{2a^3b^2}} = \sqrt{49a^3b^2}$
$= \sqrt{49a^2b^2a} = 7ab\sqrt{a}$

16. $\sqrt{9x} + 3 = 18$
$\sqrt{9x} = 15$
$\left(\sqrt{9x}\right)^2 = 15^2$
$9x = 225$
$x = 25$
Check: $\sqrt{9x} + 3 = 18$
$\sqrt{9(25)} + 3 \stackrel{?}{=} 18$
$\sqrt{225} + 3 \stackrel{?}{=} 18$
$15 + 3 \stackrel{?}{=} 18$
$18 = 18$

17. $\sqrt{192x^{13}y^5} = \sqrt{64x^{12}y^4 \cdot 3xy} = 8x^6y^2\sqrt{3xy}$

18. $2a\sqrt{2ab^3} + b\sqrt{8a^3b} - 5ab\sqrt{ab}$
$= 2a\sqrt{b^2 \cdot 2ab} + b\sqrt{4a^2 \cdot 2ab} - 5ab\sqrt{ab}$
$= 2ab\sqrt{2ab} + 2ab\sqrt{2ab} - 5ab\sqrt{ab}$
$= 4ab\sqrt{2ab} - 5ab\sqrt{ab}$

19. $\left(\sqrt{a}-2\right)\left(\sqrt{a}+2\right) = a - 4$

20. $\frac{3}{2-\sqrt{5}} \cdot \frac{2+\sqrt{5}}{2+\sqrt{5}} = \frac{6+3\sqrt{5}}{4-5} = -6 - 3\sqrt{5}$

21.
$$3 = 8 - \sqrt{5x}$$
$$\sqrt{5x} = 5$$
$$\left(\sqrt{5x}\right)^2 = 5^2$$
$$5x = 25$$
$$x = 5$$
$$Check: 3 = 8 - \sqrt{5x}$$
$$3 \stackrel{?}{=} 8 - \sqrt{5(5)}$$
$$3 \stackrel{?}{=} 8 - \sqrt{25}$$
$$3 \stackrel{?}{=} 8 - 5$$
$$3 = 3$$

22.
$$\sqrt{3}\left(\sqrt{6} - \sqrt{x^2}\right) = \sqrt{18} - \sqrt{3x^2}$$
$$= \sqrt{9 \cdot 2} - \sqrt{x^2 \cdot 3}$$
$$= 3\sqrt{2} - x\sqrt{3}$$

23. $3\sqrt{a} - 9\sqrt{a} = -6\sqrt{a}$

24. $\sqrt{108} = \sqrt{36 \cdot 3} = 6\sqrt{3}$

25. $\sqrt{63} \approx 7.937$

26. $\dfrac{\sqrt{108a^7b^3}}{\sqrt{3a^4b}} = \sqrt{\dfrac{108a^7b^3}{3a^4b}} = \sqrt{36a^2b^2a} = 6ab\sqrt{a}$

27.
$$\sqrt{x} - \sqrt{x+3} = 1$$
$$\sqrt{x} = \sqrt{x+3} + 1$$
$$\left(\sqrt{x}\right)^2 = \left(\sqrt{x+3} + 1\right)^2$$
$$x = x + 3 + 2\sqrt{x+3} + 1$$
$$x = x + 4 + 2\sqrt{x+3}$$
$$-4 = 2\sqrt{x+3}$$
$$\sqrt{x+3} = -2$$

The equation has no solution because a square root cannot be equal to a negative number.

28.
$$\sqrt{n + n + 1} = 9$$
$$\sqrt{2n+1} = 9$$
$$\left(\sqrt{2n+1}\right)^2 = 9^2$$
$$2n + 1 = 81$$
$$2n = 80$$
$$n = 40$$

The smaller integer is 40.

29.
$$T = 2\pi\sqrt{\frac{L}{32}}$$
$$2.5 = 2\pi\sqrt{\frac{L}{32}}$$
$$\frac{2.5}{2\pi} = \sqrt{\frac{L}{32}}$$
$$\left(\frac{2.5}{2\pi}\right)^2 = \left(\sqrt{\frac{L}{32}}\right)^2$$
$$0.1583 = \frac{L}{32}$$
$$L \approx 5.07\,ft$$

30.
$$c^2 = a^2 + b^2$$
$$16^2 = 5^2 + b^2$$
$$256 = 25 + b^2$$
$$231 = b^2$$
$$b = \sqrt{231} \approx 15.2\,ft$$

Cumulative Review Exercises

1.
$$\left(\frac{2}{3}\right)^2\left(\frac{3}{4} - \frac{3}{2}\right) + \left(\frac{1}{2}\right)^3 = \left(\frac{4}{9}\right)\left(\frac{3}{4} - \frac{3}{2}\right) + \left(\frac{1}{4}\right)$$
$$= \left(\frac{4}{9}\right)\left(\frac{3-6}{4}\right) + \left(\frac{1}{4}\right)$$
$$= \left(\frac{4}{9}\right)\left(\frac{-3}{4}\right) + \left(\frac{1}{4}\right)$$
$$= -\frac{1}{3} + \frac{1}{4} = -\frac{4}{12} + \frac{3}{12} = -\frac{1}{12}$$

2.
$$-3[x - 2(3 - 2x) - 5x] + 2x$$
$$= -3[x - 6 + 4x - 5x] + 2x$$
$$= -3[-6] + 2x = 18 + 2x$$

3. $2x-4[3x-2(1-3x)]=2(3-4x)$

$$2x-4[3x-2+6x]=6-8x$$
$$2x-4[9x-2]=6-8x$$
$$2x-36x+8=6-8x$$
$$-34x+8=6-8x$$
$$-26x=-2$$
$$x=\frac{1}{13}$$

4. $3(x-7)\geq 5x-12$

$$3x-21\geq 5x-12$$
$$-2x\geq 9$$
$$x\leq -\frac{9}{2}$$

5. $m=\dfrac{y_2-y_1}{x_2-x_1}=\dfrac{3-(-5)}{-4-2}=\dfrac{8}{-6}=-\dfrac{4}{3}$

6. $y-y_1=m(x-x_1)$

$$y-(-3)=\frac{1}{2}[x-(-2)]$$
$$y+3=\frac{1}{2}(x+2)$$
$$y+3=\frac{1}{2}x+1$$
$$y=\frac{1}{2}x-2$$

7. $f(x)=\dfrac{5}{2}x-8$

$$f(-4)=\frac{5}{2}(-4)-8$$
$$f(-4)=-10-8=-18$$

8. 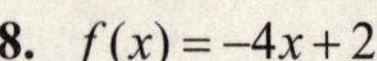 $f(x)=-4x+2$

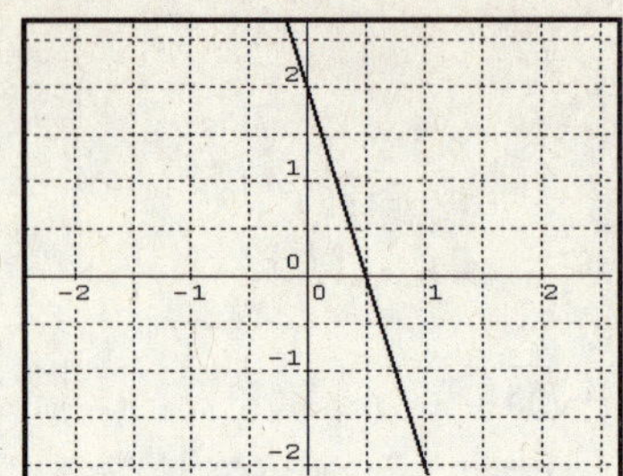

9. $2x+y<-2$

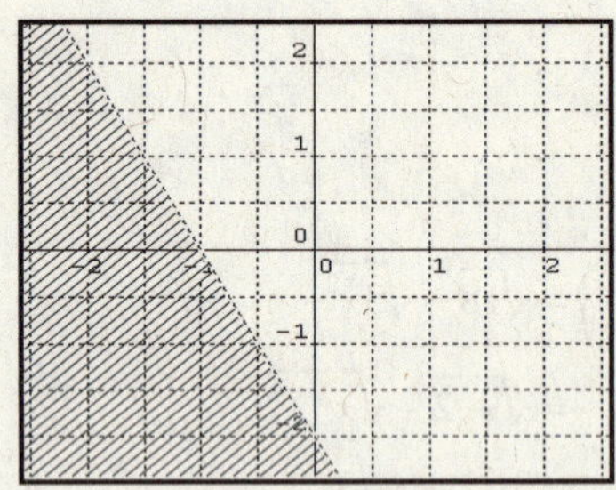

10. The solution is (2, -1).

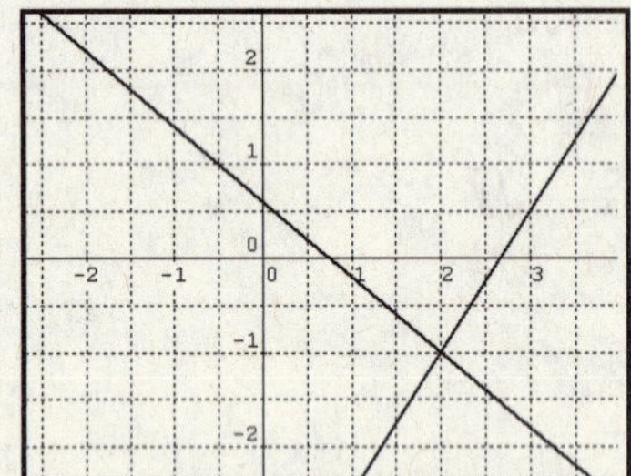

11. $4x-3y=1$

$$2x+y=3\Rightarrow y=-2x+3$$
$$4x-3(-2x+3)=1$$
$$4x+6x-9=1$$
$$10x=10$$
$$x=1$$
$$y=-2x+3$$
$$y=-2(1)+3$$
$$y=1$$

The solution is (1, 1).

12. $5x+4y=7$

$\underline{2(3x-2y)=2(13)}$

$5x+4y=7$

$\underline{6x-4y=26}$

$11x \quad =33$

$x=3$

$5x+4y=7$

$5(3)+4y=7$

$15+4y=7$

$4y=-8$

$y=-2$

13. $(-3x^2y)(-2x^3y^{-4})=6x^5y^{-3}=\dfrac{6x^5}{y^3}$

14. $\dfrac{12b^4-6b^2+2}{-6b^2}=\dfrac{12b^4}{-6b^2}-\dfrac{6b^2}{-6b^2}+\dfrac{2}{-6b^2}$

$=-2b^2+1-\dfrac{1}{3b^2}$

15. $12x^3y^2-9x^2y^3=3x^2y^2(4x-3y)$

16. $9b^2+3b-20=(3b+5)(3b-4)$

17. $2a^3-16a^2+30a=2a(a^2-8a+15)$

$=2a(a-3)(a-5)$

18. $\dfrac{3x^3-6x^2}{4x^2+4x}\cdot\dfrac{3x-9}{9x^3-45x^2+54x}$

$=\dfrac{\cancel{3}\cancel{x^2}\cancel{(x-2)}}{4\cancel{x}(x+1)}\cdot\dfrac{\cancel{3}\cancel{(x-3)}}{\cancel{9}\cancel{x}\cancel{(x-3)}\cancel{(x-2)}}$

$=\dfrac{1}{4(x+1)}$

19. $\dfrac{\left(1-\dfrac{2}{x}-\dfrac{15}{x^2}\right)x^2}{\left(1-\dfrac{9}{x^2}\right)x^2}=\dfrac{x^2-2x-15}{x^2-9}$

$=\dfrac{(x-5)(x+3)}{(x-3)(x+3)}=\dfrac{x-5}{x-3}$

20. $\dfrac{x+2}{x-4}-\dfrac{6}{(x-4)(x-3)}$

$=\dfrac{(x+2)(x-3)}{(x-4)(x-3)}-\dfrac{6}{(x-4)(x-3)}$

$=\dfrac{x^2-x-6-6}{(x-4)(x-3)}=\dfrac{x^2-x-12}{(x-4)(x-3)}$

$=\dfrac{(x-4)(x+3)}{(x-4)(x-3)}=\dfrac{x+3}{x-3}$

21. $\dfrac{x}{2x-5}-2=\dfrac{3x}{2x-5}$

$(2x-5)\left(\dfrac{x}{2x-5}-2\right)=\left(\dfrac{3x}{2x-5}\right)(2x-5)$

$x-2(2x-5)=3x$

$x-4x+10=3x$

$-3x+10=3x$

$10=6x$

$x=\dfrac{10}{6}=\dfrac{5}{3}$

22. $2\sqrt{27a}-5\sqrt{49a}+8\sqrt{48a}$

$=2\sqrt{9\cdot 3a}-5\sqrt{49a}+8\sqrt{16\cdot 3a}$

$=2(3)\sqrt{3a}-5(7)\sqrt{a}+8(4)\sqrt{3a}$

$=6\sqrt{3a}-35\sqrt{a}+32\sqrt{3a}$

$=38\sqrt{3a}-35\sqrt{a}$

23. $\dfrac{\sqrt{320}}{\sqrt{5}}=\sqrt{\dfrac{320}{5}}=\sqrt{64}=8$

24. $\sqrt{2x-3}-5=0$

$\sqrt{2x-3}=5$

$\left(\sqrt{2x-3}\right)^2=5^2$

$2x-3=25$

$2x=28$

$0=0$

$x=14$

Check: $\sqrt{2\cdot 14-3}-5\overset{?}{=}0$

$\sqrt{28-3}-5\overset{?}{=}0$

$\sqrt{25}-5\overset{?}{=}0$

$5-5\overset{?}{=}0$

$0=0$

25. $\frac{3}{8}n < -12$

$\frac{8}{3}\cdot\frac{3}{8}n < -12\cdot\frac{8}{3}$

$n < -32$

The largest integer that is less than -32 is -33.

26. $S = C + rC$

$29.40 = C + 0.20C$

$29.40 = 1.20C$

$\frac{29.40}{1.20} = C$

$C = 24.50$

The cost is $24.50.

27.

	B	·	P	=	A
Pure water	x	·	0	=	0
12% sol.	40	·	0.12	=	0.12(40)
Total	x +40	·	0.05	=	$0.05(x + 40)$

$0.12(40) = 0.05(x + 40)$

$4.8 = 0.05x + 2$

$2.8 = 0.05x$

$x = 56$

$x + 40 = 96$

56 ounces of pure water are needed.

28. Let n = the smaller integer.

$2n + 3 + n = 21$

$3n + 3 = 21$

$3n = 18$

$n = 6$

$2n + 3 = 15$

The numbers are 6 and 15.

29. Let x = the time for the larger pipe to fill the tank. Then $2x$ = the time for the smaller pipe to fill the tank. In one hour, the larger tank fills $\frac{1}{x}$ of the tank; the smaller tank fills $\frac{1}{2x}$; together they fill $\frac{1}{16}$.

$\frac{1}{x} + \frac{1}{2x} = \frac{1}{16}$

$16x\left(\frac{1}{x} + \frac{1}{2x}\right) = \frac{1}{16}(16x)$

$16 + 8 = x$

$x = 24$

$2x = 48$

Working alone, it will take the smaller pipe 48 hours to fill the tank.

30. $\sqrt{n + n + 1} = 7$

$\sqrt{2n + 1} = 7$

$\left(\sqrt{2n + 1}\right)^2 = 7^2$

$2n + 1 = 49$

$2n = 48$

$n = 24$

The smaller integer is 24.

CHAPTER 11 Quadratic Equations

Chapter 11 Prep Test

1. $b^2 - 4ac;\ a = 2, b = -3, c = -4$

$(-3)^2 - 4(2)(-4)$

$9 + 32 = 41$

2. $5x + 4 = 3$

$5x = -1$

$\frac{5x}{5} = \frac{-1}{5}$

$x = -\frac{1}{5}$

3. $x^2 + x - 12 = (x+4)(x-3)$

4. $4x^2 - 12x + 9 = (2x-3)^2$

5. Yes. It is a perfect square trinomial.

6. $\frac{5}{x-2} = \frac{15}{x}$

$5x = 15(x-2)$

$5x = 15x - 30$

$-10x = -30$

$x = 3$

7. $y = -2x + 3$

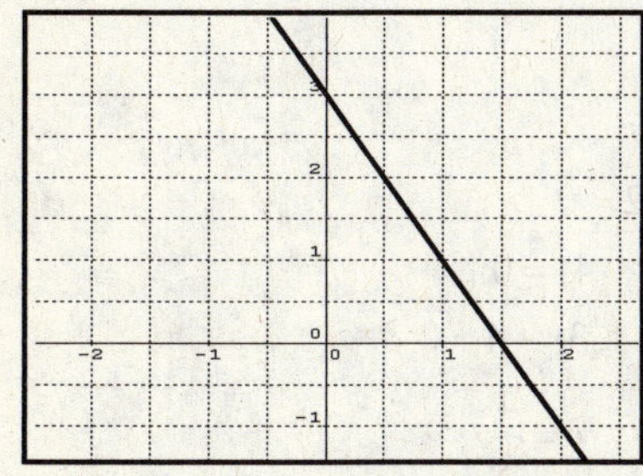

8. $\sqrt{28} = \sqrt{4 \cdot 7} = \sqrt{4} \cdot \sqrt{7} = 2\sqrt{7}$

9. $\sqrt{a^2} = |a|$

10. The distance walked in each direction is the same.

Let t = the time to get to the end of the trail.

Beginning to end: $d = 4.5t$

Back to beginning: $d = 3(2 - t)$

$4.5t = 3(2-t)$

$4.5t = 6 - 3t$

$7.5t = 6$

$t = 0.8\ hr$

$d = 4.5t = 4.5(0.8)$

$d = 3.6\ mi$

The trail is 3.6 miles long.

Section 11.1 Solving Quadratic Equations by Factoring or Taking Square Roots

1. $2x^2 - 3x + 9 = 0$ is a **quadratic** equation.

3. To solve the equation $3x^2 - 26x = 9$ by factoring, first write the equation in **standard** form.

5. $a = 3, b = -4, c = 1$

7. $a = 2, b = 0, c = -5$

9. $a = 6, b = -3, c = 0$

11. Standard form: $x^2 - 3x - 8 = 0$

13. Standard form: $x^2 - 16 = 0$

15. Standard form: $2x^2 + 12x + 13 = 0$

17. A quadratic equation is an equation that can be written in the form $ax^2 + bx + c = 0, a \neq 0$. A linear equation has no term with a degree greater than 1.

19. The proposed solution is modeled after the Principle of Zero Products. However, zero is the only number for which the principle holds.

21. $x^2 - 36 = 0$
$(x+6)(x-6) = 0$
$x+6=0 \qquad x-6=0$
$x=-6 \qquad x=6$

23. $(x+3)(x-5)=0$
$x+3=0 \qquad x-5=0$
$x=-3 \qquad x=5$

25. $x(x-7)=0$
$x=0 \qquad x-7=0$
$x=7$

27. $(2x+5)(3x-1)=0$
$2x+5=0 \qquad 3x-1=0$
$x=-\frac{5}{2} \qquad x=\frac{1}{3}$

29. $x^2+2x-15=0$
$(x+5)(x-3)=0$
$x+5=0 \qquad x-3=0$
$x=-5 \qquad x=3$

31. $z^2-4z+3=0$
$(z-3)(z-1)=0$
$z-3=0 \qquad z-1=0$
$z=3 \qquad z=1$

33. $p^2+3p+2=0$
$(p+2)(p+1)=0$
$p+2=0 \qquad p+1=0$
$p=-2 \qquad p=-1$

35. $x^2-6x+9=0$
$(x-3)(x-3)=0$
$x-3=0 \qquad x-3=0$
$x=3 \qquad x=3$

37. $6x^2-9x=0$
$3x(2x-3)=0$
$3x=0 \qquad 2x-3=0$
$x=0 \qquad x=\frac{3}{2}$

39. $r^2-10=3r$
$r^2-3r-10=0$
$(r-5)(r+2)=0$
$r-5=0 \qquad r+2=0$
$r=5 \qquad r=-2$

41. $3v^2-5v+2=0$
$(3v-2)(v-1)=0$
$3v-2=0 \qquad v-1=0$
$v=\frac{2}{3} \qquad v=1$

43. $3s^2+8s=3$
$3s^2+8s-3=0$
$(3s-1)(s+3)=0$
$3s-1=0 \qquad s+3=0$
$s=\frac{1}{3} \qquad s=-3$

45. $6r^2=12-r$
$6r^2+r-12=0$
$(2r+3)(3r-4)=0$
$2r+3=0 \qquad 3r-4=0$
$r=-\frac{3}{2} \qquad r=\frac{4}{3}$

47. $5y^2+11y=12$
$5y^2+11y-12=0$
$(y+3)(5y-4)=0$
$y+3=0 \qquad 5y-4=0$
$y=-3 \qquad y=\frac{4}{5}$

49. $9s^2-6s+1=0$
$(3s-1)(3s-1)=0$
$3s-1=0 \qquad 3s-1=0$
$s=\frac{1}{3} \qquad s=\frac{1}{3}$

51. $t^2-16=0$
$(t+4)(t-4)=0$
$t+4=0 \qquad t-4=0$
$t=-4 \qquad t=4$

53. $9z^2-4=0$
$(3z-2)(3z+2)=0$
$3z-2=0 \qquad 3z+2=0$
$z=\frac{2}{3} \qquad z=-\frac{2}{3}$

55. $\frac{3x^2}{2}=4x-2$
$2\cdot\frac{3x^2}{2}=2(4x-2)$
$3x^2=8x-4$
$3x^2-8x+4=0$
$(3x-2)(x-2)=0$
$3x-2=0 \qquad x-2=0$
$x=\frac{2}{3} \qquad x=2$

57. $\frac{2x^2}{9}+x=2$
$9\left(\frac{2x^2}{9}+x\right)=9\cdot 2$
$2x^2+9x-18=0$
$(x+6)(2x-3)=0$
$x+6=0 \qquad 2x-3=0$
$x=-6 \qquad x=\frac{3}{2}$

59. $\frac{3}{4}z^2-z=-\frac{1}{3}$
$12\left(\frac{3}{4}z^2-z\right)=12\left(-\frac{1}{3}\right)$
$9z^2-12z+4=0$
$(3z-2)(3z-2)=0$
$3z-2=0 \qquad 3z-2=0$
$z=\frac{2}{3} \qquad z=\frac{2}{3}$

61. $p+18=p(p-2)$
$p+18=p^2-2p$
$p^2-3p-18=0$
$(p-6)(p+3)=0$
$p-6=0 \qquad p+3=0$
$p=6 \qquad p=-3$

63. $s^2+5s-4=(2s+1)(s-4)$
$s^2+5s-4=2s^2-7s-4$
$0=s^2-12s$
$s(s-12)=0$
$s=0 \qquad s-12=0$
$s=12$

65. a. $a > 0, b < 0, c > 0$ tells us that both binomials in the factorization are differences, so there are **two positive** solutions.

b. $a > 0, b < 0, c < 0$ tell us that the binomials are a sum and a difference, so there is **one positive and one negative** solution.

c. $a > 0, b > 0, c < 0$ tell us that the binomials are a sum and a difference, so there is **one positive and one negative** solution.

67. To solve the equation $x^2 - 9 = 0$ by taking square roots, the first step is to add **9** to both sides of the equation.

69. $x^2 = 36$

$\sqrt{x^2} = \sqrt{36}$

$x = \pm 6$

71. $v^2 - 1 = 0$

$v^2 = 1$

$\sqrt{v^2} = \sqrt{1}$

$v = \pm 1$

73. $4x^2 - 49 = 0$

$4x^2 = 49$

$x^2 = \frac{49}{4}$

$\sqrt{x^2} = \sqrt{\frac{49}{4}}$

$x = \pm\frac{7}{2}$

75. $9y^2 = 4$

$y^2 = \frac{4}{9}$

$\sqrt{y^2} = \sqrt{\frac{4}{9}}$

$y = \pm\frac{2}{3}$

77. $16v^2 - 9 = 0$

$16v^2 = 9$

$v^2 = \frac{9}{16}$

$\sqrt{v^2} = \sqrt{\frac{9}{16}}$

$v = \pm\frac{3}{4}$

79. $y^2 - 12 = 0$

$y^2 = 12$

$\sqrt{y^2} = \sqrt{12} = \sqrt{4 \cdot 3}$

$y = \pm 2\sqrt{3}$

81. $w^2 - 24 = 0$

$w^2 = 24$

$\sqrt{w^2} = \sqrt{24} = \sqrt{4 \cdot 6}$

$w = \pm 2\sqrt{6}$

83. $(x-1)^2 = 36$

$\sqrt{(x-1)^2} = \sqrt{36}$

$x - 1 = \pm 6$

$x - 1 = 6 \qquad x - 1 = -6$

$x = 7 \qquad x = -5$

85. $2(x+5)^2 = 8$

$(x+5)^2 = 4$

$\sqrt{(x+5)^2} = \sqrt{4}$

$x + 5 = \pm 2$

$x + 5 = 2 \qquad x + 5 = -2$

$x = -3 \qquad x = -7$

87. $2(x+1)^2 = 50$

$(x+1)^2 = 25$

$\sqrt{(x+1)^2} = \sqrt{25}$

$x + 1 = \pm 5$

$x + 1 = 5 \qquad x + 1 = -5$

$x = 4 \qquad x = -6$

89. $4(x+5)^2 = 64$

$(x+5)^2 = 16$

$\sqrt{(x+5)^2} = \sqrt{16}$

$x+5 = \pm 4$

$x+5 = 4 \qquad x+5 = -4$

$x = -1 \qquad x = -9$

91. $12(x+3)^2 = 27$

$(x+3)^2 = \frac{27}{12} = \frac{9}{4}$

$\sqrt{(x+3)^2} = \sqrt{\frac{9}{4}}$

$x+3 = \pm\frac{3}{2}$

$x+3 = \frac{3}{2} \qquad x+3 = -\frac{3}{2}$

$x = -\frac{3}{2} \qquad x = -\frac{9}{2}$

93. $9(x-1)^2 - 16 = 0$

$9(x-1)^2 = 16$

$(x-1)^2 = \frac{16}{9}$

$\sqrt{(x-1)^2} = \sqrt{\frac{16}{9}}$

$x-1 = \pm\frac{4}{3}$

$x-1 = \frac{4}{3} \qquad x-1 = -\frac{4}{3}$

$x = \frac{7}{3} \qquad x = -\frac{1}{3}$

95. $(x-4)^2 - 20 = 0$

$(x-4)^2 = 20$

$\sqrt{(x-4)^2} = \sqrt{20} = \sqrt{4 \cdot 5 \cdot}$

$x-4 = \pm\, 2\sqrt{5}$

$x = 4 \pm 2\sqrt{5}$

97. $(x+1)^2 - 18 = 0$

$(x+1)^2 = 18$

$\sqrt{(x+1)^2} = \sqrt{18} = \sqrt{9 \cdot 2}$

$x+1 = 3\sqrt{2}$

$x = -1 \pm 3\sqrt{2}$

99. $2\left(z - \frac{1}{2}\right)^2 = 12$

$\left(z - \frac{1}{2}\right)^2 = 6$

$\sqrt{\left(z - \frac{1}{2}\right)^2} = \sqrt{6}$

$z - \frac{1}{2} = \pm\sqrt{6}$

$z = \frac{1}{2} \pm \sqrt{6}$

101. $(x+a)^2 = 0$

$\sqrt{(x+a)^2} = \sqrt{0}$

$x + a = 0$

$x = -a$

One solution.

103. $ax^2 - b = 0$

$ax^2 = b$

$x^2 = \frac{b}{a}$

$\sqrt{x^2} = \sqrt{\frac{b}{a}}$

$x = \pm\sqrt{\frac{b}{a}}$

Two solutions.

105. $n(n-2)=15$

$n^2-2n-15=0$

$(n-5)(n+3)=0$

$n-5=0 \qquad n+3=0$

$n=5 \qquad n=-3$

$$\begin{aligned}2n^2-7n-4&=2(5^2)-7(5)-4\\&=2(25)-35-4\\&=50-35-4=11\end{aligned}$$

$$\begin{aligned}2n^2-7n-4&=2(-3)^2-7(-3)-4\\&=2(9)+21-4\\&=18+21-4=35\end{aligned}$$

107. If 3 is a solution, then $x-3$ is a factor of the quadratic. $2x^2-5x+c=(x-3)(2x+1)$ so the other factor is $2x+1$. Setting this second factor equal to 0 yields $x=-\frac{1}{2}$.

109. $N=\frac{t^2-t}{2}$

$21=\frac{t^2-t}{2}$

$42=t^2-t$

$0=t^2-t-42$

$0=(t-7)(t+6)$

$t-7=0 \qquad t+6=0$

$t=7 \qquad t=-6$

There are 7 teams in the league.

111. $E=\frac{1}{2}mv^2$

$250=\frac{1}{2}5v^2$

$500=5v^2$

$100=v^2$

$\sqrt{100}=\sqrt{v^2}$

$\pm 10=v$

$v=10m/s$

The speed is 10 m/s.

Section 11.2 Solving Quadratic Equations by Completing the Square

1. When we square a binomial, the result is a perfect square trinomial.

3. To "complete the square" means to add to x^2+bx the constant term that will make it a perfect-square trinomial.

5. $\left[\frac{1}{2}(18)\right]^2=9^2=81$

7. $\left[\frac{1}{2}(12)\right]^2=6^2=36$

$x^2+12x+36=(x+6)^2$

9. $\left[\frac{1}{2}(10)\right]^2=5^2=25$

$x^2+10x+25=(x+5)^2$

11. $\left[\frac{1}{2}(-1)\right]^2=\left(\frac{1}{2}\right)^2=\frac{1}{4}$

$x^2-x+\frac{1}{4}=\left(x-\frac{1}{2}\right)^2$

13. $x^2+2x-3=0$

$x^2+2x+1=3+1$

$(x+1)^2=4$

$\sqrt{(x+1)^2}=\sqrt{4}$

$x+1=\pm 2$

$x+1=2 \qquad x+1=-2$

$x=1 \qquad x=-3$

15. $v^2+4v+1=0$

$v^2+4v+4=-1+4$

$(v+2)^2=3$

$\sqrt{(v+2)^2}=\sqrt{3}$

$v+2=\pm\sqrt{3}$

$v=-2\pm\sqrt{3}$

17. $x^2=4x-4$

$x^2-4x=-4$

$x^2-4x+4=-4+4$

$x^2-4x+4=0$

$(x-2)^2=0$

$\sqrt{(x-2)^2}=\sqrt{0}$

$x-2=0$

$x=2$

19. $z^2=2z+1$

$z^2-2z=1$

$z^2-2z+1=1+1$

$(z-1)^2=2$

$\sqrt{(z-1)^2}=\sqrt{2}$

$z-1=\pm\sqrt{2}$

$z=1\pm\sqrt{2}$

21. $p^2+3p-1=0$ *does not factor*

$p^2+3p=1$

$p^2+3p+\frac{9}{4}=1+\frac{9}{4}$

$\left(p+\frac{3}{2}\right)^2=\frac{13}{4}$

$\sqrt{\left(p+\frac{3}{2}\right)^2}=\sqrt{\frac{13}{4}}$

$p+\frac{3}{2}=\pm\frac{\sqrt{13}}{2}$

$p=-\frac{3}{2}\pm\frac{\sqrt{13}}{2}=\frac{-3\pm\sqrt{13}}{2}$

23. $w^2+7w=8$

$w^2+7w-8=0$

$(w+8)(w-1)=0$

$w+8=0 \quad w-1=0$

$w=-8 \quad w=1$

25. $x^2+6x+4=0$ *does not factor*

$x^2+6x+9=-4+9$

$(x+3)^2=5$

$\sqrt{(x+3)^2}=\sqrt{5}$

$x+3=\pm\sqrt{5}$

$x=-3\pm\sqrt{5}$

27. $r^2-8r+2=0$ *does not factor*

$r^2-8r=-2$

$r^2-8r+16=-2+16$

$(r-4)^2=14$

$\sqrt{(r-4)^2}=\sqrt{14}$

$r-4=\pm\sqrt{14}$

$r=4\pm\sqrt{14}$

29. $t^2-3t=-2$

$t^2-3t+2=0$

$(t-2)(t-1)=0$

$t-2=0 \quad t-1=0$

$t=2 \quad t=1$

31. $w^2-3w-5=0$ *does not factor*

$w^2-3w=5$

$w^2-3w+\frac{9}{4}=5+\frac{9}{4}$

$\left(w-\frac{3}{2}\right)^2=\frac{29}{4}$

$\sqrt{\left(w-\frac{3}{2}\right)^2}=\sqrt{\frac{29}{4}}$

$w-\frac{3}{2}=\pm\frac{\sqrt{29}}{2}$

$w=\frac{3}{2}\pm\frac{\sqrt{29}}{2}=\frac{3\pm\sqrt{29}}{2}$

33. $x^2 - x - 1 = 0$ *does not factor*

$$x^2 - x = 1$$
$$x^2 - x + \frac{1}{4} = 1 + \frac{1}{4}$$
$$\left(x - \frac{1}{2}\right)^2 = \frac{5}{4}$$
$$\sqrt{\left(x - \frac{1}{2}\right)^2} = \sqrt{\frac{5}{4}}$$
$$x - \frac{1}{2} = \pm\frac{\sqrt{5}}{2}$$
$$x = \frac{1}{2} \pm \frac{\sqrt{5}}{2} = \frac{1 \pm \sqrt{5}}{2}$$

35. $y^2 + 10y - 7 = 0$ *does not factor*

$$y^2 + 10y + 25 = 7 + 25$$
$$(y+5)^2 = 32$$
$$\sqrt{(y+5)^2} = \sqrt{32} = \sqrt{16 \cdot 2}$$
$$y + 5 = \pm 4\sqrt{2}$$
$$y = -5 \pm 4\sqrt{2}$$

37. $s^2 + 3s + 1 = 0$ *does not factor*

$$s^2 + 3s + \frac{9}{4} = -1 + \frac{9}{4}$$
$$\left(s + \frac{3}{2}\right)^2 = \frac{5}{4}$$
$$\sqrt{\left(s + \frac{3}{2}\right)^2} = \sqrt{\frac{5}{4}}$$
$$s + \frac{3}{2} = \pm\frac{\sqrt{5}}{2}$$
$$s = -\frac{3}{2} \pm \frac{\sqrt{5}}{2} = \frac{-3 \pm \sqrt{5}}{2}$$

39. $t^2 - t - 4 = 0$ *does not factor*

$$t^2 - t + \frac{1}{4} = 4 + \frac{1}{4}$$
$$\left(t - \frac{1}{2}\right)^2 = \frac{17}{4}$$
$$\sqrt{\left(t - \frac{1}{2}\right)^2} = \sqrt{\frac{17}{4}}$$
$$t - \frac{1}{2} = \pm\frac{\sqrt{17}}{2}$$
$$t = \frac{1}{2} \pm \frac{\sqrt{17}}{2} = \frac{1 \pm \sqrt{17}}{2}$$

41. $2t^2 - 3t + 1 = 0$

$$(t-1)(2t-1) = 0$$
$$t - 1 = 0 \qquad 2t - 1 = 0$$
$$t = 1 \qquad t = \frac{1}{2}$$

43. $2r^2 + 5r = 3$

$$2r^2 + 5r - 3 = 0$$
$$(2r-1)(r+3) = 0$$
$$2r - 1 = 0 \qquad r + 3 = 0$$
$$r = \frac{1}{2} \qquad r = -3$$

45. $4v^2 - 4v - 1 = 0$ *does not factor*

$$\frac{1}{4}(4v^2 - 4v) = \frac{1}{4} \cdot 1$$
$$v^2 - v + \frac{1}{4} = \frac{1}{4} + \frac{1}{4}$$
$$\left(v - \frac{1}{2}\right)^2 = \frac{1}{2}$$
$$\sqrt{\left(v - \frac{1}{2}\right)^2} = \sqrt{\frac{1}{2}}$$
$$v - \frac{1}{2} = \pm\frac{1}{\sqrt{2}} = \pm\frac{\sqrt{2}}{2}$$
$$v = \frac{1}{2} \pm \frac{\sqrt{2}}{2} = \frac{1 \pm \sqrt{2}}{2}$$

47. $4z^2 - 8z = 1$

$\frac{1}{4}(4z^2 - 8z) = \frac{1}{4}(1)$

$z^2 - 2z + 1 = \frac{1}{4} + 1$

$(z-1)^2 = \frac{5}{4}$

$\sqrt{(z-1)^2} = \sqrt{\frac{5}{4}}$

$z - 1 = \frac{\pm\sqrt{5}}{2}$

$z = 1 \pm \frac{\sqrt{5}}{2} = \frac{2 \pm \sqrt{5}}{2}$

49. $3y - 5 = (y-1)(y-2)$

$3y - 5 = y^2 - 3y + 2$

$y^2 - 6y + 7 = 0$

$y^2 - 6y + 9 = -7 + 9$

$(y-3)^2 = 2$

$\sqrt{(y-3)^2} = \sqrt{2}$

$y - 3 = \pm\sqrt{2}$

$y = 3 \pm \sqrt{2}$

51. $\frac{x^2}{4} - \frac{x}{2} = 3$

$4\left(\frac{x^2}{4} - \frac{x}{2}\right) = 4(3)$

$x^2 - 2x + 1 = 12 + 1$

$(x-1)^2 = 13$

$\sqrt{(x-1)^2} = \sqrt{13}$

$x - 1 = \pm\sqrt{13}$

$x = 1 \pm \sqrt{13}$

53. $\frac{2x^2}{3} = 2x + 3$

$\frac{2x^2}{3} - 2x = 3$

$\frac{3}{2}\left(\frac{2x^2}{3} - 2x\right) = \frac{3}{2}(3)$

$x^2 - 3x + \frac{9}{4} = \frac{9}{2} + \frac{9}{4}$

$\left(x - \frac{3}{2}\right)^2 = \frac{27}{4}$

$\sqrt{\left(x - \frac{3}{2}\right)^2} = \sqrt{\frac{27}{4}}$

$x - \frac{3}{2} = \frac{\pm 3\sqrt{3}}{2}$

$x = \frac{3}{2} \pm \frac{3\sqrt{3}}{2} = \frac{3 \pm 3\sqrt{3}}{2}$

55. $\frac{x}{3} + \frac{3}{x} = \frac{8}{3}$

$3x\left(\frac{x}{3} + \frac{3}{x}\right) = 3x\left(\frac{8}{3}\right)$

$x^2 + 9 = 8x$

$x^2 - 8x + 16 = -9 + 16$

$(x-4)^2 = 7$

$\sqrt{(x-4)^2} = \sqrt{7}$

$x - 4 = \pm\sqrt{7}$

$x = 4 \pm \sqrt{7}$

57. Since $\sqrt{5}$ is less than 3, both solutions will be negative.

59. $y^2+3y=5$

$y^2+3y+\frac{9}{4}=5+\frac{9}{4}$

$\left(y+\frac{3}{2}\right)^2=\frac{29}{4}$

$\sqrt{\left(y+\frac{3}{2}\right)^2}=\sqrt{\frac{29}{4}}$

$y+\frac{3}{2}=\pm\frac{\sqrt{29}}{2}$

$y=-\frac{3}{2}\pm\frac{\sqrt{29}}{2}$

$y=-1.5\pm 2.693$

$y=-1.5+2.693 \qquad y=-1.5-2.693$

$y\approx 1.193 \qquad y\approx -4.193$

61. $2z^2-3z=7$

$\frac{1}{2}(2z^2-3z)=\frac{1}{2}(7)$

$z^2-\frac{3}{2}z+\frac{9}{16}=\frac{7}{2}+\frac{9}{16}$

$\left(z-\frac{3}{4}\right)^2=\frac{65}{16}$

$\sqrt{\left(z-\frac{3}{4}\right)^2}=\sqrt{\frac{65}{16}}$

$z-\frac{3}{4}=\pm\frac{\sqrt{65}}{4}$

$z=\frac{3}{4}\pm\frac{\sqrt{65}}{4}=0.75\pm 2.016$

$z=0.75+2.016 \qquad z=0.75-2.016$

$z\approx 2.766 \qquad z\approx -1.266$

63. $4x^2+6x-1=0$

$\frac{1}{4}(4x^2+6x)=\frac{1}{4}(1)$

$x^2+\frac{3}{2}x+\frac{9}{16}=\frac{1}{4}+\frac{9}{16}$

$\left(x+\frac{3}{4}\right)^2=\frac{13}{16}$

$\sqrt{\left(x+\frac{3}{4}\right)^2}=\sqrt{\frac{13}{16}}$

$x+\frac{3}{4}=\pm\frac{\sqrt{13}}{4}$

$x=-\frac{3}{4}\pm\frac{\sqrt{13}}{4}=-0.75\pm 0.901$

$x=-0.75+0.901 \qquad x=-0.75-0.901$

$x\approx 0.151 \qquad x\approx -1.651$

65. If $(x+6)^2=9$, then $x + 6$ must be equal to 3 or -3.

67. $b^2-6b+7=0$

$b^2-6b+9=-7+9$

$(b-3)^2=2$

$\sqrt{(b-3)^2}=\sqrt{2}$

$b-3=\pm\sqrt{2}$

$b=3\pm\sqrt{2}$

$2b^2=2\left(3\pm\sqrt{2}\right)^2$

$=2(9\pm 6\sqrt{2}+2)$

$=2(11\pm 6\sqrt{2})$

$=22\pm 12\sqrt{2}$

69. $\sqrt{2x+7}-4=x$

$\sqrt{2x+7}=x+4$

$\left(\sqrt{2x+7}\right)^2=(x+4)^2$

$2x+7=x^2+8x+16$

$0=x^2+6x+9$

$0=(x+3)^2$

$x+3=0$

$x=-3$

71. $\frac{x-2}{3}+\frac{2}{x+2}=4$

$3(x+2)\left(\frac{x-2}{3}+\frac{2}{x+2}\right)=3(x+2)4$

$(x+2)(x-2)+6=12(x+2)$

$x^2-4+6=12x+24$

$x^2-12x+36=22+36$

$(x-6)^2=58$

$\sqrt{(x-6)^2}=\sqrt{58}$

$x-6=\pm\sqrt{58}$

$x=6\pm\sqrt{58}$

73. $x^2-2x-11=0$

$x^2-2x+1=11+1$

$(x-1)^2=12$

$\sqrt{(x-1)^2}=\sqrt{12}=\sqrt{4\cdot 3}$

$x-1=\pm 2\sqrt{3}$

$x=1\pm 2\sqrt{3}$

$a=1+2\sqrt{3}$

$a^2=(1+2\sqrt{3})^2=1+4\sqrt{3}+12$

$a^2=13+4\sqrt{3}$

$b=1-2\sqrt{3}$

$b^2=(1-2\sqrt{3})^2=1-4\sqrt{3}+12$

$a^2+b^2=1+4\sqrt{3}+12+1-4\sqrt{3}+12$

$a^2+b^2=26$

Section 11.3 Solving Quadratic equations by Using the Quadratic Formula

1. The solutions are $\frac{1+\sqrt{13}}{2}$ and $\frac{1-\sqrt{13}}{2}$.

3. True

5. False. Standard form is $4x^2-3x-9=0$.

7. The quadratic formula is $x=\frac{-b\pm\sqrt{b^2-4ac}}{2a}$, where *a*, *b*, *c* and *x* are from the standard form of a quadratic equation, $ax^2+bx+c=0$. *a* is the coefficient of x^2, *b* is the coefficient of *x*, *c* is the constant, and *x* is the unknown.

9. To write the equation $x^2=4x+2$ in standard form, subtract 4*x* and 2 from each side of the equation. The resulting equation is $x^2-4x-2=0$. Then $a=1$, $b=-4$, and $c=-2$.

11. $z^2+6z-7=0$

$a=1 \quad b=6 \quad c=-7$

$z=\frac{-b\pm\sqrt{b^2-4ac}}{2a}$

$z=\frac{-6\pm\sqrt{6^2-4(1)(-7)}}{2(1)}$

$z=\frac{-6\pm\sqrt{36+28}}{2}=\frac{-6\pm\sqrt{64}}{2}$

$z=\frac{-6\pm 8}{2}$

$z=\frac{-6+8}{2} \qquad z=\frac{-6-8}{2}$

$z=1 \qquad z=-7$

13. $w^2=3w+18$

$w^2-3w-18=0$

$a=1 \quad b=-3 \quad c=-18$

$w=\frac{-b\pm\sqrt{b^2-4ac}}{2a}$

$w=\frac{-(-3)\pm\sqrt{(-3)^2-4(1)(-18)}}{2(1)}$

$w=\frac{3\pm\sqrt{9+72}}{2}=\frac{3\pm\sqrt{81}}{2}$

$w=\frac{3\pm 9}{2}$

$w=\frac{3+9}{2} \qquad w=\frac{3-9}{2}$

$w=6 \qquad w=-3$

15. $t^2 - 2t = 5$

$t^2 - 2t - 5 = 0$

$a = 1 \quad b = -2 \quad c = -5$

$t = \dfrac{-b \pm \sqrt{b^2 - 4ac}}{2a}$

$t = \dfrac{-(-2) \pm \sqrt{(-2)^2 - 4(1)(-5)}}{2(1)}$

$t = \dfrac{2 \pm \sqrt{4+20}}{2} = \dfrac{2 \pm \sqrt{24}}{2}$

$t = \dfrac{2 \pm 2\sqrt{6}}{2} = 1 \pm \sqrt{6}$

17. $t^2 + 6t - 1 = 0$

$a = 1 \quad b = 6 \quad c = -1$

$t = \dfrac{-b \pm \sqrt{b^2 - 4ac}}{2a}$

$t = \dfrac{-6 \pm \sqrt{6^2 - 4(1)(-1)}}{2(1)}$

$t = \dfrac{-6 \pm \sqrt{36+4}}{2} = \dfrac{-6 \pm \sqrt{40}}{2}$

$t = \dfrac{-6 \pm 2\sqrt{10}}{2} = -3 \pm \sqrt{10}$

19. $w^2 + 3w - 5 = 0$

$a = 1 \quad b = 3 \quad c = -5$

$w = \dfrac{-b \pm \sqrt{b^2 - 4ac}}{2a}$

$w = \dfrac{-3 \pm \sqrt{3^2 - 4(1)(-5)}}{2(1)}$

$w = \dfrac{-3 \pm \sqrt{9+20}}{2}$

$w = \dfrac{-3 \pm \sqrt{29}}{2}$

21. $w^2 = 4w + 9$

$w^2 - 4w - 9 = 0$

$a = 1 \quad b = -4 \quad c = -9$

$w = \dfrac{-b \pm \sqrt{b^2 - 4ac}}{2a}$

$w = \dfrac{-(-4) \pm \sqrt{(-4)^2 - 4(1)(-9)}}{2(1)}$

$w = \dfrac{4 \pm \sqrt{16+36}}{2} = \dfrac{4 \pm \sqrt{52}}{2}$

$z = \dfrac{4 \pm 2\sqrt{13}}{2} = 2 \pm \sqrt{13}$

23. $p^2 - p = 0$

$p(p-1) = 0$

$p = 0 \qquad p - 1 = 0$

$p = 1$

25. $4t^2 - 4t - 1 = 0$

$a = 4 \quad b = -4 \quad c = -1$

$t = \dfrac{-b \pm \sqrt{b^2 - 4ac}}{2a}$

$t = \dfrac{-(-4) \pm \sqrt{(-4)^2 - 4(4)(-1)}}{2(4)}$

$t = \dfrac{4 \pm \sqrt{16+16}}{8} = \dfrac{4 \pm \sqrt{32}}{8}$

$t = \dfrac{4 \pm 4\sqrt{2}}{8} = \dfrac{1 \pm \sqrt{2}}{2}$

27. $4t^2 - 9 = 0$

$(2t-3)(2t+3) = 0$

$2t - 3 = 0 \qquad 2t + 3 = 0$

$t = \dfrac{3}{2} \qquad t = -\dfrac{3}{2}$

29. $3x^2 - 6x + 2 = 0$

$a = 3 \quad b = -6 \quad c = 2$

$$x = \frac{-b \pm \sqrt{b^2 - 4ac}}{2a}$$

$$x = \frac{-(-6) \pm \sqrt{(-6)^2 - 4(3)(2)}}{2(3)}$$

$$x = \frac{6 \pm \sqrt{36 - 24}}{6} = \frac{6 \pm \sqrt{12}}{6}$$

$$x = \frac{6 \pm \sqrt{4 \cdot 3}}{6} = \frac{6 \pm 2\sqrt{3}}{6}$$

$$x = \frac{2(3 \pm \sqrt{3})}{6} = \frac{3 \pm \sqrt{3}}{3}$$

31. $3t^2 = 2t + 3$

$3t^2 - 2t - 3 = 0$

$a = 3 \quad b = -2 \quad c = -3$

$$t = \frac{-b \pm \sqrt{b^2 - 4ac}}{2a}$$

$$t = \frac{-(-2) \pm \sqrt{(-2)^2 - 4(3)(-3)}}{2(3)}$$

$$t = \frac{2 \pm \sqrt{4 + 36}}{6} = \frac{2 \pm \sqrt{40}}{6}$$

$$t = \frac{2 \pm \sqrt{4 \cdot 10}}{6} = \frac{2 \pm 2\sqrt{10}}{6}$$

$$t = \frac{2(1 \pm \sqrt{10})}{6} = \frac{1 \pm \sqrt{10}}{3}$$

33. $2y^2 + 3 = 8y$

$2y^2 - 8y + 3 = 0$

$a = 2 \quad b = -8 \quad c = 3$

$$y = \frac{-b \pm \sqrt{b^2 - 4ac}}{2a}$$

$$y = \frac{-(-8) \pm \sqrt{(-8)^2 - 4(2)(3)}}{2(2)}$$

$$y = \frac{8 \pm \sqrt{64 - 24}}{4} = \frac{8 \pm \sqrt{40}}{4}$$

$$y = \frac{8 \pm \sqrt{4 \cdot 10}}{4} = \frac{8 \pm 2\sqrt{10}}{4}$$

$$y = \frac{2(4 \pm \sqrt{10})}{4} = \frac{4 \pm \sqrt{10}}{2}$$

35. $3t^2 = 7t + 6$

$3t^2 - 7t - 6 = 0$

$(3t + 2)(t - 3) = 0$

$3t + 2 = 0 \qquad t - 3 = 0$

$t = -\frac{2}{3} \qquad t = 3$

37. $3y^2 - 4 = 5y$

$3y^2 - 5y - 4 = 0$

$a = 3 \quad b = -5 \quad c = -4$

$$y = \frac{-b \pm \sqrt{b^2 - 4ac}}{2a}$$

$$y = \frac{-(-5) \pm \sqrt{(-5)^2 - 4(3)(-4)}}{2(3)}$$

$$y = \frac{5 \pm \sqrt{25 + 48}}{6} = \frac{5 \pm \sqrt{73}}{6}$$

39. $3x^2 = x + 3$

$3x^2 - x - 3 = 0$

$a = 3 \quad b = -1 \quad c = -3$

$$x = \frac{-b \pm \sqrt{b^2 - 4ac}}{2a}$$

$$x = \frac{-(-1) \pm \sqrt{(-1)^2 - 4(3)(-3)}}{2(3)}$$

$$x = \frac{1 \pm \sqrt{1 + 36}}{6} = \frac{1 \pm \sqrt{37}}{6}$$

41. $5d^2 - 2d - 8 = 0$

$a = 5 \quad b = -2 \quad c = -8$

$$d = \frac{-b \pm \sqrt{b^2 - 4ac}}{2a}$$

$$d = \frac{-(-2) \pm \sqrt{(-2)^2 - 4(5)(-8)}}{2(5)}$$

$$d = \frac{2 \pm \sqrt{4 + 160}}{10} = \frac{2 \pm \sqrt{164}}{10}$$

$$d = \frac{2 \pm \sqrt{4 \cdot 41}}{10} = \frac{2 \pm 2\sqrt{41}}{10}$$

$$d = \frac{2(1 \pm \sqrt{41})}{10} = \frac{1 \pm \sqrt{41}}{5}$$

43. $5z^2+11z=12$

$5z^2+11-12=0$

$(5z-4)(z+3)=0$

$5z-4=0 \qquad z+3=0$

$z=\dfrac{4}{5} \qquad z=-3$

45. $v^2+6v+1=0$

$a=1 \quad b=6 \quad c=1$

$$v=\frac{-b\pm\sqrt{b^2-4ac}}{2a}$$

$$v=\frac{-6\pm\sqrt{6^2-4(1)(1)}}{2(1)}$$

$$v=\frac{-6\pm\sqrt{36-4}}{2}=\frac{-6\pm\sqrt{32}}{2}$$

$$v=\frac{-6\pm\sqrt{16\cdot 2}}{2}=\frac{-6\pm 4\sqrt{2}}{2}$$

$$v=\frac{2(-3\pm 2\sqrt{2})}{2}=-3\pm 2\sqrt{2}$$

47. $4t^2-12t-15=0$

$a=4 \quad b=-12 \; c=-15$

$$t=\frac{-b\pm\sqrt{b^2-4ac}}{2a}$$

$$t=\frac{-(-12)\pm\sqrt{(-12)^2-4(4)(-15)}}{2(4)}$$

$$t=\frac{12\pm\sqrt{144+240}}{8}=\frac{12\pm\sqrt{384}}{8}$$

$$t=\frac{12\pm\sqrt{64\cdot 6}}{8}=\frac{12\pm 8\sqrt{6}}{8}$$

$$t=\frac{4(3\pm 2\sqrt{6})}{8}=\frac{3\pm 2\sqrt{6}}{2}$$

49. $9y^2+6y-1=0$

$a=9 \quad b=6 \quad c=-1$

$$y=\frac{-b\pm\sqrt{b^2-4ac}}{2a}$$

$$y=\frac{-6\pm\sqrt{6^2-4(9)(-1)}}{2(9)}$$

$$y=\frac{-6\pm\sqrt{36+36}}{18}=\frac{-6\pm\sqrt{72}}{18}$$

$$y=\frac{-6\pm\sqrt{36\cdot 2}}{18}=\frac{-6\pm 6\sqrt{2}}{18}$$

$$y=\frac{6(-1\pm\sqrt{2})}{18}=\frac{-1\pm\sqrt{2}}{3}$$

51. $6s^2-s-2=0$

$(2s+1)(3s-2)=0$

$2s+1=0 \qquad 3s-2=0$

$s=-\dfrac{1}{2} \qquad s=\dfrac{2}{3}$

53. $4p^2+16p=-11$

$4p^2+16p+11=0$

$a=4 \quad b=16 \quad c=11$

$$p=\frac{-b\pm\sqrt{b^2-4ac}}{2a}$$

$$p=\frac{-16\pm\sqrt{16^2-4(4)(11)}}{2(4)}$$

$$p=\frac{-16\pm\sqrt{256-176}}{8}$$

$$p=\frac{-16\pm\sqrt{80}}{8}=\frac{-16\pm\sqrt{16\cdot 5}}{8}$$

$$p=\frac{-16\pm 4\sqrt{5}}{8}=\frac{4\left(-4\pm\sqrt{5}\right)}{8}$$

$$p=\frac{-4\pm\sqrt{5}}{2}$$

55. $4x^2 = 4x + 11$

$4x^2 - 4x - 11 = 0$

$a = 4 \quad b = -4 \quad c = -11$

$x = \frac{-b \pm \sqrt{b^2 - 4ac}}{2a}$

$x = \frac{-(-4) \pm \sqrt{(-4)^2 - 4(4)(-11)}}{2(4)}$

$x = \frac{4 \pm \sqrt{16 + 176}}{8} = \frac{4 \pm \sqrt{192}}{8}$

$x = \frac{4 \pm \sqrt{64 \cdot 3}}{8} = \frac{4 \pm 8\sqrt{3}}{8}$

$x = \frac{4(1 \pm 2\sqrt{3})}{8} = \frac{1 \pm 2\sqrt{3}}{2}$

57. $9v^2 = -30v - 23$

$9v^2 + 30v + 23 = 0$

$a = 9 \quad b = 30 \quad c = 23$

$v = \frac{-b \pm \sqrt{b^2 - 4ac}}{2a}$

$v = \frac{-30 \pm \sqrt{30^2 - 4(9)(23)}}{2(9)}$

$v = \frac{-30 \pm \sqrt{900 - 828}}{18} = \frac{-30 \pm \sqrt{72}}{18}$

$v = \frac{-30 \pm \sqrt{36 \cdot 2}}{18} = \frac{-30 \pm 6\sqrt{2}}{18}$

$v = \frac{6(-5 \pm \sqrt{2})}{18} = \frac{-5 \pm \sqrt{2}}{3}$

59. $\frac{x^2}{2} - \frac{x}{3} = 1$

$6\left(\frac{x^2}{2} - \frac{x}{3}\right) = 6(1)$

$3x^2 - 2x - 6 = 0$

$a = 3 \quad b = -2 \quad c = -6$

$x = \frac{-b \pm \sqrt{b^2 - 4ac}}{2a}$

$x = \frac{-(-2) \pm \sqrt{(-2)^2 - 4(3)(-6)}}{2(3)}$

$x = \frac{2 \pm \sqrt{4 + 72}}{6} = \frac{2 \pm \sqrt{76}}{6}$

$x = \frac{2 \pm \sqrt{4 \cdot 19}}{6} = \frac{2 \pm 2\sqrt{19}}{6}$

$x = \frac{2(1 \pm \sqrt{19})}{6} = \frac{1 \pm \sqrt{19}}{3}$

61. $\frac{2x^2}{5} = x + 1$

$5 \cdot \frac{2x^2}{5} = 5(x + 1)$

$2x^2 - 5x - 5 = 0$

$a = 2 \quad b = -5 \quad c = -5$

$x = \frac{-b \pm \sqrt{b^2 - 4ac}}{2a}$

$x = \frac{-(-5) \pm \sqrt{(-5)^2 - 4(2)(-5)}}{2(2)}$

$x = \frac{5 \pm \sqrt{25 + 40}}{4} = \frac{5 \pm \sqrt{65}}{4}$

63. $\frac{x}{5}+\frac{5}{x}=\frac{12}{5}$

$5x\left(\frac{x}{5}+\frac{5}{x}\right)=5x\left(\frac{12}{5}\right)$

$x^2+25=12x$

$x^2-12x+25=0$

$a=1 \quad b=-12 \quad c=25$

$x=\frac{-b\pm\sqrt{b^2-4ac}}{2a}$

$x=\frac{-(-12)\pm\sqrt{(-12)^2-4(1)(25)}}{2(1)}$

$x=\frac{12\pm\sqrt{144-100}}{2}=\frac{12\pm\sqrt{44}}{2}$

$x=\frac{12\pm\sqrt{4\cdot 11}}{2}=\frac{12\pm 2\sqrt{11}}{2}$

$x=\frac{2\left(6\pm\sqrt{11}\right)}{2}=6\pm\sqrt{11}$

65. True

67. $x^2-2x-21=0$

$a=1 \quad b=-2 \quad c=-21$

$x=\frac{-b\pm\sqrt{b^2-4ac}}{2a}$

$x=\frac{-(-2)\pm\sqrt{(-2)^2-4(1)(-21)}}{2(1)}$

$x=\frac{2\pm\sqrt{4+84}}{2}=\frac{2\pm\sqrt{88}}{2}=\frac{2\pm 9.3808}{2}$

$x=\frac{2+9.3808}{2} \qquad x=\frac{2-9.3808}{2}$

$x=\frac{11.3808}{2} \qquad x=\frac{-7.3808}{2}$

$x=5.690 \qquad x=-3.690$

69. $s^2-6s-13=0$

$a=1 \quad b=-6 \quad c=-13$

$s=\frac{-b\pm\sqrt{b^2-4ac}}{2a}$

$s=\frac{-(-6)\pm\sqrt{(-6)^2-4(1)(-13)}}{2(1)}$

$s=\frac{6\pm\sqrt{36+52}}{2}=\frac{6\pm\sqrt{88}}{2}=\frac{6\pm 9.3808}{2}$

$s=\frac{6+9.3808}{2} \qquad s=\frac{6-9.3808}{2}$

$s=\frac{15.3808}{2} \qquad s=\frac{-3.3808}{2}$

$s=7.690 \qquad s=-1.690$

71. $2p^2-7p-10=0$

$a=2 \quad b=-7 \quad c=-10$

$p=\frac{-b\pm\sqrt{b^2-4ac}}{2a}$

$p=\frac{-(-7)\pm\sqrt{(-7)^2-4(2)(-10)}}{2(2)}$

$p=\frac{7\pm\sqrt{49+80}}{4}=\frac{7\pm\sqrt{129}}{4}=\frac{7\pm 11.358}{4}$

$p=\frac{7+11.358}{4} \qquad p=\frac{7-11.358}{4}$

$p=\frac{18.358}{4} \qquad p=\frac{-4.358}{4}$

$p=4.589 \qquad p=-1.089$

73. $4z^2+8z-1=0$

$a=4 \quad b=8 \quad c=-1$

$z=\frac{-b\pm\sqrt{b^2-4ac}}{2a}$

$z=\frac{-8\pm\sqrt{8^2-4(4)(-1)}}{2(4)}=\frac{-8\pm\sqrt{64+16}}{8}$

$z=\frac{-8\pm\sqrt{80}}{8}=\frac{-8\pm 8.944}{8}$

$z=\frac{-8+8.944}{8} \qquad z=\frac{-8-8.944}{8}$

$z=\frac{0.944}{8} \qquad z=\frac{-16.944}{8}$

$z=0.118 \qquad z=-2.118$

75. $5v^2 - v - 5 = 0$

$a = 5 \quad b = -1 \quad c = -5$

$$v = \frac{-b \pm \sqrt{b^2 - 4ac}}{2a}$$

$$v = \frac{-(-1) \pm \sqrt{(-1)^2 - 4(5)(-5)}}{2(5)}$$

$$v = \frac{1 \pm \sqrt{1+100}}{10} = \frac{1 \pm \sqrt{101}}{10} = \frac{1 \pm 10.0499}{10}$$

$$v = \frac{1+10.050}{10} \qquad v = \frac{1-10.050}{10}$$

$$v = \frac{11.050}{10} \qquad v = \frac{-9.050}{10}$$

$$v = 1.105 \qquad v = -0.905$$

77. $x^2 - 6x = 14$

$x^2 - 6x - 14 = 0$

$a = 1 \quad b = -6 \quad c = -14$

$$x = \frac{-b \pm \sqrt{b^2 - 4ac}}{2a}$$

$$x = \frac{-(-6) \pm \sqrt{(-6)^2 - 4(1)(-14)}}{2(1)}$$

$$x = \frac{6 \pm \sqrt{36+56}}{2} = \frac{6 \pm \sqrt{92}}{2}$$

$$x = \frac{6 \pm \sqrt{4 \cdot 23}}{2} = \frac{6 \pm 2\sqrt{23}}{2}$$

$$x = \frac{2\left(3 \pm \sqrt{23}\right)}{2} = 3 \pm \sqrt{23}$$

$(3+\sqrt{23}) - (3-\sqrt{23})$

$3 + \sqrt{23} - 3 + \sqrt{23} = 2\sqrt{23}$

79. $\dfrac{x+2}{3} - \dfrac{4}{x-2} = 2$

$$3(x-2)\left(\frac{x+2}{3} - \frac{4}{x-2}\right) = 3(x-2)2$$

$(x-2)(x+2) - 12 = 6(x-2)$

$x^2 - 4 - 12 = 6x - 12$

$x^2 - 6x - 4 = 0$

$a = 1 \quad b = -6 \quad c = -4$

$$x = \frac{-b \pm \sqrt{b^2 - 4ac}}{2a}$$

$$x = \frac{-(-6) \pm \sqrt{(-6)^2 - 4(1)(-4)}}{2(1)}$$

$$x = \frac{6 \pm \sqrt{36+16}}{2} = \frac{6 \pm \sqrt{52}}{2}$$

$$x = \frac{6 \pm \sqrt{4 \cdot 13}}{2} = \frac{6 \pm 2\sqrt{13}}{2}$$

$$x = \frac{2(3 \pm \sqrt{13})}{2} = 3 \pm \sqrt{13}$$

81. A quadratic equation is an equation of the form $ax^2 + bx + c = 0$ where $a \neq 0$. The equation $0x^2 + 3x + 4 = 0$ is not quadratic because $a = 0$. The quadratic formula is used only for solving quadratic equations. Therefore, the equation $0x^2 + 3x + 4 = 0$ cannot be solved by the quadratic formula.

83. Sum of solutions: $(2+\sqrt{7}) + (2-\sqrt{7}) = 4 = -b$

Product of solutions: $(2+\sqrt{7})(2-\sqrt{7}) = 4 - 7 = -3 = c$

Yes

85. Sum of solutions:

$(-4+\sqrt{15})+(-4-\sqrt{15})=-8\neq -b$

Solutions: $x=\dfrac{-b\pm\sqrt{b^2-4ac}}{2a}$

$$x=\frac{-(-8)\pm\sqrt{(-8)^2-4(1)(-14)}}{2(1)}$$

$$x=\frac{8\pm\sqrt{64+56}}{2}=\frac{8\pm\sqrt{120}}{2}$$

$$x=\frac{8\pm\sqrt{4\cdot 30}}{2}=\frac{8\pm 2\sqrt{30}}{2}$$

$$x=4\pm\sqrt{30}$$

Section 11.4 Complex Numbers

1. $i^2=-1$

3. $\sqrt{-7}=i\sqrt{7}$

5. Conjugate: $6-5i$

7. Conjugate: $2+8i$

9. Multiply by: $\dfrac{1+6i}{1+6i}$

11. a. The real part of the complex number $5-8i$ is **5.**

b. The imaginary part of the complex number $5-8i$ is **-8.**

13. $\sqrt{-81}=\sqrt{-1}\sqrt{81}=i\cdot 9=9i$

15. $5\sqrt{-49}=5\sqrt{-1}\sqrt{49}=5i\cdot 7=35i$

17. $\sqrt{-13}=\sqrt{-1}\sqrt{13}=i\sqrt{13}$

19. $\sqrt{-75}=\sqrt{-1}\sqrt{75}=\sqrt{-1}\sqrt{25\cdot 3}$
$=\sqrt{-1}\sqrt{25}\sqrt{3}=5i\sqrt{3}$

21. $-4\sqrt{-54}=-4\sqrt{-1}\sqrt{54}=-4\sqrt{-1}\sqrt{9\cdot 6}$
$=-4\sqrt{-1}\sqrt{9}\sqrt{6}=-4\cdot 3i\sqrt{6}$
$=-12i\sqrt{6}$

23. $a-\sqrt{-b^2}=a-\sqrt{-1}\sqrt{b^2}=a-ib=a-bi$

25. $5+\sqrt{-4}=5+2i$

27. $-7-2\sqrt{-36}=-7-2(6i)=-7-12i$

29. $5+3\sqrt{-48}=5+3\sqrt{-1}\sqrt{16\cdot 3}$
$=5+3i(4)\sqrt{3}=5+12i\sqrt{3}$

31. $7-6\sqrt{-28}=7-6\sqrt{-1}\sqrt{4\cdot 7}$
$=7-6i(2\sqrt{7})=7-12i\sqrt{7}$

33. $\dfrac{6+5\sqrt{-4}}{2}=\dfrac{6+5(2i)}{2}=\dfrac{6+10i}{2}=\dfrac{6}{2}+\dfrac{10i}{2}=3+5i$

35. $\dfrac{-6-2\sqrt{-54}}{6}=\dfrac{-6-2\sqrt{-1}\sqrt{9\cdot 6}}{6}$
$=\dfrac{-6-2i(3\sqrt{6})}{6}=\dfrac{-6-6i\sqrt{6}}{6}$
$=\dfrac{-6}{6}-\dfrac{6i\sqrt{6}}{6}=-1-i\sqrt{6}$

37. $\dfrac{-2-4\sqrt{-90}}{5}=\dfrac{-2-4\sqrt{-1}\sqrt{9\cdot 10}}{5}$
$=\dfrac{-2-4i(3\sqrt{10})}{5}=\dfrac{-2-12i\sqrt{10}}{5}$
$=-\dfrac{2}{5}-\dfrac{12\sqrt{10}}{5}i$

39. $(6-8i)+(-2+3i)=(6+(-2))+(-8+3)i=4-5i$

41. $3i+7i=10i$

43. $-2i - 7i = -9i$

45. $6i + (7 + 3i) = 7 + 9i$

47. $-2i - (8 - 9i) = -2i - 8 + 9i = -8 + 7i$

49. $(7 + 4i) + (6 - 2i) = 13 + 2i$

51. $(5 - 8i) - (3 + 2i) = 2 - 10i$

53. $(-10 + 4i) - (-2 + 6i) = -8 - 2i$

55. $(8 + 4i) + (-2 - 4i) = 6$

57. $(-2 + 11i) - (-2 + 11i) = 0$

59. False. $m = a - bi$
$n = c + di$
$a = c, b = d : m = c - di$
$n = c + di$
$m - n = (c - di) - (c + di)$
$m - n = c - di - c - di = -2di$

61. Negative. The imaginary part of $m + n$ is $-b + d$. Since $b > d$, the quantity $-b + d$ is negative.

63. $4i(7i) = 28i^2 = 28(-1) = -28$

65. $4i \cdot 5i = 20i^2 = -20$

67. $(-2i)(9i) = -18i^2 = 18$

69. $5i(2 + 4i) = 10i + 20i^2 = 10i - 20 = -20 + 10i$

71. $(3 + 2i)(4 + 5i) = 12 + 23i + 10i^2$
$= 12 + 23i - 10 = 2 + 23i$

73. $(-3 + 2i)(1 + 3i) = -3 - 7i + 6i^2$
$= -3 - 7i - 6 = -9 - 7i$

75. $(3 - 7i)(-2 - 5i) = -6 - i + 35i^2$
$= -6 - i - 35 = -41 - i$

77. $(2 + 5i)^2 = 4 + 20i + 25i^2$
$= 4 + 20i - 25 = -21 + 20i$

79. $(1 - 2i)^2 = 1 - 4i + 4i^2 = 1 - 4i - 4 = -3 - 4i$

81. $(2 - 5i)(-5 + 2i) = -10 + 29i - 10i^2$
$= -10 + 29i + 10 = 29i$

83. $(2 + 4i)(2 - 4i) = 4 + 16 = 20$

85. $(4 - i)(4 + i) = 16 + 1 = 17$

87. False. $(ai)(bi) = abi^2 = -ab$, whose sign changes depending on the signs of a and b.

89. $\frac{2}{3i} = \frac{2}{3i} \cdot \frac{i}{i} = \frac{2i}{3i^2} = \frac{2i}{-3} = -\frac{2i}{3}$

91. $\frac{9}{2i} = \frac{9}{2i} \cdot \frac{i}{i} = \frac{9i}{2i^2} = \frac{9i}{-2} = -\frac{9i}{2}$

93. $\frac{4}{-7i} = \frac{4}{-7i} \cdot \frac{i}{i} = \frac{4i}{-7i^2} = \frac{4i}{7}$

95. $\frac{6 + i}{i} = \frac{6 + i}{i} \cdot \frac{i}{i} = \frac{6i + i^2}{i^2} = \frac{6i - 1}{-1} = 1 - 6i$

97. $\frac{-5 - 9i}{-i} = \frac{-5 - 9i}{-i} \cdot \frac{i}{i} = \frac{-5i - 9i^2}{-i^2}$
$= \frac{-5i + 9}{1} = 9 - 5i$

99. $\dfrac{6-4i}{2i}=\dfrac{6-4i}{2i}\cdot\dfrac{i}{i}=\dfrac{6i-4i^2}{2i^2}=\dfrac{6i+4}{-2}=-2-3i$

101. $\dfrac{4-10i}{6i}=\dfrac{4-10i}{6i}\cdot\dfrac{i}{i}=\dfrac{4i-10i^2}{6i^2}$

$=\dfrac{4i+10}{-6}=-\dfrac{10}{6}-\dfrac{4}{6}i=-\dfrac{5}{3}-\dfrac{2}{3}i$

103. $\dfrac{4-15i}{-6i}=\dfrac{4-15i}{-6i}\cdot\dfrac{i}{i}=\dfrac{4i-15i^2}{-6i^2}$

$=\dfrac{4i+15}{6}=\dfrac{15}{6}+\dfrac{4}{6}i=\dfrac{5}{2}+\dfrac{2}{3}i$

105. $\dfrac{-4}{1-2i}=\dfrac{-4}{1-2i}\cdot\dfrac{1+2i}{1+2i}=\dfrac{-4-8i}{1+4}$

$=\dfrac{-4-8i}{5}=-\dfrac{4}{5}-\dfrac{8}{5}i$

107. $\dfrac{12i}{2-2i}=\dfrac{12i}{2-2i}\cdot\dfrac{2+2i}{2+2i}$

$=\dfrac{24i+24i^2}{4+4}=\dfrac{24i-24}{8}$

$=-3+3i$

109. $\dfrac{12+i}{1-2i}=\dfrac{12+i}{1-2i}\cdot\dfrac{1+2i}{1+2i}=\dfrac{12+25i+2i^2}{1+4}$

$=\dfrac{12+25i-2}{5}=\dfrac{10+25i}{5}$

$=2+5i$

111. $\dfrac{-18+13i}{1+4i}=\dfrac{-18+13i}{1+4i}\cdot\dfrac{1-4i}{1-4i}$

$=\dfrac{-18+85i-52i^2}{1+16}=\dfrac{-18+85i+52}{17}$

$=\dfrac{34+85i}{17}=2+5i$

113. $\dfrac{5+15i}{3+4i}=\dfrac{5+15i}{3+4i}\cdot\dfrac{3-4i}{3-4i}=\dfrac{15+25i-60i^2}{9+16}$

$=\dfrac{15+25i+60}{25}=\dfrac{75+25i}{25}$

$=3+i$

115. **False**

117. $y=\dfrac{-b\pm\sqrt{b^2-4ac}}{2a}$

$=\dfrac{-(-3)\pm\sqrt{(-3)^2-4(1)(-8)}}{2(1)}$

119. $z^2+25=0$

$z^2=-25$

$\sqrt{z^2}=\sqrt{-25}$

$z=\pm5i$

121. $3x^2+54=-54$

$3x^2=-108$

$x^2=-36$

$\sqrt{x^2}=\sqrt{-36}$

$x=\pm6i$

123. $z^2+12=-28$

$z^2=-40$

$\sqrt{z^2}=\sqrt{-40}=\sqrt{-1}\sqrt{4\cdot10}$

$z=\pm2i\sqrt{10}$

125. $4z^2-13=-21$

$4z^2=-8$

$z^2=-2$

$\sqrt{z^2}=\sqrt{-2}$

$z=\pm i\sqrt{2}$

127. $(x-3)^2+81=0$

$(x-3)^2=-81$

$\sqrt{(x-3)^2}=\sqrt{-81}$

$x-3=\pm9i$

$x=3\pm9i$

129. $(z+4)^2+18=0$

$(z+4)^2=-18$

$\sqrt{(z+4)^2}=\sqrt{-18}=\sqrt{-1}\sqrt{9\cdot 2}$

$z+4=\pm 3i\sqrt{2}$

$z=-4\pm 3i\sqrt{2}$

131. $(x-5)^2+17=9$

$(x-5)^2=-8$

$\sqrt{(x-5)^2}=\sqrt{-8}=\sqrt{-1}\sqrt{4\cdot 2}$

$x-5=\pm 2i\sqrt{2}$

$x=5\pm 2i\sqrt{2}$

133. $(z-1)^2+25=9$

$(z-1)^2=-16$

$\sqrt{(z-1)^2}=\sqrt{-16}$

$z-1=\pm 4i$

$z=1\pm 4i$

135. $z^2-2z+5=0$

$z^2-2z=-5$

$z^2-2z+1=-5+1$

$(z-1)^2=-4$

$\sqrt{(z-1)^2}=\sqrt{-4}$

$z-1=\pm 2i$

$z=1\pm 2i$

137. $x^2+4x=-8$

$x^2+4x+4=-8+4$

$(x+2)^2=-4$

$\sqrt{(x+2)^2}=\sqrt{-4}$

$x+2=\pm 2i$

$x=-2\pm 2i$

139. $x^2+10=2x$

$x^2-2x=-10$

$x^2-2x+1=-10+1$

$(x-1)^2=-9$

$\sqrt{(x-1)^2}=\sqrt{-9}$

$x-1=\pm 3i$

$x=1\pm 3i$

141. $x^2+2x+13=0$

$x^2+2x=-13$

$x^2+2x+1=-13+1$

$(x+1)^2=-12$

$\sqrt{(x+1)^2}=\sqrt{-12}=\sqrt{-1}\sqrt{4\cdot 3}$

$x+1=\pm 2i\sqrt{3}$

$x=-1\pm 2i\sqrt{3}$

143. $4x^2+12x+25=0$

$a=4 \quad b=12 \quad c=25$

$$x=\frac{-b\pm\sqrt{b^2-4ac}}{2a}$$

$$x=\frac{-12\pm\sqrt{12^2-4(4)(25)}}{2(4)}$$

$$x=\frac{-12\pm\sqrt{144-400}}{8}=\frac{-12\pm\sqrt{-256}}{8}$$

$$x=\frac{-12\pm 16i}{8}=\frac{-12}{8}\pm\frac{16}{8}i$$

$$x=-\frac{3}{2}\pm 2i$$

145. Since a is positive, the equation will yield **two imaginary solutions**.

147. $2i+5i=7i$

149. $(1-i)(1+i)(2-3i)=(1+1)(2-3i)$

$=2(2-3i)=4-6i$

151. $(1+i)^3 = (1+i)^2(1+i)$
$= (1+2i+i^2)(1+i) = 2i(1+i)$
$= 2i+2i^2 = 2i-2 = -2+2i$

153. $(4-3i)(1+2i)+(3-2i)(2-i)$
$= 4+5i-6i^2+6-7i+2i^2$
$= 10-2i-4i^2 = 14-2i$

155. $(x-1)(x+1)+5=0$
$x^2-1+5=0$
$x^2+4=0$
$x^2=-4$
$\sqrt{x^2}=\sqrt{-4}$
$x=\pm 2i$

157. $(x+4)(x+6)+4=0$
$x^2+10x+24+4=0$
$x^2+10x+28=0$
$x^2+10x=-28$
$x^2+10x+25=-28+25$
$(x+5)^2=-3$
$\sqrt{(x+5)^2}=\sqrt{-3}$
$x+5=\pm i\sqrt{3}$
$x=-5\pm i\sqrt{3}$

159. $(x-2)(x-4)+4=0$
$x^2-6x+8+4=0$
$x^2-6x=-12$
$x^2-6x+9=-12+9$
$(x-3)^2=-3$
$\sqrt{(x-3)^2}=\sqrt{-3}$
$x-3=\pm i\sqrt{3}$
$x=3\pm i\sqrt{3}$

161. $4x^2+20=0$
$4(i\sqrt{5})^2+20\stackrel{?}{=}0$
$4\cdot i^2\cdot 5+20\stackrel{?}{=}0$
$20i^2+20\stackrel{?}{=}0$
$-20+20\stackrel{?}{=}0$
$0=0$

$4x^2+20=0$
$4(-i\sqrt{5})^2+20\stackrel{?}{=}0$
$4\cdot i^2\cdot 5+20\stackrel{?}{=}0$
$20i^2+20\stackrel{?}{=}0$
$-20+20\stackrel{?}{=}0$
$0=0$

163. a. $i^4 = i^2\cdot i^2 = (-1)(-1) = 1$

b. $i^5 = i^2\cdot i^2\cdot i = (-1)(-1)i = i$

c. $i^6 = i^2\cdot i^2\cdot i^2 = (-1)(-1)(-1) = -1$

d. $i^7 = i^2\cdot i^2\cdot i^2\cdot i = (-1)(-1)(-1)i = -i$

e. $i^{18} = i^9\cdot i^9 = i\cdot i = i^2 = -1$

f. $i^{27} = i^9\cdot i^9\cdot i^9 = i\cdot i\cdot i = i^2\cdot i = -1\cdot i = -i$

g. $i^{36} = i^9\cdot i^9\cdot i^9\cdot i^9 = i\cdot i\cdot i\cdot i = i^2\cdot i^2 = (-1)(-1) = 1$

165. Linear
$2x=3x-2$
$-x=-2$
$x=2$

167. Linear
$4(x-2)=12$
$4x-8=12$
$4x=20$
$x=5$

Section 11.5 Graphing Quadratic Equations in Two Variables

1. The graph of an equation in two variables is called a **parabola**.

3. The graph opens up because the coefficient of the x^2 term is positive.

5. Find the y-intercept by letting $x = 0$ and solving for y.

7. **Up.** The coefficient of x^2 is positive.

9. **Down.** The coefficient of x^2 is negative.

11. **Up.** The coefficient of x^2 is positive.

13. **a.** $a = -1 \quad b = 0 \quad c = 9$

    b. Because $a < 0$, the parabola opens **down**.

    c. $f(x) = -x^2 + 9$
    $f(2) = -(2)^2 + 9$
    $f(2) = -4 + 9 = 5$

15. $f(x) = x^2 - 2x + 1$
    $f(3) = 3^2 - 2 \cdot 3 + 1$
    $f(3) = 9 - 6 + 1 = 4$

17. $f(x) = 4 - x^2$
    $f(-3) = 4 - (-3)^2$
    $f(-3) = 4 - 9 = -5$

19. $f(x) = -x^2 + 5x - 6$
    $f(-4) = -(-4)^2 + 5(-4) - 6$
    $f(-4) = -16 - 20 - 6 = -42$

21. $y = x^2$

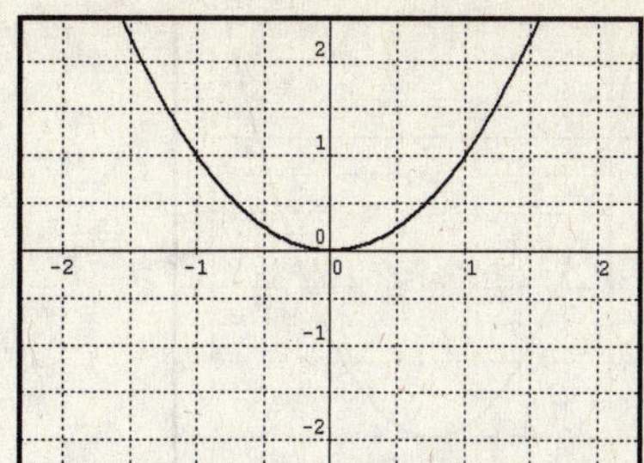

23. $y = -x^2 + 1$

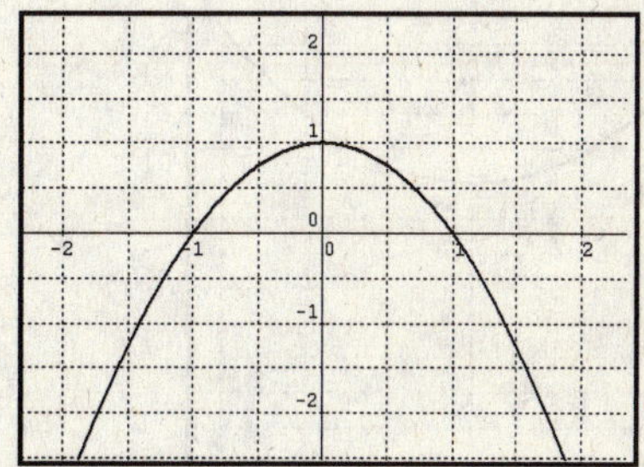

25. $y = 2x^2$

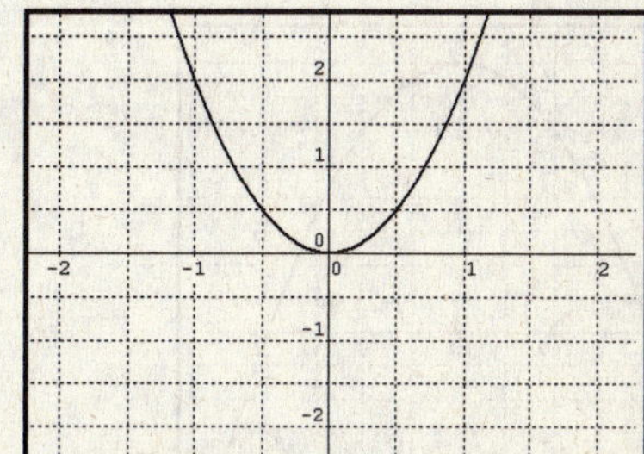

27. $y = -\frac{1}{2}x^2 + 1$

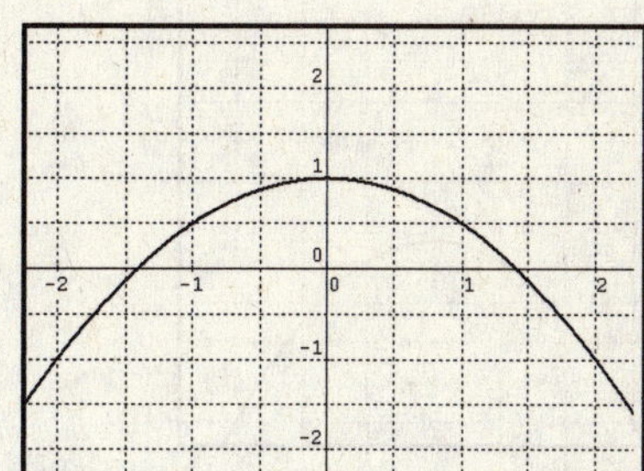

29. $y = x^2 - 4x$

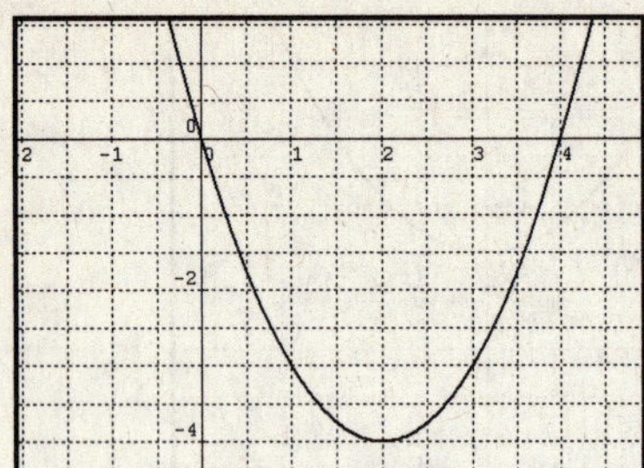

31. $y = x^2 - 2x + 3$

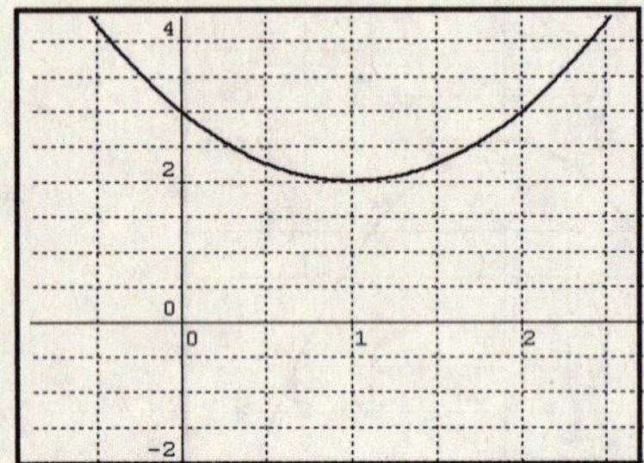

33. $y = -x^2 + 2x + 3$

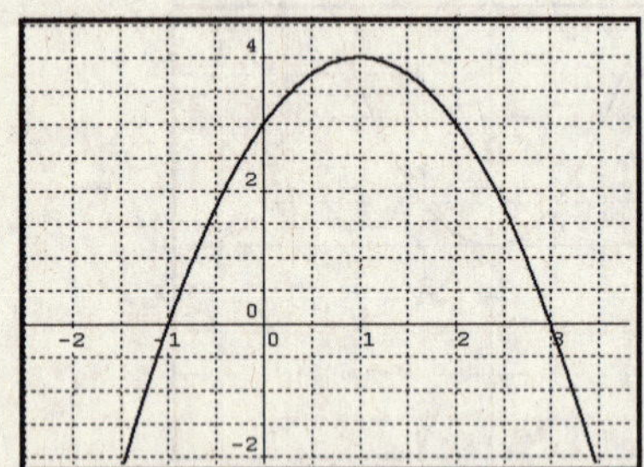

35. $y = -x^2 + 3x - 4$

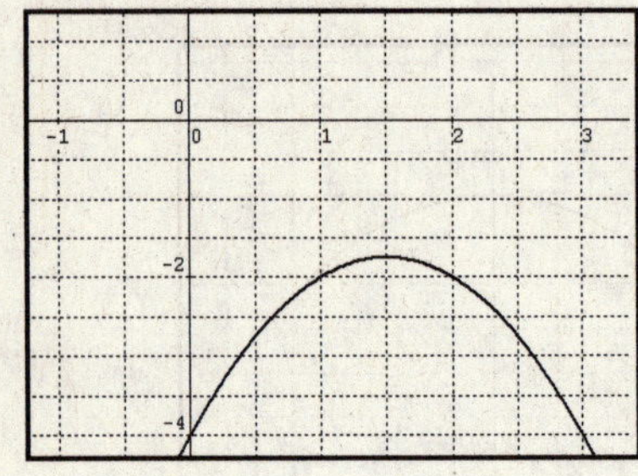

37. $y = 2x^2 + x - 3$

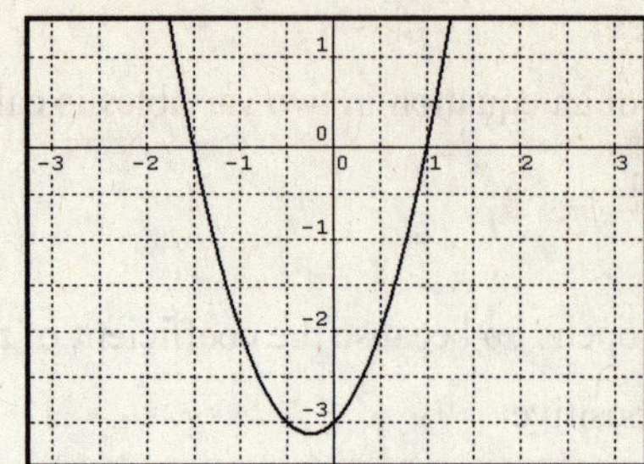

39. **False.** $a < 0$ means that the parabola opens downward.

41. x-intercepts: $0 = x^2 - 5x + 6$

$0 = (x-2)(x-3)$

$x - 2 = 0 \qquad x - 3 = 0$

$x = 2 \qquad x = 3$

$(2,0) \qquad (3,0)$

y-intercept: $y = 0^2 - 5 \cdot 0 + 6$

$y = 6$

$(0,6)$

43. x-intercepts: $0 = 9 - x^2$

$0 = (3-x)(3+x)$

$3 - x = 0 \qquad 3 + x = 0$

$x = 3 \qquad x = -3$

$(3,0) \qquad (-3,0)$

y-intercept: $f(0) = 9 - 0^2$

$f(0) = 9$

$(0,9)$

45. x-intercepts:
$$0 = x^2 + 2x - 6$$
$$x^2 + 2x + 1 = 6 + 1$$
$$(x+1)^2 = 7$$
$$\sqrt{(x+1)^2} = \sqrt{7}$$
$$x + 1 = \pm\sqrt{7}$$
$$x = -1 \pm \sqrt{7}$$
$$(-1+\sqrt{7}, 0) \quad (-1-\sqrt{7}, 0)$$

y-intercept:
$$y = 0^2 + 2 \cdot 0 - 6$$
$$y = -6$$
$$(0, -6)$$

47. x-intercepts:
$$0 = 2x^2 - x - 3$$
$$0 = (x+1)(2x-3)$$
$$x + 1 = 0 \qquad 2x - 3 = 0$$
$$x = -1 \qquad x = \frac{3}{2}$$
$$(-1, 0) \qquad \left(\frac{3}{2}, 0\right)$$

y-intercept:
$$y = 2 \cdot 0^2 - 0 - 3$$
$$y = -3$$
$$(0, -3)$$

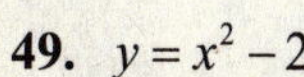

49. $y = x^2 - 2$

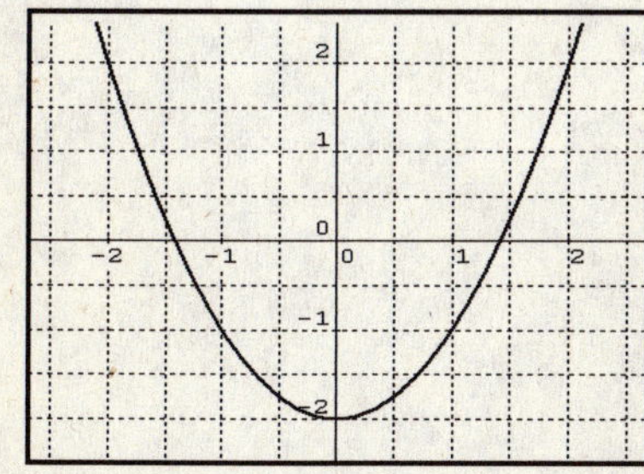

51. $y = x^2 + 2x$

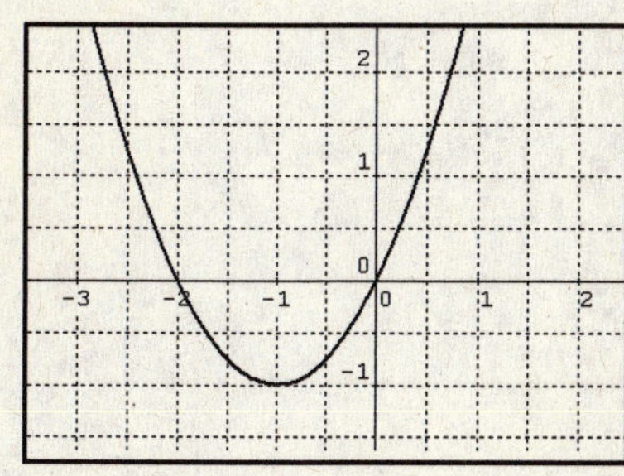

53. $y = \frac{1}{2}x^2 - x$

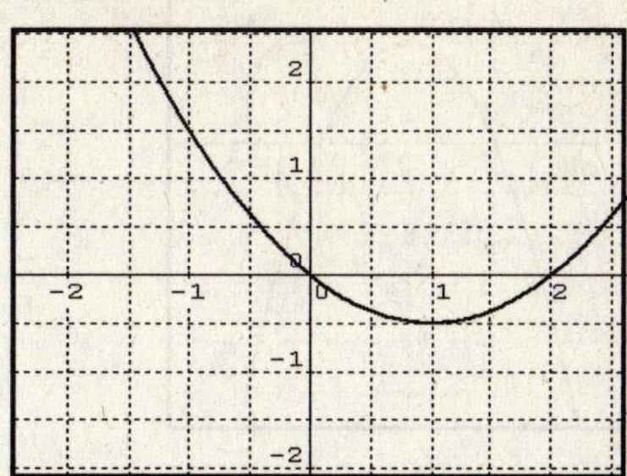

55. $y = x^2 - x - 2$

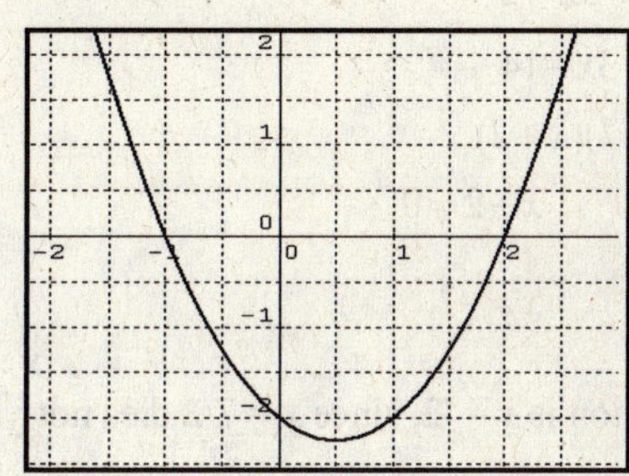

57. $y = -x^2 - 2x - 1$

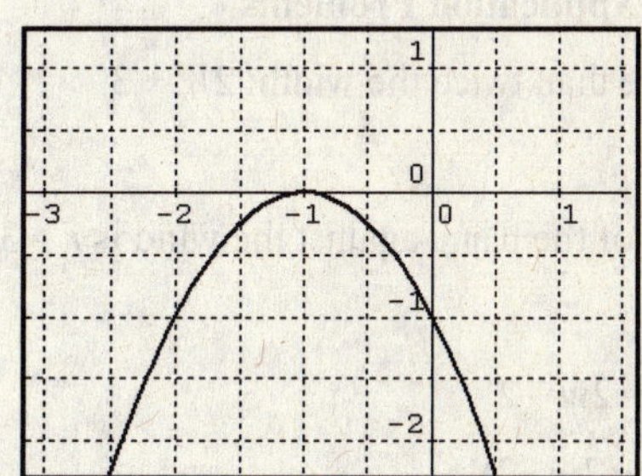

59. Linear

61. Neither

63. Quadratic

65.
a. (-2, 0) and (2, 0)
b. (0, -4)
c. a is positive
d. (0, -4)
e. The y-axis
f. When x = -1, y = 3.

67. Sample graph: $y=-x^2+6x-8$

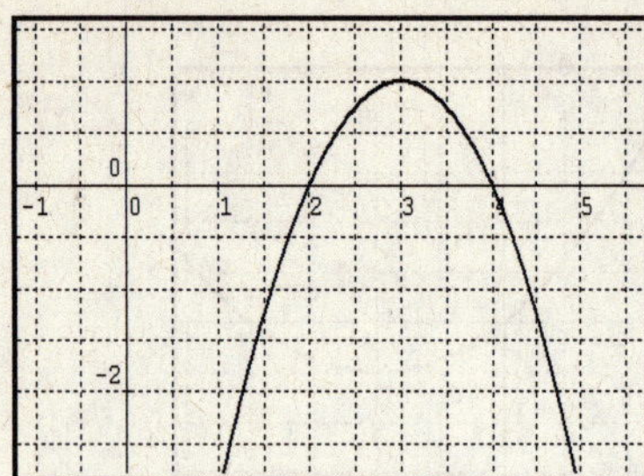

69. $y=2x^2-3x-2$

$12=2x^2-3x-2$

$0=2x^2-3x-14$

$0=(2x-7)(x+2)$

$2x-7=0 \qquad x+2=0$

$x=\frac{7}{2} \qquad x=-2$

The solution is x = -2, since x = 7/2 dies not yield a point in quadrant II.)

Section 11.6 Application Problems

1. Three more than twice the width: $2W+3$

3. The speed of the plane against the wind is $r-30$.

5. a. length = $2w-2$

b. area = $w(2w-2)$

c. $w(2w-2)=84$

7. Let b = the base. Then the height = $2b+2$.

$20=\frac{1}{2}b(2b+2)$

$40=b(2b+2)$

$40=2b^2+2b$

$0=2b^2+2b-40$

$0=b^2+b-20$

$0=(b+5)(b-4)$

$b+5=0 \qquad b-4=0$

$b=-5 \qquad b=4$

$2b+2=10$

The base measures 4m and the height measures 10 m.

9. Let w = the width. Then the length = $w+2$.

$24=w(w+2)$

$24=w^2+2w$

$0=w^2+2w-24$

$0=(w+6)(w-4)$

$w+6=0 \qquad w-4=0$

$w=-6 \qquad w=4$

$w+2=6$

The width is 4 ft and the length is 6 ft.

11. Let t = the time for the one drain. Then the time for the other drain = $t+16$.

$\frac{1}{t}+\frac{1}{t+16}=\frac{1}{6}$

$6t(t+16)\left(\frac{1}{t}+\frac{1}{t+16}\right)=6t(t+16)\cdot\frac{1}{6}$

$6(t+16)+6t=t(t+16)$

$6t+96+6t=t^2+16t$

$12t+96=t^2+16t$

$0=t^2+4t-96$

$0=(t+12)(t-8)$

$t+12=0 \qquad t-8=0$

$t=-12 \qquad t=8$

$t+16=24$

The faster drain takes 8 min and the other drain takes 24 min.

13. Let w = the width. Then the length = $2w$.

$w(2w)=5000$

$2w^2=5000$

$w^2=2500$

$w=\pm 50$

$w=50$

$2w=100$

The width is 50 ft and the length is 100 ft.

15. Let t = the time for one engine.

Then $t+6$ = the time for the other engine.

$\frac{1}{t}+\frac{1}{t+6}=\frac{1}{4}$

$4t(t+6)\left(\frac{1}{t}+\frac{1}{t+6}\right)=\frac{1}{4}(4t)(t+6)$

$4(t+6)+4t=t(t+6)$

$4t+24+4t=t^2+6t$

$8t+24=t^2+6t$

$0=t^2-2t-24$

$0=(t-6)(t+4)$

$t-6=0 \qquad t+4=0$

$t=6 \qquad t=-4$

$t+6=12$

The slower engine takes 6 hours and the other engine takes 12 hours.

17. The rate of the plane was 100 mph.

	Distance	÷	Rate	=	Time
Against the wind	375	÷	r - 25	=	$\frac{375}{r-25}$
With the wind	375	÷	r + 25	=	$\frac{375}{r+25}$

$\frac{375}{r-25}=\frac{375}{r+25}+2$

$(r-25)(r+25)\left(\frac{375}{r-25}\right)=(r-25)(r+25)\left(\frac{375}{r+25}+2\right)$

$375(r+25)=375(r-25)+(r-25)(r+25)(2)$

$375r+9375=375r-9375+2r^2-1250$

$375r+9375=2r^2+375r-10{,}625$

$0=2r^2-20{,}000$

$r^2=10{,}000$

$r=\pm 100$

$r=100$

19. The rate for the first 150 miles was 50 mph.

	Distance	÷	Rate	=	Time
Part 1	150	÷	r	=	$\frac{150}{r}$
Part 2	35	÷	r - 15	=	$\frac{35}{r-15}$

$$\frac{150}{r}+\frac{35}{r-15}=4$$
$$r(r-15)\left(\frac{150}{r}+\frac{35}{r-15}\right)=4r(r-15)$$
$$150(r-15)+35r=4r^2-60r$$
$$150r-2250+35r=4r^2-60r$$
$$185r-2250=4r^2-60r$$
$$0=4r^2-245r+2250$$
$$0=(4r-45)(r-50)$$
$$4r-45=0 \qquad r-50=0$$
$$r=\frac{45}{4} \qquad r=50$$

21. $h=48t-16t^2$

$$32=48t-16t^2$$
$$16t^2-48t+32=0$$
$$16(t^2-3t+2)=0$$
$$t^2-3t+2=0$$
$$(t-2)(t-1)=0$$
$$t-2=0 \qquad t-1=0$$
$$t=2 \qquad t=1$$

The arrow is 32 ft above the ground after 1 sec and after 2 sec.

23. $A(t)=0.005\pi t^2$

$$10=0.005\pi t^2$$
$$\frac{10}{0.005\pi}=t^2$$
$$\sqrt{\frac{10}{0.005\pi}}=\sqrt{t^2}$$
$$t\approx\pm25$$

It will take about 25 days.

25. The year will be 2017.

$$y=0.03x^2+0.36x+34.6$$
$$50=0.03x^2+0.36x+34.6$$
$$0=0.03x^2+0.36x-15.4$$
$$0=3x^2+36x-1540$$
$$x=\frac{-b\pm\sqrt{b^2-4ac}}{2a}$$
$$x=\frac{-36\pm\sqrt{36^2-4(3)(-1540)}}{2(3)}$$
$$x=\frac{-36\pm\sqrt{1296+18480}}{6}=\frac{-36\pm\sqrt{19776}}{6}$$
$$x=\frac{-36\pm140.63}{6}$$
$$x=\frac{-36+140.63}{6}=\frac{104.63}{6}\approx17$$

The year will be 2017.

27. The ball has been in the air about 1.58 seconds.\

$$h=-16t^2+24t+4$$
$$2=-16t^2+24t+4$$
$$16t^2-24t-2=0$$
$$8t^2-12t-1=0$$
$$t=\frac{-b\pm\sqrt{b^2-4ac}}{2a}$$
$$t=\frac{-(-12)\pm\sqrt{(-12)^2-4(8)(-1)}}{2(8)}$$
$$t=\frac{12\pm\sqrt{144+32}}{16}=\frac{12\pm\sqrt{176}}{16}$$
$$t=\frac{12\pm13.3}{16}$$
$$t=\frac{12+13.3}{16}=\frac{25.3}{16}\approx1.58$$

29. $s=-16t^2+88t+1$

$0=-16t^2+88t+1$

$t=\frac{-b\pm\sqrt{b^2-4ac}}{2a}$

$t=\frac{-88\pm\sqrt{88^2-4(-16)(1)}}{2(-16)}$

$t=\frac{-88\pm\sqrt{7744+64}}{-32}=\frac{-88\pm\sqrt{7808}}{-32}$

$t=\frac{-88\pm 88.36}{-32}$

$t=\frac{0.36}{-32}$ $\quad t=\frac{-176.36}{-32}$

$t=-0.011$ $\quad t=5.511$

The ball hits the ground after 5.5 seconds.

31. **False.** The time working together will be less than the time for either pipe working alone.

33. Let a = one leg. Then $b=2a-1$.

$c^2=a^2+b^2$

$(\sqrt{13})^2=a^2+(2a-1)^2$

$13=a^2+4a^2-4a+1$

$0=5a^2-4a-12$

$0=(5a+6)(a-2)$

$5a+6=0$ $\quad a-2=0$

$a=-\frac{6}{5}$ $\quad a=2$

$2a-1=3$

The legs measure 2 cm and 3 cm.

35. $V=\frac{1}{3}\pi r^2h$

$800=\frac{1}{3}\pi r^2(12)$

$\frac{3\cdot 800}{12\pi}=r^2$

$\sqrt{\frac{200}{\pi}}=\sqrt{r^2}$

$\pm 7.98=r$

The radius is 7.98 cm.

37. $V=2(x-4)(x-4)$

$512=2(x^2-8x+16)$

$256=x^2-8x+16$

$0=x^2-8x-240$

$x^2-8x=240$

$x^2-8x+16=240+16$

$(x-4)^2=256$

$\sqrt{(x-4)^2}=\sqrt{256}$

$x-4=\pm 16$

$x=4\pm 16$

$x=4+16=20$

The cardboard measures 20 in by 20 in.

39. **a.** $12=2L+2W$

b. $A=LW$

c. $12=2L+2W$

$12-2W=2L$

$6-W=L$

$A=LW$

$A=(6-W)W$

$A=-W^2+6W$

d. The largest area is when the rectangle is a square with side length of 3 ft.

Chapter 11 Review Exercises

1. $b^2-16=0$

$(b-4)(b+4)=0$

$b-4=0$ $\quad b+4=0$

$b=4$ $\quad b=-4$

2. $x^2-x-3=0$

$x=\frac{-b\pm\sqrt{b^2-4ac}}{2a}$

$x=\frac{-(-1)\pm\sqrt{(-1)^2-4(1)(-3)}}{2(1)}$

$x=\frac{1\pm\sqrt{1+12}}{2}=\frac{1\pm\sqrt{13}}{2}$

3. $x^2-3x-5=0$

$$x=\frac{-b\pm\sqrt{b^2-4ac}}{2a}$$

$$x=\frac{-(-3)\pm\sqrt{(-3)^2-4(1)(-5)}}{2(1)}$$

$$x=\frac{3\pm\sqrt{9+20}}{2}=\frac{3\pm\sqrt{29}}{2}$$

4. $49x^2=25$

$49x^2-25=0$

$(7x-5)(7x+5)=0$

$7x-5=0 \qquad 7x+5=0$

$x=\frac{5}{7} \qquad x=-\frac{5}{7}$

5. $y=-\frac{1}{4}x^2$

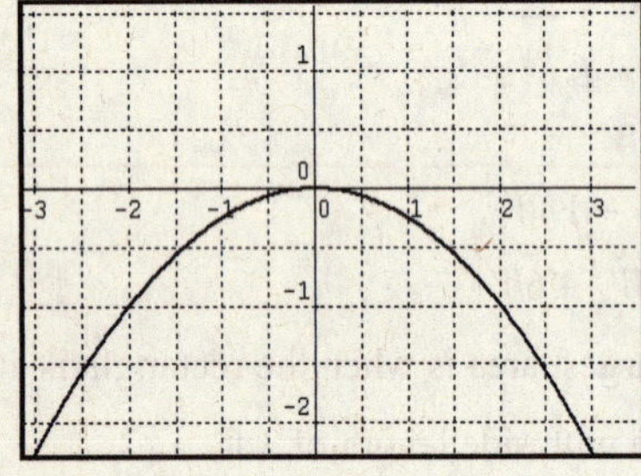

6. $y=-3x^2$

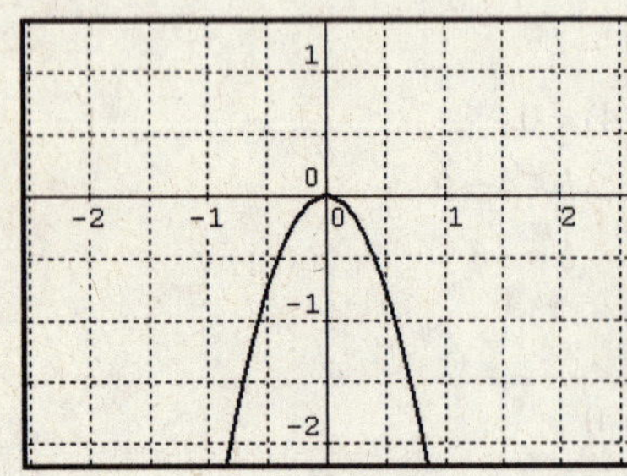

7. $3\sqrt{-28}=3\sqrt{-1}\sqrt{4\cdot 7}=3i\cdot 2\sqrt{7}=6i\sqrt{7}$

8. $6x(x+1)=x-1$

$6x^2+6x-x+1=0$

$6x^2+5x+1=0$

$(3x+1)(2x+1)=0$

$3x+1=0 \qquad 2x+1=0$

$x=-\frac{1}{3} \qquad x=-\frac{1}{2}$

9. $4y^2+9=0$

$4y^2=-9$

$y^2=-\frac{9}{4}$

$\sqrt{y^2}=\sqrt{-\frac{9}{4}}$

$y=\pm\frac{3}{2}i$

10. $5x^2+20x+12=0$

$$x=\frac{-b\pm\sqrt{b^2-4ac}}{2a}$$

$$x=\frac{-20\pm\sqrt{20^2-4(5)(12)}}{2(5)}$$

$$x=\frac{-20\pm\sqrt{400-240}}{10}=\frac{-20\pm\sqrt{160}}{10}$$

$$x=\frac{-20\pm4\sqrt{10}}{10}=\frac{2(-10\pm2\sqrt{10})}{10}$$

$$x=\frac{-10\pm2\sqrt{10}}{5}$$

11. $(-4+i)+(6-5i)=(-4+6)+(1-5)i=2-4i$

12. $x^2-x=30$

$x^2-x-30=0$

$(x-6)(x+5)=0$

$x-6=0 \qquad x+5=0$

$x=6 \qquad x=-5$

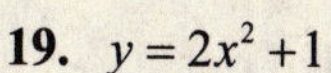

13. $6x^2 + 13x - 28 = 0$

$(3x-4)(2x+7) = 0$

$3x - 4 = 0 \qquad 2x + 7 = 0$

$x = \frac{4}{3} \qquad x = -\frac{7}{2}$

14. $x^2 = 40$

$\sqrt{x^2} = \sqrt{40}$

$x = \pm\sqrt{4 \cdot 10}$

$x = \pm 2\sqrt{10}$

15. $(-1+2i) - (2-3i) = (-1-2) + (2+3)i = -3 + 5i$

16. $x^2 - 2x - 10 = 0$

$x^2 - 2x = 10$

$x^2 - 2x + 1 = 10 + 1$

$(x-1)^2 = 11$

$\sqrt{(x-1)^2} = \sqrt{11}$

$x - 1 = \pm\sqrt{11}$

$x = 1 \pm \sqrt{11}$

17. $x^2 - 12x + 27 = 0$

$(x-9)(x-3) = 0$

$x - 9 = 0 \qquad x - 3 = 0$

$x = 9 \qquad x = 3$

18. $(x-7)^2 = 81$

$\sqrt{(x-7)^2} = \sqrt{81}$

$x - 7 = \pm 9$

$x = 7 \pm 9$

$x = 7 + 9 \qquad x = 7 - 9$

$x = 16 \qquad x = -2$

19. $y = 2x^2 + 1$

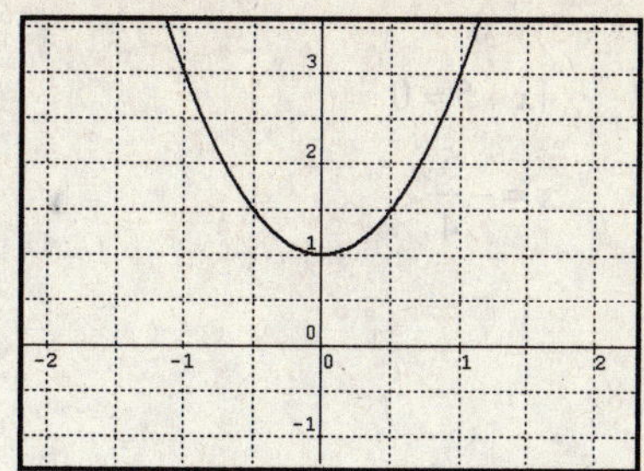

20. $y = \frac{1}{2}x^2 - 1$

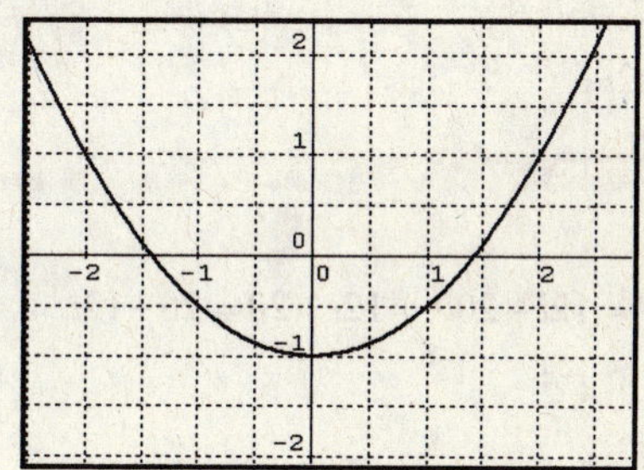

21. $(y+4)^2 - 25 = 0$

$(y+4)^2 = 25$

$\sqrt{(y+4)^2} = \sqrt{25}$

$y + 4 = \pm 5$

$y + 4 = 5 \qquad y + 4 = -5$

$y = 1 \qquad y = -9$

22. $4x^2 + 16x = 7$

$x^2 + 4x = \frac{7}{4}$

$x^2 + 4x + 4 = \frac{7}{4} + 4$

$(x+2)^2 = \frac{23}{4}$

$\sqrt{(x+2)^2} = \sqrt{\frac{23}{4}}$

$x + 2 = \pm\frac{\sqrt{23}}{2}$

$x = -2 \pm \frac{\sqrt{23}}{2} = \frac{-4 \pm \sqrt{23}}{2}$

23. $24x^2 + 34x + 5 = 0$

$(6x+1)(4x+5) = 0$

$6x + 1 = 0 \qquad 4x + 5 = 0$

$x = -\dfrac{1}{6} \qquad x = -\dfrac{5}{4}$

24. $x^2 = 4x - 8$

$x^2 - 4x = -8$

$x^2 - 4x + 4 = -8 + 4$

$(x-2)^2 = -4$

$\sqrt{(x-2)^2} = \sqrt{-4}$

$x - 2 = \pm 2i$

$x = 2 \pm 2i$

25. $4i(3-5i) = 12i - 20i^2 = 12i + 20 = 20 + 12i$

26. $25(2x^2 - 2x + 1) = (x+3)^2$

$50x^2 - 50x + 25 = x^2 + 6x + 9$

$49x^2 - 56x + 16 = 0$

$(7x-4)^2 = 0$

$7x - 4 = 0$

$x = \dfrac{4}{7}$

27. $\left(x - \dfrac{1}{2}\right)^2 = \dfrac{9}{4}$

$\sqrt{\left(x - \dfrac{1}{2}\right)^2} = \sqrt{\dfrac{9}{4}}$

$x - \dfrac{1}{2} = \pm\dfrac{3}{2}$

$x = \dfrac{1}{2} \pm \dfrac{3}{2}$

$x = \dfrac{1}{2} + \dfrac{3}{2} \qquad x = \dfrac{1}{2} - \dfrac{3}{2}$

$x = 2 \qquad x = -1$

28. $(-2+4i)(3-i) = -6 + 2i + 12i - 4i^2$

$= -6 + 14i + 4 = -2 + 14i$

29. $x^2 + 7x = 3$

$x^2 + 7x - 3 = 0$

$x = \dfrac{-b \pm \sqrt{b^2 - 4ac}}{2a}$

$x = \dfrac{-7 \pm \sqrt{7^2 - 4(1)(-3)}}{2(1)}$

$x = \dfrac{-7 \pm \sqrt{49 + 12}}{2} = \dfrac{-7 \pm \sqrt{61}}{2}$

30. $12x^2 + 10 = 29x$

$12x^2 - 29x + 10 = 0$

$(12x - 5)(x - 2) = 0$

$12x - 5 = 0 \qquad x - 2 = 0$

$x = \dfrac{5}{12} \qquad x = 2$

31. $4(x-3)^2 = 20$

$(x-3)^2 = 5$

$\sqrt{(x-3)^2} = \sqrt{5}$

$x - 3 = \pm\sqrt{5}$

$x = 3 \pm \sqrt{5}$

32. $x^2 + 8x - 3 = 0$

$x^2 + 8x = 3$

$x^2 + 8x + 16 = 3 + 16$

$(x+4)^2 = 19$

$\sqrt{(x+4)^2} = \sqrt{19}$

$x + 4 = \pm\sqrt{19}$

$x = -4 \pm \sqrt{19}$

33. 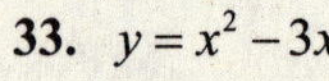
$y = x^2 - 3x$

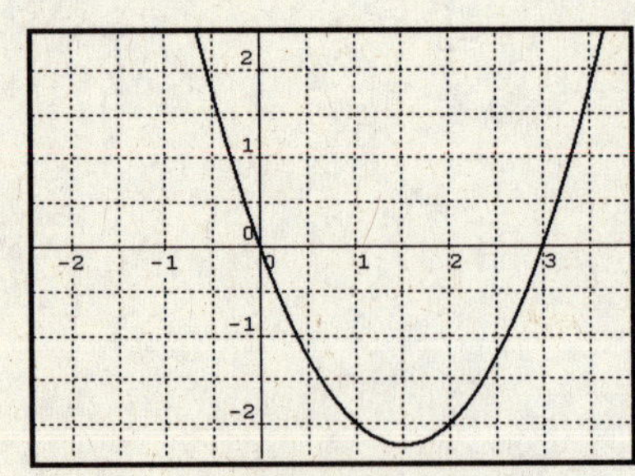

34. $y = x^2 - 4x + 3$

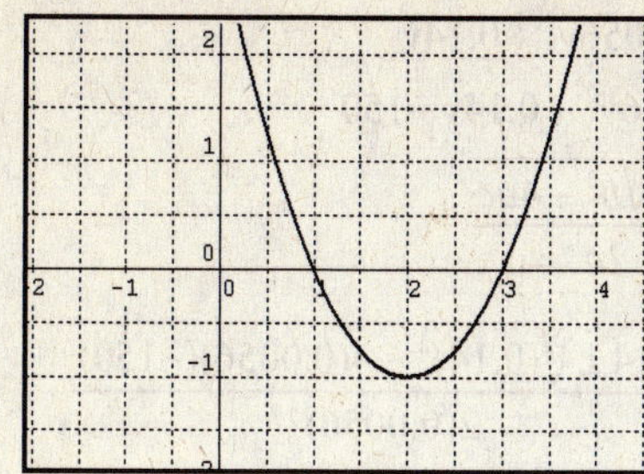

35. $(x+9)^2 = x+11$

$x^2 + 18x + 81 = x + 11$

$x^2 + 17x + 70 = 0$

$(x+10)(x+7) = 0$

$x + 10 = 0 \qquad x + 7 = 0$

$x = -10 \qquad x = -7$

36. $\dfrac{3-i}{2-4i} = \dfrac{3-i}{2-4i} \cdot \dfrac{2+4i}{2+4i} = \dfrac{6+10i-4i^2}{4+16}$

$= \dfrac{10+10i}{20} = \dfrac{1}{2} + \dfrac{1}{2}i$

37. $x^2 + 6x + 12 = 0$

$x^2 + 6x = -12$

$x^2 + 6x + 9 = -12 + 9$

$(x+3)^2 = -3$

$\sqrt{(x+3)^2} = \sqrt{-3}$

$x + 3 = \pm i\sqrt{3}$

$x = -3 \pm i\sqrt{3}$

38. $x^2 + 6x - 2 = 0$

$x^2 + 6x = 2$

$x^2 + 6x + 9 = 2 + 9$

$(x+3)^2 = 11$

$\sqrt{(x+3)^2} = \sqrt{11}$

$x + 3 = \pm\sqrt{11}$

$x = -3 \pm \sqrt{11}$

39. $18x^2 - 52x = 6$

$18x^2 - 52x - 6 = 0$

$2(9x^2 - 26x - 3) = 0$

$2(9x+1)(x-3) = 0$

$9x + 1 = 0 \qquad x - 3 = 0$

$x = -\dfrac{1}{9} \qquad x = 3$

40. $2x^2 + 5x = 1$

$2x^2 + 5x - 1 = 0$

$x = \dfrac{-b \pm \sqrt{b^2 - 4ac}}{2a}$

$x = \dfrac{-5 \pm \sqrt{5^2 - 4(2)(-1)}}{2(2)}$

$x = \dfrac{-5 \pm \sqrt{25+8}}{4} = \dfrac{-5 \pm \sqrt{33}}{4}$

41. $y = -x^2 + 4x - 5$

42. $\dfrac{2-6i}{2i} = \dfrac{2-6i}{2i} \cdot \dfrac{i}{i} = \dfrac{2i - 6i^2}{2i^2} = \dfrac{2i+6}{-2} = -3 - i$

43. $2x^2 + 5 = 7x$

$2x^2 - 7x + 5 = 0$

$(2x-5)(x-1) = 0$

$2x - 5 = 0 \qquad x - 1 = 0$

$x = \dfrac{5}{2} \qquad x = 1$

44. $y = 4 - x^2$

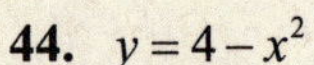

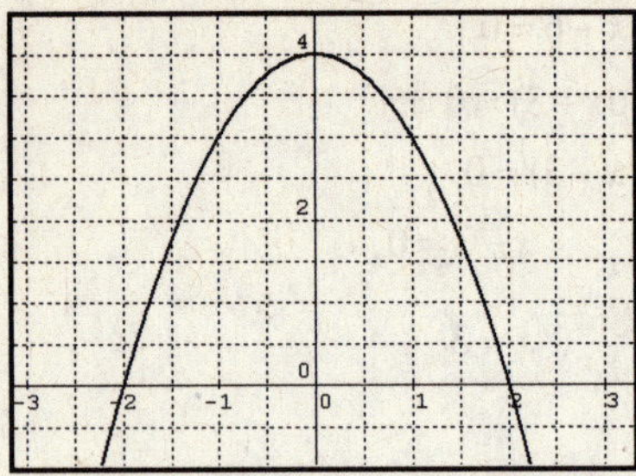

45. The rate of the balloon is 25 mph.

	Distance	÷	Rate	=	Time
Against the wind	60	÷	r - 5	=	$\frac{60}{r-5}$
With the wind	60	÷	r + 5	=	$\frac{60}{r+5}$

$$\frac{60}{r-5} = \frac{60}{r+5} + 1$$
$$(r-5)(r+5)\frac{60}{r-5} = (r-5)(r+5)\left(\frac{60}{r+5} + 1\right)$$
$$60(r+5) = 60(r-5) + (r-5)(r+5)$$
$$60r + 300 = 60r - 300 + r^2 - 25$$
$$0 = r^2 - 625$$
$$625 = r^2$$
$$\sqrt{625} = \sqrt{r^2}$$
$$r = \pm 25$$

46. $A = \frac{1}{2}bh$

$$20 = \frac{1}{2}b(2b+2)$$
$$40 = 2b^2 + 2b$$
$$0 = 2b^2 + 2b - 40$$
$$0 = b^2 + b - 20$$
$$0 = (b+5)(b-4)$$
$$b + 5 = 0 \qquad b - 4 = 0$$
$$b = -5 \qquad b = 4$$
$$2b + 2 = 10$$

The base is 4 m and the height is 10 m.

47. $d = 0.0056v^2 + 0.14v$

$$150 = 0.0056v^2 + 0.14v$$
$$0 = 0.0056v^2 + 0.14v - 150$$
$$v = \frac{-b \pm \sqrt{b^2 - 4ac}}{2a}$$
$$v = \frac{-0.14 \pm \sqrt{(0.14)^2 - 4(0.0056)(-150)}}{2(0.0056)}$$
$$v = \frac{-0.14 \pm \sqrt{3.3796}}{0.0112} = \frac{-0.14 \pm 1.8384}{0.0112}$$
$$v = \frac{-0.14 + 1.8384}{0.0112} = 151.6$$

The maximum speed is 151.6 km/h.

48. The rate of the boat is 5 mph.

	Distance	÷	Rate	=	Time
Against the current	12	÷	r - 1	=	$\frac{12}{r-1}$
With the current	12	÷	r + 1	=	$\frac{12}{r+1}$

$$\frac{12}{r-1} + \frac{12}{r+1} = 5$$
$$(r-1)(r+1)\left(\frac{12}{r-1} + \frac{12}{r+1}\right) = 5(r-1)(r+1)$$
$$12(r+1) + 12(r-1) = 5(r^2 - 1)$$
$$12r + 12 + 12r - 12 = 5r^2 - 5$$
$$0 = 5r^2 - 24r - 5$$
$$0 = (5r+1)(r-5)$$
$$5r + 1 = 0 \qquad r - 5 = 0$$
$$r = -\frac{1}{5} \qquad r = 5$$

49. $h = 32t - 16t^2$

$12 = 32t - 16t^2$

$16t^2 - 32t + 12 = 0$

$4t^2 - 8t + 3 = 0$

$(2t-3)(2t-1) = 0$

$2t - 3 = 0 \qquad 2t - 1 = 0$

$t = \frac{3}{2} \qquad t = \frac{1}{2}$

The object is 12 ft above the floor at $\frac{1}{2}$ s and at $\frac{3}{2}$ s.

50. Let t = the time for the larger drain.

$\frac{1}{t} + \frac{1}{t+8} = \frac{1}{3}$

$3t(t+8)\left(\frac{1}{t} + \frac{1}{t+8}\right) = 3t(t+8)\frac{1}{3}$

$3(t+8) + 3t = t(t+8)$

$3t + 24 + 3t = t^2 + 8t$

$6t + 24 = t^2 + 8t$

$0 = t^2 + 2t - 24$

$0 = (t+6)(t-4)$

$t + 6 = 0 \qquad t - 4 = 0$

$t = -6 \qquad t = 4$

$t + 8 = 12$

It takes the larger drain 4 hours and it takes the smaller drain 12 hours.

Chapter 11 Test

1. $3(x+4)^2 - 60 = 0$

$3(x+4)^2 = 60$

$(x+4)^2 = 20$

$\sqrt{(x+4)^2} = \sqrt{20}$

$x + 4 = \pm\sqrt{4 \cdot 5}$

$x + 4 = \pm 2\sqrt{5}$

$x = -4 \pm 2\sqrt{5}$

2. $2x^2 + 8x = 3$

$x^2 + 4x = \frac{3}{2}$

$x^2 + 4x + 4 = \frac{3}{2} + 4$

$(x+2)^2 = \frac{11}{2}$

$\sqrt{(x+2)^2} = \sqrt{\frac{11}{2}}$

$x + 2 = \pm\frac{\sqrt{11}}{\sqrt{2}} \cdot \frac{\sqrt{2}}{\sqrt{2}}$

$x = -2 \pm \frac{\sqrt{22}}{2} = \frac{-4 \pm \sqrt{22}}{2}$

3. $3x^2 + 7x = 20$

$3x^2 + 7x - 20 = 0$

$(3x-5)(x+4) = 0$

$3x - 5 = 0 \qquad x + 4 = 0$

$x = \frac{5}{3} \qquad x = -4$

4. $4\sqrt{-45} = 4\sqrt{-1}\sqrt{9 \cdot 5} = 4i \cdot 3\sqrt{5} = 12i\sqrt{5}$

5. $x^2 + 4x - 16 = 0$

$x^2 + 4x = 16$

$x^2 + 4x + 4 = 16 + 4$

$(x+2)^2 = 20$

$\sqrt{(x+2)^2} = \sqrt{20} = \sqrt{4 \cdot 5}$

$x + 2 = \pm 2\sqrt{5}$

$x = -2 \pm 2\sqrt{5}$

6. 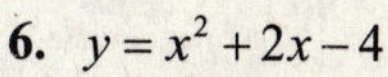
$y = x^2 + 2x - 4$

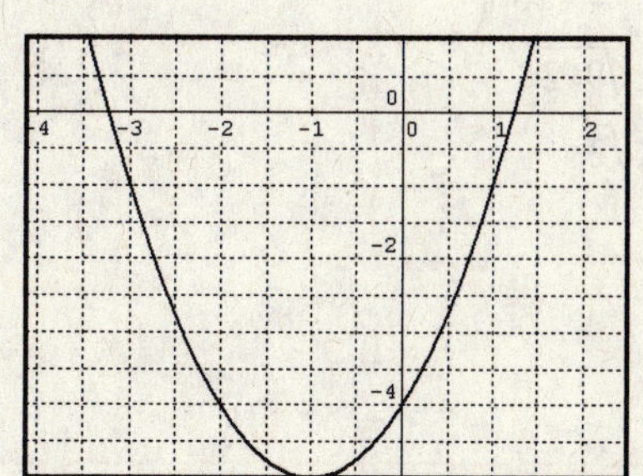

7. $x^2+4x+2=0$
$x^2+4x=-2$
$x^2+4x+4=-2+4$
$(x+2)^2=2$
$\sqrt{(x+2)^2}=\sqrt{2}$
$x+2=\pm\sqrt{2}$
$x=-2\pm\sqrt{2}$

8. $(-5+6i)-(3-7i)=(-5-3)+(6+7)i=-8+13i$

9. $2x^2-5x-3=0$
$(2x+1)(x-3)=0$
$2x+1=0 \qquad x-3=0$
$x=-\frac{1}{2} \qquad x=3$

10. $2x^2-6x+1=0$
$x=\frac{-b\pm\sqrt{b^2-4ac}}{2a}$
$x=\frac{-(-6)\pm\sqrt{(-6)^2-4(2)(1)}}{2(2)}$
$x=\frac{6\pm\sqrt{36-8}}{4}=\frac{6\pm\sqrt{28}}{4}$
$x=\frac{6\pm\sqrt{4\cdot 7}}{4}=\frac{6\pm 2\sqrt{7}}{4}=\frac{3\pm\sqrt{7}}{2}$

11. $6i(5-4i)=30i-24i^2=30i+24=24+30i$

12. $2(x-5)^2=36$
$(x-5)^2=18$
$\sqrt{(x-5)^2}=\sqrt{18}$
$x-5=\pm\sqrt{9\cdot 2}$
$x=5\pm 3\sqrt{2}$

13. $x^2-6x-5=0$
$x^2-6x=5$
$x^2-6x+9=5+9$
$(x-3)^2=14$
$\sqrt{(x-3)^2}=\sqrt{14}$
$x-3=\pm\sqrt{14}$
$x=3\pm\sqrt{14}$

14. $(-5+2i)(4-i)=-20+5i+8i-2i^2$
$=-20+13i+2=-18+13i$

15. $x^2-5x=2$
$x^2-5x-2=0$
$x=\frac{-b\pm\sqrt{b^2-4ac}}{2a}$
$x=\frac{-(-5)\pm\sqrt{(-5)^2-4(1)(-2)}}{2(1)}$
$x=\frac{5\pm\sqrt{25+8}}{2}=\frac{5\pm\sqrt{33}}{2}$

16. $6x^2-17x=-5$
$6x^2-17x+5=0$
$(3x-1)(2x-5)=0$
$3x-1=0 \qquad 2x-5=0$
$x=\frac{1}{3} \qquad x=\frac{5}{2}$

17. $x^2+3x-7=0$
$x=\frac{-b\pm\sqrt{b^2-4ac}}{2a}$
$x=\frac{-3\pm\sqrt{3^2-4(1)(-7)}}{2(1)}$
$x=\frac{-3\pm\sqrt{9+28}}{2}=\frac{-3\pm\sqrt{37}}{2}$

18. $\frac{3-i}{2i}=\frac{3-i}{2i}\cdot\frac{i}{i}=\frac{3i-i^2}{2i^2}=\frac{3i+1}{-2}=-\frac{1}{2}-\frac{3}{2}i$

19. $x^2-8x+17=0$

$x^2-8x=-17$

$x^2-8x+16=-17+16$

$(x-4)^2=-1$

$\sqrt{(x-4)^2}=\sqrt{-1}$

$x-4=\pm i$

$x=4\pm i$

20. $y=x^2-2x-3$

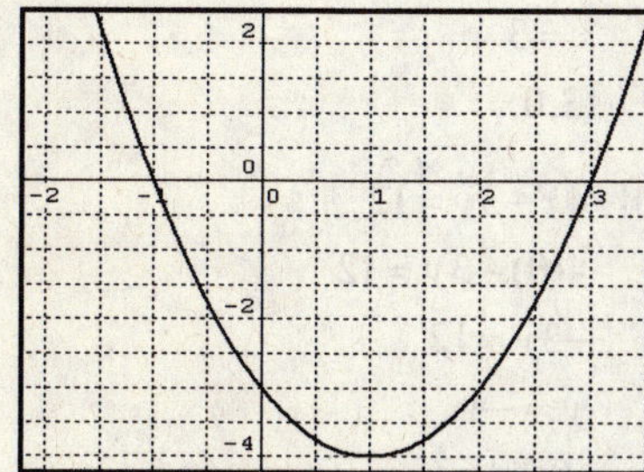

21. $\frac{4-2i}{1-i}=\frac{4-2i}{1-i}\cdot\frac{1+i}{1+i}=\frac{4+2i-2i^2}{1+1}$

$=\frac{4+2i+2}{2}=\frac{6+2i}{2}=3+i$

22. $A=w(2w-2)$

$40=2w^2-2w$

$0=2w^2-2w-40$

$0=w^2-w-20$

$0=(w-5)(w+4)$

$w-5=0 \qquad w+4=0$

$w=5 \qquad w=-4$

$2w-2=8$

The rectangle is 5 ft by 8 ft.

23. The rate of the boat was 11 mph.

	Distance	÷	Rate	=	Time
Against the current	60	÷	r - 1	=	$\frac{60}{r-1}$
With the current	60	÷	r + 1	=	$\frac{60}{r+1}$

$\frac{60}{r-1}=\frac{60}{r+1}+1$

$(r-1)(r+1)\frac{60}{r-1}=(r-1)(r+1)\left(\frac{60}{r+1}+1\right)$

$60(r+1)=60(r-1)+(r-1)(r+1)$

$60r+60=60r-60+r^2-1$

$0=r^2-121$

$r^2=121$

$\sqrt{r^2}=\sqrt{121}$

$r=\pm 11$

24. Let n, $n+2$ and $n+4$ be consecutive odd integers.

$n^2+(n+2)^2+(n+4)^2=83$

$n^2+n^2+4n+4+n^2+8n+16-83=0$

$3n^2+12n-63=0$

$n^2+4n-21=0$

$(n+7)(n-3)=0$

$n+7=0 \qquad n-3=0$

$n=-7 \qquad n=3$

$n+2=-5 \qquad n+2=5$

The middle integer is 5 or -5.

25. The rate for the last 8 miles was 4 mph.

	Distance	÷	Rate	=	Time
Part I	7	÷	r	=	$\frac{7}{r}$
Part II	8	÷	r - 3	=	$\frac{8}{r-3}$

$$\frac{7}{r}+\frac{8}{r-3}=3$$
$$r(r-3)\left(\frac{7}{r}+\frac{8}{r-3}\right)=3r(r-3)$$
$$7(r-3)+8r=3r^2-9r$$
$$7r-21+8r=3r^2-9r$$
$$15r-21=3r^2-9r$$
$$0=3r^2-24r+21$$
$$0=r^2-8r+7$$
$$0=(r-7)(r-1)$$
$$r-7=0 \qquad r-1=0$$
$$r=7 \qquad r=1$$
$$r-3=4 \qquad r-3=-2$$

Cumulative Review Exercises

1. $2x-3[2x-4(3-2x)+2]-3$
$$=2x-3[2x-12+8x+2]-3$$
$$=2x-3[10x-10]-3$$
$$=2x-30x+30-3$$
$$=-28x+27$$

2. $-\frac{3}{5}x=-\frac{9}{10}$
$$-\frac{5}{3}\left(-\frac{3}{5}x\right)=\frac{-5}{3}\left(-\frac{9}{10}\right)$$
$$x=\frac{3}{2}$$

3. $2x-3(4x-5)=-3x-6$
$$2x-12x+15=-3x-6$$
$$-10x+15=-3x-6$$
$$-7x=-21$$
$$x=3$$

4. $2x-3(2-3x)>2x-5$
$$2x-6+9x>2x-5$$
$$11x-6>2x-5$$
$$9x>1$$
$$x>\frac{1}{9}$$

5. x-intercepts: $4x-3y=12$
$$4x-3(0)=12$$
$$4x=12$$
$$x=3$$
$$(3,0)$$

y-intercept: $4x-3y=12$
$$4(0)-3y=12$$
$$-3y=12$$
$$y=-4$$
$$(0,-4)$$

6. $y-y_1=m(x-x_1)$
$$y-2=-\frac{4}{3}(x-(-3))$$
$$y-2=-\frac{4}{3}(x+3)$$
$$y=-\frac{4}{3}x-4+2$$
$$y=-\frac{4}{3}x-2$$

7. Domain: {-2, -1, 0, 1, 2}
Range: {-8, -1, 0, 1, 8};Function

8. $f(x)=-3x+10$
$$f(-9)=-3(-9)+10$$
$$f(-9)=27+10=37$$

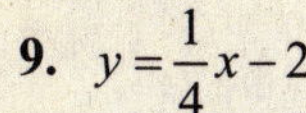

9. $y = \frac{1}{4}x - 2$

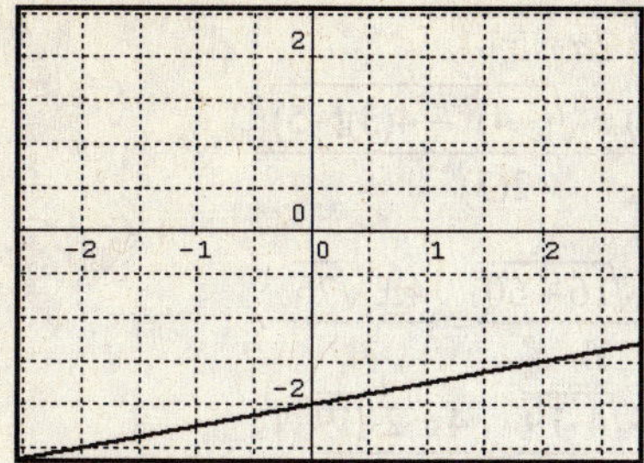

10. $2x - 3y > 6$

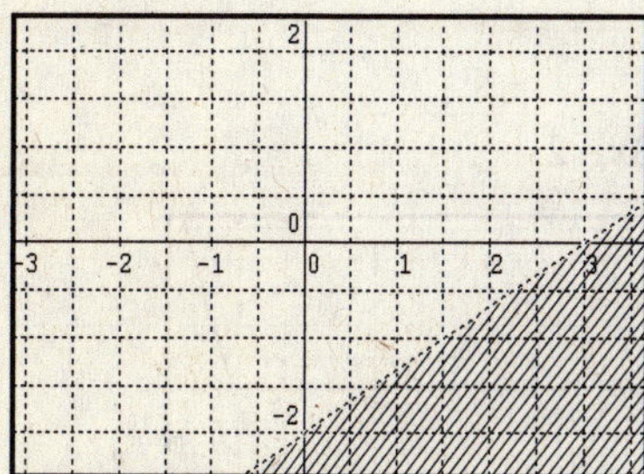

11. $3x - y = 5$

$y = 2x - 3$

$3x - (2x - 3) = 5$

$3x - 2x + 3 = 5$

$x + 3 = 5$

$x = 2$

$y = 2x - 3 = 4 - 3 = 1$

The solution is (2, 1).

12. $3x + 2y = 2$

$5x - 2y = 14$

$8x = 16$

$x = 2$

$3(2) + 2y = 2$

$2y = -4$

$y = -2$

The solution is (2, -2).

13. $\frac{(2a^{-2}b)^2}{-3a^{-5}b^4} = \frac{4a^{-4}b^2}{-3a^{-5}b^4} = -\frac{4}{3}a^{-4-(-5)}b^{2-4}$

$= -\frac{4}{3}ab^{-2} = -\frac{4a}{3b^2}$

14.
$$\begin{array}{r} x+2 \\ x-2\overline{)x^2+0x-8} \\ \underline{x^2-2x} \\ 2x-8 \\ \underline{2x-4} \\ -4 \end{array}$$

$(x^2 - 8) \div (x - 2) = x + 2 - \frac{4}{x-2}$

15. $4y(x-4) - 3(x-4) = (x-4)(4y-3)$

16. $3x^3 + 2x^2 - 8x = x(3x^2 + 2x - 8)$

$= x(3x-4)(x+2)$

17. $\frac{3x^2-6x}{4x-6} \div \frac{2x^2+x-6}{6x^2-24x} = \frac{3x^2-6x}{4x-6} \cdot \frac{6x^2-24x}{2x^2+x-6}$

$= \frac{3x(x-2)}{\cancel{2}_1(2x-3)} \cdot \frac{\overset{3}{\cancel{6}}x(x-4)}{(2x-3)(x+2)} = \frac{9x^2(x-2)(x-4)}{(2x-3)^2(x+2)}$

18. $\frac{x}{2(x-1)} - \frac{1}{(x-1)(x+1)} = \frac{x(x+1)-1(2)}{2(x-1)(x+1)}$

$= \frac{x^2+x-2}{2(x-1)(x+1)} = \frac{(x+2)(x-1)}{2(x-1)(x+1)} = \frac{x+2}{2(x+1)}$

19. $\frac{\left(1 - \frac{7}{x} + \frac{12}{x^2}\right)}{\left(2 - \frac{1}{x} - \frac{15}{x^2}\right)} \cdot \frac{x^2}{x^2} = \frac{x^2-7x+12}{2x^2-x-15}$

$= \frac{(x-3)(x-4)}{(x-3)(2x+5)} = \frac{x-4}{2x+5}$

20. $\dfrac{x}{x+6}=\dfrac{3}{x}$

$x^2=3(x+6)$

$x^2=3x+18$

$x^2-3x-18=0$

$(x-6)(x+3)=0$

$x-6=0 \qquad x+3=0$

$x=6 \qquad x=-3$

Both answers check.

21. $\left(\sqrt{a}-\sqrt{2}\right)\left(\sqrt{a}+\sqrt{2}\right)=a-2$

22. $3=8-\sqrt{5x}$

$\sqrt{5x}=5$

$\left(\sqrt{5x}\right)^2=5^2$

$5x=25$

$x=5$

23. $2x^2-7x=-3$

$2x^2-7x+3=0$

$(2x-1)(x-3)=0$

$2x-1=0 \qquad x-3=0$

$x=\dfrac{1}{2} \qquad x=3$

24. $3(x-2)^2=36$

$(x-2)^2=12$

$\sqrt{(x-2)^2}=\sqrt{12}$

$x-2=\pm\sqrt{4\cdot 3}$

$x=2\pm 2\sqrt{3}$

25. $3x^2-4x-5=0$

$x=\dfrac{-b\pm\sqrt{b^2-4ac}}{2a}$

$x=\dfrac{-(-4)\pm\sqrt{(-4)^2-4(3)(-5)}}{2(3)}$

$x=\dfrac{4\pm\sqrt{16+60}}{6}=\dfrac{4\pm\sqrt{76}}{6}$

$x=\dfrac{4\pm\sqrt{4\cdot 19}}{6}=\dfrac{4\pm 2\sqrt{19}}{6}$

$x=\dfrac{2(2\pm\sqrt{19})}{6}=\dfrac{2\pm\sqrt{19}}{3}$

26. $y=x^2-3x-2$

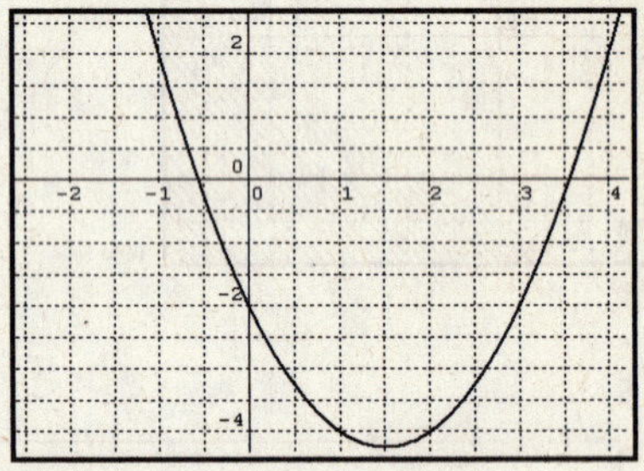

27. $A=PB$

$5.22=0.0725B$

$\dfrac{5.22}{0.0725}=B$

$\$72=B$

The book cost $72.

28. The mixture costs $4.50 per pound.

	Amount	·	Unit cost	=	Total
Cashews	20	·	7	=	7(20)
Peanuts	50	·	3.50	=	3.5(50)
Total	70	·	x	=	$70x$

$7(20)+3.5(50)=70x$

$140+175=70x$

$315=70x$

$\$4.50=x$

29. $\frac{100}{215}=\frac{100+x}{752.50}$
$75250=215(100+x)$
$75250=21500+215x$
$53750=215x$
$250=x$
250 additional shares are needed.

30. The rate of the plane is 200 mph and the rate of the wind is 40 mph.

	Distance	=	Rate	·	Time
Against the wind	720	=	$r - w$	·	4.5
With the wind	720	=	$r + w$	·	3

$4.5(r-w)=720$
$3(r+w)=720$
$r-w=160$
$r+w=240$
$2r=400$
$r=200$
$r-w=160$
$200-w=160$
$w=40$

31. $70+91+85+77+x\geq 400$
$323+x\geq 400$
$x\geq 77$
A grade of 77 or higher will give a minimum of 400 points.

32. $c^2=a^2+b^2$
$c^2=30^2+10^2$
$c^2=900+100$
$c^2=1000$
$c=\sqrt{1000}=31.62$
The wire is 31.62 m long.

33. $A=w(2w+24)$
$2106=2w^2+24w$
$0=2w^2+24w-2106$
$0=w^2+12w-1053$
$w^2+12w=1053$
$w^2+12w+36=1053+36$
$(w+6)^2=1089$
$\sqrt{(w+6)^2}=\sqrt{1089}$
$w+6=\pm 33$
$w=-6\pm 33$
$w=-6+33 \qquad w=-6-33$
$w=27 \qquad w=-39$
$2w+24=78$
The tennis court measures 27 ft by 78 ft.

Final Exam

1. $-|-3|=-3$

2. $-15-(-12)-3=-15+12+(-3)$
$=-3+(-3)$
$=-6$

3. $\frac{1}{8}\cdot\left(\frac{100}{1}\%\right)=\frac{100}{8}\%=12.5\%$

4. $-2^4\cdot(-2)^4=-16\cdot 16=-256$

5. $-7-\frac{12-15}{2-(-1)}\cdot(-4)=-7-\frac{-3}{3}\cdot(-4)$
$=-7-(-1)\cdot(-4)$
$=-7+1\cdot(-4)$
$=-7+(-4)$
$=-11$

6. $\dfrac{a^2-3b}{2a-2b^2}; a=3, b=-2$

$$\frac{3^2-3(-2)}{2(3)-2(-2)^2}=\frac{9+6}{6-8}=\frac{15}{-2}=-\frac{15}{2}$$

7. $6x-(-4y)-(-3x)+2y=6x+4y+3x+2y$
$=9x+6y$

8. $(-15z)\left(-\frac{2}{5}\right)=-\frac{\overset{3}{\cancel{15}}}{1}\left(-\frac{2}{\underset{1}{\cancel{5}}}\right)z=6z$

9. $-2[5-3(2x-7)-2x]=-2[5-6x+21-2x]$
$=-2[-8x+26]$
$=16x-52$

10. $20=-\frac{2}{5}x$

$$-\frac{5}{\underset{1}{\cancel{2}}}\cdot\overset{10}{\cancel{20}}=\left(-\frac{\overset{1}{\cancel{5}}}{\underset{1}{\cancel{2}}}\right)\left(-\frac{\overset{1}{\cancel{2}}}{\underset{1}{\cancel{5}}}x\right)$$

$-50=x$

11. $4-2(3x+1)=3(2-x)+5$
$4-6x-2=6-3x+5$
$-6x+2=-3x+11$
$-6x+2-2=-3x+11-2$
$-6x=-3x+9$
$-6x+3x=-3x+3x+9$
$-3x=9$
$\frac{-3x}{-3}=\frac{9}{-3}$
$x=-3$

12. $A=PB$
$A=0.19(80)$
$A=15.2$

13. $4-x\ge 7$
$4-x-4\ge 7-4$
$-x\ge 3$
$\frac{-x}{-1}\le\frac{3}{-1}$
$x\le -3$

14. $2-2(y-1)\le 2y-6$
$2-2y+2\le 2y-6$
$-2y+4\le 2y-6$
$-2y\le 2y-10$
$-4y\le -10$
$\frac{-4y}{-4}\ge\frac{-10}{-4}$
$y\ge\frac{5}{2}$

15. $(x_1,y_1)=(-1,-3);\ (x_2,y_2)=(2,-1)$

$$m=\frac{y_2-y_1}{x_2-x_1}=\frac{-1-(-3)}{2-(-1)}=\frac{-1+3}{2+1}=\frac{2}{3}$$

16. $(x_1,y_1)=(3,-4);\ m=-\frac{2}{3}$
$y-y_1=m(x-x_1)$
$y-(-4)=-\frac{2}{3}(x-3)$
$y+4=-\frac{2}{3}x+2$
$y=-\frac{2}{3}x-2$

17. $y=-\frac{1}{2}x-3$

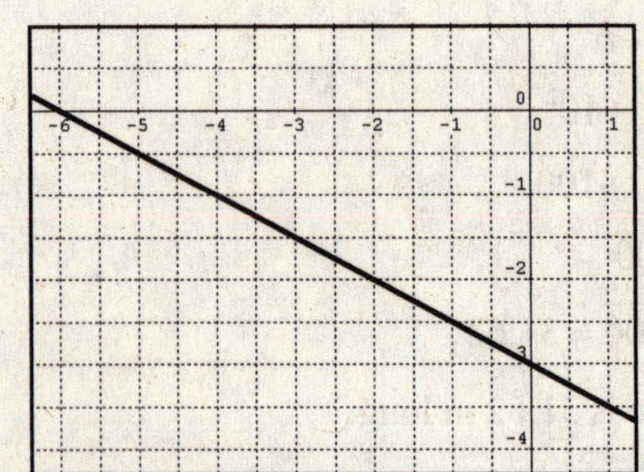

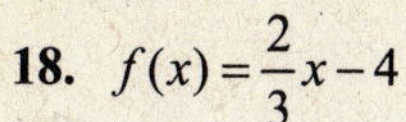

18. $f(x) = \frac{2}{3}x - 4$

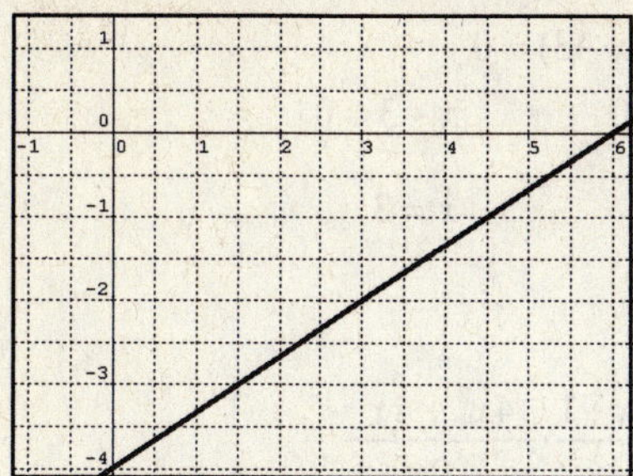

19. $f(x) = -x + 5$

$f(-6) = -(-6) + 5 = 11$

$f(-3) = -(-3) + 5 = 8$

$f(0) = 0 + 5 = 5$

$f(3) = -3 + 5 = 2$

$f(6) = -6 + 5 = -1$

Range: {-1, 2, 5, 8, 11}

20. $3x - 2y \geq 6$

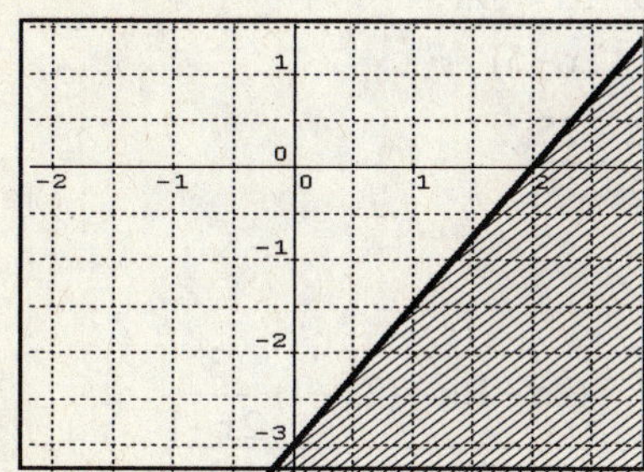

21.

$$\begin{aligned} y &= 4x - 7 \\ y &= 2x + 5 \\ 4x - 7 &= 2x + 5 \\ 4x &= 2x + 12 \\ 2x &= 12 \\ x &= 6 \end{aligned} \qquad \begin{aligned} y &= 4x - 7 \\ y &= 4(6) - 7 \\ y &= 24 - 7 \\ y &= 17 \end{aligned}$$

The solution is (6, 17).

22.

$$\begin{aligned} 4x - 3y &= 11 \\ 2x + 5y &= -1 \end{aligned}$$

$$\begin{aligned} 4x - 3y &= 11 \\ -2(2x + 5y) &= -2(-1) \end{aligned}$$

$$\begin{aligned} 4x - 3y &= 11 \\ -4x - 10y &= 2 \end{aligned}$$

$$\begin{aligned} -13y &= 13 \\ y &= -1 \\ 4x - 3y &= 11 \\ 4x - 3(-1) &= 11 \\ 4x + 3 &= 11 \\ 4x &= 8 \\ x &= 2 \end{aligned}$$

The solution is (2, -1).

23. $(2x^2 - 5x + 1) - (5x^2 - 2x - 7)$

$2x^2 - 5x + 1 - 5x^2 + 2x + 7$

$-3x^2 - 3x + 8$

24. $(-3xy^3)^4 = (-3)^4 x^4 (y^3)^4 = 81x^4 y^{12}$

25.

$$\begin{array}{r} 3x^2 - x - 2 \\ 2x + 3 \\ \hline 9x^2 - 3x - 6 \\ 6x^3 - 2x^2 - 4x \phantom{{}-6} \\ \hline 6x^3 + 7x^2 - 7x - 6 \end{array}$$

26. $\dfrac{(-2x^2y^3)^3}{(-4x^{-1}y^4)^2} = \dfrac{-8x^6y^9}{16x^{-2}y^8} = -\dfrac{x^8y}{2}$

27. $(4x^{-2}y)^3(2xy^{-2})^{-2} = 4^3x^{-6}y^3 \cdot 2^{-2}x^{-2}y^4$

$= \dfrac{64}{4}x^{-8}y^7 = \dfrac{16y^7}{x^8}$

28. $\dfrac{12x^3y^2-16x^2y^2-20y^2}{4xy^2}$

$=\dfrac{12x^3y^2}{4xy^2}-\dfrac{16x^2y^2}{4xy^2}-\dfrac{20y^2}{4xy^2}$

$=3x^2-4x-\dfrac{5}{x}$

29. $(5x^2-2x-1)\div(x+2)=5x-12+\dfrac{23}{x+2}$

$$\begin{array}{r|l} & 5x-12 \\ x+2 & 5x^2-2x-1 \\ & \underline{5x^2+10x} \\ & -12x-1 \\ & \underline{-12x-24} \\ & 23 \end{array}$$

30. $0.000000039=3.9\times10^{-8}$

31. $2a(4-x)-6(x-4)=2a(4-x)+6(4-x)$

$=(4-x)(2a+6)$

$=2(4-x)(a+3)$

32. $x^2-5x-6=(x-6)(x+1)$

33. $2x^2-x-3=(2x-3)(x+1)$

34. $6x^2-5x-6=(3x+2)(2x-3)$

35. $8x^3-28x^2+12x=4x(2x^2-7x+3)$

$=4x(2x-1)(x-3)$

36. $25x^2-16=(5x-4)(5x+4)$

37. $75y-12x^2y=3y(25-4x^2)$

$=3y(5-2x)(5+2x)$

38. $2x^2=7x-3$

$2x^2-7x+3=0$

$(2x-1)(x-3)=0$

$2x-1=0 \qquad x-3=0$

$x=\dfrac{1}{2} \qquad x=3$

39. $\dfrac{2x^2-3x+1}{4x^2-2x}\cdot\dfrac{4x^2+4x}{x^2-2x+1}$

$=\dfrac{\overset{1}{\cancel{(2x-1)}}\,\overset{1}{\cancel{(x-1)}}}{\underset{1}{\cancel{2x}}\,\underset{1}{\cancel{(2x-1)}}}\cdot\dfrac{\overset{2}{\cancel{4x}}(x+1)}{\underset{1}{\cancel{(x-1)}}(x-1)}$

$=\dfrac{2(x+1)}{x-1}$

40. $\dfrac{5}{x+3}-\dfrac{3x}{2x-5}=\dfrac{5(2x-5)}{(x+3)(2x-5)}-\dfrac{3x(x+3)}{(x+3)(2x-5)}$

$=\dfrac{10x-25}{(x+3)(2x-5)}-\dfrac{3x^2+9x}{(x+3)(2x-5)}$

$=\dfrac{10x-25-3x^2-9x}{(x+3)(2x-5)}$

$=\dfrac{-3x^2+x-25}{(x+3)(2x-5)}$

41. $\dfrac{x-\dfrac{3}{2x-1}}{1-\dfrac{2}{2x-1}}=\dfrac{(2x-1)x-(2x-1)\dfrac{3}{2x-1}}{(2x-1)1-(2x-1)\dfrac{2}{2x-1}}$

$=\dfrac{2x^2-x-3}{2x-1-2}=\dfrac{2x^2-x-3}{2x-3}$

$=\dfrac{(2x-3)(x+1)}{2x-3}=x+1$

42. $\dfrac{5x}{3x-5}-3=\dfrac{7}{3x-5}$

$(3x-5)\dfrac{5x}{3x-5}-3(3x-5)=\dfrac{7}{3x-5}(3x-5)$

$5x-9x+15=7$

$-4x+15=7$

$-4x=-8$

$x=2$

The solution checks.

43. $a = 3a - 2b$

$a - 3a = 3a - 3a - 2b$

$-2a = -2b$

$a = b$

44. $\sqrt{49x^6} = 7x^3$

45. $2\sqrt{27a} + 8\sqrt{48a} = 2\sqrt{9 \cdot 3a} + 8\sqrt{16 \cdot 3a}$

$= 2 \cdot 3\sqrt{3a} + 8 \cdot 4\sqrt{3a}$

$= 6\sqrt{3a} + 32\sqrt{3a}$

$= 38\sqrt{3a}$

46. $\dfrac{\sqrt{3}}{\sqrt{5}-2} = \dfrac{\sqrt{3}}{\sqrt{5}-2} \cdot \dfrac{\sqrt{5}+2}{\sqrt{5}+2} = \dfrac{\sqrt{15}+2\sqrt{3}}{5-4}$

$= \dfrac{\sqrt{15}+2\sqrt{3}}{1}$

$= \sqrt{15}+2\sqrt{3}$

47. $\sqrt{x+4} - \sqrt{x-1} = 1$

$\sqrt{x+4} = \sqrt{x-1} + 1$

$\left(\sqrt{x+4}\right)^2 = \left(\sqrt{x-1}+1\right)^2$

$x + 4 = x - 1 + 2\sqrt{x-1} + 1$

$x + 4 = x + 2\sqrt{x-1}$

$4 = 2\sqrt{x-1}$

$2 = \sqrt{x-1}$

$2^2 = \left(\sqrt{x-1}\right)^2$

$4 = x - 1$

$5 = x$

The solution checks.

48. $(x-3)^2 = 7$

$\sqrt{(x-3)^2} = \sqrt{7}$

$x - 3 = \pm\sqrt{7}$

$x = 3 \pm \sqrt{7}$

49. $4x^2 - 2x - 1 = 0$

$a = 4 \quad b = -2 \quad c = -1$

$x = \dfrac{-b \pm \sqrt{b^2 - 4ac}}{2a}$

$x = \dfrac{-(-2) \pm \sqrt{(-2)^2 - 4(4)(-1)}}{2(4)}$

$x = \dfrac{2 \pm \sqrt{4+16}}{8}$

$x = \dfrac{2 \pm \sqrt{20}}{8} = \dfrac{2 \pm 2\sqrt{5}}{8}$

$x = \dfrac{2(1 \pm \sqrt{5})}{2 \cdot 4} = \dfrac{1 \pm \sqrt{5}}{4}$

50. $y = x^2 - 4x + 3$

51. Let x = the number.

$2x + 3(x-2) = 2x + 3x - 6 = 5x - 6$

52. $A = PB$

$2400 = 0.80B$

$\dfrac{2400}{0.80} = \dfrac{0.80B}{0.80}$

$3000 = B$

The original value is \$3000.

53. $d = kw$

$16 = k(36)$

$\frac{16}{36} = \frac{4}{9} = k$

$d = kw$

$d = \frac{4}{9}w$

$d = \frac{4}{\cancel{9}_1} \cdot \cancel{45}^5$

$d = 20$

The dosage is 20 mg.

54. Let x = the measure of the second angle.

Then $x + 10$ = the measure of the first angle.

And $x + 10 + 10 = x + 20$ = the measure of the third angle.

$$x + x + 10 + x + 20 = 180$$
$$3x + 30 = 180$$
$$3x = 150$$
$$x = 50°$$
$$x + 10 = 60°$$
$$x + 20 = 70°$$

The angles measure 50°, 60° and 70°.

55. $C = \$900$; $S = \$1485$

$$S = C + rC$$
$$1485 = 900 + 900r$$
$$1485 - 900 = 900 + 900r - 900$$
$$585 = 900r$$
$$\frac{585}{900} = \frac{900r}{900}$$
$$0.65 = r$$

The markup rate is 65%.

56.

	P	·	r	=	I
Amount at 8%	3000	·	0.08	=	0.08(3000)
Amount at 11%	x	·	0.11	=	$0.11x$

$$0.08(3000) + 0.11x = 0.10(3000 + x)$$
$$240 + 0.11x = 300 + 0.10x$$
$$0.01x = 60$$
$$x = 6000$$

$6000 must be invested at 11%.

57.

	A	·	C	=	V
Peanuts	4	·	2	=	8
Walnuts	2	·	5	=	10
Mixture	6	·	x	=	$6x$

$$8 + 10 = 6x$$
$$18 = 6x$$
$$3 = x$$

The mixture costs $3 per pound.

58.

	A	·	r	=	Q
60% solution	20	·	0.60	=	20(0.6)
20% solution	30	·	0.20	=	30(0.2)
Mixture	50	·	x	=	$50x$

$$20(0.6) + 30(0.2) = 50x$$
$$12 + 6 = 50x$$
$$18 = 50x$$
$$\frac{18}{50} = \frac{50x}{50}$$
$$0.36 = x$$

The mixture is 36% acid.

59.

	r	·	t	=	d
First leg	r	·	1	=	r
Second leg	$2r$	·	1.5	=	$2r(1.5)$

$$r + 2r(1.5) = 860$$
$$r + 3r = 860$$
$$4r = 860$$
$$r = 215$$

The plane traveled 215 km during the first hour.

60.

	r	·	t	=	d
With current	$r + c$	·	2.5	=	50
Against current	$r - c$	·	5	=	50

$$2.5(r + c) = 50$$
$$5.0(r - c) = 50$$
$$r + c = 20$$
$$r - c = 10$$
$$2r = 30$$
$$r = 15$$
$$r - c = 10$$
$$15 - c = 10$$
$$c = 5$$

The speed of the boat is 15 mph and the speed of the current is 5 mph.

61. Let w = the width. Then $w + 5$ = the length.

$$length \cdot width = area$$
$$(w+5)w = 50$$
$$w^2 + 5w - 50 = 0$$
$$(w+10)(w-5) = 0$$
$$w + 10 = 0 \qquad w - 5 = 0$$
$$w = -10 \qquad w = 5$$
$$w + 5 = 10$$

The width is 5 m and the length is 10 m.

62. $\dfrac{2oz}{15oz} = \dfrac{n}{120oz}$

$$15n = 2(120)$$
$$15n = 240$$
$$n = 16$$

16 oz of dye are required.

63.

	r	·	t	=	
Chef	1	·	t	=	t
Apprentice	$\dfrac{1}{1.5}$	·	t	=	$\dfrac{t}{1.5}$

$$t + \frac{t}{1.5} = 1$$
$$1.5\left(t + \frac{t}{1.5}\right) = 1(1.5)$$
$$1.5t + t = 1.5$$
$$2.5t = 1.5$$
$$t = 0.6$$

The job will take 0.6 hr if they work together.

64. $a^2 + b^2 = c^2$

$8^2 + b^2 = 14^2$

$64 + b^2 = 196$

$b^2 = 132$

$b = 11.5cm$

65.

	d	$\div$	r	$=$	t
With wind	500	$\div$	$225 + w$	$=$	$\frac{500}{225+w}$
Agains t wind	500	$\div$	$225 - w$	$=$	$\frac{500}{225-w}$

$\frac{500}{225-w} = \frac{500}{225+w} + 0.5$

$2(225-w)(225+w)\left(\frac{500}{225-w} = \frac{500}{225+w} + \frac{1}{2}\right)$

$(225+w)1000 = (225-w)1000 + (225-w)(225+w)$

$225000 + 1000w = 225000 - 1000w + 50625 - w^2$

$225000 + 1000w = 275625 - 1000w - w^2$

$w^2 + 2000w - 50625 = 0$

$(w-25)(w+2025) = 0$

$w - 25 = 0 \qquad w + 2025 = 0$

$w = 25 \qquad w = -2025$

The rate of the wind is 25 mph.